Undergraduate Lecture Notes in Physics

For further volumes:
http://www.springer.com/series/8917

Undergraduate Lecture Notes in Physics (ULNP) publishes authoritative texts covering topics throughout pure and applied physics. Each title in the series is suitable as a basis for undergraduate instruction, typically containing practice problems, worked examples, chapter summaries, and suggestions for further reading.

ULNP titles must provide at least one of the following:

- An exceptionally clear and concise treatment of a standard undergraduate subject.
- A solid undergraduate-level introduction to a graduate, advanced, or non-standard subject.
- A novel perspective or an unusual approach to teaching a subject.

ULNP especially encourages new, original, and idiosyncratic approaches to physics teaching at the undergraduate level.

The purpose of ULNP is to provide intriguing, absorbing books that will continue to be the reader's preferred reference throughout their academic career.

Series Editors

Antonis Modinos

From Aristotle to Schrödinger

The Curiosity of Physics

Antonis Modinos
Emeritus Professor of Physics
National Technical University of Athens
Athens
Greece

ISSN 2192-4791 ISSN 2192-4805 (electronic)
ISBN 978-3-319-00749-6 ISBN 978-3-319-00750-2 (eBook)
DOI 10.1007/978-3-319-00750-2
Springer Cham Heidelberg New York Dordrecht London

Library of Congress Control Number: 2013943981

Printed on acid-free paper

Springer is part of Springer Science+Business Media (www.springer.com)

Raise now the third Olympus, on which Science shall be enthroned.

She is the only reality, laughterless, and yet

What smile, what gold or silver, is sweeter than her face?

Away with Olympian mists! If the heart is a miracle, the mind is its eye!

Kostis Palamas, Twelve Lays of the Gipsy

Translated by George Thomson

Preface

The romantic poet Samuel Taylor Coleridge, born in 1772, was an amateur chemist and had a lively interest in all science. He opined that: *the activity of science being necessarily performed with the passion of hope, it is poetical*. And when one looks at the collective effort of humanity to understand and describe in an intelligible manner the ways of nature, the process of doing so passing from generation to generation and from one epoch to the next, he may reasonably claim that there is an inspiring epic story second to none, poetic indeed! And yet one must admit that there is no easy way to tell this story.

There are many books (some very good) which endeavour to tell the story of physics (or some branch of physics) without using any mathematics at all. However good these books are, they miss at least half of the story. They are not good enough for the student and teacher of physics who knows that physics would not be the exact science it is without mathematics. The present book is intended primarily for the student and teacher of physics and assumes a familiarity with mathematics beyond high school, for the text to be fully appreciated. The book aims not only to say *who has done what*, but to do so in a way that contributes to a better understanding of physics and how it progresses. However, knowing that too much mathematics will be a hindrance to many readers, I chose a middle way; I use a minimum of mathematics, by which I mean the descriptive power of its symbols, but very little mathematical calculation as such, except in some of the exercises. Moreover, the mathematical concepts needed for the telling of the story are introduced along the way. They are part of the story. Neither do I aim to present a complete history of physics and all its achievements past and present. That lies beyond my ability. I tried to put together a meaningful story of the main events (ideas and facts) from Aristotle to Schrödinger as they developed historically. Though the emphasis is on the physics (observation, experiment, theory, some applications), the story is supplemented by short sections on the lives and times of the main protagonists. My story ends with the development of quantum mechanics (Chap. 13). Some important applications of quantum mechanics to atoms, molecules and solids are described in Chap. 14. Finally, in Chap. 15, devoted to the very small and the very large, I refer very briefly to recent developments (of the last 50 years or so) in nuclear particle physics and in cosmology (a comprehensive review of these developments lying beyond the scope of the present book). I included a few exercises at the end of each chapter. Some of these relate directly to formulae given in the text, and some deal with applications not dealt with in the text.

Though the book is meant primarily for the student and teacher of physics, it will be of interest, I hope, to other scientists and to historians of science, and also to curious minds of any age or profession who wish to know something about science, how it started, relatively simple, and how it developed to its present day magnificence and sophistication.

I would like to thank Andreas Kyritsakis for his valuable assistance in the preparation of the diagrams.

A. Modinos

Contents

1 **The Language of Physics** 1
1.1 Infinite Series and Sequences: From Zeno to Weierstrass 1
1.2 Geometry and Numbers. 6
Exercises 11

2 **The Dawn of Science** 13
2.1 Aristotle 13
2.2 Space, Time, and Matter 14
2.3 The Universe 16
2.4 Does Nature Act for an End?. 21
Exercises 23

3 **Astronomy Paves the Way** 25
3.1 The Astronomers of Alexandria 25
3.2 Copernicus and his Advocacy of the Heliocentric System 31
3.3 Kepler and his Contribution to Astronomy 36
Exercises 41

4 **Galileo: His Life and Work** 43
Exercises 54

5 **The Seventeenth Century: The Bloom of Science** 55
5.1 The Continuum of Numbers. Functions Descartes' Analytical Geometry 55
5.2 Some Important Experiments. 65
5.2.1 Torricelli's Barometer and Boyle's Gas Law 65
5.2.2 Snell's Law of Refraction and the Nature of Light . . . 68
5.3 The Invention of Calculus 72
Exercises 81

6 **Isaac Newton** 83
6.1 The Unhappy Childhood and His Life at Cambridge. 83
6.2 The Principia 87
6.2.1 Introduction and the Laws of Motion 87

6.2.2 Examples of Motion . . . 93
6.2.3 Gravity . . . 99
6.2.4 Galilean Relativity . . . 102
6.3 Newton's Life After the Principia. . . . 103
Exercises . . . 104

7 Classical Mechanics . . . 107
7.1 The Theory of Mechanics After Newton . . . 107
7.1.1 Angular Momentum . . . 108
7.1.2 Rotation of a Solid Body About a Fixed Axis . . . 109
7.1.3 The Simple Pendulum. . . . 112
7.2 The Vibrating String. Partial Differential Equations Fourier Series. . . . 114
7.3 Some Important Concepts of Analytical Mechanics . . . 119
7.4 Fluid Dynamics . . . 123
Exercises . . . 127

8 The Beginnings of Chemistry . . . 133
8.1 Chemistry's Relation to Alchemy . . . 133
8.2 The Pneumatic Chemists . . . 134
8.3 John Dalton and the Atomic Theory of Matter . . . 139
8.4 Avogadro' Theory and the Law of Ideal Gases . . . 143
Exercises . . . 146

9 Thermodynamics and Statistical Mechanics. . . . 147
9.1 Thermodynamics . . . 147
9.1.1 Heat . . . 147
9.1.2 Energy Conservation and the First Law of Thermodynamics . . . 149
9.1.3 Thermodynamic Functions, the Equation of State and Phase Diagrams . . . 151
9.1.4 The Second Law of Thermodynamics: Entropy . . . 156
9.2 Statistical Mechanics. . . . 165
9.2.1 The Kinetic Theory of Gases . . . 165
9.2.2 The Maxwell–Boltzmann Distribution. . . . 166
9.2.3 Gibbs' Statistical Physics. . . . 174
9.2.4 The Equipartition Theorem . . . 177
9.2.5 Fluctuations . . . 178
9.2.6 Diffusion. . . . 180
Exercises . . . 182

10 Electromagnetism. . . . 189
10.1 Early History . . . 189
10.2 Electrostatics, Steady Currents and Magnetism. . . . 190

10.2.1 Electrostatics 190
10.2.2 The Discovery of the Battery 193
10.2.3 Electric Currents and Magnetism 195
10.2.4 Electromagnetic Units 199
10.2.5 Ohm's Law 201
10.2.6 Electricity and Chemistry 203
10.3 Faraday's Electromagnetic Induction 204
10.4 Electromagnetic Waves 211
10.4.1 Confirmation of the Wave Nature of Light 211
10.4.2 Two Important Theorems Concerning Vector Fields 214
10.4.3 Maxwell and His Equations 219
10.4.4 Electromagnetic Fields in Material Media 231
Exercises 236

11 Cathode Rays and X-rays 243
11.1 Cathode Rays 243
11.2 The Discovery of X-rays 244
11.3 The Discovery of the Electron 247
11.4 X-ray Diffraction 249
11.5 Electron Emission and Electronic Valves 254
11.5.1 Thermionic Emission 254
11.5.2 Field Emission 257
Exercises 257

12 Einstein's Theory of Relativity 261
12.1 Einstein's Early Life and Studies 261
12.2 The Special Theory of Relativity 263
12.2.1 Preliminaries 263
12.2.2 Derivation of the Lorentz Transformation 268
12.2.3 Length Contraction and Time Dilation 269
12.2.4 The Law of Motion in Special Relativity 270
12.3 On the Invariance of Physical Laws 272
12.3.1 The Minkowski 4-Dimensional Spacetime 272
12.3.2 Tensors and Relativistic Invariance 275
12.4 The General Theory of Relativity 281
12.4.1 The Principle of General Relativity 281
12.4.2 Curvilinear Tensors 285
12.4.3 Einstein's Equation 287
12.5 The Famous Scientist 292
Exercises 294

13 Quantum Mechanics 297
13.1 The Quantization of Light 297
13.1.1 Black Body Radiation and Planck's Law 297
13.1.2 The Photoelectric Effect 304
13.1.3 The Einstein Coefficients 306
13.1.4 The Laser 308
13.2 Atomic Structure 309
13.2.1 Thomson's Static Model 309
13.2.2 Rutherford's Atom 310
13.2.3 Bohr's Model 313
13.2.4 The Discovery of Spin 318
13.3 The Wave-Particle Duality 319
13.4 Formal Quantum Mechanics 321
13.4.1 Schrödinger's Equation 321
13.4.2 The Stationary States of a Bound Particle 326
13.4.3 Heisenberg's Uncertainty Principle 331
13.4.4 Scattering States and the Tunnelling Phenomenon 337
13.4.5 Transitions 342
13.4.6 The Relation Between Classical and Quantum Mechanics 343
Exercises 344

14 Atoms, Molecules and Solids 355
14.1 The One-Electron Atom 355
14.1.1 Orbital Angular Momentum 355
14.1.2 Stationary States of a Particle in a Central Field 360
14.1.3 Stationary States of the Hydrogen Atom 363
14.2 Electron Spin 367
14.2.1 Fine Structure of the Energy Levels of Hydrogen 367
14.2.2 Formal Description of the Electron Spin 369
14.3 The Radiation Field and its Interaction with Matter 373
14.3.1 The Linear Harmonic Oscillator 373
14.3.2 Quantum-Mechanical Description of the Radiation Field 374
14.3.3 Interaction of the Radiation Field with Matter 376
14.4 Indistinguishableness of Same Particles and Quantum Statistics 378
14.4.1 The Indistinguishableness of Same Particles 378
14.4.2 Bosons 380
14.4.3 Fermions 384
14.5 The Many-Electron Atom 386
14.5.1 The Periodic Table of the Elements 386
14.5.2 The Self-consistent Field 391
14.5.3 The Energy Levels of a Many-Electron Atom 394

14.6 Molecules 397
14.6.1 The Born-Oppenheimer Approximation 397
14.6.2 Diatomic Molecules: Electronic Terms 399
14.6.3 Nuclear Motion of Diatomic Molecules 403
14.6.4 Polyatomic Molecules 405
14.7 Solids 406
14.7.1 Symmetric Cells of Real and Reciprocal Lattices 406
14.7.2 One-electron States in Crystalline Solids 408
14.7.3 Metals 411
14.7.4 Ferromagnetic Metals 418
14.7.5 Semiconductors 420
14.7.6 p-n Junctions and Transistors 424
14.7.7 Lattice Vibrations 425
14.7.8 Superconductivity 430
Exercises 435

15 The Very Small and the Very Large 443
15.1 Nuclear Physics 443
15.1.1 Radioactivity 443
15.1.2 Nuclear Reactions and The Discovery of The Neutron 448
15.1.3 Nuclear Fission and the First Nuclear Reactor 450
15.1.4 Nuclear Models 453
15.1.5 Elementary Particles and the Standard Model 453
15.2 Dirac's Theory of the Electron and Quantum Electrodynamics 458
15.2.1 Dirac's Theory of the Electron 458
15.2.2 Quantum Electrodynamics 466
15.3 Cosmology 483
15.3.1 Galaxies 483
15.3.2 Generation and Death of Stars 484
15.3.3 The Earth 487
15.3.4 The Universe: Expanding Universe 488
Exercises 489

Appendix A: Notes on Mathematics 491

Appendix B: Physical Constants 503

Subject Index 505

Name Index 513

Chapter 1
The Language of Physics

1.1 Infinite Series and Sequences: From Zeno to Weierstrass

I shall begin by asking you to look carefully at the following equation. You need only know the fractions one learns at school.

$$1+\frac{1}{2}+\frac{1}{4}+\frac{1}{8}+\frac{1}{16}+\cdots=2 \tag{1.1}$$

The three dots on the left of the equation imply that the *series* has no end: the sum is made up by an *infinite* number of terms. We note that every term in the series is equal to the half of the preceding term: the 1/4 is equal to the half of 1/2, the 1/8 is equal to the half of 1/4, the 1/16 is equal to the half of 1/8 and so on. Having this in mind we can write down (include in the calculation of the sum) as many terms as we like, thousands or millions of terms. With every term that we include in it the sum comes closer to 2, but no matter how many terms we include in it the sum will not equal 2 *exactly*.

You can more easily convince yourself that the above statement is true by doing the following: On a straight line mark two points to correspond to the numbers 1 and 2 respectively, as on a ruler. You can if you so wish think of the line segment between the points as the unit of length. Then mark on the line the point corresponding to the sum of the first two terms of the series, namely (1 + 1/2), and note that this lies to the left of 2 by a half (1/2) of the unit length. And the point corresponding to the sum of the first three terms, namely (1 + 1/2 + 1/4) lies to the left of 2 by a quarter (1/4) of the unit length. Similarly note that as more terms, 1/8, 1/16 and so on, are included in the sum, the point on the straight line corresponding to it lies to the left of 2 by one eighth (1/8) of the unit length, and then by one sixteenth (1/16) of the unit length and so on.

We are therefore convinced that as the numbers in the sequence:

$$1/2,\ 1/4,\ 1/8,\ 1/16, \ldots \tag{1.2}$$

A. Modinos, *From Aristotle to Schrödinger*,
Undergraduate Lecture Notes in Physics, DOI: 10.1007/978-3-319-00750-2_1,

are added to unity one by one, as indicated in Eq. (1.1), the difference between the sum on the left of Eq. (1.1) and 2 is diminished to 1/2, 1/4, 1/8, 1/16, and so on. At each step of the procedure the difference between the sum and 2 is halved. Therefore by including a sufficient number of terms in the summation this difference can be made as small as we want it, but it can not be extinguished. What then does the equality sign in the above expression mean?

For practical purposes, i.e. in calculations of one kind or the other, it is always permissible to substitute for a required number another that is practically but not exactly the same with the one required. It is, for example, quite common to approximate fractions, such as 2/3, by a decimal number even when this can not be done exactly. In the case of 2/3, depending on the accuracy you require you can put (this is what your calculator will probably give you) $2/3 = 0.6666667$, or if you require more accuracy you can put $2/3 = 0.6666666667$, where the last digit (7) replaces a non ending sequence of sixes (as in $2/3 = 0.666666666\ldots$). It is clear that in this sense and for practical purposes a number can be replaced by a series, as in Eq. (1.1), provided we keep sufficiently many terms to obtain the required accuracy. And in fact this is how great pioneers of science from Archimedes (who lived in Syracuse in the third century BC) to Newton (who was born in 1642) understood the limiting process implied by Eq. (1.1), but they certainly felt the clumsiness of this approach and would hesitate to put an exact equality sign in Eq. (1.1), as we have done.

We must now explain how this equality is to be properly understood. What we need is a *formal* criterion according to which one can decide unambiguously whether the equality in Eq. (1.1) holds or not, and by the way clarify in what sense this equality is to be understood. This has been provided by the German mathematician Karl Wilhelm Weierstrass who was born in 1815 in Westphalia. He was in his mid-thirties and working as a teacher in a village school when he came up with his formal criterion of convergence of a mathematical expression to a limit. His criterion applies to a variety of cases, but here we shall introduce it in relation to Eq. (1.1).

Let us choose a number (any number) smaller than 2 and as near to it as we like, then it is evident that if we include sufficiently many terms in the sum on the left of Eq. (1.1) we shall obtain a value for it between the chosen number and 2 (and for that matter closer to 2, if we want it so). It is clearly evident that there is no number other than 2 that bears this relation to the series on the left of Eq. (1.1). Therefore we can say, according to Weierstrass, that the series on the left of Eq. (1.1) converges to 2. It is in this limiting sense that the sum on the left of Eq. (1.1) equals 2. Sometimes (the rigorous mathematician would prefer that) we say that 2 is the limit of the said series.

Weierstrass's criterion is often stated as follows: A mathematical expression (it could be the series on the left of Eq. (1.1), or the sequence of numbers in expression (1.2) above, or some other expression) converges to a limit if the difference between this limit and the value of the given expression as it develops becomes smaller than ε, where ε (the Greek letter epsilon) denotes a quantity which can be arbitrarily small.

For example: we can be certain that the sequence of numbers given by expression (1.2) converges to zero (0), because by going far enough into the given sequence we can always get a fraction smaller than ε, however small the value of the latter.

In some cases it may not be that easy to establish the convergence of an expression to a limit, but this need not concern us here. The important thing is that thanks to Weierstrass we have a criterion by which we can establish by algebraic means (using the rules of arithmetic) whether an expression converges to a limit or not. And that is all we need.

Let me rewrite the series of Eq. (1.1) as follows:

$$1+\frac{1}{2}+\left(\frac{1}{2}\times\frac{1}{2}\right)+\left(\frac{1}{2}\times\frac{1}{2}\times\frac{1}{2}\right)+\left(\frac{1}{2}\times\frac{1}{2}\times\frac{1}{2}\times\frac{1}{2}\right)+\cdots=\frac{1}{1-\frac{1}{2}} \quad (1.3)$$

Once we have written our series in this form, we are tempted to ask: for which other values of ζ may the following be true?

$$1+\zeta+(\zeta\times\zeta)+(\zeta\times\zeta\times\zeta)+(\zeta\times\zeta\times\zeta\times\zeta)+\ldots=\frac{1}{1-\zeta} \quad (1.4a)$$

By applying Weierstrass's criterion we find (see exercise 1.1) that the series on the left of Eq. (1.4a), known as the geometric series, converges to the limit on the right of Eq. (1.4a) for $-1<\zeta<+1$, i.e. for values of ζ between -1 and +1.

We may therefore conclude that the series of Eq. (1.1) is a special case of the geometric series.

We often write an infinite series more compactly, as an infinite sum. For example, Eq. (1.4a) will be written as:

$$\sum_{n=0}^{\infty}\zeta^n=\frac{1}{1-\zeta} \quad (1.4b)$$

where $\zeta^0=1, \zeta^1=\zeta, \zeta^2=\zeta\times\zeta, \zeta^3=\zeta\times\zeta\times\zeta$, and so on.

It is worth remembering that the application of Weierstrass's criterion to the series of Eq. (1.1), or equivalently to the sequence of expression (1.2) is the final chapter of a story which began a long time ago. The said sequence appears (written in words of course and not in present day numerals) in Zeno's paradox of motion. Zeno was a student of Parmenides at the Elea school of philosophy in southern Italy at the beginning of the 5th century BC. The Eleatics believed in the permanence of the world and the impossibility of change. And it is perhaps in his desire to ridicule the idea of motion and change that Zeno describes a 'race' between Achilles, the fast-foot hero of the Iliad, and a tortoise in the following manner: Let the tortoise be a unit of length ahead of Achilles at the beginning of the race, and let it be that with every unit of time that passes Achilles halves the distance that separates him from the tortoise. The distance between Achilles and the tortoise will then be given by *one* unit of length at the beginning of the race, by *one half* of the unit of length when *one* unit of time has elapsed, by *one quarter* of a unit of length when *two* units of time have elapsed, and so on. The distance separating Achilles from the tortoise will be successively smaller, the division (halving) of this distance repeated *ad infinitum* (to infinity), but it will never be extinguished: the tortoise will be for ever ahead of Achilles.

Zeno's paradox was not taken lightly by his contemporaries (as one might think today) because 'motion' was not properly defined at that time. Aristotle, who took it upon himself to clarify the vocabulary of emerging science in his treatise on *Physics*, refers to Zeno's paradox pointing out the fallacy in his argument, which comes down to the fact that the speed by which Achilles runs in Zeno's race can not be constant but must be reduced successively if the result that Zeno requires is to be obtained.

Much more interesting from our point of view is the use that Aristotle makes of division (bisection) ad infinitum, of which the sequence of expression (1.2) is a representation, in his discussion of infinity. Aristotle devotes a whole chapter to infinity in his *Physics*, admitting that the problem of infinity is a difficult one with *difficulties arising whether one assumes that it exists or does not exist.*[1] After carefully arguing that there is no body (a sensible physical thing) which is *actually* infinite, he points out that *to assume that the infinite does not exist in any way leads obviously to many impossible consequences: there will be a beginning and an end of time, a magnitude will not be divisible into magnitudes, number will not be infinite. If, then, in view of the above considerations, neither alternative is possible, an arbiter must be called in; and clearly there is a sense in which infinity exists and another in which it does not.* Aristotle proposes that infinity has *potential existence*. He then clarifies what he means by this term with a number of examples whereby it becomes clear that *the infinite has in general this mode of existence: one thing is being taken after another, and each thing that is taken is always finite, but always different. In other words,* Aristotle concludes, *a quantity is infinite if it is such that we can always take a part outside what has been already taken.*

We note in passing that the infinite series of Eq. (1.1) and the infinite sequence of expression (1.2) do have this mode of existence. For however many terms we include in the sum of Eq. (1.1), or list in the sequence of expression (1.2), there is more of them outside what we have already included in the sum or listed in the sequence.

In his consideration of infinity Aristotle makes a distinction between infinity in relation to physics and infinity in relation to numbers, and in both cases he considers the infinitely small (whether it exists or not) and the infinitely large or infinitely many (whether they exist or not).

[1] Here and throughout the book when quoting from another book, the quoted passage is written in italics. In relation to Aristotle: the quoted passages were extracted from his text on *Physics*, as translated by R. P. Hardie and R. K. Gaye, or from *On the Heavens*, as translated by J. L. Stocks. Both works are included in: *The Basic Works of Aristotle*, Edited by Richard McKeon, New York: Random House, 1941; 2001 Modern Library Paperback Edition.

According to Aristotle time is infinite (he assumes that there will always be another day), but there can not be a physical (material) body or entity (including space itself) which is spatially infinite because nothing can extend beyond the heavens (the heavens being the sphere enclosing the universe). On the other hand a continuous body (and every material body dense or rare is continuous according to Aristotle) can be divided (bisected) ad infinitum (this is where the sequence of expression (1.2) appears implicitly in his argument), so there is *no limit* to smallness. It is worth noting in this respect that Aristotle could not and would not refer to *zero* as the limit of bisection ad infinitum. Such a number as *zero* did not exist (see below) and the nearest concept to it for an ancient mathematician or physicist would be *nothingness*. Moreover, and more importantly, according to Aristotle's definition, infinitely small meant something which however small can be further divided.

In Aristotle's analysis there is no mention of a limit of a convergent sequence or series, which in any sense resembles the concept of the limit as defined by Weierstrass. However, it is worth noting that Antiphon a philosopher and mathematician who was born may be a hundred years before Aristotle (he was a contemporary of Socrates) proposed a version of division ad infinitum (or to be precise division after division sufficiently many times) as a way of estimating the circumference of a circle: inscribe a square in it (see Fig. 1.1a), by drawing isosceles triangles on the sides of the square inscribe an octagon in the circle, repeat the process again and again, each time making isosceles triangles on the sides of the last drawn polygon. *Eventually* the sides of the inscribed polygon will be so small that they will coincide with the circumference of the circle.

Aristotle (in his *Physics*) refers to Antiphon's idea in order to point out that Antiphon's polygon will not ever match the circumference of the circle because there is no end to the division that Antiphon prescribes. Aristotle is of course right, but he misses the important point which is: by carrying the process of division far enough we can get as close to the circumference of the circle as we want to or

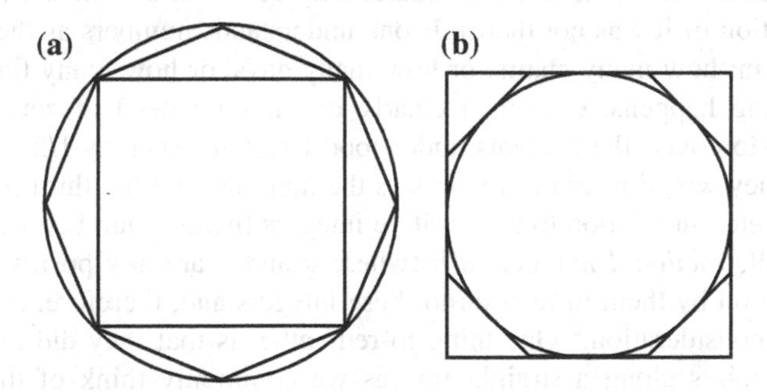

Fig. 1.1 **a** The inscribed square has a diagonal equal to the diameter of the *circle*. **b** The side of the *square* enclosing the *circle* equals the diameter of the *circle*. Replacing the *squares* by *polygons* of sufficiently many sides, Archimedes obtained the circumference of a circle with sufficient accuracy: it lies between the circumferences of the inscribed and enclosing *polygons*

require (for the purpose of the intended calculation). It was left to Archimedes (about a hundred years later) to appreciate Antiphon's idea, and to use it to calculate the ratio, π, of the circumference to the diameter of a circle with impressive accuracy (see Fig. 1.1).

1.2 Geometry and Numbers

Aristotle uses bisection ad infinitum as well in his discussion of infinity in relation to numbers in order to demonstrate that contrary to what is the case for matter and the physical space, there is no upper bound to numbers *because it is always possible to think of a larger number: for the number of times a number can be bisected is infinite.* The mathematicians need not worry about the finiteness of physical space, Aristotle writes, for: *In point of fact they do not need the infinite and do not use it. They postulate only that the finite straight line may be produced as far as they wish. Hence for the purposes of proof, it will make no difference to them whether its existence will be in the sphere of real magnitudes.* And in this he is right: the mathematical world and the physical world are not one and the same, although in some sense the one may be a good approximation to the other.

As to the existence of a lower bound to numbers, there is obviously one, writes Aristotle, and this naturally is: *unity.*

The present day reader is naturally surprised by this statement. It appears at first sight that Aristotle contradicts himself, after what he has written in relation to bisection ad infinitum. This is not so, for according to the ancients, numbers are only the natural (the counting) numbers:

$$1, 2, 3, 4, 5, 6, \text{and so on.} \tag{1.5}$$

And they would usually denote these numbers by words: unity, duet, trio, quartet, etc., not by numerals as we do today.

We note that zero (0) is not included among the natural numbers (it was not there, the notion of it was not there). If one understands numbers as the means of counting (as in: how many chairs, or how many days, or how many times this or the other thing happens, etc.) then clearly one has no need of zero (it is not *necessary*). Moreover, the ancients understood fractions, such as 1/2 and 1/3 and 1/4 (which they would refer to in words as the half part and the third part and the quarter part) etc., in relation to the positive integers (natural numbers). And in the same way, all *fractional* numbers a/b (where a and b are any positive integers) were understood by them in relation to these integers and, therefore, not meriting independent consideration.[2] One thing to remember is that they did not thing of numbers as points along a straight line as we commonly think of them today. A point on a line segment, according to the ancients, divides the segment into two

---

[2] See, e.g., D. Fowler, *The mathematics of Plato's academy*, Oxford University Press, 1999.

parts (smaller segments), the lengths of which relate or are to each other as (for example) the numbers 2 and 3 are to each other and then, of course, the length of the shorter part would relate to the length of the whole as 2 relates to 5, and the length of the longer part would relate to that of the whole as 3 relates to 5. The ancients certainly knew how to add and multiply fractions (or ratios of numbers). And because geometry deals with line segments, with ratios of their lengths and the angles between them when they cross, they were able to develop geometry to the highest standard (as evidenced by Euclid's books of geometry), not hindered by this way of looking at the fractional numbers, downgrading them so to speak in relation to the natural numbers. And of course, when they needed to, for practical purposes, they could measure any length by noting its ratio to a chosen (more or less arbitrarily) unit of length.

The fractional numbers referred to above belong to the wider set of *rational* numbers by which, now days, we mean all numbers *a*/*b*, where *a* can be any positive or negative integer or zero, and *b* can be any positive or negative integer. We note also that, negative numbers, like fractions, did not exist independently according to the ancients. One could always subtract a number (fractions included) from a greater number and that is all they needed.

In the early days of Greek mathematics it was assumed that the ratio of any two segments of length can always be expressed as: a/b where a and b are positive integers. And they were certainly amazed when they discovered that this is not so. The story is worth telling: it begins with Pythagoras's theorem which states that (see Appendix A): the square of the hypotenuse of an orthogonal triangle is equal to the sum of the squares of its two other sides.

The man who presumably proved this theorem was born some time around 570 BC on the Aegean island of Samos. The son of a merchant, Pythagoras was well educated and travelled widely before creating his own school at Croton in southern Italy. The Pythagoreans were fascinated with mathematics and often attributed metaphysical qualities to numbers, especially those from one (which signified the Oneness of the mind, to ten (which signified perfection). However, Pythagoras's legacy to mankind rests with his famous theorem. In any case, using Pythagoras's theorem one could show that *there are line segments whose ratio can not be written as a ratio a/b of positive integers.*

This is how it goes: we consider the orthogonal triangle of Fig. 1.2, the two sides of which have the same length (which we have taken as the unit of length). According to Pythagoras's theorem the square of the hypotenuse of this triangle is equal to: $1^2 + 1^2 = 2$. Therefore the ratio of the hypotenuse's length to that of its sides is equal to the square root of two ($\sqrt{2}$). It can be shown (see Appendix A) that the $\sqrt{2}$ can not be written as a ratio of integers. It is, therefore, in today's terminology, an *irrational* (not a rational) number.

In other words: the hypotenuse of the triangle of Fig. 1.2 and the length of its sides are not commensurate: there is not a unit of length in respect of which the lengths of all three of its sides are rational numbers. We can of course choose the unit of length so that the length of the hypotenuse of the triangle of Fig. 1.2 is a

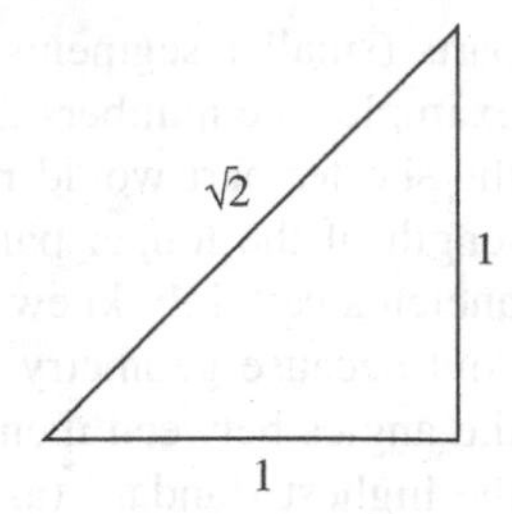

Fig. 1.2 The ratio of the hypotenuse's length to that of its sides is equal to $\sqrt{2}$

rational number (unity, for example) but then the length of its sides will not be given by a rational number.

It is said that when the Pythagoreans discovered the irrationality of the $\sqrt{2}$, they were so shocked that they kept it secret for a long while. They could not deny its existence: it was the ratio of the hypotenuse of a triangle (that of Fig. 1.2) to one of its sides, but they could not associate an (exact) rational number with it. And of course the $\sqrt{2}$ would not be the only irrational number. One can easily see, for example, that multiplying the irrational $\sqrt{2}$ with any rational number produces another irrational.

Today we can calculate the $\sqrt{2}$ in decimal form with any accuracy we require. In a first approximation we may write: $\sqrt{2} \approx 1.41$ $[(1.41)^2 = 1.9881]$; we can do better by including more digits: $\sqrt{2} \approx 1.4142$ $[(1.4142)^2 = 1.9999788]$; and we can improve on that by including further digits: $\sqrt{2} \approx 1.414213$ $[(1.414213)^2 = 1.9999984]$; and so on until we have the accuracy we want. But, of course, however many digits we include in our calculation we will not get the $\sqrt{2}$ *exactly*. The ancients did not have the advantage of our decimal system of numerals (see below), but they developed their own way of approximating irrational numbers. A method of doing so, that of continued fractions, probably due to Eudoxos, who was a student of Plato and a brilliant mathematician, is worth mentioning. According to this method the $\sqrt{2}$ is given by an (infinitely) continued fraction, which in present day notation can be written as follows:

$$\sqrt{2} = \cfrac{1}{2 + \cfrac{1}{2 + \cfrac{1}{2 + \cfrac{1}{2 + \cfrac{1}{2 + \frac{1}{2}}}}}} \tag{1.6}$$

The above fraction can be continued by replacing 2 at the bottom of the fraction by 2 + 1/2, and go on repeating this process ad infinitum, each time replacing the 2 at the bottom by 2 + 1/2. The interested reader can verify the validity of this method by comparing the value of successive values of the continued fraction to the decimal value of $\sqrt{2}$.

It turns out that while an irrational number can never be known exactly (in the sense that a rational number or a decimal number with a finite number of non-zero digits is known exactly) we can always know it (by some means or other) with any degree of accuracy we require. In other words we know it as the (approachable in the sense of Weierstrass) limit of a sequence or series. For example: the ratio of the circumference of a circle to its diameter, the very important π is an irrational number and can be known only in this way. We have already seen how Archimedes obtained a very reasonable value for this irrational via a sequence of geometrical constructions. We now know that there are a number of series converging to π, some faster than others. Here is one of them, not a fast one but the simplest I could find:

$$1 - 1/3 + 1/5 - 1/7 + 1/9 - 1/11 + \cdots = \pi/4 \tag{1.7}$$

One possible reason as to why algebra (the branch of mathematics that deals with numbers themselves rather than in relation to geometry) has not been advanced by the Greeks may be due to the fact that their way of denoting numbers was indeed cumbersome (see Footnote 2). The Greeks would often use symbols to denote numbers, especially in accountancy, but their way of writing numbers did not have the simplicity and effectiveness of present day numerals. In the Attic system, believed to be the earliest, the number 1 was denoted by a bar: I, 2 was denoted by two bars: II, *three* and *four* were denoted by three bars: III and four bars: IIII, respectively. The number 5 was denoted by the capital Greek letter Π (from $\Pi\varepsilon\nu\tau\acute{\alpha}\varsigma$, the Greek word for five); the number 10 was denoted by Δ (from $\Delta\varepsilon\kappa\acute{\alpha}\varsigma$); 100 would be denoted by H; 1,000 by X ($X\iota\lambda\iota\acute{\alpha}\varsigma$); 10000 by M ($M\upsilon\rho\iota\acute{\alpha}\varsigma$). By putting such symbols together they would write down any other number. For example: 233 would be written as $HH\Delta\Delta\Delta$III. The position of the symbols in the row representing a number had no significance, so 233 could also be written as III$\Delta\Delta\Delta HH$ or as any other arrangement of the same symbols. Sometimes they would use superscripts to shorten the representation of a number, writing, for example, H^{Π} instead of $HHHHH$. And with time they introduced more letters to denote large numbers, but their representation of numbers remained cumbersome nevertheless. Simple 'mechanical' calculators had to be invented (by Archimedes among others) to facilitate calculations involving addition and multiplication of large numbers. Roman numerals, familiar to us by inscriptions we see on monuments and on the walls of churches and public buildings, were equally cumbersome. With a few exceptions, as in *IX* meaning 9 (*I* which denotes 1 being subtracted from *X* which stands for 10) whereas *XI* means 11, the position of the symbols in the row that represents a number has no significance in Roman numerals, as was the case with the Attic numerals. And there lies the weakness of both systems in relation to the decimal system we use today to represent numbers. The shape of the symbols we use (1, 2, 3, 4, 5, 6, 7, 8, 9) is not by itself important. What is important is the fact that: in the row that represents a number the digit in the first column (from right to left) means *units*, the one in the second column means *tens*, that in the third column means *hundreds* and so on. This is what makes addition and multiplication of numbers relatively easy and straightforward. This, the Hindu-Arabic system of numerals, was introduced to the West by Leonardo of Pisa in a book he wrote in the year 1202 AD. It is generally accepted that this way of writing numbers originated in India and was taken over by Arab philosophers and scientists, and it is from the latter that Leonardo of Pisa came to know of it. To begin with, 0 (zero) was not among the digits employed in the representation of a number, so when a number had no (say) *tens* or *hundreds* in it, the corresponding columns would be empty as demonstrated in the addition of 403 to 649 shown below (where I have used dots to indicate the position of the columns):

$$\begin{array}{cccc} \cdot & \cdot & \cdot & \cdot \\ & 6 & 4 & 9 \\ + & 4 & & 3 \\ \hline 1 & & 5 & 2 \end{array}$$

However, empty columns could easily go unnoticed by tired accountants, and it is perhaps to avoid such errors that 0 was first introduced (to denote a vacant column) into our way of representing numbers. One may assume that by the end of the 13th century the decimal way of writing numbers as we know it today was complete, and this together with the use of exponents to write large numbers, as in 10^3 (= 1,000) or 10^6 (= 1,000,000), etc., made calculations involving large numbers much easier.

However, one must not assume that the ancient mathematicians shied away from large numbers. Archimedes actually calculated the number of grains of sand one would need to fill up the 'universe' for no other reason, it seems, than to demonstrate that one can deal with very large numbers if need be, and that therefore to talk about an upper limit to numbers (as some people apparently did at that time) was unfounded. Archimedes assumed the Aristotelian model of the universe (see next chapter): a grand sphere with its centre at the centre of a motionless spherical earth, the stars orbiting about it fixed on the revolving outer sphere of the universe. Using estimates by Aristarchus (who was a contemporary of Archimedes working in Alexandria, where the two of them probably met) concerning the perimeter of the earth and the distance of the sun from the earth, and assuming that the fixed stars on the outer sphere of the universe were not much further away than the sun, he showed geometrically that the size of the universe was not larger than a myriad (ten thousand) times the size of the earth. He then went on to estimate an upper bound to the number of poppy seeds (spherical with a diameter equal to 1/40 of a finger's breadth) needed to fill up this universe. Finally assuming that each poppy seed contained at most a myriad grains of sand, he was able to calculate the uppermost number of grains of sand that can be in the universe. In order to deal with the very large numbers involved in his calculations, Archimedes introduced a classification of numbers in order of increasing magnitude as follows: he described the numbers up to a myriad of myriads (a hundred million or 10^8) as the *first period*. The myriad of myriads was taken as the unit for the *second period* which in turn included numbers up to 10^{16}, which was taken as the unit for the *third period*, and so on. By using these larger 'units' of numbers (more or less like we do today when dealing with big numbers) he concluded that the number of grains of sand one would need to fill up the universe was not higher than 1,000 units of the seventh period (1,000 times 10^{56}; or 10…0, the 1 followed by 59 zeros).

Archimedes presented his results, in a book entitled *Sand-reckoner*, to King Gelon of Syracuse who was apparently an enlightened person interested in such matters.

Of course, Archimedes's universe is very small compared to the actual solar system, and minutely small compared to the universe as we know it today. It is also much smaller than Aristarchus's model of the universe which was known to Archimedes. Aristarchus put the sun and the (fixed) stars on a spherical surface whose radius was much larger than the radius of the spherical universe employed by Archimedes. In Aristarchus's model the earth moves around the sun in an orbit whose radius is minute compared to the spherical surface of the stars. Aristarchus's model is clearly nearer to the real thing but, of course, Archimedes was not in a position to know it. The astronomical data at his disposal were consistent with Aristotle's model which was the simpler (an advantage) of the two models, with 'measurable' parameters, unlike that proposed by Aristarchus which contained many unknowable parameters. One must also remember that in working out the answer to a rather artificial problem, Archimedes was not much interested in the problem per second, he only wanted to show that one could deal with very large numbers in a systematic way, if the need for that arose.[3]

Exercises

1. Prove Eq. (1.4b).

   Proof: Let

$$\sum\nolimits_n \equiv 1 + z + z^2 + \cdots + z^n; \text{ then: } \sum\nolimits_{n+1} \equiv 1 + z + z^2 + \cdots + z^n + z^{n+1} = 1 + z\sum\nolimits_n \tag{1.8}$$

   We also have:

$$\sum\nolimits_{n+1} \equiv \sum\nolimits_n + z^{n+1} \tag{1.9}$$

From (1.8) and (1.9) we obtain: $1 + z\sum_n = \sum_n + z^{n+1}$; Therefore:

$$(1 - z)\sum\nolimits_n = 1 - z^{n+1} \tag{1.10}$$

In the limit $n \to \infty$, $z^{n+1} \to 0$ when $-1 < z < +1$, and Eq. (1.10) reduces to Eq. (1.4b).

[3] For a comprehensive review of the mathematics underlying the development of physics from antiquity to the present day, see: R. Penrose, *The Road to Reality – A Complete Guide to the Laws of the Universe*, Jonathan Cape, London, 2004.

2. Find the limit of $f(k) = \frac{4k^2+k-2}{8k^2-2k}$ as $k \to \infty$.
   Hint: divide the numerator and denominator of $f(k)$ by k^2.
   Answer: 1/2.
3. Verify by direct calculation that the continued fraction of Eq. (1.6) converges to $\sqrt{2}$.
4. Verify by direct calculation that the series of Eq. (1.7) converges to $\pi/4$.

Chapter 2
The Dawn of Science

2.1 Aristotle

Aristotle was the first great scientist. The fact that he was also a great philosopher, the founder of logic, and a great literary critic, need not concern us here. There were, of course, other philosopher-scientists before and during his time, and we shall refer to some of them, as indeed Aristotle does in his own books. But Aristotle is one of the two great scientists of antiquity, the other being Archimedes, whose work defined and guided scientific endeavour for centuries.

Before I summarize for you his view of the world, let me tell you something about the man himself: Aristotle was born in 384 BC in Stagira in northern Greece, the son of a court physician at the court of king Amynthas II, the father of Philip the Great. When he was eighteen years old, he went to Athens to study at the Academy of Plato (who was a student of Socrates). This school was founded by Plato two years before Aristotle's arrival there, and was already established as a great centre of learning when Aristotle came to Athens (and remained a great school for nine hundred years until 529 AD). Aristotle stayed in Athens, in close association with the Academy, for twenty years, until Plato's death in 348 BC. After Plato's death, he moved for a period to Mysia in Asia Minor (present-day Turkey), probably for political reasons, and then to Mitylene on the island of Lesbos, where he engaged among other things in biological research. In 343 BC he returned to Macedonia at the invitation of king Philip to tutor Alexander, which he did until 340 BC. A year after Philip's death, in 335 BC, he was back in Athens. There he created his own school, the Lyceum, where he taught for twelve years. And then he had to go again.

It seems that Aristotle's connection with the Macedonian royal family made him some enemies in republican Athens, although there is no evidence that he was sympathetic to Alexander's imperial ambitions and, remarkably, in his political writings there is scant mention, if at all, of Alexander's achievements! In any case, when Alexander died in 323 BC, Aristotle left Athens to avoid charges of impiety (similar to those advanced against Socrates some seventy years earlier). He moved

A. Modinos, *From Aristotle to Schrödinger*,
Undergraduate Lecture Notes in Physics, DOI: 10.1007/978-3-319-00750-2_2,

to Chalcis in Evvia under the protection of Antipater, viceroy to Alexander, where he died in 322 BC.

2.2 Space, Time, and Matter

Aristotle's notion of space is not essentially different from the way most scientists think of it today. He points out, to begin with, that we thought of *place* in the first instance only because there is *motion with respect to place*. Place is a non portable vessel: *When what is within a thing which is moved, is moved and changes its place, as a boat on a river, what contains plays the part of a vessel rather than that of place. Place on the other hand is rather what is motionless: so it is rather the whole river that is place, because as a whole is motionless.*

Aristotle's understanding of *time* is impressively modern: We perceive time through change or motion, and if change was not there, or if we were not conscious of it, time would not exist either. In a way, he writes, the flow of time resembles the motion of a body along its trajectory; so that, so to speak, the 'now' corresponds to the body that is carried along, as time corresponds to the motion, both of them varying continuously. So it is that 'now' separates the past (what happened before) from the future (what will happen after). Aristotle claims that: time as perceived and counted by man would not exist without soul reason. What would exist and does exist independently of the observer is movement of which time is an attribute. In his own words: ... *Hence time is either movement or something that belongs to movement. Since then it is not movement, it must be the other. .... If nothing but soul, or in soul reason, is qualified to count, there would not be time unless there were soul, but only that of which time is an attribute, i.e. movement can exist without soul, and the before and after are attributes of movement, and time is these qua numerable.*

He then introduces the concept of a clock as follows (in summary): In the same way that everything is measured by some one thing homogeneous with it, units by a unit, horses by a horse, so time is measured by a unit of time. And he observes that from all kinds of locomotion the circular movement (at constant speed) is the most appropriate in this respect and therefore one can choose the period of one such motion as the unit of time. And he adds: *This is why time is thought to be the movement of the sphere* (of the heavens), viz. *because the other movements are measured by this, and time by this movement.* The revolution of the heavens (see Sect. 2.2) defined a universal time, everywhere the same, its movement unchanged. Later scientists became accustomed to thinking of time in this way, as if it had an existence of its own beyond the clock, and lost sight of Aristotle's initial, and the more profound, insight, namely that *time has no existence separate from the clock's movement.* A view of time more readily adapted to the present-day notion of space–time introduced by Einstein in 1905.

In relation to *matter,* it must be said, that Aristotle and the other scientists of antiquity, did ask the 'right' questions (by which I mean the kind of questions that defined science ever since) but, by necessity, they could only speculate about its true nature.

There were the atomists, of whom Empedocles and Democritus were the more prominent, who postulated that: all matter consists of (invisible) minute particles, of the same substance but differing in shape, some are spheres others regular tetrahedra, there may be other shapes. And always *there is vacuum* (empty space) between the atoms of any substance (gas, liquid or solid). They claimed that only matter constituted in this way can be divided, as matter obviously does. Moreover, according to the atomists the (observed) different properties of matter (earth, water, air, fire) derive ultimately from the atoms they are made of: how many of this shape and how many of that or the other shape are in a unit mass (or unit volume) of the material. And matter changes from one form to another (as when air becomes water on cooling or the other way around when heated) by changing the ratio of, say, the spherical to the tetrahedral atoms per unit mass of the material.

And there were those, Aristotle was the most influential of them, who opposed the atomic structure of matter on various grounds. Aristotle believed that matter is always continuous, even in its rarest form (air or ether), and that there is no void (vacuum) either about or within material bodies. And, apparently, there is no vacuum beyond the earth's atmosphere either: a continuous ether-like substance fills the entire universe according to Aristotle. This is so, he claims, for otherwise one body could not influence (hinder or assist) the natural motion of another body. As it obviously happens when, for example, air hinders the downward motion of a falling body, or pushes it upward. And by the same token the motion we impart on a body by pushing on one of its sides would not be transferred to the other parts of the body if there were no continuity between the parts of the body (if there were voids within the body). What Aristotle is saying (in today's terminology) is that there is no action at a distance (a body large or small can not exert a force on another body if the two are not in contact). This is a view that prevailed until Newton's theory of gravity, who, it is said, was perplexed by the fact that bodies distant from each other attract each other nevertheless, and he did not exclude the possibility that the gravitational force was mediated by some unseen medium. And later when the electromagnetic waves were discovered people thought that their propagation implied the existence of a continuous 'ether' extending over all space. It was only after direct evidence for the existence of atomic like particles was obtained, at about the beginning of the twentieth century, that the atomic theory prevailed.

Aristotle, like the atomists, accepted that matter can change from one form to another, but can not be created out of nothing. While rejecting the atomic theory of matter (as contrary to the facts as he perceived them), he does not reject the idea that all matter may derive from a small number of basic elements (e.g. earth, water, air, fire). This may be so, he says. On the other hand, observing the great variety of properties of different forms of matter, he does not exclude the possibility that matter exists in infinite many variations, in the way that light exists, varying continuously in colour and shade.

2.3 The Universe

Aristotle's model of the universe was based on the apparent movement of the moon around the earth and similarly of the sun and of the fixed stars (fixed on the revolving outer surface of a finite spherical universe), as seen (by the naked eye) from earth. Aristotle explains (in his treatise *On the heavens*) that his view of a universe rotating about a spherical and motionless earth at its centre is *supported by the contributions of mathematicians to astronomy, since the observations made as the shapes change by which the order of the stars is determined are fully accounted on the hypothesis that the earth lies at the centre* (see Fig. 2.1).

Aristotle postulates that: it is *in the nature* of the heavenly bodies (the moon, the sun and the fixed stars) to move as they move (rotating about the earth), the way it is *in the nature* of all material things (on earth) which possess heaviness (all elements except fire) to come to rest on the surface of the earth and stay there except if they are hindered or forced to do otherwise by some other body.

Perhaps the most interesting aspect of Aristotle's model of the universe is that which relates to the falling of bodies, what we now know and refer to as gravity, and the way he connected this process with the formation of the earth. The basic idea of his theory was very beautiful and simple: The heaviness of bodies (the gravitational field we would say today) relates, according to Aristotle, to an *intrinsic* property of universal space: The universe is spherical with a centre, and it distinguishes between the two directions: the one towards the centre of the universe and its opposite (away from the centre), and it is *in the nature* of all matter (and especially earthy matter which is the heaviest) to fall towards the *centre of the universe*, unless it is constrained by some agent (another body). And this is how the earth came to be: by parts of earthy matter falling towards the centre of the universe from all directions and accumulating there, the heaviest bits nearer the centre, the lighter ones resting on the heavier ones. And this is why the earth has a spherical shape, as it was formed by different pieces coming from everywhere towards the centre.

The spherical shape of the earth, a view originally suggested by Anaxagoras, is confirmed by observations, Aristotle tells us. In his own words: *How else would eclipses of the moon show segments shaped as we see them? The outline is always curved: and since it is the interpolation of the earth that makes the eclipse, the form of this line will be formed by the earth's surface, which is therefore spherical. Again, our observations of the stars make it evident, not only that the earth is circular, but also that it is a circle of no great size. For quite a small change of position to south or north causes a manifest alteration of the horizon. … Indeed there are some stars seen in Egypt and in the neighbourhood of Cyprus which are not seen in the northerly regions; and stars which in the north are never beyond the range of observation, in those regions rise and set* (see Fig. 2.2).

Aristotle concludes his passage on the size and shape of the earth by noting that the mathematicians of his time who tried to calculate the earth's circumference obtained a figure of 400,000 states (9,987 miles, which is almost twice its actual

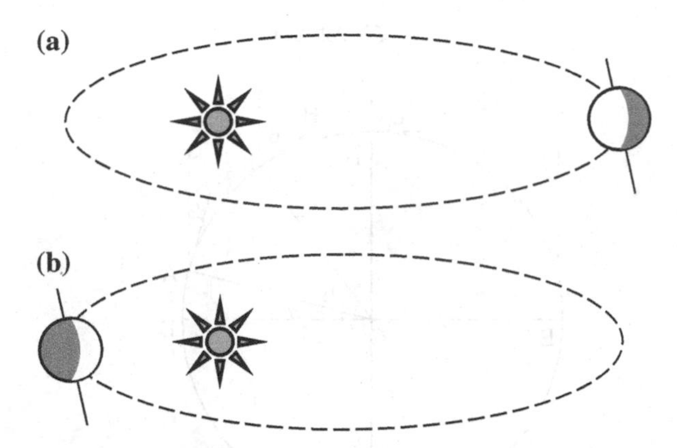

Fig. 2.1 We now know that the earth orbits the sun on an elliptical path (with the sun at one of the ellipse's foci. On the 4th of July the earth is at the aphelion, the point in the orbit farthest from the sun (1.0167 astronomical units (AU) away from it) (**a**). On the 3rd of January the earth is at the perihelion, the point in the orbit nearest to the sun (0.9833 AU away from it) (**b**). The AU (149,597,870 km) equals the mean distance to the sun. At the same time the earth rotates about its axis. The earth's equatorial plane (normal to the axis of rotation) makes an angle of 23.5° with the plane of its orbital motion. This inclination of the earth does not change as it orbits the sun: its axis of rotation points permanently to the Pole Star (Polaris). The rotation of the earth about its axis defines the 24 h day-night cycle, while the orbit of the earth defines the cycle of seasons. In the Northern Hemisphere, we have summer when the North Pole is tilted towards the sun (**a**), and winter when it points away from the sun (**b**). Because of the earth's revolution about its axis (always pointing to the Pole Star), in the Northern Hemisphere the stars appear to move from east to west in relation to the permanently fixed Pole Star. The earth completes a revolution about its axis in 24 h exactly, but because of the earth's orbital motion the stars appear to complete a revolution about the Pole Star in 23 h and 56 min and 4 s. After this discrepancy was noticed, astronomers were able to tell the time at night with reasonable accuracy. One may think that the angular distance between two stars will be different when viewed from two different positions on the earth's orbit around the sun, but such differences are minute and unobservable by the naked eye because of the vast distances to even the nearer stars. Compared to these distances the orbit of the earth around the sun is practically reduced to a point. The ancients believed that the earth was motionless at the centre of the universe. The fixed stars were attached to a revolving outer sphere of the universe, revolving on an axis through the centre of the earth and pointing permanently to the Pole Star. The apparent revolution of the fixed stars was for them a real one (its period: 23 h and 56 min and 4 s). Similarly, the sun was fixed on a revolving sphere revolving on an axis through the centre of the earth. The revolution of the solar sphere about its axis was responsible for the 24 h day-night cycle. It was further assumed that the axis of revolution of the solar sphere was inclined at 23.5° to the axis of revolution of the fixed stars, and rotated about it in a spin -top like rotation (always at 23.5°). The period of this rotation defined the solar year and was responsible of the cycle of seasons. The ancients' geocentric model of the universe was compatible with the astronomical data available to them

value of 5,400 miles), which implied according to Aristotle, that the mass of the earth compared to that of the stars is not of great size. Apparently the ancients were aware that the stars are very large (probably much larger than the sun) but seem small due to their much greater distance from the earth.

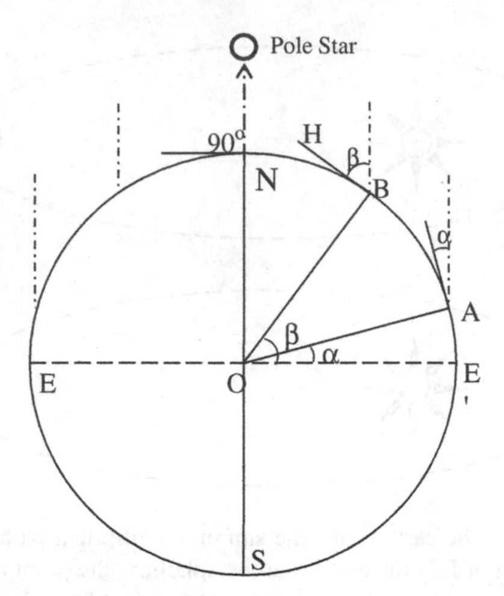

Fig. 2.2 Consider two places *A* and *B* on the meridian (the half of a great circle on the surface of the earth from the north pole (*N*) to the south pole (*S*) through a place (*E*) on the equator. The vertical lines show the direction in which the Pole Star appears at various latitudes in the Northern Hemisphere. The latitude of a place *A is* by definition the angle (α) that the radius *OA* (from the centre of the earth to *A*) makes with the equator (*EOE*). It varies from 0° (at the equator) to 90° (north) at the north pole (N). We see that the latitude of a place is equal to the angle that the direction to the Pole Star makes with the horizon of the place, and it can, therefore be determined by measuring this angle. It is also clear that as the earth rotates about its axis (*ON*), or equivalently as the heavens revolve about the Pole Star, only those stars will come into view at place *A* that enter for a time period the space above the horizon of the place which is of course determined by the latitude (α) of the place. This explains why certain stars can be seen (they rise and set) at place *A* but not at place *B*

Aristotle's initial conception of gravity as a process directly related to the geometry of space (to the spherical shape of the Universe he postulated) and his use of it to account for the spherical shape of the earth was certainly very clever, many would say a brilliant idea. Unfortunately, it was not followed (as one would expect by today's way of thinking) by careful observations of falling bodies that would clarify the process further.

The first step towards a correct description of gravity starts from the fact that, in the absence of air resistance different bodies released simultaneously from rest and from the same height reach the ground at the same time. Could Aristotle and his contemporaries establish this fact? One can not, of course, get rid of air resistance, but if the suspicion entered one's mind that air resistance played a hindering role in the falling of bodies, he should at least try (so the present-day scientist thinks) to design experiments to take account of this effect, to separate it out in some way. One could, for example, let bodies of *different* mass, but all sufficiently heavy and of nearly the same shape (in which case air resistance would be small compared to gravity and nearly the same for all of them), fall from the same high point to the ground, and compare their times of flight. One would then discover that different bodies reach the ground not exactly at the same time, but very nearly at the same time. And this would have lead, probably, to further experiments.

Aristotle did know (it is evident in his writings) that air affects the motion of falling bodies, but he did not suspect that this might be the reason why some bodies fall faster than others. And he did not do, and neither did his contemporaries, any experiments which would help them see things more clearly, and to measure things, when possible with the means at their disposal. Instead of isolating, so to speak, the falling process, they looked for it in phenomena as they happened, under a variety of conditions which could not be described in simple terms, as the falling process would be in isolation.

Aristotle noted that while all bodies (except fire) fall towards the centre of the earth when in the air, this was not the case when in water or some other media, and to account for this he introduced the idea of relative heaviness: Only earth (earthly matter) is absolutely heavy because *it* always rests at the bottom of things, and only fire is absolutely light because only *it* moves always upward. The rest possess heaviness in some media (and sink in them) and lightness in other media (in which case they float or rest at the top of them): *In air, for instance, a talent's weight of wood is heavier than a mina of lead, but in water the wood is the lighter. The reason is that all the elements except fire have weight and all but earth lightness* (Aristotle points out, in this respect, that even air possesses heaviness because *a bladder when inflated weighs more than when is empty*). And he goes on to say that the flowing or sinking of bodies in air or water depends on the ratio of the forces, the one (its weight) pulling it down (towards the centre of the earth) and that of the disruption-resisting surface of the medium (larger in water than in air because the latter divides easily). All this may be correct but it does not amount to a deeper understanding of gravity. Moreover, while many of his observations concerning falling bodies are correct, some are simply wrong (contrary to what is actually happening).

Ancient science was held back, did not go as far as it could possibly go, by the lack of experimentation. That of course changed with Archimedes, who was not only a great mathematician and theorist but a competent experimentalist as well, the equal, many would say, in his combined talents, with Galileo and Newton. We do not know much about his person except that he was born about 287 BC, lived most of his life in the Greek city of Syracuse on the island of Sicily, and died there (killed by a soldier) during the Roman attack on the city in 212 BC. The historian Polybius wrote that Archimedes helped to design catapults so accurate in their range that the Romans gave up their first attempt to invade the city by the sea. That may be an exaggeration by Polybius, because Archimedes could not have advanced so far in his understanding of ballistics, but he was certainly interested in engineering. He is known to have devised, among other things, a pump, the (so called) Archimedian screw. And he was certainly a great mathematician. We have already described (see Sect. 1.2) how he dealt with big numbers and how he was able to calculate the ratio of the circumference of a circle to its diameter using the infinitesimals of the calculus to be (though he did not name them so). And he discovered many of the characteristic properties of line shapes such as the parabola and studied the shapes of certain solids, namely paraboloids of revolution (whose vertical cross section in any plane is a parabola).

His work in physics was concerned with statics that branch of mechanics which deals with stationary bodies at equilibrium. He wrote books *On Centres of Gravity* and *On the Lever*. Of course people of his time had a vague idea of the centre of gravity of a body as the point at which the weight of the body appears to act (from the use of the balance in weighing and by observing the positions of rest of symmetrical and asymmetrical bodies suspended from a point), but it was Archimedes who gave a precise definition of it in terms of the distribution of the mass of the body, which allowed its precise determination by geometrical methods, which could then be verified by accurate measurements. Archimedes was able to calculate the centre of gravity of bodies (with a uniform mass density) of various shapes including paraboloids (the geometry of which he had already worked out), and no doubt he and his coworkers checked the results with experiments. Similarly, people new a lot about the lever at the practical level before Archimedes, but it was he who formulated in a precise mathematical manner the conditions for equilibrium, which could then be checked by precise measurements. And he was the first scientist to grasp the idea that a pure substance has measurable properties which uniquely characterize it: he proclaimed that gold could be distinguished from other metals by its weight (per unit volume), but his idea was not taken further by other scientists until much later. Archimedes's most celebrated work in physics is undoubtedly the *principle* that bears his name. It was elaborated in his book *On Floating Bodies* and appears in practically every introductory text on physics ever since.

Archimedes's principle states (in today's terminology) that: when a solid body is immersed in a fluid, the fluid exerts on the body a force opposing its weight and equal in magnitude to the weight of the volume of the fluid that has been displaced by the body (Clearly the volume of the displaced fluid is less than the volume of the body when the body is partially immersed in the fluid [it floats], and equals the volume of the body when it is completely immersed).

Archimedes based his theoretical derivation of the above principle on the following well defined assumptions[1]:

a. The various parts of the fluid lie evenly and are continuous and that which is under less pressure is driven along by that under greater pressure.
b. Each part of the fluid is under pressure from the fluid perpendicularly above it.
c. A body born upwards in a fluid is doing so along the perpendicular through its centre of gravity.

We note that (a) is in effect a statement concerning the establishment of equilibrium in a fluid. (b) is his way of saying that the fluid is in a gravitational field, and therefore at equilibrium the upward pressure of the fluid equals in

[1] A. D. Ritchie, *History and Methods of the Sciences*, The Edinburgh University Press, 1965; page 83.

magnitude the downward pressure of gravity. (c) identifies the upward motion of the body with that of its centre of gravity (It should be said that Archimedes's arguments imply that the fluid is incompressible, which is of course true at small and moderate pressures).

There is no doubt that Archimedes thought like a modern scientist: He based his theory on well defined assumptions, worked through the mathematics (geometry) to a final conclusion, and checked its validity by experiments, which he must have done.[2]

After Archimedes the Greek era of scientific inquiry came to an end. The Romans were more interested in practical engineering. And when the Roman Empire eventually fell, Christian Europe turned away from science altogether, until the Renaissance and Galileo's time.

2.4 Does Nature Act for an End?

Empedocles and Anaxagoras claimed that physical processes occur by necessity: if a certain A is there, then a certain B will follow. The business of the physicist according to these philosophers (and most scientists of today would agree with them) is to describe this process (and therefore be able to predict the B that follows A). The scientist asks no further questions. It is not his business to ask the philosophical question: *why this is so* and what is the *first* cause of all that happens? But Aristotle, the philosopher scientist, wishes to know the first cause: Could it be that certain things started by chance, as Empedocles and Anaxagoras apparently believed? Aristotle rejects this point of view. According to him, *nature belongs to the class of causes which act for the sake of something.* In the following extract from his *Physics* (Book II, Chap. 8) he dismisses the point of view of Anaxagoras and Empedocles, and states his own quite clearly. In reading this extract, one should bear in mind that both Aristotle and his opponents believed that the same laws operate in biological systems as in ordinary non-living matter. It is worth noting, in this respect, that Anaxagoras and Empedocles believed in a kind of Darwinian-like evolution of living organisms through selection: by the survival of the fittest from creatures that came to be formed more or less by chance. In their case, as in Democritus's atomic theory of matter, they appear to have guessed correctly. But their theories were merely guesses, games of the mind, unsupported by observations of any kind, and Aristotle had no difficulty in rejecting them. He goes about it as follows:

A difficulty presents itself: why should not nature work, not for the sake of something, nor because it is better so, but just as the sky rains, not in order to make the corn grow, but of necessity? What is drawn up must cool and what has been cooled must become water and descend, the result of this being that the corn

[2] R. Netz and W.Noel, *The Archimedes Codex*, Phoenix, U.K., 2007.

grows. Similarly, if a man's crop is spoiled on the threshing-floor, the rain did not fall for the sake of this—in order that the crop might be spoiled—but that result just followed. Why then should it not be the same with the parts in nature, e.g. that our teeth should come up of necessity—the front teeth sharp, fitted for tearing, the molars broad and useful for grinding down the food—since they did not arise for this end, but it was merely a coincident result; and so with all other parts in which we suppose that there is purpose? Wherever then all the parts came about just what they would have been if they had come to be for an end, such things survived, being organized spontaneously in a fitting way; whereas those which grew otherwise perished and continue to perish, as Empedocles says his 'man-faced ox-progeny' did.

Such are the arguments (and others of the kind) which may cause difficulty on this point. Yet it is impossible that this (what Anaxagoras and Empedocles claimed to be the case) *should be the true view. For teeth and all other natural things either invariably or normally come about in a given way; but of not one of the results of chance or spontaneity is this true. We do not ascribe to chance or mere coincidence the frequency of rain in winter, but frequent rain in the summer we do; nor heat in the hot days, but only if we have it in winter. If then, it is agreed that things are either the result of coincidence or for an end, and these can not be the result of coincidence or spontaneity, it follows that they must be for an end; and that such things are due to nature even the champions of the theory which is before us would agree. Therefore, action for an end is present in things which come to be and are by nature.*

This is most obvious in the animals other than man: they make things neither by art nor after inquiry or deliberation. Wherefore people discuss whether it is by intelligence or by some other faculty that these creatures work, spiders, ants and the like. By gradual advance in this direction we come to see clearly that in plants too that is produced which is conducive to the end—leaves, e.g. grow to provide shade for the fruit. If then it is both by nature and for an end that the swallow makes its nest and the spider its web, and plants grow leaves for the sake of the fruit and send their roots down (not up) for the sake of nourishment, it is plain that this kind of course is operative in things which come to be and are by nature. And since 'nature' means two things, the matter and the form, of which the latter is the end, and since all the rest is for the sake of the end, the form must be the cause in the sense of 'that for the sake of which'.

Aristotle admits that 'matter' is there and *necessary* in the sense that without it nothing can be. But what comes to be, he insists, comes to be for an end. We can put it like this: when B comes to be after a necessarily preexistent A, it does so not as a necessary sequence to the existence of A, but rather the contrary is true: A exists for the sake of B. Bricks do not exist because there is clay. The contrary is true: clay exists for the sake of bricks, which in turn exist for the sake of building houses, and houses exist for the sake of human habitation, and so on. 'That for the sake of which' is what drives nature onward, the arbiter of change, according to Aristotle.

It is absurd, he says, *to suppose that purpose is not present because we do not observe the agent of deliberating. Art does not deliberate. If the ship building art were in the wood it would produce the same results by* ***nature****. If, therefore, purpose is present in art, it is present also in nature. The best illustration is a doctor doctoring himself: nature is like that.*

This doctrine of Aristotle and the implication that flows from it: that there is a deliberating mind behind the workings of nature, was very conveniently adopted by the Popes and the ministers of the Catholic Church, who for many centuries used Aristotle's authority to discourage, when they did not ban it outright, proper scientific inquiry. It is indeed ironic that the works of an inquiring genius were reduced in this way by power-hungry bishops and the lazy minds of fanatics. And especially so, when there is nothing in Aristotle's writings to suggest that the deliberating agent behind the workings of nature that he had in mind, is the God of the Christians, as they claimed.

Exercises

2.1 A cylindrical rod of uniform mass density (ρ) comes to equilibrium when 2/3 of it is immersed in water. What is the value of ρ?
Note: $\rho_{water} = 1$ g/cm^3.
Answer: $\rho = 2/3$ g/cm^3.

2.2 Sea water has a density of 1.03 g/cm^3, and ice has a density of 0.92 g/cm^3. Evaluate the percentage of the volume of an iceberg that is above the water.
Answer: 11 %.

2.3 The true weight of a body is the one obtained in vacuum. If a body of volume V is weighed in air on a balance, using weights of density ρ, show that its true weight (W) is given by:

$$W = W^* + (V - \frac{W^*}{\rho g})\rho_{air} g$$

Where W^* is the weight shown by the balance and g is the acceleration due to gravity (9.80 m/s^2, the same for all bodies; see Chap. 6).

Chapter 3
Astronomy Paves the Way

3.1 The Astronomers of Alexandria

Alexandria, the city created by Alexander the Great on the Mediterranean coast of Egypt, became a lively centre of learning during the centuries following Alexander's death and remained so till Roman times. Outstanding geometricians, among them the great Euclid, lived and worked there, and so did the outstanding astronomers of this era. And there was obviously communication between the two. The astronomers would analyse their astronomical data using the geometricians' theorems. Astronomers and geometricians were convinced that the postulates of (Euclidian) geometry reflected the reality of physical space, which they assumed was one and the same over the entire universe.

We have seen how ancient astronomers came to believe that the earth was a sphere at the centre of the universe. In the third century BC Eratosthenes estimated the circumference of the earth by measuring the latitude difference (angle α in Fig. 3.1) of two places, namely Alexandria and Syene (present day Aswan), nearly on the same meridian and separated by a known and sufficiently large distance. He did so by measuring the lengths of the shadows of two vertical rods, the one at Syene and the other at Alexandria, at the same time: noon of the midsummer (solstice) day. At Syene the shadow was zero, implying that sunlight was incident normally on Syene, this being so because Syene's latitude (23.5°) equals the angle by which the equator plane is tilted relative to the orbital plane of the earth (see caption of Fig. 2.1). At Alexandria the length of the rod's shadow implied that the sunlight was incident at a certain angle (α) from the vertical. By measuring this angle, Eratoshenes had effectively measured the difference in latitude between Syene and Alexandria (see Fig. 3.1). And because he knew the arc length corresponding to this angle (this being the known distance between Alexandria and Syene), he could calculate the length of the complete great circle (the circumference) and the diameter of the earth. Eratoshenes's method was used subsequently by other astronomers who tried to better his estimate of the earth's diameter.

A. Modinos, *From Aristotle to Schrödinger*,
Undergraduate Lecture Notes in Physics, DOI: 10.1007/978-3-319-00750-2_3,

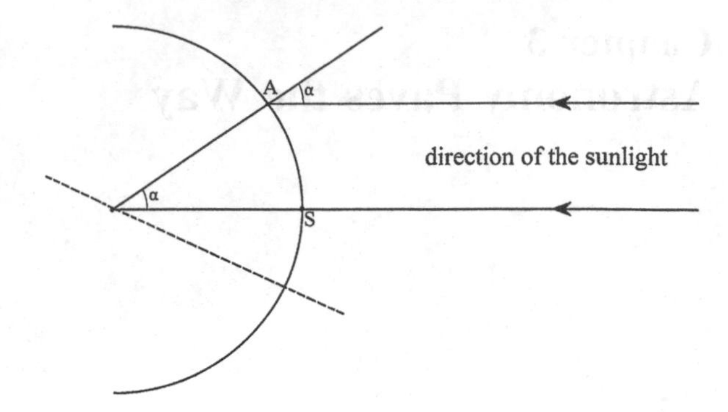

Fig. 3.1 Eratoshenes's way of measuring the circumference of the earth. The angle α has been exaggerated in the figure for the sake of clarity. The *broken line* denotes the equator

And then they looked at the moon and the sun. Aristarchus estimated the size of the moon by observing its passage through the shadow of the earth during a lunar eclipse (see Fig. 3.2). Evidently the moon was smaller than the earth and much nearer to it than it was to the sun. He noticed that the entire duration of the eclipse (when at least part of the moon was not visible) was about two times the interval during which the entire moon was eclipsed. Assuming it to be exactly two, and replacing the conical shape of the earth's shadow by a cylinder with a diameter equal to that of the earth (which is not a very bad approximation when the sun is at a much greater distance from the earth than the moon) one finds that the diameter of the moon must be less than half of the diameter of the earth. It appears (see Table 3.1) that Aristarchus believed the diameter of the moon to be 9/25 (0.36) of that of the earth (to be compared with its modern value of 0.27).[1]

The distance of the moon was more difficult to determine. Hipparchus, who lived in the second century BC, went about it as follows: He noticed that a total solar eclipse of the sun as viewed from place A (Hellespont) was only a partial eclipse when viewed from place B (Alexandria). This is so because the cone defined by the point of observation and the size of the moon (which determines what can not be seen from the given point) is different at each place. In Fig. 3.3 this cone (which we shall call the shadow cone) is shown by solid lines at place A and by broken lines at place B. One can see that the sun is entirely within the shadow cone of A, whereas a part of the sun is outside the shadow cone of place B and therefore visible from B. Now if one assumes (as Hipparchus did): (a) that the angles subtended by the discs of the sun and moon are equal when the sun is totally eclipsed, (b) that the visible (from B) part is indeed one fifth of the sun's disc, (c) that the distance between A and B is known (it could be easily estimated), then one can determine the distance to the moon (from the earth), if one already knows the diameter of the moon (estimates of which existed at the time of Hipparchus).

---

[1] A. D. Ritchie, *History and Methods of the Sciences*, The Edinburgh University Press, 1965; page 62. For more on the methods of physics from antiquity to modern times (including many references to original texts) see: G. Holton and S. G. Brush, *Introduction to Concepts and Theories in Physical Science*, 2nd Edition, Princeton University Press, 1985.

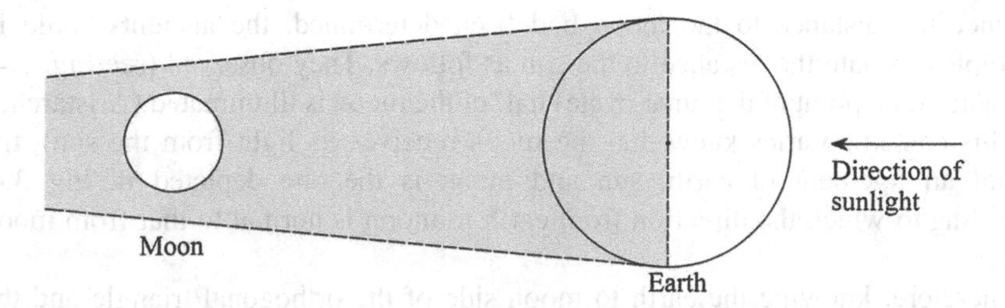

Fig. 3.2 The moon crossing the shadow of the earth during a lunar eclipse

Table 3.1 All numbers are multiples of the Earth's diameter

	Moon Mean Distance	Moon Diameter	Sun Mean Distance	Sun Diameter
Aristarchus	9.5	0.36	180	6.75
Hipparchus	33.667	0.33	1,245	12.334
Posidonius	26.2	0.157	6,545	39.25
Ptolemy	29.5	0.29	605	5.5
Modern	30.2	0.27	11,726	108.9

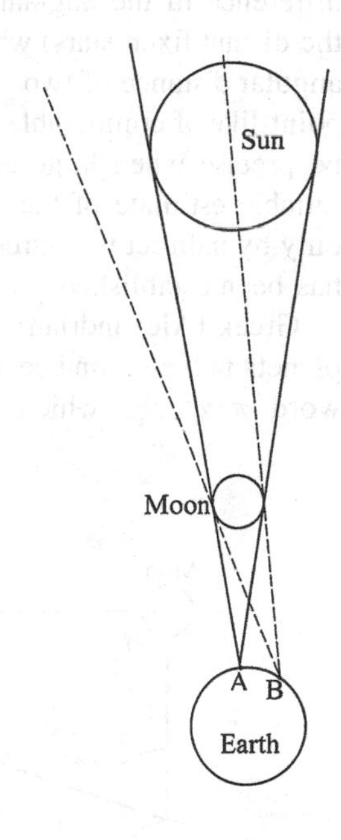

Fig. 3.3 A full eclipse of the sun at place A is only a partial eclipse at point B

We need not go through the trigonometry required to prove the above. Estimates of the (mean) distance of the moon obtained in this way by Hipparchus and other ancient astronomers are given in Table 3.1.

Once the distance to the moon had been determined, the ancients could in principle estimate the distance to the sun as follows. They observed (see Fig. 3.4) that when (at a point of the lunar cycle) half of the moon is illuminated (Aristarchus and his contemporaries knew that the moon receives its light from the sun), the spatial arrangement of earth, sun and moon is the one depicted in Fig. 3.4, according to which the direction from earth to moon is normal to that from moon to sun.

Therefore, knowing the earth to moon side of the orthogonal triangle and the angle (α) between the moon and the sun, fixes the triangle completely, and thereby one obtains the distance to the sun. The real difficulty arises in the measurement of α, which the ancients could not obtain with the required accuracy because of the very large optical errors involved. We now know that α equals 89° and 50 min, and that the distance to the moon is 30.2 times the diameter of the earth. The values obtained by the ancient astronomers for α were generally smaller (Aristarchus believed it to be 87°), and consequently their estimates of the distance to the sun, shown in Table 3.1, were not so good, except perhaps that of Posidonius (and this may be due to luck rather than anything else). And it was equally or more difficult for them to measure the size (the diameter) of the sun (see Table 3.1).

We should note in this respect that measurements of distances of bodies in the solar system and of the nearer stars outside it depend on the *parallax*. This is the difference in the angular position of a body (the direction it is seen relative to the distant fixed stars) when the position of the observer is changed. However, the angular distance of two luminous bodies can be measured precisely when both are point-like of comparable and not so bright luminosity. Such measurements can not be precise when large and very luminous bodies, like the sun, are involved. A reliable estimate of the sun's distance became possible in the 18th century (and only by indirect measurements) when the Newtonian theory of the planets' motion has been established.

Greek (Alexandrian) astronomy reached its peak with Ptolemy's work on the planets in the second century after Christ. The word *planet* derives from the Greek word πλανήτης which means *wanderer*. They were called so because they

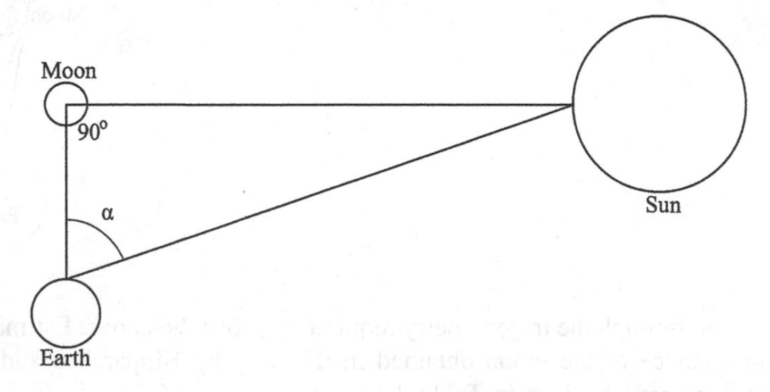

Fig. 3.4 Spatial arrangement of earth, moon and sun, when half of the moon is illuminated

appeared to follow complicated paths relative to the fixed stars. And yet the ancient astronomers, by the way of careful observations (by the naked eye assisted by the simplest of instruments) and a thorough analysis of the data, made up a geometrical model which described well the planets' celestial paths.

We now know that there are eight planets (there is also Pluto, which some say ought to be included in the list, and other staff in the solar system which need not concern us here). The planets orbit the sun on elliptical paths (not far removed from the circular). The size of the orbit (the average distance of the planet from the sun) increases in the following order: Mercury is the nearest to the Sun, then come Venus, Earth, Mars, Jupiter, Saturn, Uranus, and finally Neptune (the most distant from the Sun). [We note that at the time of Ptolemy, Saturn, Uranus and Neptune had not been discovered. They could not be seen without a telescope]. Knowing the orbit (its position at any time) of a planet other than the Earth, as well as that of the Earth, and taking into account the rotation of the earth about its axis, we can at any time (and relatively easily) determine the angular position of the other planet relative to the fixed stars, as seen by an observer from a certain place on the earth. The resulting (apparent) orbit of the other planet about the earth can look in some instances very strange indeed. We can understand why this so by reference to Fig. 3.5. When Mars and Earth find themselves in the relative position shown in (a), Mars will appear practically motionless because the earth is moving straight away from it. Similarly in (b): Mars appears stationary because the earth is moving directly towards it. And it may also appear that Mars temporarily reverses its movement, as shown in (c), when the earth overtakes Mars. Of course, in reality the two planets move with practically constant speeds all the time. The broken line in Fig. 3.5c represents the ecliptic (the apparent orbit of the sun about the earth). The proximity of the ecliptic to the orbit of Mars shows that the orbital plane of Mars and that of the earth are very close to each other. And the same is true for the rest of the planets. Ptolemy accepted the Aristotelian model of the universe with the earth at its centre and the distant stars fixed on the revolving outer sphere of the universe. He then put the moon, the sun and the planets in orbits about the earth as follows: the moon is the nearest to Earth, then come Mercury and after it Venus, then comes the sun, and after the sun come Mars, Jupiter and Saturn.

He put Mercury and Venus between the earth and the sun because they, like the moon, sometimes crossed in front of the sun, while Mars, Jupiter and Saturn never did. They would sometimes be seen in front of a fixed star, but that was understandable in view of the far greater distance of the latter from Earth. He was then able to describe, with relatively good accuracy, the observed orbits of the moon and the sun about the earth (these were circular). And he was able to describe, equally well, the apparent orbits of the planets about the earth which were not circular. He was able to fit the data (from observations by himself and other astronomers), assuming that each planet moves (with a constant speed) on an *epicycle*: a circle whose centre moves (with a constant speed) on a greater circle about the earth. The resulting path is an epicycloid (so called); an example is shown in Fig. 3.6. Ptolemy also allowed for the centres of circular orbits to be displaced, to a small degree, from the centre of the earth, when he could not

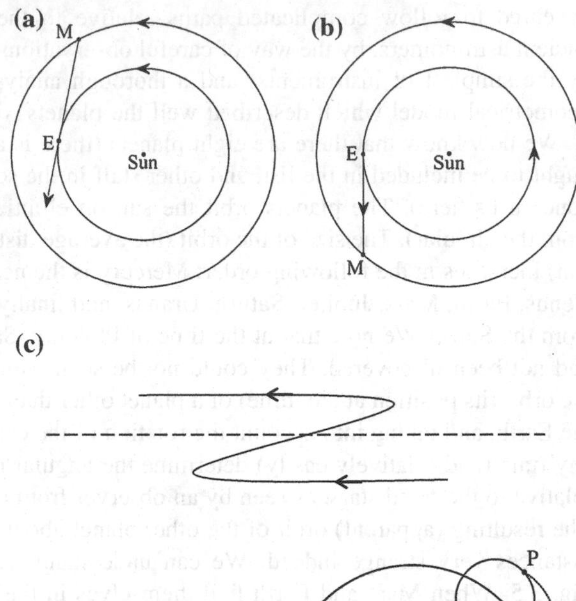

Fig. 3.5 Motion of Mars relative to Earth (schematic): In (**a**) and (**b**) Mars appears stationary when viewed from Earth. In (**c**) Mars appears to temporarily reverse the direction of its motion when the earth overtakes it

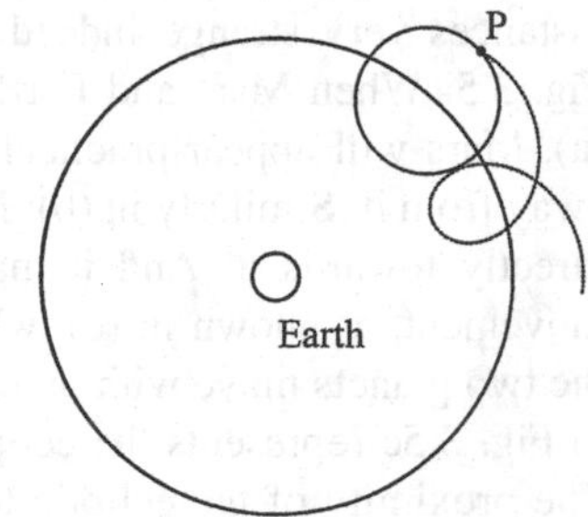

Fig. 3.6 Planetary motion according to Ptolemy

otherwise fit the data. This in effect meant the abandonment of the doctrine that all circles or spheres were centred on the earth.

Although Ptolemy's model is wrong (the planets do not orbit the earth) and laborious, it provided an effective means of presenting the raw data, and as such it was useful to later astronomers. After him it was taken for granted (until Kepler's time) that epicycloids described well the paths of the planets around the earth. And Copernicus, who came to believe that the planets actually orbit the sun, was convinced that the orbits of the planets around the sun should be epicycloids.

Although all observations were made by the naked eye, the astronomers of Alexandria were able to fix the direction of observation with sufficient accuracy using the Astrolabe, an instrument they developed themselves, which mimicked a spherical universe with the earth at its centre. The great circles at the meridian and the equator (the equatorial plane at the appropriate angle to the ecliptic) defined the reference frame by which the angular positions of the fixed stars were known and, in turn, those of the planets, as a function of time.

And for the measurement of time they would rely on the revolving heavens. The crossings of the meridian by recognizable fixed stars were a most reliable clock (of local time) at night. During the day, they would normally use an ordinary sundial (like those seen today in some old gardens) to tell the time, by following the shadow of a pointer on a horizontal base. The pointer was aligned with the earth's axis (by aiming it to the Pole Star at night), which meant that the projection of the pointer on the horizontal plane corresponded to noon time (when the sun crosses the meridian and is therefore at its highest point on the sky). The angle between the pointer and the horizontal plane, and therefore the length of its shadow, would of course vary with the latitude of the place (see Fig. 2.2).

3.2 Copernicus and his Advocacy of the Heliocentric System

Nicholas Copernicus was born in Torum on the banks of the river Vistula in 1473. He was the son of immigrant parents from the region of present day Silesia. At some stage his father's family must have been merchants of copper, hence the name of Koppenigk which his son changed to the Latin Copernicus. Nicholas Copernicus was born at the time when science was taking its first timid steps after a very long sleep. The printing press was invented by Johan Gutenberg about the time of his birth, and this meant that the number of people who could read and write was growing year by year. And along with it, learning through experiment, the hallmark of modern science, was gradually coming out of the shadows.

The telescope had not yet been discovered. Glass making, known to the Romans, had gone into decline, and even glass windows were very expensive to produce and could only be afforded by the very rich, during most of the Middle Ages (the time between the Dark Ages that followed the collapse of the Roman Empire and the Renaissance). However, spectacles had been known since the late thirteenth century. It is probable that they were invented by an English friar, Roger Bacon, at about 1262. But it was about 20 years later that the Florentine Salvinus de Amatus began producing spectacles for the public, because experimentation on lenses, as in everything else, was discouraged by the Church at the time. Bacon actually applied to the Pope requesting his permission to write a book about the advantages of experimental methods, and he got it on the condition that he would send the finished book to the pope in secrecy. He obviously needed the pope's permission to write such a book, fearing he might be accused of interfering with God's intentions by his endeavour to facilitate a man's sight with the use of spectacles. But he did not in the end avoid punishment for his labours. In 1277 he was imprisoned for two years by his Order (the Franciscans) for "novelties" in his teachings. He apparently entertained the idea of producing lenses that would make it possible to see distant objects as if they were near, and lenses that would make it possible to see minute objects which were invisible to the unassisted eye. The telescope was eventually discovered by the Dutchman Hans Lippershey in 1608, and perfected by Galileo in 1609. The reason it took three and a half centuries after

Bacon to make the telescope is an obvious one: fear. The innovator who could see distant objects as if they were near, and tiny things that could not be seen by the unassisted eye, using instruments that he devised, could easily be accused of being himself an instrument of Satan and severely punished.

Copernicus was well educated by the standards of his time. His father could pay for it, and when his father died he was supported by his uncle who was a man of the Church. At the age of 18, Copernicus became a student at Krakow University, which was founded in 1364 and was by this time a renowned centre of learning. There he studied, among other things, mathematics, and read the works of Aristotle and other philosopher scientists of antiquity (as was common in the established universities of his time). It was probably at Krakow that he first read Ptolemy's book on astronomy. After graduating from Krakow, the young Copernicus was planning to become a canon (a kind of monk) at the cathedral of Frauenburg, where his uncle had already been established as a bishop. In this way he would have a job and financial security for life. As there was no post immediately available, his uncle sent him to Bologna to continue his studies in law. Once there, Copernicus studied also Astronomy with professor Novara, and it is known that Novara had strong doubts about Ptolemy's model of the universe. He returned to Frauenberg in 1500 where both he and his brother became canons of the cathedral. In 1501 Copernicus was back in Italy. He stayed there for two years sharing his time between Padua and Ferrara. In Ferrara he studied law (and got his doctorate in 1503), and in Padua he studied medicine although he did not get a degree. Students of medicine at that time would study the texts of Hippocrates and Galen (who lived in the second century), would learn about natural drugs and their use in the treatment of certain illnesses and they would perhaps practice some minor surgery (primitive at the time as it was performed with little knowledge of anatomy and without anesthetic). Copernicus was probably not much interested in medicine, but like many clergymen of that time he needed to know a bit of medicine to be able to help his parishioners in the absence of a doctor. He was, nevertheless, a man of many interests and varied talents, a renaissance man in the manner of his great contemporary Leonardo Da Vinci. He could paint well. A 17th century copy of a self portrait of his exists to this day at Krakow's Jagiellonian University. And he found the time to translate from Greek to Latin a collection of 85 brief poems, called «Epistles», by the 7th century Byzantine historian Theophylact Simocatta.

On returning to Poland, Copernicus joined his ageing uncle, bishop Lucas Waczennode, at his castle in Ermland. He served the bishop as his personal doctor, secretary and adviser on matters of economics and administration. During this time he made no astronomical observations but began to develop his heliocentric theory, and actually wrote a *Little Commentary* on it, and gave hand written copies of it to friends he thought he could trust. In the year 1512 his uncle died and Copernicus moved back to Frauenburg to finally take up his post as canon, living in the precinct of the cathedral with 16 other canons.

In the year 1513 Copernicus was asked to prepare a proposal for calendar reform. It had been assumed that there were exactly 365 ¼ days in a year, when in reality the true year is slightly less. Julius Caesar implemented his own calendar by adding an extra day every four years. The resulting small error in the Julian Calendar resulted in the actual calendar being approximately 10 days in advance of what it should be, at the time of Copernicus. This meant that the beginnings of the seasons were misplaced, which had obvious consequences in agriculture at a time when the farmers relied on the calendar for the sowing of plants and other activities. And the Church needed to accurately know the age of the moon on the 21st of March (which was taken as the first day of the year), in order to fix the date of Easter Sunday and other religious events. Copernicus promised Pope Leo that he would try to determine the true length of the year.

This was a difficult undertaking, not least because there seem to be two different years. One was the solar year: the time it takes for the sun to return to a position where it is the same number of degrees above the horizon of a certain place as it had been previously at, for example, midday in midsummer. The length of this (the solar) year equals 365 days, 5 h, 48 min and 46 s. One would expect (see caption to Fig. 2.1) that at the end of the solar year the fixed stars would be at exactly the same positions on the night sky as they were at the beginning of it. But it turns out that this occurs 20 min and 24 s later. In other words: the length of the year as determined by the apparent revolution of the fixed stars, known as the sidereal year, is 20 min and 24 s longer than the solar year. So that, after many years the dates of an event given in solar years and sidereal years respectively will be very different. We now know that the difference between the solar year and the sidereal year derives from the fact that the direction of the earth's rotation axis is nearly but not exactly constant (see Fig. 3.7). It seems that Hipparchus of Alexandria suspected the non constancy of the rotation axis of the earth (it would be the axis of rotation of the heavenly sphere in his case) by comparing his own data with those obtained by Babylonian astronomers 200 years before him. And Copernicus would have known of this assertion. In any case, whatever the cause of the discrepancy between the solar and the sidereal year, it meant that one had to decide which of the two to take as the unit year and stick to it in order to obtain a reliable calendar. From a scientist's point of view it makes no difference, the one is as good as the other. But there were other considerations to be taken into account in relation to this matter. It would seem that for agriculture a calendar based on the solar year is preferable, but the Church might prefer the sidereal year as the basis for the dates of religious events. In any case, Copernicus would like to know as accurately as possible the length of both years. He set out to do so and much more. Copernicus would determine the orbits of the planets as well as those of the sun and the moon as accurately as was possible using the data obtained by earlier astronomers, which he would check and occasionally supplement with his own observations.

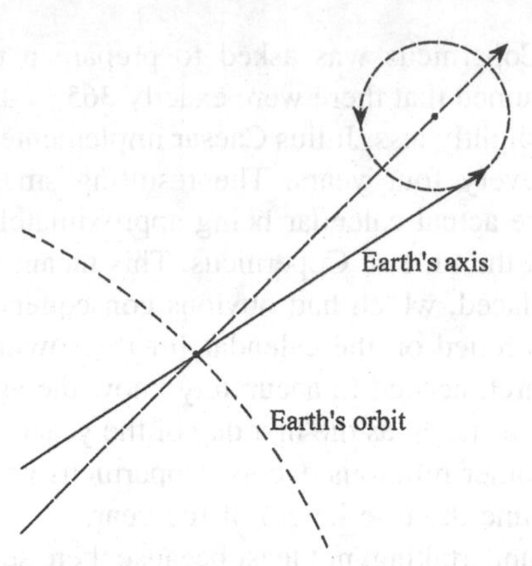

Fig. 3.7 The direction of the earth's axis of rotation changes in the manner of a spinning top as indicated by the broken circle, but the angle it makes with the orbital plane of the earth remains constant. Now the rotation axis points to the Pole Star (changing its direction extremely slowly), in 12,000 years it will point to the star Vega, and in 26,000 years it will again point to the Pole Star, and will go on repeating the cycle every 26,000 years. This so called precessional motion of the earth is well accounted for by the combined gravitational pull of the earth by the sun and the moon. In any case, since we define the meridian by aligning to the direction of the Pole Star, we observe an apparent rotation of the heavens in consequence of the above motion of the earth's axis

After years of meticulous analysis and hard work he concluded that the data were consisted with a heliocentric model of the universe.[2] By putting together the motions of the earth (the rotation about its axis and its orbit around the sun) with those of the other planets (around the sun), he could fit the available astronomical data. Of equal importance was the fact that, in Copernicus's model, the fixed stars (in the outermost region of the universe) were at rest, they were not moving. He realized also that the earth and the (other) planets went around the sun in egg-shaped orbits, not in circles as he expected.

Copernicus believed that the sun, being the giver of light and heat, ought to be at the centre of the solar system, installed there by God. And he similarly believed that the planets orbited in paths that God ordained for them, and these, he thought, would be circles as the most perfect. If not circles, then they ought to be combinations of circles in the manner of Ptolemy. He therefore tried and tried again to fit his data to increasingly complex epicycloids (we use the term to denote any

---

[2] According to Copernicus the sun is not only at the centre of the solar system but also at the centre of the universe (with the fixed stars in the outermost celestial sphere). The centre of the earth is, according to Copernicus, only the centre of gravity and of the lunar sphere. He also asserted that the distance between the earth and the sun is negligibly small compared to the distance between the sun and the fixed stars. And that explains the absence of parallax (difference in the apparent position of two stars on the celestial sphere when viewed from different positions, in this case due to the orbiting of the earth. A very small such parallax was observed (using a powerful telescope) in 1838.

combination of circles not only that of Fig. 3.6). Apparently Copernicus never tried to fit his data to ellipses. And when after much effort over many years he failed to describe the orbits by epicycloids he began to suspect the data he used in his calculations. And there was another fact which worried him: the angular distances between the fixed stars appeared the same when viewed from widely different points on the presumed orbit of the earth around the sun. He could explain it by assuming that the fixed stars were far away from the solar system, but he could not be certain about it. And finally, being a religious man, he must have been to some degree at least reluctant to contradict the doctrine of his church. He began to doubt his own theory. He certainly kept his ideas to himself and published nothing.

Fortunately, in the year 1539 Copernicus found a coworker, George Rheticus, who was at the age of 25 already a professor of astronomy at the University of Wittenberg and a protestant. Copernicus gave his data and calculations to Rheticus who devoted himself to checking and refining these calculations, until eventually he managed to fit the data to very complex epicycles. [This would not be the last time in the history of physics when reluctance to give up cherished ideas led to unnecessary complexity]. Copernicus could now tell the world about his heliocentric model of the universe. To avoid the expected controversy, Copernicus considered publishing his astronomical tables without explaining the underlying heliocentric model. But, eventually, encouraged by Rheticus and his friend, the bishop Giese, he allowed Rheticus to publish an initial, popularized account entitled: *Narratio Prima* (First Account), outlining the ideas of the heliocentric system. Rheticus prepared the manuscript and in February of 1540 presented it to the printers in Danzig. Copies of the *Narratio* were sent to interested persons in Protestant and Catholic regions of Europe. And there was a second edition of the book printed by an admirer in Basle. Many among those who read the *Narratio* wanted him to publish the complete work and joined his friends in urging him to do so. Copernicus was convinced and the work was eventually published in 1542. Its title: *De Revolutionibus Orbium Coelestium Libri VI* (Six Books Concerning the Revolution of the Spheres). The book was published as Copernicus was dying. He was shown the book but he could not read it as he had already suffered a stroke. Which was in a way a blessing to him, because the publisher worried about the consequences of its publication, and probably in order to protect his author, changed the preface to the book, without asking Copernicus, and instead of saying that the book's aim was to affirm the heliocentric system, he presented it (the heliocentric system) as no more than a mathematical device in the calculation and prediction of astronomical events (presented in the Tables). Of course those who mattered, the astronomers and the mathematicians got the real message of the book.

Strangely, Copernicus failed to acknowledge in his book the important contribution that Rehticus made to the content and the presentation of the book. Rehticus was understandably deeply hurt by this omission. He eventually lost his interest in astronomy. He lost his job at the University following a drunken homosexual assault on a young student, and ended his life as an ordinary physician rather than as the renowned astronomer he expected to become.

Copernicus's book was officially banished by the Catholic Church 60 years after his death, by which time his ideas became well known to astronomers and others.[3]

3.3 Kepler and his Contribution to Astronomy

Johannes Kepler was born in 1571 in the Stuttgart region of Germany. His father left his family when Johannes was five years old, and probably died in the 80 years War in the Netherlands. His maternal grandfather was an inn keeper, and it seems that he took good care of his daughter's family. After completing his primary and secondary school education in the Württemberg state-run education system, Keppler studied at the University of Tübingen. He read theology, philosophy, mathematics and astronomy. He must have been an admirer of Aristotle. In later years he would describe his own work as « a supplement of Aristotle's *On the Heavens* », and as « an excursion into Aristotle's *Metaphysics* ». He was also interested in astrology and he certainly enjoyed casting horoscopes for his fellow students. At the end of his studies he got a job as teacher of mathematics and astronomy at the Protestant school in Graz (in Austria).

Kepler's admiration for Aristotle did not stop him from becoming a Copernican. He believed, like Copernicus, that the sun is the principal source of motive power in the universe and as such ought to be at the centre of it. In his first book on astronomy, the *Mysterium Cosmographicum* (The Cosmographic Mystery), published in 1597, Kepler looks at Copernicus's universe with Pythagorean eyes. He argued that the planetary orbits were placed around the sun in accordance with certain geometrical structures (which he endeavoured to discover) made up of the five Platonic solids (tetrahedron, cube, octahedron, dodecahedron, and icosahedron) inscribed and circumscribed by spheres. And he suggested that the above, God's geometrical plan of the universe, was itself a representation of the divine: the Sun (at the centre of the universe) represents the Father, the stellar sphere at the outermost of the universe represents the Son, and the space between represents the Holy Spirit. In the first manuscript of his book there was also a chapter (which he deleted at the request of the senate of the University of Tübingen) where he cited biblical passages which, he claimed, supported heliocentrism.

In 1600, after refusing to convert to Catholicism, Kepler had to leave Graz. Interestingly, during his last months in Graz, Kepler wrote an essay, *In Terra inest virtus quae Lunam ciet* (There is a force in the earth which causes the moon to move), suggesting that the moon's motion is caused by some (quasi-spiritual) force exerted on it by the earth. It must be understood that, when Kepler says that the motion of the moon is caused by the said force, what he means is that if this force was not there the moon would be at rest (not moving), and similarly if the

[3] For more on Copernicus see: I. Crow, *Copernicus*, Tempus Publishing Ltd, 2003.

force disappeared the moon would stop at the point it were at the moment the force vanished. Kepler dedicated this essay to Archduke Ferdinand hoping, perhaps, for a position as mathematician to his court. But this did not happen. Kepler and his wife (Barbara Müller) would go to Prague. They had two children but both had died in infancy. They would later have a daughter and two sons.

In Prague, Kepler was offered a job by the astronomer Tycho Brahe. Kepler would help him complete the *Rudolphine Tables* (of astronomical data) he was working on at the time. And when Brahe died unexpectedly, in October 1601, Kepler took over his post as imperial mathematician with the responsibility to complete the unfinished *Rudolphine Tables.* The work he did during his 11 years as imperial mathematician turned out to be of great importance to the development of astronomy. This work went beyond the *Rudolphine Tables,* and had nothing to do with his other official duties which included providing astrological advice to the emperor and the casting of detailed horoscopes for allies and foreign leaders. And one may add that his salary for all the services he provided was hardly adequate, which contributed to his quarrels with his wife who was sick and unhappy in Prague.

His research which culminated in his book *Astronomia Nova* (A New Astronomy) began with the analysis of Brahe's data on the orbits of Earth and Mars. He calculated these orbits many times, but he could not devise a geometrical model for them that would fit Brahe's data within the accuracy of Brahe's observations (two arc minutes, which is 2/60 of a degree). That was until he had his epiphany. It sprang from his work in optics. He assumed to begin with (this assumption is an extension of the one he postulated in relation to the moon's motion around the earth) that the planets are kept in motion by a motive force emanating from the sun. He then supposed that this motive power weakens with the distance from the sun, the way the intensity of the light emitted from a point source weakens with the distance from the source, causing the planet to move faster or slower as it moves closer or further away from the sun. This led him to a close examination of the data on the aphelion and perihelion of Earth and Mars after which he proposed a formula according to which a planet's rate of motion is inversely proportional to its distance from the sun. He subsequently reformulated this rule, mainly in order to facilitate its verification over the entire orbit of a planet, as follows: *Planets sweep out equal areas in equal times*, now well known as Kepler's second law of planetary motion (see Fig. 3.8). Kepler then tried to fit the data on Mars to an egg-shaped orbit using this law. After forty or so unsuccessful attempts he at last (in 1605) tried an ellipse, which he overlooked for so long because (so he said) it was so simple that he assumed that it would have been discovered by earlier astronomers. But when he discovered that an ellipse fitted the data on Mars, he immediately concluded that: *All planets move in elliptical paths*, which became Kepler's first law of planetary motion. As he had no calculation assistants he did not at this stage extend his analysis to the other planets. He had nevertheless enough material for his *Astronomia Nova* which he completed by the end of 1605. But the publication of the book was delayed until 1609, because of legal disputes with Brahe's heirs concerning the use of Brahe's data which they claimed as their property.

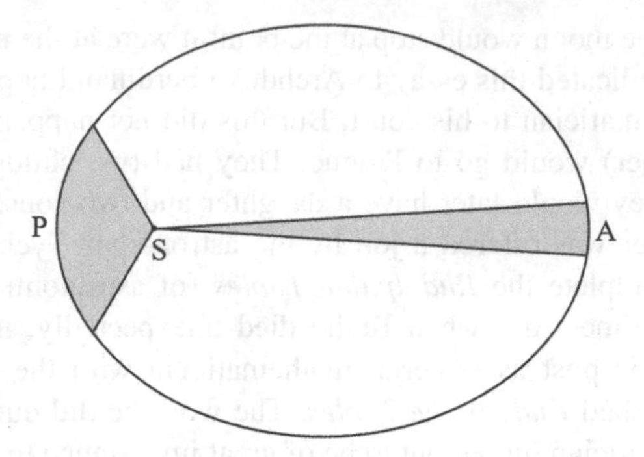

Fig. 3.8 The planet's orbit is an ellipse with the sun (*S*) at one of the two foci of the ellipse. At the perihelion (*P*) the planet moves faster than at the aphelion (*A*). The area (*shaded*) swept by the planet in a given time-interval at P, is equal to the area (*shaded*) swept by the planet in an equal time-interval at A. A more detailed description of an ellipse is given later (in Fig. 5.4)

In 1610, a year after the publication of *Astronomia Nova*, Galileo discovered, using his own telescope, four satellites orbiting Jupiter. After the publication of his findings under the title *Sidereus Nuncius* (Starry Messenger) he sent a copy of his work to Kepler, who responded enthusiastically with a short published reply, entitled *Dissertadio cum Nuncio Sidereo* (Conversations with the Starry Messenger), endorsing Galileo's observations. In contrast, Galileo never published his reactions to Kepler's *Astronomia Nova,* which was a disappointment to Kepler. Apparently, Galileo would not give up the epicycles of Copernicus in favour of the ellipses of Kepler.

During the time he was analyzing Brahe's data on Mars, Kepler did also some pioneering work in optics, the results of which he published (in 1604) in a book entitled *Astronomiae Pars Optica* (The Optical Part of Astronomy). There he described such things as the reflection of light by flat and curved mirrors, he established the inverse-square law (which states that the intensity of the light emitted from a point source weakens proportionally to the inverse of the square of the distance from the source), and he dealt with issues related to atmospheric refraction which applied to all astronomical observations. He also discussed the optics of the human eye. He realized (the first to do so) that images are projected inverted by the eye's lens onto the retina, to be subsequently corrected, as he put it, *in the hollows of the brain due to the activity of the soul*. Kepler's interest in optics stayed with him, and soon after Galileo's discovery of the refractive telescope he proposed an improved version of it (the Keplerian telescope), which is still in use in small observatories (Fig. 3.9).

Kepler left Prague for Linz in 1612, soon after his wife died of Hungarian spotted fever. In Linz he took a post as teacher and district mathematician. And he, of course, continued with his work on the *Rudolphine Tables*. After a difficult period in his life (his son Friedrich died at the age of six), made more difficult by the political and religious upheaval of the times, he settled into a new marriage

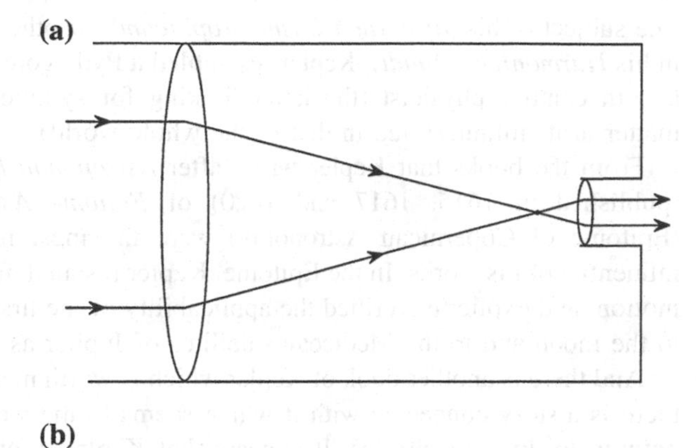

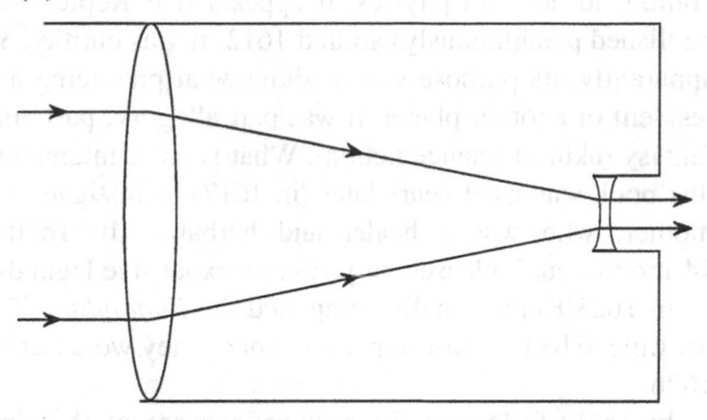

Fig. 3.9 Schematic diagram of a telescope. It basically consists of two lenses, separated by an adjustable distance, in a tube. The first (*convergent*) lens collects the light from the observed object, and the resulting image is magnified by a second lens (*the eyepiece*). Kepler used a convergent eyepiece (**a**). Galileo used a diverging lens as eyepiece (**b**)

with the much younger Susanna Reuttinger, which appears to have been much happier than his first marriage.

In the years after *Astronomia Nova*, Kepler worked more on the preparation of the *Rudolphine Tables*, but he did find the time to write a number of other books as well.

He believed that *the geometrical things have provided the creator with the model for decorating the whole world.* His book *Harmonices Mundi* (Harmonies of the World), published in 1619, begins with an exploration of regular polygons and regular solids. He then goes on to consider the harmonies of music, meteorology and astrology. He writes, in relation to the latter, that the tones made by the souls of heavenly bodies interact with human souls. Kepler, though he thought that most rules and practices of astrology were *evil smelling dung*, he would say that *there was also perhaps a good little grain in it.* The final part of *Harmonices Mundi*, and the most important, deals with planetary motions and in particular with the relation between the orbital velocity of a planet and its distance from the sun. Using Brahe's data and his own analysis of the data he proposed that: *The squares of the periodic times are to each other as the cubes of the mean distances* (from the sun), which became Kepler's third law of planetary motion.

It is fair to say that in his search for harmonies in the physical world, whether that concerned the arrangement of the planets' orbits in the space around the sun (the subject of his *Mysterium Cosmographicum*), or other aspects of it (dealt with in his *Harmonices Mundi*), Kepler resembled a Pythagorean and, at the same time, a 20th century physicist (the latter looking for symmetries in the structure of matter at its minutest and in that of the whole world).

From the books that Kepler wrote after *Astronomia Nova,* the three volumes (published in 1615, 1617 and 1620) of *Epitome Astronomia Copernicanae* (Epitome of Copernican Astronomy) were the most important, and the more influential of his works. In the Epitome, Kepler restated his three laws of planetary motion, and explicitly verified the applicability of the first two laws to all planets, to the moon and to the Medicean satellites of Jupiter as well.

And there is another book of Kepler which is worth mentioning, mostly because there is a story connected with it which seems to me worth telling though it has nothing to do with physics. It appears that Kepler wrote this book (which was published posthumously) around 1612. It was entitled *Somnium* (The Dream) and, apparently, its purpose was to show what practicing astronomy would mean to a resident of another planet. It was part allegory, part autobiography and part sheer fantasy (akin to science fiction). What is more interesting is, that his manuscript of this book was used years later (in 1617) to instigate a witchcraft trial against his mother, who was a healer and herbalist. In 1620 she was imprisoned for 14 months, and released only after an expensive legal defense drawn up by Kepler.

In 1623 Kepler finally completed the *Rudolphine Tables* which was thought at the time to be his most important work. They were published at his own expense in 1626.

In 1628, following the military success of the Emperor Ferdinand's armies under General Wallenstein, Kepler became an official adviser to Wallenstein, and would occasionally write horoscopes for him. He died during a visit to Regensburg, in November 1630, and was buried there. He had written his own epitaph[4]:

Mensus eram coelos, nunc terrae metior umbras
Mens coelestis erat, corporis umbra iacet.

It translates as follows:

I measured the skies, now the shadows I measure
Skybound was the mind, earthbound the body rests.

Although Kepler's laws were not immediately accepted (they were ignored not only by Galileo, but by Rene Descartes as well), several astronomers did test some of Kepler's predictions, though not always successfully. For example, in 1631 Pierre Gassendi observed for the first time a transit of Mercury across the face of the sun on the date predicted by Kepler, but failed to observe the transit of Venus some time later due to inaccuracies in the *Rudolphine Tables*. Similarly, Kepler's elliptical orbits were only gradually accepted, as was his idea of attractive forces (though not the quasi-spiritual ones postulated by him) between the sun and

[4] A. Koestler, *The Sleepwalkers*, page 427.

the planets. However, his work was fully appreciated only when Newton used Kepler's (empirical) laws of planetary motion to establish his own theory of planetary motion in the context of universal gravity.

Exercises

3.1 Consider Fig. 3.3. Denote the angle at A by a (in radians). Derive an approximate formula for the distance (D) of the moon from the earth, in terms of the diameter (M) of the moon and the distance (L) from A to B, subject to the assumptions made by Hipparchus (see text).
Note: a and L could be measured and M was known.
Answer: $D \approx \frac{5}{6\alpha}(M + L)$.

3.2 Consider Fig. 3.4. Convince yourself that knowing the earth to moon distance and the angle α uniquely determines the earth to sun distance.

Chapter 4
Galileo: His Life and Work

Galileo Galilei was born in Pisa in 1564, the year Michelangelo died. In the same year Shakespeare was born. His father, Vincenzo Galilei, was a talented musician with unorthodox views on music, and a generally independent mind. In one of his books (*Dialogo della musica antica e della moderna*) he wrote[1]: *those who try to prove an assertion by relying simply on authority act very absurdly.* And he apparently had the same attitude in matters of philosophy and religion. There is no doubt that in this respect Galileo took after his father. His mother, had no other interests beyond her family, financial security, and social advancement. Unfortunately Vincenzo Galilei, in spite of his good efforts, never earned enough money to satisfy all the needs of his family. Galileo was the first and, for a while, the only child of the family. In time he had six siblings but only three survived to adulthood (two sisters and a brother). Galileo learned his first letters at the local primary school at Pisa. When, in 1575, his family moved to Florence, Galileo was tutored by a local teacher in Latin, Greek and Rhetoric, but this was only for a few months. He continued his studies (as a resident student) at the monastery of St. Mary of Vallomprosa (very near to Florence). He stayed at the monastery for three years, and it seems that at some stage he entertained the idea of becoming a priest. His father, however, would have nothing of it. He took his son back home claiming that the monks allowed him to fall ill with an eye infection. Galileo returned to Pisa where he was taught at the famous boarding school of Sapienza. Two years later, the 17 years old Galileo began a degree course in medicine at the University of Pisa. By this time he was already a polymath. Thanks to his father he was an accomplished lutist, and he was also a very good painter. Above all he was extremely intelligent and he knew it. There is no doubt that his insistence on seeing things his own way often appeared as arrogance to many of his fellow students and to some of his teachers.

Galileo soon discovered that he was much more interested in mathematics than medicine. He began skipping lectures in medicine to attend lectures in mathematics. Instead of reading Galen, he read Euclid and Archimedes. Archimedes

[1] M. White, Galileo ANTICHRIST, (Phoenix), 2009. The interested reader will find references to the original texts in White's book.

A. Modinos, *From Aristotle to Schrödinger*,
Undergraduate Lecture Notes in Physics, DOI: 10.1007/978-3-319-00750-2_4,

would be for him the model scientist, combining mathematics with observation and experiment to discover the laws of nature. His parents allowed him (they were paying for his studies) to pursue his studies in mathematics on the condition that he would at the same time continue to completion his degree course in medicine. Galileo agreed but in fact he stopped attending medical lectures altogether.

According to Viviani (a student of Galileo and his first biographer), the student Galileo made his first scientific discovery in 1583, as a result of his observation of the swinging movement of an oil lamp, hanging from the ceiling of the church, under the influence of air currents. He noticed that, when the lamp swung through a longer arc it moved faster than when it swung through a shorter arc, and established, using his own pulse to measure time, that the period of the swinging was the same for any length of the arc. When at home he made a pendulum, by tying a metal bob to a string, and measured its period as accurately as he could, using his pulse (there was not a better way of doing it at that time). He confirmed that the period of the oscillation did not vary with the arc length of the oscillation. It did not change as the latter became gradually smaller during the pendulum's motion. He further established, by varying the mass (weight) of the bob and the length of the string, that: the period of the oscillation does not depend on the mass of the bob but it does depend on length of the string.[2] Galileo realized that the pendulum could be turned into a practical device for measuring time and, being a student at the medical school at the time, he actually wrote a document suggesting that, using the pendulum one could make a device for measuring a patient's pulse. He submitted his paper together with a prototype of the device (which he called *pulsilogium*) to the authorities of the University. The authorities of the University, acknowledged its value but attributed the invention to «the Medical Faculty».

Evidently, Galileo was not much liked by his teachers at the University. So when, in 1584, Galileo was in need of a scholarship, they refused him one. By this time his family was in a dire economic situation and his father could not support him anymore. Galileo left the University without a degree. For a period he earned some money (for him and in order to help his family) by teaching mathematics to children of wealthy families in preparation for university courses. At the same time he was looking for ways to advance his own scientific career. In 1586, working at home (he lived with his family), he constructed a *little balance* (basically a bar suspended from a wire) which could measure small weights with greater accuracy than existing devices. And he wrote a document on it, which he hoped would help him in his quest for a university job. During this period (1586–1587) Galileo applied for a job at the Universities of Bologna, Siena, Padua, Pisa and Florence, but he failed in all these attempts. Then, towards the end of 1588, he was invited to give a talk at the Academy of Florence that changed his career prospects dramatically.

The Academy of Florence was at the time a nexus for humanists and thinkers who would gather to discuss various topics and to listen to invited speakers.

[2] We now know that the period of such a pendulum increases in proportion to the square root of the length of the pendulum.

A topic that interested them at the time related to the dimensions and location of Dante's inferno. The president of the Academy, Senator Baccio Valori, decided to ask a natural philosopher to give a lecture on this topic. Why he choose the little known Galileo, we shall never know. Perhaps none of the established professors was willing to talk on the subject and the task fell on Galileo by default. The topic of the talk may appear ridiculous to us, but that was not so for the thinkers of that time, who thought that Dante was writing about the real Hell using information given to him by God. Galileo assumed that much and proceeded to estimate the size of Hell as follows[3]: *Let us speculate on the size of Lucifer, he said, there is a relation between the size of Dante and the size of Nimrod in the pit of Hell, and in turn between Nimrod and the arm of Lucifer. Therefore if we know Dante's size, and Nimrod's size we can deduce the size of Lucifer.* Using clues from Dante's poem he estimated that Lucifer was nearly two thousand yards tall. And he went on to demonstrate using some mathematics that, Hell had the shape of a cone with its apex at the centre of the earth, and taking about one twelfth of its volume. His audience, which included many of the most important citizens of Tuscany, was greatly impressed. Within three months of his talk at the Academy of Florence, the grand duke offered him a three-year contract for the post of professor of mathematics at the University of Pisa. He took his post in the autumn of 1589.

The University of Pisa was at the time much smaller than and not as reputed as the Universities of Bologna and Padua. Most of its students studied law and only a few of them graduated. And only a very small number amongst them attended Galileo's lectures (based mostly on Aristotle, Euclid and Archimedes). He gave tutorials to small groups, conventional lectures and occasionally public lectures. Galileo taught every day except Sunday, and all that for a very small salary. Apparently his unconventional behaviour (he refused to wear an academic gown which, he claimed, was only a disguise of intellectual inadequacy), and his independent thinking on scientific matters did not make him many friends among his colleagues. He was still a very young man and often felt more comfortably among his students. Like most young men of his time, he was fond of the tavern, and soon discovered the pleasures of gambling and whoring.

Among his fellow professors was a philosopher, Girolamo Boro, who a few years before Galileo's arrival at Pisa decided to test experimentally Aristotle's assertion that, heavier (more massive bodies) fall to the ground much faster than bodies of lesser mass. Boro concluded (in his treatise) that Aristotle was right, and that balls made from the same material when dropped from the same height landed at different times depending on their mass. Galileo, who had already done some experiments with falling bodies, asserted that Boro (and Aristotle) were wrong. But nobody took him seriously (he was after all a young professor arguing against Aristotle's wisdom and Boro's experience). Whether (as the story goes) Galileo let different objects fall from the leaning Tower of Pisa or gathered his data in a more practical way, by 1590 he had sufficient evidence to conclude that: all bodies fall at

[3] see Footnote 1.

the same speed irrespective of their physical nature or their mass. He wrote a short treatise, *De Motu,* presenting his experimental results and his conclusions in a fashion that would stay with him (and perfected in the process) in all his subsequent writings. He would set up a conversation (in the manner of a Platonic dialogue) between an enlightened individual, asking the questions and evaluating the answers, and two scientists representing alternative explanations, the one would be that of Galileo, the other that of his opponents (the Aristotelians in *De Motu*). Galileo was aware that bodies (in air) were not falling with *exactly* the same speed but he suspected that this was due to some secondary cause which could be set aside, to be clarified at a later stage. This is the mark of genius: to separate (guess) the essential from the inessential and have the intellectual courage to declare it as such in the face of established 'truth' or 'wisdom'. Galileo would later write (in his *Discourses on Two New Sciences*): *Aristotle says that a hundred pound ball falling from a height of a hundred cubits hits the ground before a one-pound ball has fallen one cubit. I say they arrive at the same time. You (my critics) find on making the test, that the larger ball beats the smaller one by two inches. Now behind those two inches you want to hide Aristotle's ninety nine cubits.*

We now know that in a vacuum tube a feather falls as fast as a sphere of lead, but to know this we had to wait until the invention of the vacuum pump (not long after Galileo's death).

Galileo never published *De Motu.* In the years that followed its conception, he carried out some exquisite experiments that clarified further the motion of falling bodies, the results of which he published towards the end of his life in the *Discourses on Two New Sciences.* I shall describe them a bit later.

In 1592 Galileo finally succeeded (although he was not much liked in academia, he had some influential friends in high society) in obtaining an important academic job in a city where he felt comfortable. He was appointed to the post of Professor of Mathematics at the University of Padua with a salary three times that of Pisa. Padua was at the time a prosperous town within the Republic of Venice, at a time when Venice was the most liberal of the catholic regions (free to a large degree from the Pope's jurisdiction and the hated Inquisition). Unfortunately Galileo's father had died in 1591, and Galileo was now the only one in his family earning a salary. His sister was about to be married and expecting a dowry, and his brother Michelangelo who at 17 left the family to settle in Padua was, as always, in need of money. In spite of his good salary Galileo could not make ends meet. He took on private students but that did not help much. He decided to become an entrepreneur scientist.

He constructed a thermometer, but it was not good enough for commercial exploitation. In 1593 he invented a device for raising water from the earth *most easily at little expense*, to use his own words, but he only managed to sell just one such device. Apparently, he never tried, or if he tried he never succeeded, to construct a reliable time measuring device based on his discovery of the pendulum.[4]

---

[4] The first to construct a working clock using a pendulum was the Danish physicist Christian Huygens, who also published (in 1673) a book on the subject: *Horologium Oscillatorium.*

Eventually (in 1597) he found success with the invention of the 'military compass'. This had nothing to do with the ordinary compass used in navigation. It was a primitive slide rule that one could use to perform any simple arithmetical operation. It was a success: mathematicians, accountants, artillery soldiers, every one loved it. Galileo employed a craftsman, Marc Antonio Mazzoleni, and the two of them developed a production method that resulted in accurate and nice looking devices at a cost which brought them a good profit. By the turn of the century, Galileo made enough money from his military compass to pay his debts and buy a beautiful three-storey house close to the palazzo of his friend Gianvincenzo Pinelli.

It was at the house of Pinelli, who was apparently a rich play boy but had at the same time a free inquiring mind, that Galileo met Marina Gamba, a prostitute who became his mistress and the mother of his three children[5]: Virginia born in 1600, Livia two years later, and his son Vincenzo in 1606. Galileo did not at any stage formally acknowledge them as his children, but always took good care of them and of their mother, providing for them, and securing for them all the opportunities that society would allow. Marina and the children lived in a modest house next to his residence and apparently saw each other regularly if not daily. His mother, of course, did not like Marina, but she surely loved her grandchildren and had them stay with her quite often. That does not mean that Galileo's children did not suffer the inconvenience and, to some degree at least, the shame that the society of the time attached to illegitimate children (especially those born to middle-class parents). And yet all the evidence suggests that his love for them, and their love for him were true and constant.

Galileo, now financially secured, could return to pure research. When he did so in 1600, he was concerned with two different areas of science: Mechanics which he began studying in Pisa, and Astronomy.

Mechanics: The experiments he conducted in Pisa showed that bodies fall equal distances in equal times, but one could not tell from those experiments how the speed of the body changed during the fall. Did the speed change continuously or in successive but distinct installments? And if the change was continuous, was it also a uniform one (the speed increasing by the same amount in a unit of time)? The difficulty Galileo had in answering these questions arose from the fact that, the increase in the speed of the falling body was too rapid to be measured accurately by using his own pulse or other means at his disposal. Galileo went forward as follows: He *assumed* that the motion of a ball rolling downwards on an inclined plane (with a well smoothed surface to minimize friction) is *qualitatively* the same with that of a freely falling body.[6] And he set out to study with meticulous care *this* motion.

---

[5] see Footnote 1.

[6] Strictly speaking, here and throughout this chapter, by *motion of the rolling sphere* we actually mean *motion of the centre of mass of the rolling sphere*. Because of the *rolling* aspect of the motion, extrapolating the results obtained for rolling spheres on an inclined plane to free fall, is not at all an obvious proposition. It is fully explained only in the context of Newton's mechanics (see exercise 7.12).

Galileo made a grove on the planar surface of a beam of wood, about 6 m long, and covered it with a thin layer of the right material to make it as smooth as possible. Holding the beam at an angle to the ground, he let (one at a time) balls of different mass density (and therefore of different weight) roll down the groove from a certain height to the ground. Each time, he measured the time it took the ball to roll from the chosen height to the ground, using the following device: At the bottom of a large vessel which he filled with water he attached a narrow tube through which water would flow into a smaller vessel during the descent of the ball. The weight of the water collected in this way provided a measure of the time of descent. He repeated the experiment starting the balls from a different height (with the beam inclined at the same angle to the ground). After many such experiments he concluded with certainty that (with the beam inclined at the same angle):

The ratio of the distances two balls travel (in their descent from their initial position on the beam to the ground) *is equal to the ratio of the squares of the times of their travel*. (4.1)

We can write the above statement in modern notation (see next chapter) as follows:

$$S = kT^2 \tag{4.2}$$

where S denotes the distance travelled and T the time of travel, and k is a constant (a number with appropriate 'dimensions' attached to it; if, for example, T is measured in seconds and S in meters, then k must be expressed in (meters per (second)^2). [We remember that when two symbols representing different quantities appear next to each other, a multiplication is implied]. Equation (4.2) tells us that, if in a unit of time the ball has traveled a certain length (L), in two units of time it will travel four times L, in three units of time it will travel nine times L, and so on. It must be clearly understood that, as long as we do not attach a specific value to the constant k, Eq. (4.2) tells us no more than Galileo's statement (4.1). Galileo's statement is not less mathematical than Eq. (4.2) because it is expressed in words. It is also clear that because of the way he measured times in his experiments, Galileo could not attach a reliable value to the constant k. What he established beyond any doubt, is that this constant does *not* depend on the mass of the ball or on its initial position (height) on the beam. And he realized, of course, that when the beam was inclined at a higher angle to the ground, the balls were rolling faster to the ground, which means that the constant k in Eq. (4.2) increases with this angle.[7] After he established that much, Galileo wanted to know, what the statement (4.1) implied for the acceleration of a rolling ball.

[7] In the experiments of Galileo the inclined plane could not be very steep, in order to secure a rolling motion of the ball (avoid sliding which complicates the motion).

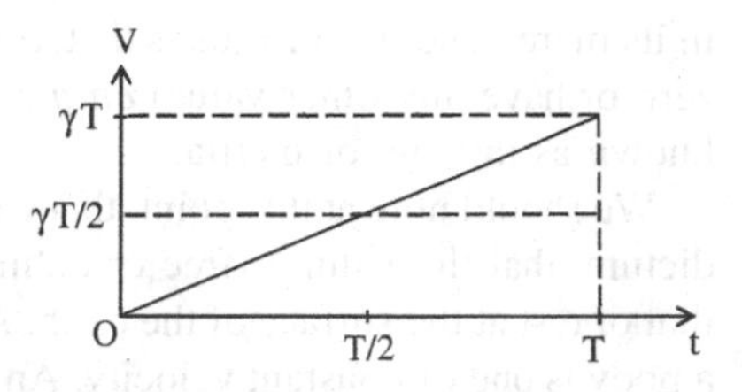

Fig. 4.1 In uniformly accelerated motion, the speed increases linearly with time

Let me summarize his analysis using the modern way of writing things mathematical. Let us assume that the acceleration of the rolling ball is constant (depending only on the inclination angle of the beam) and let us denote this constant by γ. In other words: the ball comes down the beam with a speed (V) which increases with the time (t) uniformly, as follows:

$$V = \gamma t \tag{4.3}$$

which tells us that the ball starts from rest ($V = 0$ when $t = 0$); after a unit of time has elapsed it acquires a certain speed, which is doubled after two units of time have elapsed, tripled after three units of time have elapsed, and so on. We are now accustomed to a graphical representation of Eq. (4.3), as shown in Fig. 4.1. Along the horizontal axis we measure the time (t) and along the vertical axis we measure the speed. At the start of its descent ($t = 0$) the particle is at rest ($V = 0$), at the end of its descent ($t = T$) the speed equals $V = \gamma\ T$, at half-time ($t = T/2$) the speed of the ball equals $V = \gamma\ T/2$, and so on. We can see that the *average* speed of the ball, over the time (T) of its descent, equals $\overline{V} = \gamma T/2$, which is indicated by the broken line in Fig. 4.1. From which it follows that the distance (S) travelled during its descent equals:

$$S = \frac{\gamma T}{2} T = \frac{1}{2} \gamma T^2 \tag{4.4}$$

which is the same with Eq. (4.2), with $k = \gamma/2$. The above equation shows that Galileo's observations, summarized in statement (4.1), imply that the motion (of the centre of mass) of a rolling ball on an inclined plane is uniformly accelerated. Galileo knew, of course, that a ball rolling down an inclined plane was not a free-falling body, but he 'saw' that the two were related, and from that he concluded that free fall was also a uniformly accelerated motion. Galileo did not translate the acceleration of the centre of mass of the rolling balls into ordinary units [as in: so many (meters per second) per second], and made no guess as to the value of the acceleration in free fall.

Of equal, perhaps of greater importance, is Galileo's observations relating to the rolling of balls on a horizontal plane. After many experiments he was convinced that: *A body will retain its velocity when the external causes for its acceleration (or deceleration) seize to exist.* This happens, he noted, only when a body moves on a horizontal plane: A ball set rolling on a frictionless horizontal plane will go on rolling for ever in the same direction. This 'law' of Galileo, which in its more general form states that, *a body will move with constant speed* (it can be zero or have any other value) *on a straight line if no force is acting on it*, is now known as the law of inertia.

We should note at this point, that Galileo's law of inertia contradicted Aristotle's dictum, that the natural (free or unhindered) state of a body possessing weight is that of rest at the surface of the earth. According to Galileo, the natural state of such a body is one of constant velocity. An Aristotelian could of course claim that a ball rolling on a horizontal plane would finally come to rest, because however smooth the surface of the plane some friction would remain. We now know that the subsequent development of science proved that Galileo was right to think of friction as a secondary factor (a nuisance) that one can disregard in formulating a fundamental principle. Of equal importance for the development of science was his insistence on quantifying his observations by the use of mathematics in the manner initiated by Archimedes.

Galileo worked for about three years on mechanics. And among other things, he established that the motion of a projectile occurs along a parabola. He completed these studies at the beginning of the 17th century, but many years past before he wrote a complete treatise on it, in his book *Discourses and Mathematical Demonstrations Concerning two New Sciences*, which was published in 1638, four years before he died. At the same time that he was doing his experiments on rolling balls, he became interested in astronomy. We know that by the time Kepler published his *Mysterium Cosmographicum*, Galileo was a Copernican. In 1609, having heard of Lippeshey's telescope (how it was made up of two lenses with an adjustable distance between them), Galileo hurried to produce his own version of it (see Fig. 3.9), so as to usurp any business opportunities that the new instrument could provide. And these were obvious. One need only mention its usefulness to the navy, the army, and to leisure activities such as hunting. It is said that Galileo managed, through his friends, to delay Lippershey's visit to Venice (the Duchman was in Padua on his way there) until he had completed his own instrument. And he surely succeeded to be the first to present a telescope to the doge of Venice, who immediately took a great interest in the device. The doge doubled Galileo's salary, turned his professorship into a lifetime appointment, and offered him a cash reward equal to a year's salary. Apparently, Galileo did not have any theoretical knowledge of optics at the time and never did. He was able to make a telescope through practical know-how (he made the lenses himself) and improved Lippeshey's device using his natural talent. It must also be said that he never felt guilty about having stolen (to a degree at least) Lippeshey's design.

Galileo was hoping that the doge would allow him to concentrate on his researches and that he would not have to teach anymore, but the doge insisted that he should continue to lecture at the University of Padua. By the end of 1609 Galileo, who had been a conscientious but often frustrated teacher (he once said *that one can not teach anything to anybody, one could only help those who wanted to learn by themselves*), could not bear his teaching duties any more, so when the opportunity arose (in 1610) he moved to a new position in Florence. He would be the Chief Mathematician and Philosopher to the Grand Duke of Tuscany. His salary would be the same as that of Padua, and he would be exempted from any teaching duties. What he did not realize at the time was that, by leaving the relatively democratic Republic of Venice to settle in monarchical Florence, he lost the protection that Venice could offer him against the Inquisition and the reactionaries of his time. His mistress did not follow him to Florence, but his children did. In Florence they would spend more time with their grandmother Giulia (Galileo's mother).

While still in Padua, Galileo observed the moon through his telescope and found that it was not the *perfect sphere* that most philosophers thought it to be, that it had craters and mountains not unlike those on earth. And he also discovered four planets of Jupiter, as we have already noted. The book he wrote detailing his observations (Conversations with the Starry Messenger) was well received throughout Europe, and as a result Galileo became well known.

In 1611, after settling in Florence, Galileo began observing the planet Venus. In due course, he suggested that, the variation of the planet's brightness over the year can be explained if one assumes that, Earth and Venus orbit the sun (as Copernicus had suggested), that Venus is illuminated by the sun (like the moon) and that its apparent brightness depends on its position relative to both Sun and Earth. There were, of course, alternative suggestions (by Tycho Brahe and others) as to the cause of the variation in the brightness of Venus, based on the geocentric model of the universe. Over the following years Galileo made many other observations. He observed, for example, that the Galaxy which appears continuous to the naked eye is in fact a collection of very many stars. And he made worthwhile observations of sunspots, in cooperation with his assistant Filippo Salviatti, which unfortunately led him into trouble. This had nothing to do with sun rays damaging his eyes. A former student of his, the Benedictine monk Benedeth Castelli had devised a technique of letting the sun fall onto a white card placed at an appropriate point near the eyepiece of the telescope which could then be observed without harm to the observer's eyes. It was trouble of a different kind. According to Galileo the sunspots could appear or disappear unexpectedly. Galileo did not have a sure explanation for these spots, only a tentative one: they originated from within the sun, they appeared and disappeared because the sun was rotating about its axis, and possibly for other reasons, perhaps accidentally. It so happened that during the same period, a Jesuit priest, Father Christopher Scheiner, who was a professor of mathematics at Ingolstadt (in Bavaria), was also observing the sunspots, and he explained them away by assuming that they were caused by stars coming in front of the sun as the latter orbited the earth. At first sight that was no more than an academic dispute. In any case before publishing his book on the subject, *History and Demonstrations about Sunspots and their Properties*, Galileo wrote to Cardinal Conti, the then Prefect of the Roman Inquisition, admitting that his theory was anti-Aristotelian, but what he did not know and needed to know was whether it was actually anti-doctrinal. The cardinal replied that it could not be anti-doctrinal, as there was no mention of it in the Bible, but that he was certain the spots were due to stars coming in front of the sun. Eventually, Galileo's publisher obtained the required papal permission and the book was published in 1613. The book was received with much interest, but also with great hostility by his opponents, especially as it was written in the vernacular allowing ordinary people to read his progressive ideas. Galileo described (in a later book entitled *Ansayer*) the reception of his book on the sunspots as follows: *How many men attacked my letters on sunspots, and under what disguises! The book should have opened the mind's eye for admirable speculation. Instead it met with scorn and derision.*

The ecclesiastical authorities were becoming increasingly aware that Copernican ideas could damage orthodoxy (though these ideas had not been declared by then heretical). What worried them most was that Copernican ideas were becoming the catalyst of independent thought; people like Galileo were dangerous to them. In December 1614 a Dominican priest, Tomaso Caccin, attacked Galileo openly (in a sermon) as an antichrist. And he went further. He requested and obtained an interview with the Pope, and put to him the case against Galileo, which rested, of course on the latter's Copernicanism. Eventually, Galileo had to go to Rome to explain himself. He tried to convince the bishops of the Church that his ideas, and Copernicanism, did not contradict the Gospels. It did not help. Eventually, the proceedings were formalized in a document, signed by the Pope Paul V, which concluded as follows[8]: *... the Father Commisionary, in the presence of a notary and witnesses, is to issue him* (Galileo) *an injunction to abstain completely from teaching or defending this doctrine and opinion or from discussing it; and further if he should not acquiesce, he is to be imprisoned.*

On the 26th of February 1616 Galileo agreed to the demands of the Pope and promised not to write or teach in defense of Copernicanism.

Eight days after Galileo was officially silenced, Copernicus's *De Revolutionibus Orbium Coelestium* was taken out of circulation.

Galileo, of course, continued his deliberations on Copernicanism. *I do not feel obliged to believe*, he said, *that the same God who endowed us with sense, reason and intellect has intended us to forgo their use.* And in due course he composed a book on the matter: *Dialogo Sopra i Due Massimi Sistemi del Mondo* (Dialogue on the Two Principal World Systems) which was published in 1632.

In his book Galileo argues in favour of the Copernican System using a variety of arguments. He points out, to begin with, that the assertion made by the Aristotelians, that every celestial body orbits about the earth, has been shown to be wrong. His discovery of the Jupiter's satellites proves that *not* every celestial body orbits about the earth. He goes on to clarify misconceptions that arise in relation to the rotation of the earth: A stone falling from the highest point of a very tall tower hits the ground at the base of the tower. If the earth were rotating, his opponent argues, the stone ought to land at some distance from the tower, left behind by the earth's motion. Not so, argues Galileo, because the stone retains the velocity (parallel to the surface of the earth) it had at the moment of its release, due to inertia. The horizontal displacement due to this velocity (denoted by ***A*** in the diagram of Fig. 4.2) is added to the vertical displacement due to the fall (denoted by ***B***) to produce the observed displacement (denoted by ***C***).

---

[8] see Footnote 1.

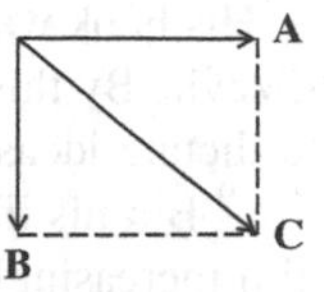

Fig. 4.2 Displacement *A* is added to displacement B to produce displacement *C*

Finally, he argues that other assertions made by the Aristotelians in defense of the geocentric system, though compatible with the available astronomical data, are far fetched, unappealing to the scientist's mind. It makes more sense, he argues, to assume that the small earth rotates about its axis in 24 h than to assume that the whole celestial sphere rotates about the earth in the same period. Moreover, there is something awkward in the geocentric system, he claims, in the way the periods of motion of celestial bodies change with the distance from the earth: we go from the $27\frac{1}{3}$ days of the moon's period, to the 30 years period of the most distant planet, and then we fall back to a mere 24 h period for the fixed stars on the outer sphere of the universe. Whereas in the heliocentric system we go progressively from the $27\frac{1}{3}$ days of the moon's period, to the 30 years period of the most distant planet, and from there to the infinitely large period of the fixed stars (meaning that these stars are practically static) in the outermost region of the universe. It looks much better, more reasonable.

We need not list here all of Galileo's arguments in favour of the heliocentric model. And, of course, he was fully aware that none of them constituted a proof of its reality. They were good reasons for preferring it (tentatively) to the geocentric model. The Copernican System eventually won, when it acquired a firm foundation in Newton's theory of universal gravity.

The publication of his book triggered an attack on Galileo by his old enemies. They discovered the document of 1616 that warned him against teaching or defending Copernicanism. This document made it possible for the Inquisition to bring Galileo to trial for heresy with an apparent breach of a ruling of the Church. And they did so in 1633. However, Galileo had some powerful defenders who managed to commute what could have been a capital sentence to house imprisonment for life.

During the six months before his confinement at his house in the village of Arcetri (in Tuscany), while he was the guest of his friend and supporter, the Archbishop of Siena, he wrote his most important book *Discourses and Mathematical Demonstrations Concerning Two New Sciences*, based on the work he had done over a lifetime. The first *New Science* dealt with his experiments and conclusions relating to the motion of bodies (the rolling balls on the inclined plane, etc., the law of inertia and the related mathematical analysis). The second *New Science* was concerned with the microscopic properties of material bodies. Galileo had experimented on the strength of metals and other materials, on the elasticity and cohesion of matter, on the viscosity of fluids, and other such things. And he drew original conclusions about these properties which he quantified mathematically. These were instrumental in the development of the study of matter in the years after him.

His book was published a few years later (in 1638) by the Duch publisher Louis Elzevir. By this time Galileo was ill and almost blind. He continued to work, and to dictate ideas to his devoted assistants Torricelli and Viviani, to the end of his life.[9] But his illnesses (arthritis among them) and the isolation from society made him increasingly bitter about his confinement and the way he had been treated by the Church.

The last months of his life were miserable. He was very ill and totally blind. Galileo died on 8 January 1642, at the age of 77, in the room in which he had spent most days of the last eight years of his life. With him, at the hour of his death, were his son Vincenzo and his faithful student Viviani.

Exercises

4.1. A particle moves on a straight line with an average velocity of 4 m/s for 3 min, and then with an average velocity of 5 m/s for 2 min.
(a) What is the total displacement of the particle during this time?
(b) What is the average velocity of the particle over this time?
Answer: (a) 1,320 m; (b) 4.4 m/s.

4.2. Sketch: (a) a velocity versus time graph showing how the velocity of a particle thrown vertically upward changes with the time. (b) a position versus time graph for the same motion.

4.3. Summarize for yourself the advantages of the Copernican system in relation to the geocentric system of the universe.

4.4. You may wish to elaborate on the following statement: Archimedes established, through his work on levers and on solid bodies in liquids, a mathematical theory for static bodies in interaction with each other. Galileo made the first important step towards a theory of the dynamics of motion of solid bodies. Both made use of concepts introduced or clarified by Aristotle. All three of them owe a lot to the work of the ancient mathematicians and astronomers.

[9] Evangelista Torricelli, then in his early thirties, would later become grand-ducal mathematician at the Tuscany Court. He is rememberd for his discovery of the barometer.

Chapter 5
The Seventeenth Century: The Bloom of Science

5.1 The Continuum of Numbers. Functions. Descartes' Analytical Geometry

After the acceptance of zero as a number, following the adoption of the decimal way of writing numbers (see Sect. 1.2), it became gradually common to think of numbers as points on a straight line: the positive numbers extending, in order of increasing magnitude, from zero (0) to infinity ($+\infty$) on the right, and the negative numbers extending from zero to 'minus infinity' ($-\infty$) on the left, as shown in Fig. 5.1.

We know that the rational numbers (the numbers: a/b, where a can be any negative or positive integer or zero, and b any negative or positive integer) do not make up a continuous straight line. We can see that this is so by noting that, any two rational numbers, a/b and c/d, however close they are to each other, have the rational number $(a/b + c/d)/2$ between them, with small but finite line-segments between it and the numbers a/b and c/d. Using a decimal representation does not make any difference (decimal numbers with a finite number of non-zero digits are, indeed, rational numbers). There is only one way out of this difficulty: to assume that if a point on the straight line of Fig. 5.1 does not have a rational number corresponding to it, then there is an irrational number corresponding to it (see Sect. 1.2). Richard Dedekind, a German mathematician who lived in the 19th century, put it like this: Let there be a *cut*, he says, which divides the continuum of real numbers in two, with all those on the left of the *cut* being smaller than those on the right. When the *knife-edge* of the cut does not *hit* a rational number but falls between, it defines an irrational number.

The reader will note that in the caption of Fig. 5.1 and in Dedekind's statement (above) the totality of rational and irrational numbers, extending from $-\infty$ to $+\infty$, have been described as the *real* numbers. This can be taken as a mere definition, but it can also be taken to imply that these numbers are directly related to nature. And in this respect one may wonder: Are the negative numbers real in this sense? Do they have an independent existence? Why can we not describe them simply as

A. Modinos, *From Aristotle to Schrödinger*,
Undergraduate Lecture Notes in Physics, DOI: 10.1007/978-3-319-00750-2_5,

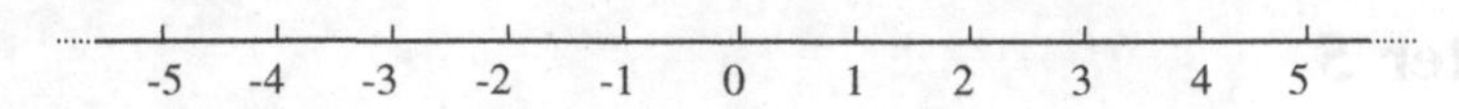

Fig. 5.1 The continuum of real numbers represented by a straight line extending from $-\infty$ to $+\infty$

'quantities to be subtracted' or as points on a direction opposite to that of the positive axis (defined by the positive numbers extending from zero to $+\infty$)? The fact is that negative physical quantities (and therefore, numbers) *do exist independently*; nature wants it this way. As we shall see, the elementary units of matter (the electron, the proton, the photon, etc.) have a certain mass and a certain electrical charge, and the electrical charge can be positive, zero, or negative. And this tells us that a negative quantity can be as real as a positive one, and the same applies, of course, to zero (0). We shall later see (in relation to quantum mechanics), that the real numbers are not the only numbers with a direct (intimate) relation to nature.

When thinking about the real numbers one may ask: are there more or fewer irrational numbers than there are rational ones? The question may not be unreasonable, but it can not be answered so easily. We can answer it in a meaningful way after we introduce the concepts necessary for that purpose. We begin with the sequence (5.1a) of the natural numbers:

$$1, 2, 3, 4, 5, \ldots \tag{5.1a}$$

$$1, 4, 9, 16, 25, \ldots \tag{5.1b}$$

The numbers in sequence (5.1b) are the squares of the terms above them. The following are immediately obvious[1]: All the numbers in the sequence of the squares appear in the sequence of the counting (natural) numbers, but not the other way round. We can say that the square numbers are a subset of (are contained in) the set of natural numbers. And yet, equally obviously, there is a one-to-one correspondence between the members of the two sets (the square numbers and the counting numbers). Galileo concluded, rightly, that the attributes 'equal', 'greater' or 'less' are not applicable to infinite sets (containing infinite many members). In fact such sets always have subsets with infinite many members in them.

We say that a set (of numbers or other 'objects', which we call its elements or its members) is a *countable* set, if its members can be put into a one-to-one correspondence with the counting (natural) numbers. The square numbers of sequence (5.1b) is a countable set (like every other infinite subset of the natural numbers). What is (at first) surprising is that the rational numbers are also a

[1] Galileo was the first to point out the one-to-one correspondence between the terms of the two sequences of (5.1). This and a discussion relating to the one-to-one correspondence between the points of line-segments of different length appear in his book Discourses and Mathematical Demonstrations Concerning Two New Sciences.

countable set.[2] On the other hand, the *irrational numbers* are obviously *not* a countable set: they can not be put into a one-to-one correspondence with the counting numbers. In summary: if one asks whether there are more or fewer irrational numbers than there are rational ones, the most we can say in answer is: there is infinitely many of either of them, but while the rational numbers are countable, the irrational ones are not. Moreover, we can easily see that the irrational numbers within a segment of the real numbers, however small, is non countable: it can not be put into a one-to-one correspondence with the counting numbers.

However, as Galileo observed, the points (real numbers) of a line- segment of the continuum of real numbers can be put into a one-to-one correspondence with the points of another line-segment of the continuum of real numbers of the same or *different* length. Galileo's arguments (based on a variety of geometrical constructions) are summarized in Fig. 5.2. The length of the circumference of the inner circle in Fig. 5.2 is smaller than that of the outer circle. But, obviously, there is a one-to-one correspondence between the points on the inner circle and those on the outer circle. This is, of course, only an example of what is possible more generally. The possibility (born out of this example) of a one-to-one correspondence between points (numbers) belonging to different segments of the continuum of real numbers, lies at the heart of modern science. We have made use of it, already, in Eqs. (4.2–4.4) of the previous chapter. Take Eq. (4.3): it tells us that to every point t (a number with units of time, e.g. seconds, attached to it) between $t = 0$ and $t = T$, *there corresponds* a point $V = \gamma\, t$ (a number with units of speed attached to it) between 0 and $\gamma\, T$. We say that V is a *function* of t described by Eq. (4.3). A *function* is a device by which a number (with or without any units attached to it; units of length, time, etc.) is turned into another number (with appropriate units attached to it as required). In general when a quantity x is turned into a quantity y according to a well defined rule, we say that y is a function of x and write it formally as follows:

---

[2] To see this, imagine a net of knots arranged in rows and columns, as in a fishing net spread on a plane. The positive rational numbers are given by a/b, where 'a' and 'b' are any two positive integers, and can therefore be arranged on the net, 'a' deciding the row and 'b' the column, if we allow the same number to appear over more than one knots. For example: 1/1 (= 1) will be on (first row-first column) but also on (nth row-nth column) where n is any integer (n/n = 1). Similarly 2/3 will be on (second row-third column) but also on ((2n)th row-(3n)th column) etc. Now if the knots are a countable set, the positive rational numbers (which correspond to a subset of the knots) are also a countable set. We can put the knots of the net into a one-to-one correspondence with the counting numbers by first counting the knots on the perimeter of the smallest square on the net (one of its corners being on (first row-first column)), go on to knots on the next square, then to knots on the square after that, and so on. We are, in this way convinced that the knots are a countable set, and therefore the positive rational numbers are a countable set. And the same, of course applies to the negative rational numbers. Now if two different sets of numbers are countable, their union is countable (to see this assume a correspondence between the members of the first set with the even numbers and a correspondence between the members of the second set with the odd numbers). We therefore conclude that the rational numbers are a countable set.

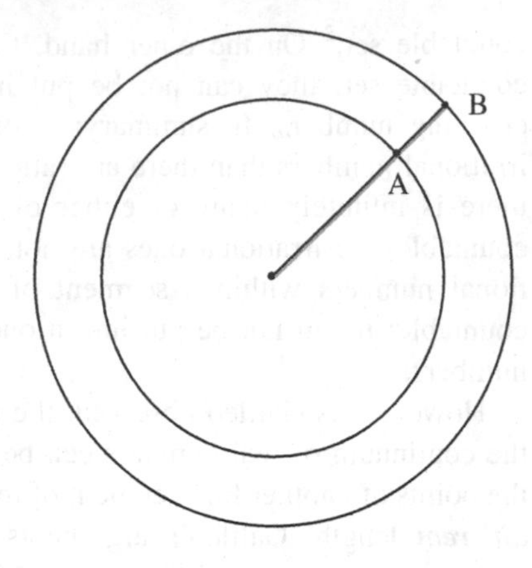

Fig. 5.2 To each point *A* on the *inner circle* corresponds a point *B* on the *outer circle*

$$y = f(x) \tag{5.2}$$

Usually, the function is described by an equation, as in Eqs. (4.2–4.4). But this is not always the case. A function can be expressed in tabular form. And more often, it can be expressed in graphical form, as in Fig. 4.1. The graphical representation of functions relates to analytical geometry, a branch of mathematics, invented by Descartes. So before I tell you more about functions, let me tell you a little about Descartes and analytical geometry.

Rene Descartes, a contemporary of Kepler and Galileo was born in the Touraine region of France in 1596. The son of an aristocratic family, he studied law at the University of Poitiers, intending to follow a military career. And for a while he entered the service of Prince Maurice who ruled the Netherlands. But army life did not suit him. He became interested in mathematics and natural philosophy. He returned to France for a short period, travelled regularly, spending time in Bohemia, Hungary, and Germany. Eventually, at the age of *32*, he settled in the Netherlands, where he stayed for most of the rest of his life. In 1649 Descartes went to Sweden to tutor Queen Christina, but within a few months of being there, he caught a severe cold and died.

Descartes, though a devout catholic who often referred to God in his writings, was an original thinker (living in protestant Netherlands he was never in danger from the Inquisition), who believed that religion's domain is that of morality, not science. He was a polymath who dealt with fundamental problems of science. In his book *Principia Philosophiae* (Principles of Philosophy), he tried to explain the observed motions of the celestial bodies in terms of an ethereal fluid-like substance, existing over the entire universe. The motion of this ethereal fluid, acquired at the moment of its creation and indestructible, could be locally turbulent; and when a celestial body lies within such a region it is forced into a rotational motion about its axis by the rotating fluid. This is, for example, why the earth rotates about its axis. Moreover, a celestial body can by its motion, propagated through the ether, influence the motion of another celestial body, with *no need of action at a distance*. Descartes' theory was too complicated, was proved wrong by Newton (it could, for example, never explain Kepler's laws of planetary motion), but it was, nevertheless, a step in the right direction in the following sense. Descartes attributed the motions of celestial bodies to entirely mechanical causes, without any reference to metaphysical causes of any kind. In contrast to Kepler who did not exclude the possibility that, the 'force' exerted by the earth on the moon had a metaphysical nature or component (see Sect. 3.3). However, Descartes' most important contribution to science, and the one he is remembered by, is his discovery of analytical geometry. It appears in his book entitled *Discourse on the Scientific Method* which was published in 1637. The book is a long one, but its most significant parts are in its three appendices: on optics (see Sect. 5.2), on meteorology, and on mathematics. It is in the latter one that analytical geometry was introduced.

The idea that a point in two-dimensional space (we assume it to be a plane surface) can be fixed by two coordinates was of course known at the time of Descartes, it was explicitly used in maps since, at least, the time of Ptolemy. Everyone is now used to reading the position of a place (*A*) on a map by reading its coordinates (so many kilometers to the east from the centre of London, and so many kilometers to the north). In the same way (see Fig. 5.3) we can fix the position of any point on a plane, with respect to a coordinate system defined by an origin (*0*), it corresponds to the centre of London in our map example, and two directions normal to each other (but otherwise arbitrarily chosen), which we shall call the x-axis and the y-axis respectively (they correspond to the west-to-east and the south-to-north directions in our map example).

Descartes used this manner of representing a point in two dimensional space (a Euclidian plane), to describe a geometrical shape, like that of a circle or an

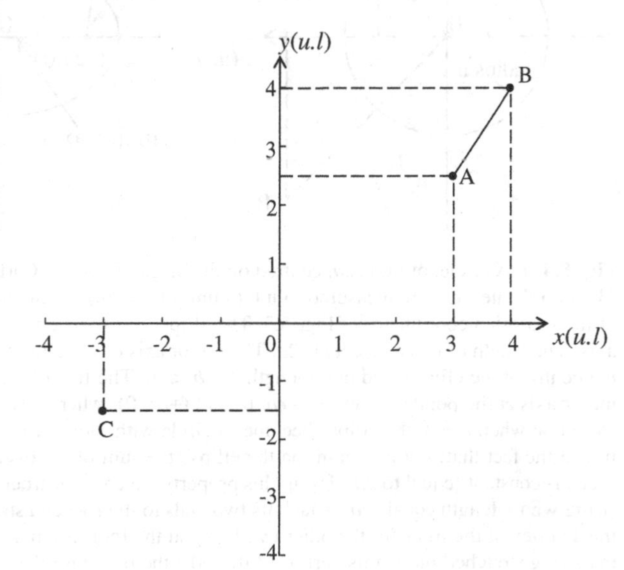

Fig. 5.3 Cartesian coordinates (so called in honour of Descartes) in a two dimensional (Euclidian) space. The unit of length (u.l.) can be anything from a mm (or smaller) to a km (or larger). The choice of unit is analogous to the choice of scale of an ordinary map. The coordinates (measured in u.l.) of point C are: $(x_C, y_C) = (-3, -1.5)$, those of point A are: $(x_A, y_A) = (3, 2.5)$, and those of point B: $(x_B, y_B) = (4, 4)$. The line AB describes a path between points A and B

ellipse, analytically; which means using algebra rather than geometrical constructions. The best way to demonstrate his method is through examples.

Consider a circle of radius a centred on the origin (0) of a Cartesian system of coordinates (x,y) as shown in Fig. 5.4a. We can inscribe a circle of radius a on the plane about the origin (0) as follows: holding one end of a string of length a at the origin, we turn the stretched string through an angle of 360°: the line that the other end of the string inscribes on the plane is the required circle. It follows from the way the circle was constructed that: the distance of any point on the circle from 0 equals a. It is also evident that any point on the plane whose distance from 0 equals a, lies on the circle. And if a point has a distance from 0 other than a then it does not lie on the circle. Now, following Descartes, we note that (according to Pythagoras's theorem, see Appendix A) the square of the distance of a point with coordinates (x, y) from the origin of the coordinates equals $x^2 + y^2$ whether it lies on the circle (like point A in Fig. 5.4a), or not (like point B); and because we put the origin of the coordinates at the centre of the circle, $x^2 + y^2$ is also equal to the square of the distance of the point (x, y) from the centre of the circle. Therefore, for a point (x, y) on the circle: $x^2 + y^2 = a^2$ (for example: $x_A{}^2 + y_A{}^2 = a^2$), and for any other point we obtain: $x^2 + y^2 < a^2$ where $<$ means *smaller than*, or $x^2 + y^2 > a^2$,

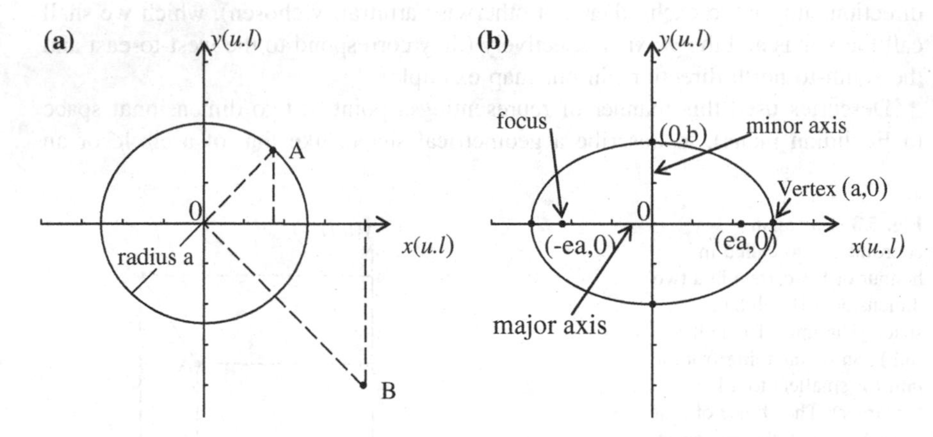

Fig. 5.4 **a** A circle, of radius a, centred on the origin (0) of the Cartesian coordinates of Fig. 5.3. The coordinates x, y are measured in u.l. (a unit of length). **b** An ellipse centred on the origin (0) of the Cartesian coordinates of Fig. 5.3. The ellipse has two vertices, one at each end of its major axis. The length of this axis equals $2a$. The minor axis of the ellipse is normal to the major axis at the centre of the ellipse, and has a length $2b$ $(b < a)$. The foci of the ellipse are two points on the major axis at the points: $(x, y) = (-ea, 0)$ and $(+ea, 0)$, where e is the eccentricity of the ellipse. Note that when $e = 0$ the ellipse becomes a circle with radius $a = b$. The importance of the foci lies in the fact that: For any point on the ellipse the sum of the distances from that point to each focus is constant (equal to $2a$). Using this property we can construct the ellipse as follows: Take a string whose length equals $2a$. Attach its two ends to the foci and stretch it to form a triangle. The third vertex of the triangle (the other two being at the foci) is a point on the ellipse. Now keeping the string stretched move this vertex a full 360°: the path inscribed on the plane by this vertex is an ellipse with the given major axis and the given eccentricity

where > means *greater than* (for example: $x_B^2 + y_B^2 > a^2$). Therefore, the equation:

$$\frac{x^2 + y^2}{a^2} = 1 \qquad (5.3)$$

describes the circle shown in Fig. 5.4a in the sense that, any point on the circle (its (*x*, *y*) coordinates) satisfies this equation, and any point not on the circle does not.

Now let me write down (we need not bother about its derivation here; see exercise 5.1) the equation for the ellipse shown in Fig. 5.4b. It is:

$$\frac{x^2}{a^2} + \frac{y^2}{b^2} = 1 \qquad (5.4)$$

The above equation describes the ellipse of Fig. 5.4b in the sense that, any point on the ellipse [its coordinates (*x*, *y*)] satisfies the above equation, and any point not on the ellipse does not. We note that when the eccentricity of the ellipse, defined by $e = \sqrt{(1 - b^2/a^2)}$, equals zero, we have $b = a$, and Eq. (5.4) reduces to that of a circle [Eq. (5.3)].

It is clear that if we are given the *x* coordinate of a point on the circle, we can determine, through Eq. (5.3), its *y* coordinate. That is $y = +\sqrt{(a^2 - x^2)}$ for the upper half ($y > 0$) of the circle, and $y = -\sqrt{(a^2 - x^2)}$ for the lower half ($y < 0$) of the circle. And the same applies in relation to an ellipse.

Let me demonstrate the usefulness of analytical geometry through the following two examples:

First example: Assume that a particle is moving on the *x*−*y* plane of Fig. 5.3, and that we know its coordinates [*x*(*t*), *y*(*t*)] at any moment of time $t \geq 0$ (measured from some initial moment ($t = 0$) onwards). If *x*(*t*), *y*(*t*) satisfy Eq. (5.4) at any moment of time the particle moves on an ellipse centred on the origin (0) of the coordinates with a major axis equal to 2*a* and a minor axis equal to 2*b*. We came to this conclusion without drawing a single diagram!

Second example: Let the coordinates of a particle moving on the x−y plane of Fig. 5.3 be: [*x*(*t*), *y*(*t*)] = (3 + *t*, 2.5 + 1.5 t) in appropriate units of length and time. It follows that initially ($t = 0$) the particle is at [*x*(0), *y*(0)] = (3, 2.5) which is point A on the figure; and at time $t = 1$ the particle is at [*x*(1), *y*(1)] = (4, 4) which is point *B* on the figure. And we can tell what the value of *y* is for a given value of *x* (in the region: $3 \leq x \leq 4$). We find (you may want to prove this by yourself):

$$y = 1.5\,x - 2, \text{ with } x \text{ in: } 3 \leq x \leq 4 \qquad (5.5)$$

The above equation defines the *y*-coordinate of a point as a function of its *x*-coordinate. The graphical representation of this function is given by the straight line between point *A* and point *B* in Fig. 5.3.

The graphical representation of functions is a generalization of the above procedure. Consider the function $y = f(x)$ of Eq. (5.2): *x* may be any quantity measured (in appropriate units) along the *x*-axis of a 'Cartesian' system of coordinates, while the corresponding to it value of y, given by $y = f(x)$, is measured (in appropriate units) along the y-axis.

Let me close this section with the description of three functions which will be very useful to us in later chapters: $y = sin(x)$, $y = cos(x)$, often referred to as the trigonometric functions; and $y = exp(x)$, known as the exponential function. The trigonometric functions are introduced most easily by reference to the circle of Fig. 5.5a. We imagine OA turning anticlockwise: x increasing from zero (when OA lies on the positive cos-axis) to $x = \pi/2$ (when OA lies on the positive sin-axis), then to $x = \pi$ (when OA lies on the negative cos-axis), then to $x = 3\pi/2$ (when it lies on the negative sin-axis), then to $x = 2\pi$ (when it lies again on the positive cos-axis), and then it goes on, increasing to $x = 2\pi + \pi/2$ ($=5\pi/2$) (when it lies again on the positive sin-axis), and so on, completing one cycle after another, as x increases from zero (0) to $+\infty$. We define $cos(x)$ and $sin(x)$ as follows: For any (positive) value of x, $cos(x)$ is given by the ratio (a pure number) of the projection of OA on the cos-axis to OA, but since we put the length of OA equal to unity, the value of $cos(x)$ is given by the value of the projection of OA on the cos-axis (positive when the projection lies on the positive side of the cos-axis, and negative when the projection lies on the negative side of this axis). And $sin(x)$ is given, in the same way, by the projection of OA on the sin-axis. The values of $cos(x)$ and $sin(x)$ for negative values of x are obtained from the following relations:

$$cos(-x) = cos(x), \quad sin(-x) = -sin(x) \tag{5.6}$$

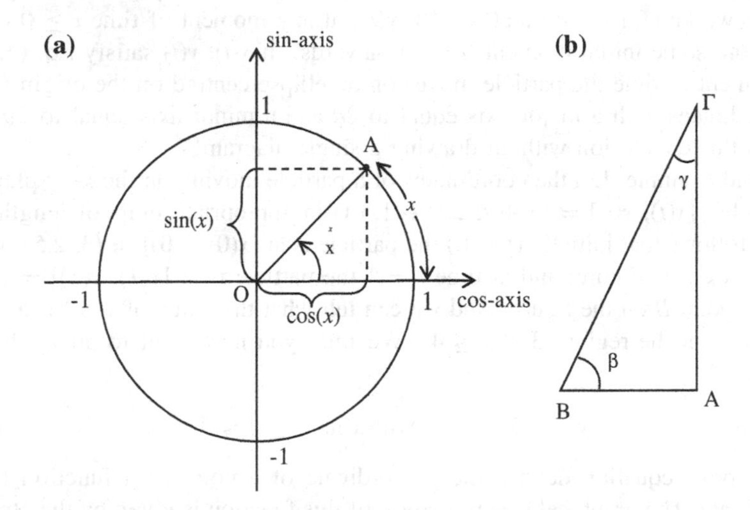

Fig. 5.5 **a** The projection of OA on the cos-axis defines $cos(x)$; the projection of OA on the sin-axis defines $sin(x)$. We have taken the radius of the circle as the unit of length ($OA = 1$), which means that the angle subtended at the centre of the circle, measured in radians, equals the length of the arc. We have: 2π radians $= 360°$; 1 radian $= 57.296°$. **b** We can see that the following relations apply to any orthogonal triangle: $BA = B\Gamma cos(\beta) = B\Gamma sin(\gamma)$; $A\Gamma = B\Gamma cos(\gamma) = B\Gamma sin(\beta)$

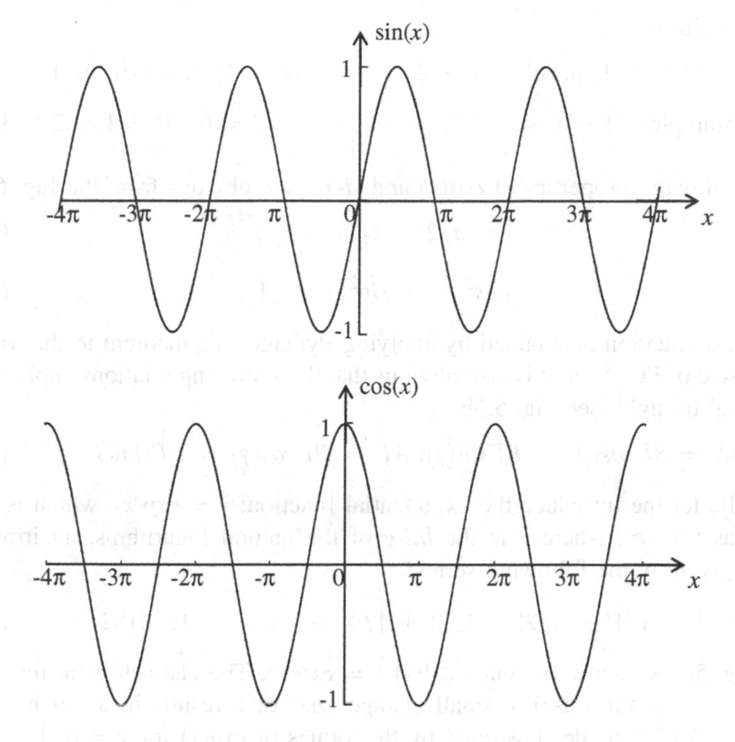

Fig. 5.6 Graphical representation of $y = sin(x)$ and $y = cos(x)$

Graphical representations of $cos(x)$ and $sin(x)$ obtained in this manner are shown in Fig. 5.6.

While a graphical representation of $cos(x)$ and $sin(x)$ demonstrates clearly the qualitative features of these functions, it is not very practical when numerical values are required. Fortunately, our understanding of infinite series (see Sect. 1.1) provides a most useful tool for this purpose. It turns out that the following infinite series converge to $sin(x)$ and $cos(x)$ respectively, for any value of x. Of course, the number of terms we include in the sum of the series to obtain good accuracy depends on the value of x. We shall know that we have included a sufficient number of terms, if by adding more, the value of the sum does not change within the accuracy required (by which we mean that the first five or ten or more decimal points, as required, do not change). Of course, you can always use a calculator to obtain $cos(x)$ and $sin(x)$. The calculator does the summation of the series for you. We have:

$$sin(x) = x - x^3/3! + x^5/5! - x^7/7! + \ldots \tag{5.7}$$

$$cos(x) = 1 - x^2/2! + x^4/4! - x^6/6! + \ldots \tag{5.8}$$

By definition:

$$0! = 1,\ 1! = 1, \text{and}\ n! = 1 \times 2 \times \cdots \times (n-1) \times n\ \text{for}\ n > 1 \qquad (5.9)$$

For example: $2! = 1 \times 2 = 2$, $3! = 1 \times 2 \times 3 = 6$, $4! = 1 \times 2 \times 3 \times 4$, etc.

The following properties of *cos*(x) and *sin*(x) are obvious from the Fig. 5.5a:

$$cos(\pi/2 - x) = sin(x) \qquad (5.10)$$

$$cos^2(x) + sin^2(x) = 1 \qquad (5.11)$$

The last equation is obtained by applying Pythagoras's theorem to the triangle in the circle of Fig. 5.5a. It is also obvious that the following relations apply to any orthogonal triangle [see Fig. 5.5b]:

$$BA = B\Gamma cos(\beta) = B\Gamma sin(\gamma);\ A\Gamma = B\Gamma cos(\gamma) = B\Gamma sin(\beta) \qquad (5.12)$$

Finally, let me introduce the exponential function: $y = exp(x)$, which is often written as $y = e^x$, where e is the *base* of the natural logarithms, an irrational number given by the following series:

$$e = 1 + 1/1! + 1/2! + 1/3! + 1/4! + \cdots = 2.71828182\ldots \qquad (5.13)$$

In Fig. 5.7 we show (schematically) $y = exp(x)$. The characteristic feature of this function is: for $x > 0$, a small change (Δx) of x results in a much greater change (Δy) of y, as demonstrated by the values of $exp(x)$ for $x = 0, 1, 2, 3, 4$. Similarly, for negative values of x, $exp(x)$ tends very quickly to zero as x becomes more negative. This is, of course, what we expect, since

$$e^{-x} = \frac{1}{e^x} \qquad (5.14)$$

The numerical value of $exp(x)$, for any value of x, is obtained from the following infinite series:

$$e^x = 1 + x/1! + x^2/2! + x^3/3! + x^4/4! + \ldots \qquad (5.15)$$

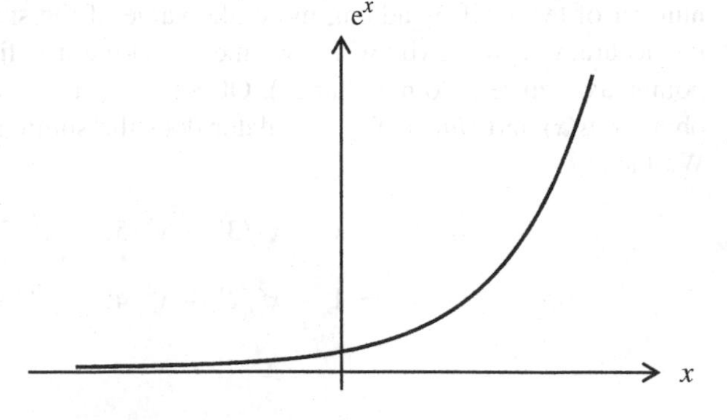

Fig. 5.7 Graphical representation of $y = e^x$. We have: $e^0 = 1$, $e^1 = 2.718$, $e^2 = 7.389$, $e^3 = 20.085$, $e^4 = 54.598$. The values of e^x for $x = -1, -2, -3, -4$, can be obtained from the preceding values using Eq. (5.14)

The comments made in relations to Eqs. (5.7 and 5.8) apply to Eq. (5.15) as well.

The exponent, x, in e^x is known as the (natural) logarithm of e^x. We can define the function: $y = log(x)$ quite generally as follows: y is the number which gives: $e^y = x$. [*logx* is often written as *lnx*.]

5.2 Some Important Experiments

5.2.1 *Torricelli's Barometer and Boyle's Gas Law*

Evangelista Torricelli, whom we met as a student of Galileo towards the end of the latter's life, did the following experiment: After completely filling a glass tube with (liquid) mercury he turned it upside down into a basin of the same, doing it carefully so as to avoid air entering the tube. The result is shown in Fig. 5.8a. Mercury flows from the tube into the basin up to a point; it stops when the height of the mercury in the tube equals 0.76 m. One can safely assume, from the way the experiment is done, that there is practically no air in the emptied top region of the tube. Torricelli repeated the experiment using different tubes, and found that the height of the mercury in the tube does not depend on the diameter of the tube or its shape (see Fig. 5.8b).

Torricelli concluded correctly that what keeps the column of mercury at the observed height is the pressure of the atmospheric air, exerted on the open surface of the mercury in the basin. One can put it, slightly differently, as follows: the column of mercury in the airless tube exerts on its base (at the free surface level) a pressure equal to the one the atmospheric air was exerting prior to its removal. By this means, an instrument (what we now call Torricelli's barometer) was obtained, by which one could measure the atmospheric pressure (and its variations with time and place) by the height (in centimeters or millimeters) of a column of mercury. Torricelli published his results in 1643. Many who read his work were greatly

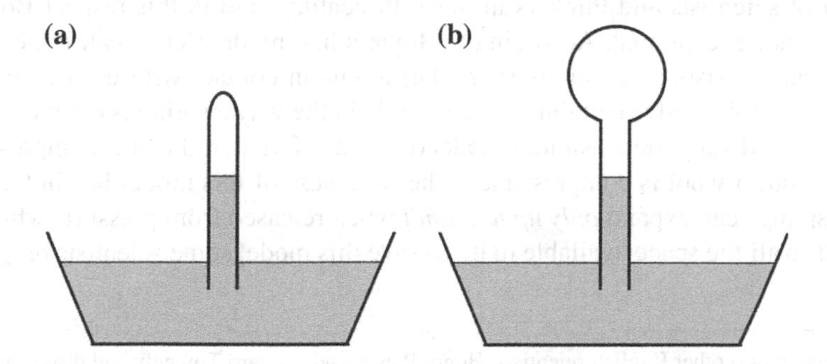

Fig. 5.8 **a** The height of the mercury in the tube is 0.76 m above the surface of the mercury in the basin. **b** This height does not depend on the shape of the tube

impressed, especially so by his assertion, that the air of the atmosphere exerted that much pressure on the bodies in it: 0.76 m of mercury is a very large pressure indeed. It is equivalent to a weight of 1.04 kg per square centimeter! In 1656 a German scientist, Otto von Guericke, proceeded to a demonstration of this pressure in a most impressive manner. He made (from bronze) two large semi-spherical shells which could close tightly on each other to form a sphere, from which he removed the air using a reasonably good vacuum pump. With the air practically removed from within the sphere, the two halves of it kept together so strongly by the atmospheric pressure (pushing them toward each other), that in order to separate them, Guericke needed eight horses tied to each of the semi-spheres in order to pull them apart!

Other scientists soon confirmed Torricelli's and Guericke's observations. One of them was Robert Boyle, a rich Englishman, who was very much interested in chemistry and science in general. Boyle was able to construct, with the assistance of fellow scientist Robert Hooke, a reasonably good vacuum pump, and was able to show that the height of the mercury column of a barometer in a vessel of thin air is reduced as the air is pumped out of the vessel, and disappears when all air has been removed from the vessel. With this experiment Boyle proved beyond any doubt that Torricelli's interpretation of his results was the correct one. Boyle went on to perform some very interesting experiments in moderately good vacuum. He showed, for example, that a candle would not burn in vacuum, and that a mouse (or some other small animal) could not live in very thin air (very low pressure). However, Boyle is remembered mostly for his discovery (in 1660) of the gas law,[3] which states that the pressure (P) of a gas is inversely proportional to its volume (V) at constant temperature. We have:

$$PV = \textit{constant}, \text{at constant temperature.} \tag{5.16}$$

What does the above law tell us about the nature of air (about the nature of a gas)? Boyle proposed two models of a gas, both of which could in principle 'explain' Eq. (5.16). Both of them assumed the existence of atoms: minute particles of matter moving in accordance to some law or other. The concept of atoms, conceived in antiquity by Empedocles and Democritus, was gradually accepted by a number of scientists and thinkers in the 17th century, and in this respect Boyle's view was not exceptional. According to Boyle's first model (let us call it the static model), air consists of elastic (spring-like) atoms in contact with each other. An aggregate of this kind is certainly compressible in the way a spring is compressible. (Torricelli had suggested a similar model of air: a soft material that is compressible the way cotton wool is compressible). The weakness of this model lies in the fact that a 'spring' can expand *only up to a limit* when released from pressure, while air expands to all the space available to it. To save this model some scientists proposed

[3] Apparently two other English scientists, Henry Power and Richard Townely had discovered the gas law in 1653. Boyle knew of their experiments and acknowledged their contribution in his monograph of 1662.

that some kind of repulsive interaction operated between the atoms pushing them away from each other even when the atoms were not in contact. However action at a distance was difficult to accept at the time without additional experimental evidence of its existence. In any case according to this model, the pressure exerted by the air on the container surface arose from atoms pushing against it as they were pushing each other. In Boyle's alternative model of a gas (let us call it the kinetic model) air consists of unconnected atoms (possibly hard particles of various sizes and shapes) moving fast and irregularly (and independently of each other) all the time; as they do so they impact often enough on the container surface and in this way exert pressure on it. This model of air is, so far, not different from that of the atomists of antiquity. But Boyle and his contemporaries would not leave it at that. They wanted a cause for this fast irregular motion. And they found one in Descartes' ethereal fluid (see Sect. 5.1). The way the motion of the ether in the universe, at the same time smooth but also at regions locally turbulent, kept the planets moving and rotating, the ethereal fluid in a vessel kept the atoms of the air in the vessel moving. Atoms in the grip of the ether's turbulences move fast and irregularly. Boyle tried to find independent experimental evidence for the existence of ether but, of course, never found one. In 1738 Daniel Bernoulli, a member of a remarkable family of Swiss mathematicians, published a book, called *Hydrodynamica*, which contains an explanation of Boyle's law in terms of a kinetic theory: the gas is made of particles which move randomly in all directions; the observed pressure is the result of collisions of these fast moving particles with the walls of the vessel. A thorough interpretation of Boyle's law (Eq. 5.16) became possible about 250 years after Boyle, when scientists understood the role of heat in the exchange of *energy* between the gas and its environment.

Robert Boyle was certainly a pioneer chemist who favoured an experimentally based chemistry, and theoretical explanations in terms of a mechanical atomic theory of matter. And yet, a recent book on Boyle,[4] reveals a more complex picture of the man. The historian Laurence Principe is convinced that Boyle, like his great contemporary Isaac Newton, had also an active and lively interest in alchemy. Here is a nice story: In 1689 Boyle managed to persuade the English parliament to remove the ban on the transmutation of metals. On hearing that, Newton suspected that Boyle had found the powder of the 'philosopher's stone' that could turn ordinary metals into gold. Whether this particular story is correct or not, there is enough evidence, according to Principe, to show that Boyle was a practicing alchemist, who believed in angels and other incorporeal beings, and that one could invoke the power of these beings through the 'philosopher's stone' not only in turning ordinary metals into gold, but also in matters spiritual (in order to support the claims of religion). This, Principe tells us, does not mean that Boyle was not an excellent scientist. He was simply a man of his time: a transition period when alchemy and chemistry could happily coexist.

[4] L. Principe, Robert Boyle: the secret alchemist, (Princeton University Press), 1998.

5.2.2 Snell's Law of Refraction and the Nature of Light

When we look at a rod partly immersed in water, the part of it in the water appears to us bent (see Fig. 5.9). This is because the light rays from a point in the water (e.g. the rod's end) change their direction when they emerge from the water into the air. The phenomenon is known as the refraction of light.[5]

The laws governing the reflection and refraction of light were first formulated in the 17th century on the basis of experimental measurements. These laws are summarised in Fig. 5.10. Snell's law of refraction states that:

$$n_1 sin(i) = n_2 sin(r) \tag{5.17}$$

where, n_1 and n_2 are the 'refractive indexes' of media '1' and '2' respectively, to be determined experimentally as follows.

We place a slab of the medium X in vacuum (in Fig. 5.10, medium '1' is replaced by vacuum, and medium '2' by the medium X).[6] We put (arbitrarily) the refractive index of vacuum equal to unity (we are permitted to do so because what matters in Eq. (5.17) is the ratio n_1/n_2 of the two indexes). So that Eq. (5.17) becomes:

$$sin(i) = n_X sin(r) \tag{5.18}$$

We measure the angles 'i' and 'r' in the laboratory. Substituting these values in the above equation we obtain n_X. Once the refractive index of a medium (air, water, glass, etc.) has been determined in this way, it can be used in conjunction with Eq. (5.17), in a variety of instances, including in the design of lenses etc. We note that in the case of the rectangular slab of Fig. 5.10 the emergent ray, in

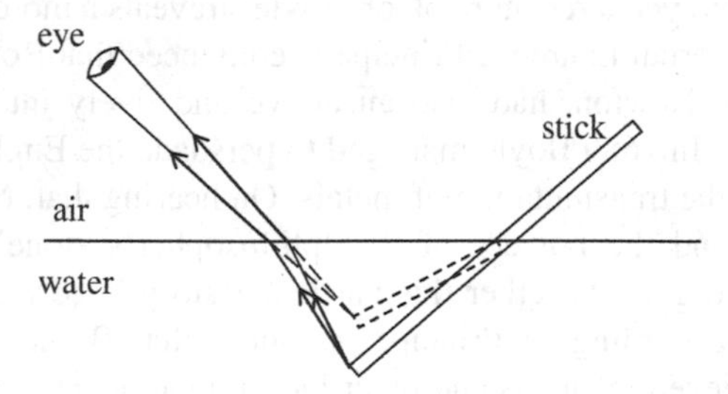

Fig. 5.9 A rod partly immersed in water appears to the eye bent, due to refraction

[5] The figure represents a cross section (a thin slice to be identified with the page) of the system under consideration. The system extends above and below the page, but the values of the physical quantities we consider do not change in the direction normal to the page. The same applies to the other figures of this section.

[6] Often, and certainly so in the 17th century, air and not vacuum was used as the reference medium. The refractive index of air (relative to vacuum) is not exactly unity and depends to some small degree on the colour (frequency) of light.

medium '3', is parallel to the incident ray. When the slab is replaced by a prism, as shown in Fig. 5.11, the emergent ray (*S*) is not parallel to the incident ray (*P*). And that is of course what we expect according to Snell's law (when applied at both interfaces).

What is indeed remarkable is what Isaac Newton discovered when he performed this experiment for the first time. He observed that, the (white) incident light emerged from the prism resolved into a continuous spectrum of coloured light, from red to violet. The red being the least deviated from the direction of the incident light. He concluded that white light is a mixture of different colours. And, of course, the resolution of the white light in the above experiment implies that the refractive index of a medium (other than vacuum) varies with the colour of the light.

It appears that the Dutchman, Willebrord Snell, who discovered the law that bears his name about the year 1620, did not publish his results. The law was published for the first time by Descartes, in his *Discourse on the Scientific Method*, without any reference to Snell. Descartes believed that, light propagates through a medium the way a mechanical disturbance propagates along a rod and rejected any

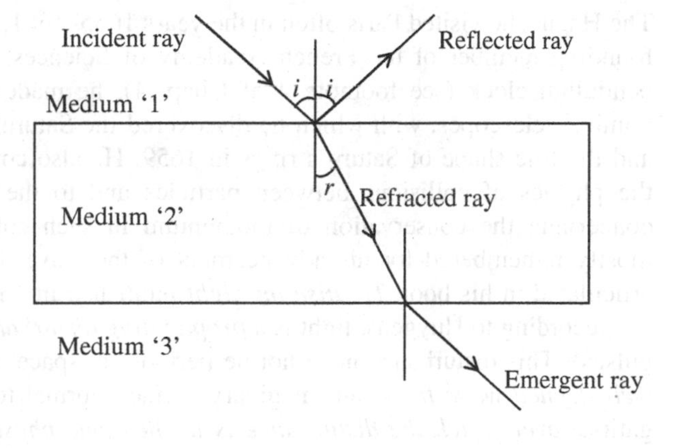

Fig. 5.10 A beam (ray) of light in medium '1' is partly reflected at the interface with medium '2' and partly refracted. A first law states that: The *incident ray*, the *reflected ray*, the *refracted ray* and the normal at the point of incidence, lie in the same plane. While the angle of reflection is always equal to the angle of incidence (i); the angle of refraction (r) is determined by Snell's law (see text). The same law applies, of course, at the interface between medium '2' and medium '3'. We assumed that $n_2 > n_1$[5]

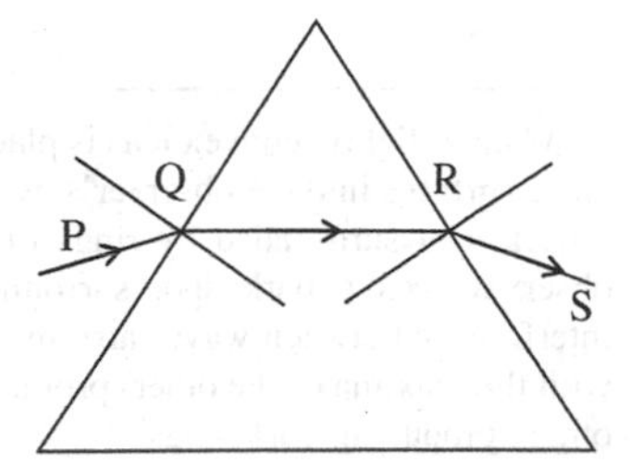

Fig. 5.11 Refraction at the interfaces of a glass prism in air

idea that light might consist of corporeal particles. He argued that such particles could not possibly enter the human eye. At the same time he believed that light propagates with 'infinitely large' speed, i.e. instantly. In any case, he suggested that when a beam of light strikes the interface of two media, it is partly reflected (in the manner that a ball is reflected by a perfectly smooth surface) and partly refracted into the second medium, and in the process it receives a 'kick' that pushes the refracted beam towards (or away) from the normal to the interface at the point of incidence. His interpretation is, by his own admission, no more than a metaphorical way of describing refraction, and as such does not tell us anything about what causes it.

Newton, on the other hand, entertained the idea that light might consist of 'particles of light'. It may seem strange that Newton who had also discovered the phenomenon of the Newton rings,[7] which is so nicely explained in terms of waves, chose to explain light in terms of 'particles of light'. He apparently believed that the then known phenomena, could in principle be explained by mechanical forces acting on these particles by matter. But he was not dogmatic in this respect and he allowed for the possibility that light might be a disturbance after all.

Of the scientists who disagreed with Newton on the nature of light, the most prominent was the Dutch astronomer and physicist Christian Huygens. Born in The Hague, he visited Paris often in the years 1655–1681, and in 1666 he became a founding member of the French Academy of Sciences. He constructed the first pendulum clock (see footnote 4 of Chap. 4), he made improvements to astronomical telescopes, with which he discovered the Saturn's satellite Titan in 1655 and the true shape of Saturn's rings in 1659. He also contributed significantly to the physics of collisions between particles and to the formulation of the law concerning the conservation of momentum in such collisions. However, he is mostly remembered for his advancement of the wave theory of light, which he articulated in his book *Treatise on Light* published in 1690.

According to Huygens, light is a *propagating disturbance* (as in a succession of pulses). This disturbance need not be periodic in space or time but *always has a well defined wave front* (an imaginary surface normal to the direction of propagation) *over which the disturbance is in the same phase*. When the disturbance (light) crosses the interface of two media (see Fig. 5.12) part of the disturbance (and therefore part of the wave front) is in one medium, and part of it in the other where it propagates with *different* speed. However, the phase at *all* points of the wave front must be the same, and this implies a shift in the direction of the refracted beam relative to the incident one. Huygen's calculation (summarised in

[7] When a slightly convex lens is placed on a flat glass plate and white light is reflected by the two close surfaces into the observer's eye at a suitable angle, the point of contact of the lens is seen as a dark spot surrounded by rings of different colours. When monochromatic light is used, the observer sees a dark spot surrounded by bright and dark rings, a phenomenon typical of interference between waves arriving at a point in phase (the maxima of the one wave coinciding with the maxima of the other) producing bright rings, or in opposite phases (the one canceling the other) producing dark rings.

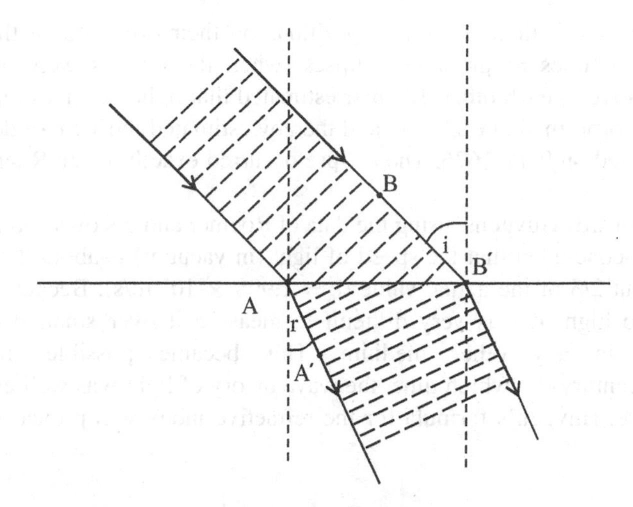

Fig. 5.12 Refraction at the interface between vacuum and medium *X*, according to Huygens (we consider monochromatic light for the sake of simplicity): We have All points on a wave front are at the same phase even when it lies partly in one medium (vacuum) and partly in another (medium *X*). Points *A* and *B* are at the same phase when the light (its wave front) reaches the interface. After certain time (*t*), point *B* propagating with speed *c* (the *speed of light in vacuum) reaches point B', and point A propagating with speed u* (the speed of light in medium *X*) reaches point *A'*. We have:. $\begin{matrix} ct = BB' = AB'cos(\pi/2 - i) = AB'sin(i) \\ ut = AA' = AB'cos(\pi/2 - r) = AB'sin(r) \end{matrix}$. Therefore: $c/u = sin(i)/sin(r)$, which is the same as: $sin(i) = (c/u)sin(r)$, which is the same with Snell's law (Eq. (5.18)) if we put $n_X = c/u$

the figure caption) tells us that this shift is in accordance with Snell' law, and it tells us, moreover, that the refractive index of a material (*X*) is given by:

$$n_X = \frac{c}{u} \tag{5.19}$$

where *c* is the speed of propagation of light in vacuum, and *u* its speed of propagation in medium *X* (which according to what we said above must also depend to some degree on the colour of the light).

Huygen's theory constituted a significant advance. Unlike the other theories of light of that time, it was not speculative; it was verifiable: it could, in principle, be proved right or wrong by measuring the speed of light in vacuum (Huygens assumed that this was very large but finite), and in some other medium such as glass or water. Unfortunately that was not possible at that time.

The first demonstration of the finiteness of the speed of light was provided by the Danish astronomer Ole Roemer. In September 1676 he announced at the Academy of Sciences in Paris, that an eclipse of a satellite of Jupiter, expected at 45 s after 5:25 pm on the 9th of November, would be observed 10 min later. He argued that a delay would be the result of the light needing more time to reach Earth from Jupiter, because the distance between the two planets was greater on this occasion due to their respective positions on their orbits about the sun. By comparing the times of previous eclipses (when the planets were at different positions relative to each other, Roemer estimated that light traverses the diameter of the earth's orbit in about 22 min, and thereby estimated the 10 min delay of the eclipse expected on 9-11-1676. The eclipse occurred exactly when Roemer said it would.

Soon afterwards Huygens, using the data of Roemer and his own estimate of the earth's orbit, concluded that the speed of light (in vacuum) is about 2×10^8 m/s, which is about 2/3 of the actual value ($c = 2.998 \times 10^8$ m/s). Because the speed of light is so high, it was very difficult to measure it over small distances, in vacuum or in any other medium. This became possible much later (in the 19th century) by which time, the wave theory of light was well established; and, of course, Huygen's formula for the refractive index was proved valid.

5.3 The Invention of Calculus

We have seen (Sect. 1.1) how Archimedes, following Antiphon's suggestion, was able to estimate the ratio of the circumference of a circle to its diameter, by inscribing a many-sided polygon to it. The idea is that when the number of sides of the polygon increases to a very large number, its perimeter becomes practically the same with the circumference of the circle; and similarly the area of the polygon becomes practically identical with the area of the circle. From this idea sprang *integral calculus*. The man who invented it, turning Archimedes's geometrical method into an efficient analytical tool for calculating the 'area' between any curve, $y = f(x)$, and the x-axis and extending from a point a on this axis to a point x (the shaded area in Fig. 5.13), was Gottfried Wilhelm Leibniz. Following Leibniz we shall call this 'area' *the integral of f(x) from point 'a' to point 'x'* and denote it as follows:

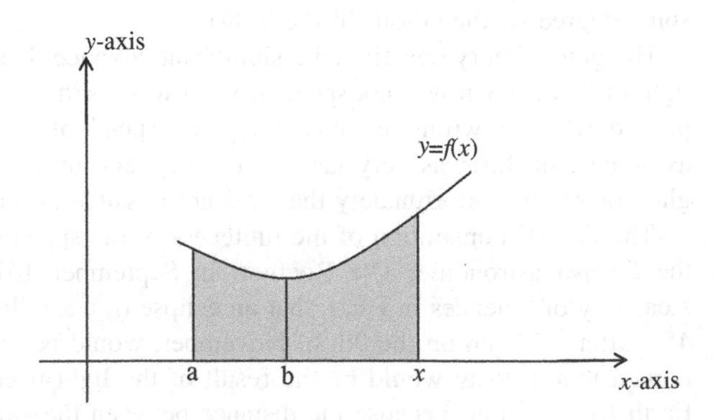

Fig. 5.13 The shaded area corresponds to the integral of expression (5.20)

$$\int_a^x f(x)dx \tag{5.20}$$

Leibniz was born in Leipzig (in Germany) in 1646. The son of a well-established lecturer in moral philosophy, he was educated, in the manner of that time, by studying the ancients, Latin philosophers and the Greeks, and he must have been well versed in the works of Aristotle, Euclid and Archimedes. Leibniz was initially more interested in philosophy than in mathematics, probably because mathematics was not taught well at his first university (Leipzig). After his first degree he travelled frequently to Frankfurt, Jena and other places, and it is probably Ehard Weigel, a professor at Jena, that excited his interest in mathematics. In 1672 he went to Paris, sent there on a diplomatic mission by his sponsor Baron Johan von Boineburg, and stayed there for four years. During these years he established many contacts with prominent mathematicians and scientists in both France and England, and his knowledge of mathematics and physics deepened. In 1673 he was made a Fellow of the Royal Society of London.

He was still in Paris when he wrote his first treatise on integral calculus, entitled: *De quadratura arithmetica circuli ellipseos et hyperbolae cujus corollarium est trigonometria sine tabulis*. But somehow this work (the longer treatise Leibniz ever wrote) was not published at that time. Before leaving Paris (in 1676) Lebniz gave the fair copy of his treatise to a friend who died before he could get it published. And when the manuscript was eventually sent to Leibniz (in Germany) it was lost in transit. In the meantime Leibniz had made further progress and never bothered to publish the original treatise, which was discovered, edited and published many years later (in 1993) by Professor Eberhard Knobloch (of the Institute of Philosophy in Berlin). Leibniz wanted to stay with the Academy in France but was not offered a job, apparently because there were already enough foreigners at the Academy. He returned to Germany, was appointed to the post of librarian and Court Councilor to the Duke of Hanover, and stayed in Hanover until his death in 1716.

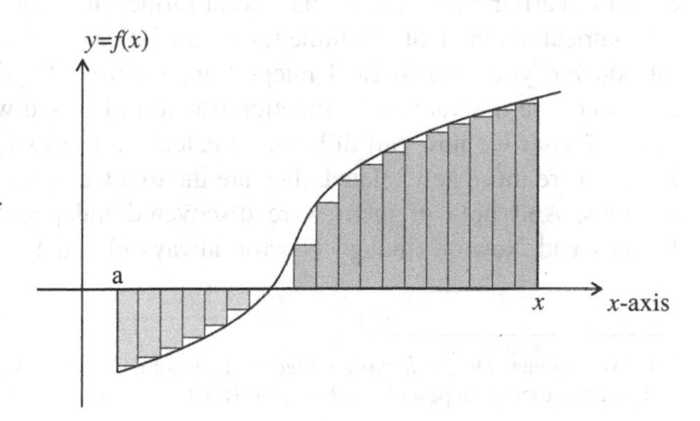

Fig. 5.14 The 'area' between the curve (defined by $y = f(x)$) and the x-axis (between point a and point x) is obtained by summing the areas (those below the x-axis taken with a negative sign) of a large number (N) of *rectangles* with the same sufficiently small breadth (Δx). Of course: $(\Delta x)N = (x-a)$

The basic idea in Leibniz's integration derives from Archimedes: he obtains the 'area' between the x-axis and a curve defined by a function $y = f(x)$, by adding together rectangles with *equal breadths of indefinite smallness*, as shown in Fig. 5.14. Enumerating the rectangles from left to right as $i = 1, 2, 3, \ldots N$, we can write the sum of the rectangles as follows (the expression on the right side of the equation is our way of writing in shorthand the sum on the left side of the equation):

$$f(x_1)\Delta x + f(x_2)\Delta x + \ldots + f(x_n)\Delta x = \sum_{i=1}^{N} f(x_i)\Delta x \qquad (5.21)$$

where Δx (a small but finite positive quantity) is the breadth of the rectangles and $f(x_i)$ the 'height' of the ith rectangle (it is determined by the value of $f(x)$ at the point where the ith rectangle touches the curve, and it can be positive or negative). We must understand Leibniz's *indefinite smallness* to mean 'not specified in advance' but chosen as small as is necessary (with a correspondingly large value of N), so that the above sum is practically identical with the integral (5.20) which stands for the exact value of the 'area' we wish to evaluate. 'Practically identical' here means 'within the accuracy we require'.

Leibniz's way of dealing with Δx (which was also Newton's way and Archimedes's way) was the cause of much worry among mathematicians. Leibniz recognized that his procedure was a bit clumsy, but he insisted that it was reliable: because one can get within any desired accuracy in matching a shape, using a large but finite number of rectangles. In his opinion 'excessive exactness' should not discourage the student's mind from other far more agreeable things by making it weary prematurely. And he added[8]: *Nothing is more alien to my mind than the scrupulous attention to minor details of some authors which imply more ostentation than utility. For they consume time, so to speak, on certain ceremonies, include more trouble than ingenuity and envelope in blind night the origin of inventions which is, as it seems to me, most prominent than the inventions themselves.*

The reader will think, and rightly so, that Leibniz's contribution (the way I described it so far) adds very little or nothing to what Archimedes had done many centuries earlier. However, Leibniz went further: he managed to express the above geometrical method of Archimedes in analytical (algebraic) terms. But before introducing you to analytical integration, I must tell you about Newton's conception of the derivative of a function (the term he used was *fluxion*), which is the basis of what we now call differential calculus. The two calculi (Differential and Integral) are intimately related, they are the two faces, so to speak, of Infinitesimal calculus. And both of them were discovered independently (more or less) by Leibniz and Newton (though Newton always claimed, wrongly according to the

[8] G. W. Leibniz, *De quadratura,* edited by E. Knobloch (Abh. Akad. Wiss., 1993).
The cited extract appears in: *Infinity*, by B. Clegg, (Constable & Robinson Ltd), 2003.

historians, that *his* was the most important contribution). Here we need not use the terms and symbols that Newton used in his development of the theory. It is easier to use the terms and symbols we use today, and these happen to be those introduced by Leibniz.

Let us consider the change $\Delta y = \Delta f$ in the value of a function, defined by $y = f(x)$, when x changes to $x + \Delta x$, where Δx is a small but finite quantity [as for the width of a rectangle in Fig. 5.14]. It is clear from Fig. 5.14, and from the functions we introduced in Sect. 5.1, [see e.g. Figs. 5.6 and 5.7] that, in general, for the same Δx the change in the value of the function, given by $\Delta f = f(x + \Delta x) - f(x)$, varies with x. It can be positive, negative or zero (if the value of y increases, decreases or does not change, when x changes to $x + \Delta x$). Let us *tentatively* define the rate of change of $f(x)$, at point x, by the ratio:

$$\frac{\Delta f}{\Delta x} = \frac{f(x + \Delta x) - f(x)}{\Delta x} \tag{5.22}$$

The question now arises: how small should Δx be? (It can not be zero, because putting $\Delta x = 0$ makes the above ratio meaningless.) Newton's answer to this question would be: evaluate the ratio for any (small) value of Δx, and then do the same for a smaller value of Δx, and then once more using an even smaller value of Δx, and go on repeating the process until the value of $\Delta f/\Delta x$ obtained in this manner does not change within the accuracy required (its value does not change up to, say, the nth decimal point). When you have reached this stable, so to speak, value of the above ratio, you have obtained the *derivative of the function f(x) at point x* within the required accuracy.

Let us, for the sake of clarity consider a specific example. Let us assume that the distance (s) travelled by a particle in time t is given by:

$$s = 2t^2 \tag{5.23}$$

where s is measured in meters and time in seconds. We could obtain the ratio $(s(t + \Delta t) - s(t))/\Delta t \equiv \Delta s/\Delta t$, which measures the speed of the particle at time t geometrically from the slope of the tangent to the curve 's versus t' at the time t, as shown schematically in Fig. 5.15. We can, in principle, draw the tangent to the

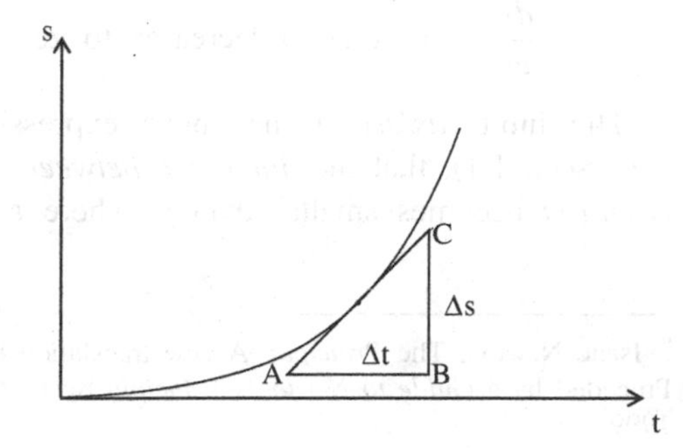

Fig. 5.15 The rate of change, $\Delta s/\Delta t$ at time t is given by the slope of the tangent to the *curve* at this moment of time

shown curve at any point of time and in this way obtain its speed at any time. In practice, however, this is a laborious process and the possibility of error is not negligible. It is much easier to obtain the speed of the particle analytically (algebraically) on the basis of Eq. (5.22).

We have [from Eq. (5.23)]:

$$\begin{aligned} s(t + \Delta t) - s(t) &= 2(t + \Delta t)(t + \Delta t) - 2t^2 = 2(t^2 + 2t\Delta t + (\Delta t)^2) - 2t^2 \\ &= 4t\Delta t + 2(\Delta t)^2 \end{aligned}$$

Therefore:

$$\frac{s(t + \Delta t) - s(t)}{\Delta t} = \frac{4t\Delta t + 2(\Delta t)^2}{\Delta t} = 4t + 2\Delta t \tag{5.24}$$

Now if Δt is very small (for example, $\Delta t = 0.001$) its square will be even smaller ($(\Delta t)^2 = 0.000001$) and can surely be neglected. Therefore the derivative of $s(t)$ (the speed of the particle) is given at any time by:

$$\frac{\Delta s}{\Delta t} = 4t \tag{5.25}$$

The advantages of the algebraic method are obvious: we have a formula which gives the speed at any moment, reliably and effortlessly in comparison with the geometric method. There remains, of course, the clumsiness of the term $2\Delta t$ and the way we got rid of it at the end of a process which began with the assumption that Δt can not be zero. What Newton and Leibniz proposed in response to this difficulty can be summarized roughly as follows: keep Δt in principle but disregard it in practice (as we have done). In the words of Newton[9]: *Quantities, and also ratios of quantities, which at any finite time constantly turn to equality, and which before the end of that time approach so close to one another that their difference is smaller than any given quantity, become ultimately equal.*

Let me add, however, that the remaining 'clumsiness' of this approach was eventually removed by *defining* the derivative of a function as a *limit*, in the manner of Weierstrass (see Sect. 1.1). In relation to our specific example this means that, the derivative of $s(t)$, to be denoted by $\frac{ds}{dt}$, must be defined as follows:

$$\frac{ds}{dt} = \text{limit (as } \Delta t \text{ decreases to zero) of } \frac{s(t + \Delta t) - s(t)}{\Delta t} \tag{5.26}$$

The limit (ds/dt) of the above expression is defined by the requirement (see Sect. 1.1), that the *difference between it and the ratio on the right of the equation* becomes smaller than ε, where ε can be arbitrarily small, when Δt

[9] Isaac Newton, The *Principia*, A new translation by I. Bernard Cohen and Anne Whitman. Preceded by a *Guide to Newton's Principia* by I. B. Cohen. (University of California Press), 1999.

becomes sufficiently small. In our example, the ratio on the right of the above equation equals $4t + \Delta t$, (according to Eq. (5.24)), so putting $ds/dt = 4t$ (according to Eq. (5.25)), makes the difference between these two quantities equal to $2\Delta t$, which can be made smaller than any ε by putting $\Delta t < \varepsilon/2$. Therefore $ds/dt = 4t$ is *formally* the derivative of $s(t) = 2t^2$.

What we said in relation to the above example applies quite generally. Therefore we can refine our definition [Eq. (5.22)] of the derivative (df/dx) of a function $f(x)$, as follows:

$$\frac{\mathrm{d}f}{\mathrm{d}x} = limit\ (as \Delta \mathrm{x}\ decreases\ to\ zero)\ of \frac{f(x+\Delta x) - f(x)}{\Delta x} \tag{5.27}$$

which is how the derivative is defined in most textbooks of mathematics.

Let me now return to integration to explain how it is related to differentiation (taking the derivative of a function), and how the latter can help in the evaluation of integrals.

We first note that the integral of expression (5.20) is a function of its upper limit x: to every x there corresponds a definite value of the integral. We denote this function as follows:

$$F_a(x) = \int_a^x f(x)dx \tag{5.28}$$

In which case $F_a(x + \Delta x)$ is given by

$$F_a(x + \Delta x) = \int_a^{x+\Delta x} f(x)dx \tag{5.29}$$

We observe (see, e.g., Fig. 5.14) that

$$F_a(x + \Delta x) - F_a(x) = f(x)\Delta x \tag{5.30}$$

which is the area of the rectangle to be added to the integral when its upper limit shifts from x to $x + \Delta x$. Therefore, when Δx is sufficiently small we obtain:

$$\frac{dF_a}{dx} = f(x) \tag{5.31}$$

This is the fundamental equation relating integration to differentiation.

Before I demonstrate the usefulness of the above equation let me clarify a simple, but important, point relating to the integral defined by Eq. (5.28). Going back to the definition of the integral (see Eq. (5.20) and Fig. 5.13), we can see that $F_a(x)$ can be written as:

$$F_a(x) = \int_a^b f(x)dx + \int_b^x f(x)dx = C + \int_b^x f(x)dx = C + F_b(x) \qquad (5.32)$$

where b is a point on the x-axis other than a (see Fig. 5.13); and C is a constant, a number corresponding to the 'area' between the curve $y = f(x)$ and the x-axis extending from point a to point b. Obviously $dF_a/dx = dF_b/dx$. Therefore (5.31) is valid whatever the value of a.

Let us now see how Eq. (5.31) helps us in the evaluation of integrals: Given a function $f(x)$, we search for a function $F(x)$ whose derivative $dF/dx = f(x)$. We call $F(x)$ the *indefinite integral* of $f(x)$. In some cases it will be easy to find $F(x)$, in other cases it will not be so easy, and in some cases it will be extremely difficult, but that need not concern us here. The important thing is that the $F(x)$ we have found in this manner satisfies the equation:

$$F(x) = \int_a^x f(x)dx \qquad (5.33)$$

where a is some point on the x-axis which we shall *not* bother to determine. In the evaluation of any *definite integral*, between a lower limit c and an upper limit d, point a cancels out; and we obtain:

$$\int_c^d f(x)dx = F(d) - F(c) \qquad (5.34)$$

And we note that the upper limit (d) can be greater or smaller than (c). In the latter case we can imagine the integration of $f(x)$ *moving* from right to left ($\Delta x < 0$). It is worth repeating that a definite integral, evaluated analytically via $F(x)$, is identical with the limit of the corresponding sum [Eq. (5.21)] as Δx tends to zero.

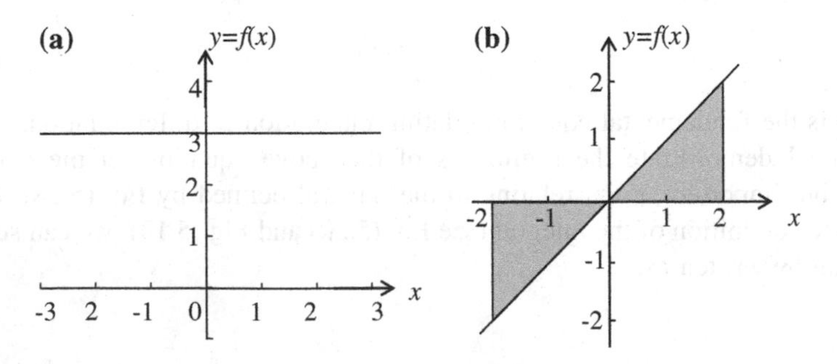

Fig. 5.16 **a** $f(x) = 3$; **b** $f(x) = x$

Usually, one denotes the indefinite integral of $f(x)$ by $\int f(x)dx$, which makes its relation to $f(x)$ obvious:

$$\int f(x)dx \equiv F(x) \tag{5.35}$$

where $F(x)$ is to be obtained in the manner I have described.

Let us now apply the above to two simple examples which will clarify what we have said above.

Example (A): We consider the function: $f(x) = 3$ (shown in Fig. 5.16a). We wish to evaluate the integrals:

$$I_1 = \int_1^3 3dx,\ I_2 = \int_3^1 3dx,\ I_3 = \int_{-1}^1 3dx$$

We search for the function $F(x)$ whose derivative $dF/dx = 3$. We can easily see that $F(x) = 3x$, because $(3(x + \Delta x) - 3x)/\Delta x = 3$. Therefore:

$I_1 = F(3) - F(1) = 9 - 3 = 6$, which is what we expect: it is the area of the rectangle between the line and the x-axis extending from $x = 1$ to $x = 3$.

$I_2 = F(1) - F(3) = 3 - 9 = -6$, which is the area of the same rectangle as for I_1 but with a negative sign which derives from the fact that Δx is negative in the direction from $x = 3$ to $x = 1$.

$I_3 = F(1) - F(-1) = 3 \times 3 - 3 \times (-3) = 9 + 9 = 18$, which again is what we expect: it is the area of the rectangle between the line (of Fig. 5.16a) and the x-axis extending from $x = -3$ to $x = 3$.

Example (B): We consider the function $f(x) = x$ (shown in Fig. 5.16b). We wish to evaluate the integrals:

$$I_1 = \int_0^2 xdx,\quad I_2 = \int_{-2}^2 xdx$$

Table 5.1 Derivatives and Indefinite integrals of frequently used functions

$f(x)$	$\frac{df}{dx}$	$\int f(x)dx$
ax^n, n = *any number*	nax^{n-1}	$\frac{a}{n+1}x^{n+1}, n \neq -1$
$\sin x$	$\cos x$	$-\cos x$
$\cos x$	$-\sin x$	$\sin x$
e^x	e^x	e^x
$\log x$	$1/x$	
$1/x$	$-1/x^2$	$\log x$

Note in relation to the first row of the table: we remember that x^{-n} is another way of writing $1/x^n$; and that $x^{1/2}$ is another way of writing $\sqrt{x}$. So that, for example, $(1/\sqrt{x})^3 = (x^{-1/2})^3 = x^{-3/2}$. Therefore: $\frac{d}{dx}\left(\frac{1}{\sqrt{x}}\right)^3 = \frac{d}{dx}x^{-3/2} = -\frac{3}{2}x^{-5/2} = -\frac{3}{2}\left(\frac{1}{\sqrt{x}}\right)^5$

We search for a function $F(x)$ whose derivative $dF/dx = x$. We try $F(x) = ax^2$ (where a is number to be determined). We have: $(F(x + \Delta x) - F(x))/\Delta x = (a(x + \Delta x)^2) - ax^2)/\Delta x = 2ax + a\ \Delta x = 2ax$ (for sufficiently small Δx). Therefore putting $a = 1/2$, we obtain:

$$\frac{d(x^2/2)}{dx} = x, therefore \int xdx = F(x) = \frac{1}{2}x^2$$

Therefore: $I_1 = F(2) - F(0) = \frac{1}{2}(2^2) - \frac{1}{2}(0^2) = 2$, which is what we expect from Fig. 5.16b: I_1 is the area of the shaded triangle above the x-axis: $(2 \times 2)/2 = 2$.

$I_2 = F(2) - F(-2) = \frac{1}{2}(2^2) - \frac{1}{2}(-2)^2 = 0$, which is again what we expect: the *negative* area of the shaded triangle below the x-axis of Fig. 5.16b cancels the positive area of the triangle above the x-axis.

Let me conclude this section with a list (Table 5.1) of the derivatives and indefinite integrals of the simple functions that we are likely to an encounter in this book. And one can, of course, convince oneself that the derivatives of the trigonometric functions given in the table agree with Fig. 5.6: the derivative of the function is positive when the function increases, negative when it decreases, and zero at the points where the tangent to the curve of the function is parallel to the x-axis, i.e. it has zero slope. [In what follows, when no ambiguity arises, we shall write *cosx*, *sinx*, etc. instead of *cos*(*x*), *sin*(*x*),etc.]

It is not our purpose to involve the reader in the technicalities of calculus. What is necessary for a good understanding of this book is an understanding of what an integral is [it is the area under the curve $f(x)$ between two limits a and b], and what the derivative of $f(x)$ is [it is the rate at which $f(x)$ changes at x]. The above list and the two rules of differentiation we give below is all we need for the purposes of this book. We first note that occasionally (when the expression for $f(x)$ is a bit long) one denotes the derivative of $f(x)$ by

$$\frac{d}{dx}f(x) \text{ instead of } \frac{df}{dx} \tag{5.36}$$

The first rule of differentiation I wish to cite gives the derivative of a function which is the *product* of two functions:

$$\frac{d}{dx}(f(x)g(x)) = f(x)\frac{dg}{dx} + g(x)\frac{df}{dx} \tag{5.37}$$

Example: $\frac{d}{dx}(ax\sin(x)) = \frac{d(ax)}{dx}\sin(x) + ax\frac{d\sin(x)}{dx} = a\sin(x) + ax\cos(x)$

The second rule gives the derivative of a function of the type: $f(q(t))$, which means that f is a function of q which is a function of t. The rule is straightforward:

$$\frac{df}{dt} = \frac{df}{dq} \times \frac{dq}{dt} \tag{5.38}$$

Example:

$$\frac{d\sin(\omega t)}{dt} = \frac{d\sin(\omega t)}{d(\omega t)} \times \frac{d(\omega t)}{dt} = \cos(\omega t) \times \omega = \omega\cos(\omega t) \quad (5.39)$$

Another example:

$$\frac{de^{-x}}{dx} = \frac{de^{-x}}{d(-x)} \times \frac{d(-x)}{dx} = e^{-x}(-1) = -e^{-x} \quad (5.40)$$

As a final example, let us obtain $\frac{d}{dx}\left(\frac{A}{r}\right)$, when $r = \sqrt{(x^2 + c)}$, where A and c are constants. We have:

$$\frac{d}{dx}\left(\frac{A}{r}\right) = \frac{dr}{dx}\frac{d}{dr}\left(\frac{A}{r}\right); \frac{dr}{dx} = \frac{d}{dx}(x^2 + c)^{1/2} = \frac{1}{2}(x^2 + c)^{-1/2}2x = \frac{x}{r}; \frac{d}{dr}\left(\frac{A}{r}\right) = -\frac{A}{r^2}$$

Therefore:

$$\frac{d}{dx}\left(\frac{A}{r}\right) = -\frac{Ax}{r^3} \quad (5.41)$$

Finally, the second, third and higher derivatives of a function, $f(x)$, are defined in similar manner:

$$\frac{d^2f}{dx^2} = \frac{d}{dx}\left(\frac{df}{dx}\right); \quad \frac{d^3f}{dx^3} = \frac{d}{dx}\left(\frac{d^2f}{dx^2}\right); \text{ and so on.} \quad (5.42)$$

Exercises

5.1 Consider an ellipse as defined in the caption of Fig. 5.4. Its eccentricity is defined by: $e = \sqrt{1 - \frac{b^2}{a^2}}$. Using this formula and the property of the ellipse, namely that the sum of the distances of any point on the ellipse to each focus equals $2a$, derive Eq. (5.4).

5.2 Using the rule: $\frac{d}{dx}ax^n = nax^{n-1}$, differentiate each term in the series representing $\cos x$ and $\sin x$ and therefore show that: $\frac{d\cos x}{dx} = -\sin x$ and $\frac{d\sin x}{dx} = \cos x$.

5.3 Using the rule: $\frac{d}{dx}ax^n = nax^{n-1}$, differentiate each term in the series representing e^x and therefore show that: $\frac{de^x}{dx} = e^x$.

5.4 Prove Eq. (5.37).
Hint: For Δx small we can write: $f(x + \Delta x) = f(x) + (df/dx)dx$ and $g(x + \Delta x) = g(x) + (dg/dx)dx$.

5.5 Find the derivative *df/dt* of: (a) $f(t) = (A + Bt)e^{-\gamma t}$; (b) $f(t) = Ae^{-\gamma t}\cos\omega t$
Answer: (a) $df/dt = Be^{-\gamma t} - \gamma (A + Bt)e^{-\gamma t}$; (b) $df/dt = -\gamma Ae^{-\gamma t}\cos\omega t - \omega Ae^{-\gamma t}\sin\omega t$.

5.6 Let $F(t) = e^{-\gamma t}(A\cos\omega t + B\sin\omega t)$. Determine A and B so that: $\frac{dF}{dt} = Ce^{-\gamma t}\cos\omega t$
Answer: $A = -\frac{\gamma C}{\omega^2+\gamma^2}$; $B = -\frac{A\omega}{\gamma}$.

5.7 (a) Convince yourself that the derivative *df/dx* of a function vanishes at those points where *f*(*x*) has a (local) maximum or minimum. Verify the above for *cos x* and *sin x*.
(b) Consider the function: $f(x) = A + B(x - x_0)^2$. Show that this function has maximum or minimum at $x = x_0$ depending on the sign of *B*.

5.8 Find the indefinite integral: $F(x) = \int \frac{dx}{(a+bx)^{3/2}}$
Hint: Put $a + bx = w$, so that $bdx = dw$, so that $F = \frac{1}{b}\int w^{-3/2}dw = -\frac{2}{b}w^{-1/2} + constant$
Answer: $F(x) = -\frac{2}{b(a+bx)^{1/2}}$+constant. We verify that: $\frac{dF}{dx} = \frac{1}{(a+bx)^{3/2}}$.

5.9 Find the indefinite integral: $F(x) = \int \sin x \cos x dx$
Hint: Put $\sin x = w$, so that $\cos x dx = dw$, so that $F = \int wdw = \frac{1}{2}w^2$+constant.
Answer: $F(x) = \frac{1}{2}\sin^2 x$+constant. We verify that: $dF/dx = \sin x \cos x$.

5.10 The mean value of $\cos^2 x$ is defined by: $\overline{\cos^2 x} = \frac{1}{2\pi}\int_0^{2\pi} \cos^2 x dx$.
Show that: $\overline{\cos^2 x} = 1/2$.
Hint: We note that: $\cos^2 x = (1 + \cos 2x)/2$.

Chapter 6
Isaac Newton

6.1 The Unhappy Childhood and His Life at Cambridge

Isaac Newton was born in Woolsthorpe, a village in Lincolnshire of England, in the year 1642, a time of civil war between the soldiers of the king and those of the parliament led by Oliver Cromwell. His father, who was an illiterate peasant but did have some property and was by no means poor, died before Isaac was born, apparently prematurely. In 1646, when Isaac was just three years old, his mother Hannah remarried. It appears that Hannah, who came from a family of respected local lower gentry, which included among its members parsons and lecturers, married the 63 years old widower Barnabas Smith for financial security, although there is no evidence to suggest that the 30 years old Hannah and her son were in need of money at that time. Barnabas Smith, who had studied at Oxford University, was the rector of North Witham, a hamlet about a mile from Woolsthorpe. It appears that over a lifetime he had collected many books on theology which he apparently never read; they were inherited by Newton and may have contributed to his interests in theology. More relevant to Hannah's decision was the fact that Smith had no children from his first marriage and that he had at the time of their marriage an independent income way above his clergyman's stipend.

When Hannah moved to North Witham to be with her new husband, the child Isaac was left behind at the Newton's manor in Woolsthorpe, where he was being looked after by his grandparents James and Margery Ayscough. Though Hannah visited regularly, the child Isaac missed his mother and resented her absence. He would later confess in one of his diaries that he hated his stepfather and desired to have his house in North Witham burnt.

Barnabas Smith died after 8 years of marriage to Hannah, during which he fathered three children, Mary, Benjamin and Hannah. After his death, the widow Hannah returned with her children to the manor in Woolsthorpe, but it was a bit too late for Isaac. A year after Hannah's return he left for Grantham (7 miles away) to attend the King's School there. According to some historians the enforced separation from his mother at an early age affected Newton's character. As a young man he kept to himself, had few friends and was generally mistrustful

A. Modinos, *From Aristotle to Schrödinger*,
Undergraduate Lecture Notes in Physics, DOI: 10.1007/978-3-319-00750-2_6,

of people and especially of women, and these characteristics he retained more or less throughout his life. At the King's School he received a good grounding in Latin, Greek and Bible studies, but had no formal mathematical training of any kind. At King's School the pupils would learn the basics of arithmetic and possibly a bit of algebra, that would be all he knew when he entered Cambridge a few years later. He was not an exceptional student, but after reading, at the age of thirteen, a book by John Bale on *The Mysteries of Nature and Art,* he developed an aptitude for building working mechanical models and gained a reputation in this respect. He could also draw and write well.

In Grantham, Newton was lodging with an apothecary's family. Mr. Clark, the apothecary, encouraged the curiosity of the inquisitive pupil who got himself involved in the production of paints and pigments, learned how to cut glass with chemicals, and other things. Newton was apparently happy during this period, and he apparently made good friends with Mr. Clark's stepdaughter, Catherine Stoner, the only female other than his mother and later his half-niece Catherine Barton to whom Newton is known to have been emotionally attached. By the autumn of 1660 Newton was preparing for Cambridge after his mother reluctantly agreed to let him go there. She would have been much happier if he had stayed with her at the manor looking after her family and the farm.

Cambridge in the 1600s had a population of about 8,000 of which about 3,000 were students. The town was dominated by an autocratic university governing body that was often inefficient and corrupt. The town was overcrowded and a dangerous place to live in. There would be prostitutes and robbers in the inns and the unlit streets. The students were of course forbidden to associate with the tradesmen of the town and were not supposed to have dealings with prostitutes, but they often did, and some of them were occasionally punished for doing so. However, the restoration of the monarchy in 1660 brought a climate of renewal. Cromwell's Protestant Commonwealth died with him in 1658. The University which remained loyal to the Crown was looking forward to a secure future.

Newton registered at Trinity College in June 1661 as a subsizar, becoming a sizar within a month or so. Subsizars and sizars were supplementing their income by serving richer students (emptying the bedpans and cleaning their rooms), and were often treated with contempt by these wealthy students, many of whom considered the University as little more than a playing ground, before moving on to an easy job and an undemanding role in society. Newton's mother paid his fees at the University (about 15 pounds per year) but offered him very little (about 10 pounds per year) for his living expenses. She could have given him more as she had an annual income of about 700 pounds, but she did not. The young Newton soon found a way to supplement his allowance, by borrowing and lending money at a profit to his fellow students, a practice he kept going for at least 2 years, until some money came his way. At nineteen Newton was about 2 years older than the average student at entry. He was a devoted Anglican with puritan ethics who believed that the acquisition of knowledge and the study of Nature were to the great glory of God. And he kept to himself.

The curriculum of the University had not changed much from that of the 1570s. To obtain a BA (Bachelor of Arts) degree all students had to reside at the University for 4 years and to attend all public lectures given by the faculty members. There was only one course. By the end of the first year the students were expected to be fluent in Latin, Greek and Hebrew. After that, students would learn theology, history, geography and science by studying well known texts, almost exclusively the Greeks and in particular Aristotle. But Newton was among the few who followed the latest developments and was reading books by Galileo, Kepler, Descartes, Tycho Brahe, Robert Boyle and others. It is fair to assume that his reading of works by these authors excited his curiosity and put him on the road to his great discoveries. It appears from his notebooks (beginning some time in 1663) that he often wrote under a heading or other, such as *Of water and salt* or *Attraction Magnetical* or *Of gravity and Levity*, the answers offered by the various authors, which he would then consider more closely.

Newton's first experiments were concerned with the nature of light and began in the summer of 1664 when Newton bought a prism at a Stourbridge fair in order to try some experiments in relation to what he had read in Descartes' *Discourse on the Scientific Method* on this subject. We have already summarized the results and the conclusions relating to the nature of light that Newton derived from these experiments (see Sect. 5.2.2). These appeared in the *Opticks,* a book he published much later (in 1704). In a letter he wrote in 1672 to Henry Oldenburgh, then Secretary of the Royal Society, he described his first experiments with the prism he bought at the Stourbridge fair as follows: *I procured me a triangular glass-prism, to try therewith the celebrated «Phenomena of Colours». And in order thereto having darkened my chamber, and made a small hole in my window-shuts, to let in a convenient quantity of the sun's light, I placed my prism at its entrance, so that it might be thereby refracted to the opposite wall. It was at first a very pleasing divertissement, to view the vivid and intense colours produced thereby. ...*

At about the same time (1664) Newton turned his attention to mathematics. When he joined the University his knowledge of mathematics was limited to simple arithmetic, some algebra and a little trigonometry. He now read Euclid's geometry which he would later use with great efficiency in his own work and he also studied Descartes' analytical geometry. Because he concentrated on his own studies and less so on the examinable material, his BA degree, which he obtained in 1665 was a second class degree. But this was enough to secure his future at the University, and this was where he wanted to be. He had already decided that his vocation was to discover the God given laws of Nature.

When the plague came he left Cambridge, sometime in the summer of 1665 and stayed in Woolsthorpe for about 2 years, and it is during these 2 years that he made his great discoveries of calculus, the laws of motion and universal gravity. Following on these discoveries he worked for 20 years to elaborate upon and complete his theories on the motion of bodies in general and of the planets in particular, which he eventually presented in the *Principia* in 1687.

On his return to Cambridge after the plague, in the spring of 1667, he completed his Master of Arts degree and thereby obtained a fellowship that secured his future at Trinity College. Newton now had a job for life and could continue to study at his leisure whatever he chose to study. The college paid him a living allowance and provided a room free of charge. He shared a flat with his fellow student John Wickins who later became his assistant. He transcribed notes and helped with the setting up of apparatus and the monitoring of investigations. The two shared a flat for many years until 1683. For a few months Newton relaxed. He visited taverns with Wickins and played an occasional game of cards. And then he withdrew again into isolated scholarship.

In October 1669 Isaac Newton, not yet 27 years old, became the second Lucasian Professor of Mathematics at Cambridge. It happened like this.

A fellow of St. John's College, Henry Lucas, one of the University's Members of Parliament, had succeeded in establishing the Lucasian Professorship of Mathematics. The post carried with it a generous renumeration of 100 pound per year and a minimum of teaching responsibilities, but the Professor was not allowed to hold any other post concurrently. The first Lucasian Professor was Isaac Barrow who was appointed to the post in 1664. Barrow, who was both a Puritan and a royalist, was a competent and versatile mathematician and philosopher, and had travelled widely in Europe for 4 years. On his return he became the Regius Professor of Greek at the Trinity College. In 1662 he moved to London to become the Gresham Professor of Geometry, and was one of the founding Fellows of the Royal Society. Barrow held the Lucasian Professorship for 5 years (1664–1669), did his best to meet his teaching responsibilities, but on most occasions he had no audience for his lectures (on mathematics and optics). Demoralized by neglect he gave up the post, which was offered to Newton, on his recommendation. The two men, though not exactly friends, had a good relation between them based on mutual respect. Barrow went on to serve for some years as Master of Trinity College before he died in 1677.

Not a single student turned up for Newton's second lecture as Lucasian Professor, and that was more or less the case there after. Occasionally he would lecture in an empty classroom for about 15 min before returning to his own studies. He lectured on mathematics, optics and on various topics that would later appear in the *Principia*. And he would deliver notes covering a minimum of ten lectures per year to the library of the University. Newton gradually reduced the times he lectured to one term a year and later, when he spent more time away from the University, he lectured even less. He had during his career three tutees (research students) but none of them did anything worth noting later in their lives.

In 1672 Newton was elected Fellow of the Royal Society (FRS). The Royal Society was created in 1648 in order *to enlarge knowledge by observation and experiment.* In those times, most announcements there were concerned with simple and often trivial observations by interested members. At the time Newton was elected to the Society, Robert Hooke was an active member of it and the Curator of Experiments. Hooke was an able man with many interests. His best work, the *Micrographia*, published in 1665, was meant as a treatise on microscopy, but it also included theoretical pieces about the nature of light among other things. Newton had read and valued the book. But when they met they immediately disliked each other. Newton was a loner, Hooke a man of society. Hooke thought Newton was lacking in good manners, and Newton thought that Hooke was vain. And may be they were both right! In any case, when Newton gave his first seminar to the Society, which concerned the nature of light, Hooke accepted the verity of Newton's prism experiments but disagreed strongly with Newton's idea of 'particles of light' insisting that light is nothing but a pulse propagating in a homogeneous medium. But apparently the way he put his case, rather than the arguments themselves, offended Newton, and what was in essence a scientific argument developed into a personal dispute between the two men which lasted well beyond that first meeting. So much so that at some stage Newton threatened to resign from the Society. In fact, as he was living in Cambridge and the meetings of the Society took place in London he very rarely attended.

In 1679 Newton returned to Woolsthorpe and stayed there for several months nursing his mother. When she died in June 1679, Newton who inherited her estate became a wealthy gentleman.

6.2 The Principia

6.2.1 Introduction and the Laws of Motion

Newton was 41 years old and almost totally grey when he was visited by Edmond Halley. Halley was a competent astronomer, he was much interested in gravity and had already discussed with Hooke the possibility that a planet's motion about the sun was due to a force (***F***) exerted on the planet by the sun, pulling the planet towards it with a strength inversely proportional to the square of the distance (*R*) between the planet and the sun ($F = constant/R^2$). When he asked Newton about it, he replied that this was indeed the case and that this led to an elliptical orbit of the planet about the sun. He had calculated it! Halley tried to convince Newton to publish his results and after much patience and effort he managed to obtain (in 1684) the manuscript of *De Motu Corporeum* in which Newton presented his 'laws of motion' and his theory of gravity in a *popular manner*. This is how he described *De Motu* when he referred to it in the *Principia,* which he wrote *in a more mathematical form so that it would not be* read *by anyone who had not mastered the principles of rational mechanics.* The great book itself was published three years later (in 1687). In his preface Newton acknowledges *the tremendous assistance he received from Halley in the publication of the book* and for putting him on the road to it: *it was he who started me off on the road to this publication.*

It is worth noting that in the *Principia* we are presented with the final product of Newton's efforts to formulate the laws of motion and of gravity. We are told nothing of his earlier attempts that failed. The reader who wishes to know something about these will find some information and further references in the *Guide to Newton's Principia* by I. Bernard Cohen (cited in Chap. 5). It must also be said, that the present day student of physics will find it extremely difficult to follow Newton's proofs of the various theorems presented in it. This is because he uses geometrical constructions (what he calls *synthesis*) in his proofs, not the calculus of functions that we described in Sect. 5.3 (what he calls the analytical method). What he takes for granted throughout the book is that any line can be approximated by a succession of straight segments of infinitesimally small length, both in relation to the evaluation of the slope of the curve (representing the rate of change of a quantity with respect to a variable) and in relation to integration. That much he states quite clearly at the beginning of Book1 of the *Principia*. It appears that though he had himself discovered (analytical) calculus, he felt more confident using the traditional methods of geometry in the development of his theories relating to the motion of bodies. In what follows we shall describe Newton's results in analytical form, except for of one or two examples where his geometrical ways are easy to follow.

The *Principia* which Newton wrote in Latin (A translation of it into English was published just before the end of his life), consists of three Books and an Introduction. In the Introduction he defines the basic concepts he will employ in his formulation of the laws of motion. His definitions of velocity and acceleration of a body are the familiar ones (as in Galileo's work) based in effect on the Aristotelian notions of space and time. It is worth noting, however, that Newton makes a distinction between true (mathematical) and apparent (common) time. In his own words: *Absolute, true, and mathematical time, in and of its own nature, without reference to anything external, flows uniformly and by another name is called duration. Relative, apparent, and common time is any sensible and external measure of duration by means of motion* (by which we must understand the motion of the stars and of clocks in general). The reader will note the difference between the above and Aristotle's view that time would not exist without motion and it can only be defined in terms of clocks.

More important in relation to actual calculation, is Newton's distinction between absolute and relative position, and similarly between absolute and relative motion. This is made clear in his Introduction by the following example: … *Thus, in a ship under sail, the relative position of a body is that place of the ship in which the body happens to be and which moves along with the ship, and relative rest is the continuous of the body in that same position. But true rest is the continuance of a body in the same position in that unmoving space in which the ship itself, along with its interior and all its contents, is moving. Therefore, if the earth is at rest, a body that is relatively at rest on the ship will move truly and absolutely with the velocity that the ship is moving on the earth. But if the earth is moving, the true and absolute motion of the body will arise partly from the true motion of the earth in unmoving space and partly from the relative motion of the ship on the earth. Further, if the body is also moving relatively on the ship, its true motion will arise partly from the true motion of the earth in unmoving space and partly from the relative motions both of the ship on the earth and of the body on the ship, and from these relative motions the relative motion of the body on the earth will arise.*

Newton demonstrates the above with an example of a ship travelling westwards with a velocity of 10 units, at a point on the earth moving eastwards with a velocity of 10,010 units. If a sailor is walking on the ship towards the east with a velocity of 1 unit, the sailor will be moving truly and absolutely in unmoving space towards the east with a velocity of 10,001 units, and relatively on the earth towards the west with a velocity of 9 units.

Of great importance is also his definition of the mass (m) of a body[1]: *it is a measure of matter that arises from its density* ρ *and volume* V *jointly* ($m = \rho V$). Newton tells us that the mass of a body is proportional to its weight and can be known by measuring its weight, but it is nevertheless something entirely different. Evidence that this is indeed the case was provided a few years earlier by the observation (noted in the report of the Richter expedition) that the weight of a body is a 'local' property, depending to some degree on the latitude of the place (its value near the North Pole being to some degree different from its value near the Equator), and therefore can not be used as a measure of a body's quantity of matter.[2]

With the above definitions in place, Newton states (as axioms) the following three *Laws of Motion of a Body.*

Law 1: *Every body perseveres in its state of being at rest or of moving uniformly straight forward, except in so far as it is compelled to change its state by forces acting on it.*

This is Galileo's law of inertia (see Chap. 4) and we need not say anything more about it here.

Law 2: *A change in motion is proportional to the force acting upon the body* (assumed to be a point-particle) *and takes place along the straight line in which that force is acting.*

This law is now commonly expressed by an equation (the notation is explained in what follows):

$$\frac{d}{dt}\boldsymbol{P} = \frac{d}{dt}(m\boldsymbol{V}) = \boldsymbol{F} \tag{6.1}$$

where m denotes the mass of the body, assumed to be a *point*-particle,[3] $\boldsymbol{V}$ is the velocity of the particle, $\boldsymbol{P} = m\boldsymbol{V}$ is the momentum of the particle, and $\boldsymbol{F}$ stands for

---

[1] Appropriate symbols in parenthesis are introduced to facilitate the presentation of the theory that may not be in the Principia.

[2] Today we take as the unit of mass the kilogram (kg). This is equal to the mass of an international prototype body, a platinum-iridium cylinder, kept by the International Bureau of Weights and Measures at Sévres, near Paris. A kilogram is equal to 1,000 g. 1 g (gram) is equal to the mass of a cubic centimeter of pure water at a temperature of 4 °C.

[3] For the sake of clarity think of a solid body of a certain mass m. Replace the body by a point particle of mass m at the centre of mass of the body.

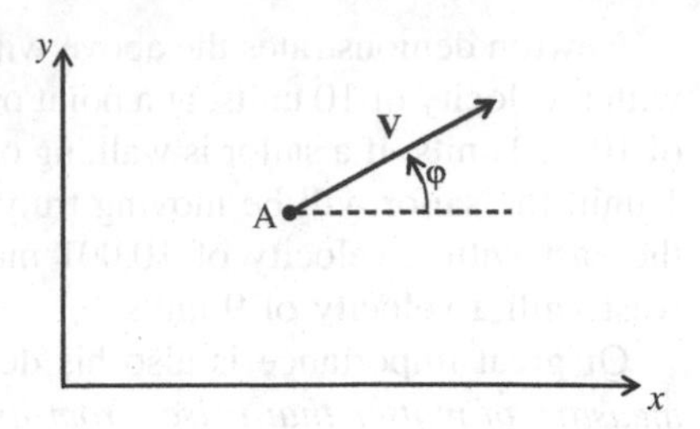

Fig. 6.1 A boat (at A) moves with velocity V at an angle (φ) to the x-axis

the force (the sum of the forces) acting on it. Eq. (6.1) tells us that: the derivative with respect to the time (i.e. the rate of change with the time) of the momentum of the particle is equal to the force acting on it.

Quantities denoted by bold letters are vectors (a vector $\boldsymbol{a}$ may also be denoted by $\vec{a}$), which means that they have a magnitude and a direction and both must be specified for the vector to be known. For example: to describe the velocity ($\boldsymbol{V}$) of a boat moving in a lake we must give its speed (so many meters per second) and its direction (it moves at an angle φ) to a specified direction (which we can take as the x-axis of a Cartesian system of coordinates; see Fig. 6.1).

The 'length' of the vector $\boldsymbol{V}$ gives the magnitude of the velocity (its speed) according to some scale (e.g., 1 cm stands for 10 m/s). In the same way to know a force $\boldsymbol{F}$ we must know not only its magnitude but also the direction along which it acts on the body. The unit of force (its magnitude) is: (kilogram $\times$ (meter/second))/second $=$ kg(m/s^2), which is in accordance with Eq. (6.1); it is now referred to as the *Newton* (N). I must now tell you, following Newton, how to sum up vector quantities.

Suppose that in a time interval (Δt) a boat is moved by its engine (in the absence of wind) from A to B (see Fig. 6.2a), and suppose that in the same time (Δt) the boat (with the engine off) moves from A to C due to wind. Then observation tells us that, if both movements happen simultaneously, the boat will move in time Δt from A to D, i.e. along the diagonal of the parallelogram formed by AB and AC. And of course the corresponding velocities add up according to the same rule as shown in Fig. 6.2b. In the same way, observation tells us, that two forces acting on a body along different directions produce the same effect as their sum obtained by the parallelogram rule as shown in Fig. 6.2c. In fact, the same rule applies to all vector quantities.

The same rule allows us to write any vector as a sum of its components along specific directions. In particular using the Cartesian system of coordinates (introduced in Sect. 5.1) we can write any vector as follows[4]:

---

[4] In Sect. 5.1 we introduced the Cartesian system of coordinates for two dimensions. The extension to 3 dimensions is obtained by adding a third axis (the z- axis) normal the xy-plane, as shown in Fig. 6.3. We note that the z-axis points in the direction a right-turning screw advances when turning from the x to the y direction. We adopt this convention throughout this book.

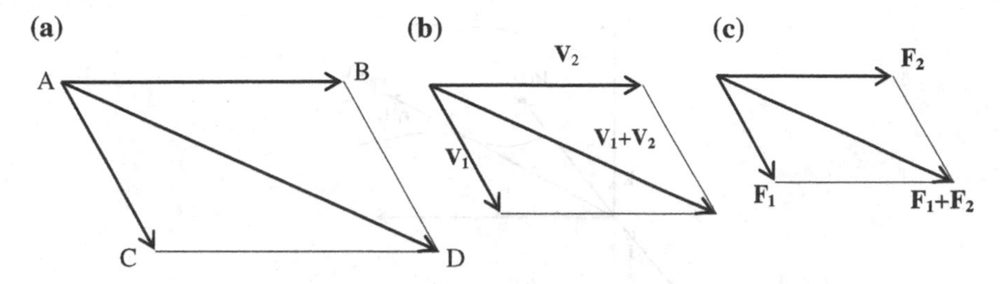

Fig. 6.2 Vector quantities add up according to the parallelogram rule

$$\boldsymbol{A} = A_x\boldsymbol{i} + A_y\boldsymbol{j} + A_z\boldsymbol{k}$$

where ***i, j, k***, are unit vectors along the *x*, *y*, and *z* coordinate axes. The magnitude *A* of the vector is of course given by: $A = \sqrt{A_x^2 + A_y^2 + A_z^2}$, obtained by using Pythagoras's theorem twice, first in the x–y plane, and then in the plane defined $A_x\boldsymbol{i} + A_y\boldsymbol{j}$ and the z-axis. Finally, we note that: $\boldsymbol{A} - \boldsymbol{B} = \boldsymbol{A} + (-\boldsymbol{B})$, where $-\boldsymbol{B}$ has the same magnitude as ***B*** but opposite direction.

In Fig. 6.3 we show the position vector

$$\boldsymbol{R}(t) = X(t)\boldsymbol{i} + Y(t)\boldsymbol{j} + Z(t)\boldsymbol{k}$$

of a particle with respect to a fixed Cartesian system of coordinates. In time Δt the particle is displaced by $\Delta\boldsymbol{R}(t)$, so its velocity $\boldsymbol{V}(t)$ is a vector with the direction of $\Delta\boldsymbol{R}(t)$, and a magnitude equal to the magnitude of $\Delta\boldsymbol{R}(t)$ divided by Δt, provided Δt is sufficiently small. In what follows we shall assume that this is always so, in which case we can write:

$$\boldsymbol{V}(t) = \Delta\boldsymbol{R}/\Delta t = d\boldsymbol{R}/dt = (dX/dt)\boldsymbol{i} + (dY/dt)_{\boldsymbol{i}}\boldsymbol{j} + (dZ/dt)\boldsymbol{k}$$

Note on differentiation of vectors: The derivative of any vector $\boldsymbol{A}(t)$ with respect to time (or any other variable) can in principle be obtained geometrically (as in Fig. 6.3), but it is more convenient to work it out in terms of its components:

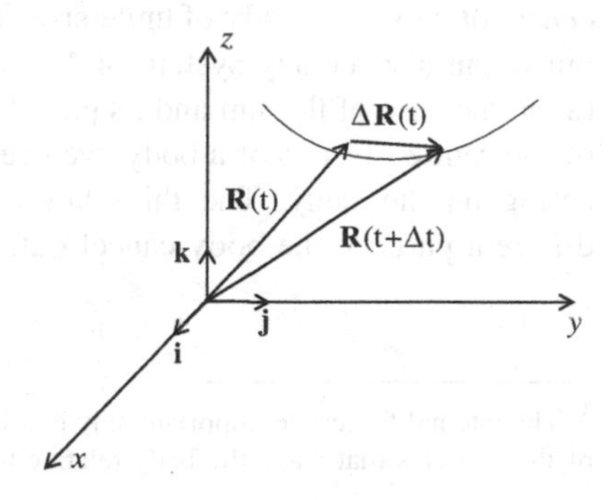

Fig. 6.3 Variation of **R**(t) with the time. $V(t) = \lim(\Delta t \to 0)\Delta\boldsymbol{R}/\Delta t$

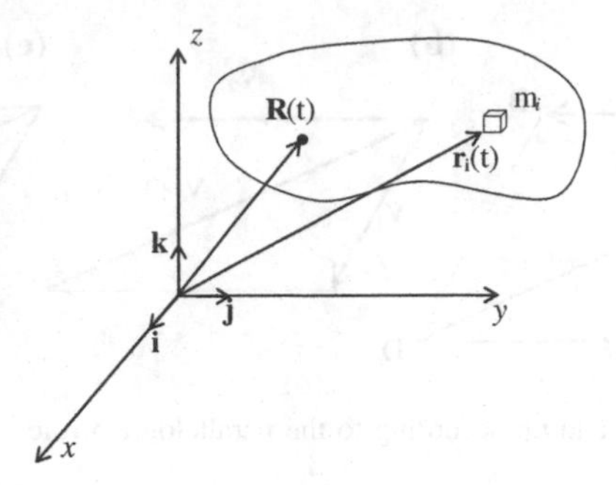

Fig. 6.4 The *i*th part of the body (the mass m_i within an infinitesimally small volume in the body) is so small that it can be treated as a point-particle. The parts of the body taken together make up the body. The relative positions of these parts are fixed for a *solid* body (such as a stone or a piece of wood), and therefore the position of the centre of mass of a solid body is fixed within the body. For example: the centre of mass of a solid sphere, whose density is constant or varies only with the distance from the centre of the sphere, coincides with the centre of the sphere

$$d\boldsymbol{A}/dt = (dA_x/dt)\boldsymbol{i} + (dA_y/dt)_y\boldsymbol{j} + (dA_z/dt)\boldsymbol{k}$$

It is clear from the above equation that the differentiation of vectors does not involve any new principle. It comes down to the evaluation of three derivatives (one for each component of the vector).

Law 3: *To any action there is always an opposite and equal* (in magnitude) *reaction; in other words the actions of two bodies upon each other are always equal and always opposite in direction.*

And by the way of example he writes: *Whatever presses or draws something is pressed or drawn just as much by it. If a finger presses a stone, the finger is also pressed by the stone. …. If some body impinging upon another body changes the motion* (the momentum) *of that body in any way by its own force, then, by the force of the other body (because of the equality of their mutual pressure), it also will in turn undergo the same change in its own motion* (momentum) *in the opposite direction.*

A most important consequence of the third law relates to the motion of the *centre of mass* of a body of finite size. The body can be a solid body (e.g., a stone), but it can also be any system of *N* particles kept together by mutual interactions (as in the case of the sun and its planets). It turns out that: to determine the motion of the centre of mass of a body, we need only take into account the *external* forces acting on the body. The third law dictates that any forces operating between different parts of the body cancel out.[5]

[5] The internal forces are important only in relation to the internal motion of the body (the motion of the particles that make the body relative to each other).

The position vector $\boldsymbol{R}(t)$ of the centre of mass of a body, of mass m, is (by definition) given by:

$$m\boldsymbol{R}(t) = \sum m_i \boldsymbol{r}_i(t) \tag{6.2a}$$

where $\boldsymbol{r}_i(t)$ is the position vector of the ith part of the body with mass m_i (see Fig. 6.4). The sum (Σ) in Eq. (6.2) includes all parts of the body, and therefore $\Sigma m_i = m$. The Cartesian coordinates (X,Y,Z) of the centre of mass of a body are correspondingly given by the following equations:

$$mX(t) = \sum m_i x_i(t), \quad mY(t) = \sum m_i y_i(t), \quad mZ(t) = \sum m_i z_i(t) \tag{6.2b}$$

Taking the time derivative of both sides of Eq. (6.2a) we obtain:

$$m\boldsymbol{V} = \sum m_i \mathbf{v}_i \quad \text{or} \quad \boldsymbol{P} = \sum \boldsymbol{p}_i \tag{6.3}$$

In words: the momentum of the body (the sum of the momenta of the different parts of the body) is equal to the mass m of the body times the velocity $\boldsymbol{V}$ of its centre of mass.

Taking the time derivative of both sides of the above equation, we obtain:

$$\frac{d}{dt}\boldsymbol{P} = \frac{d}{dt}\sum \boldsymbol{p}_i = \sum \boldsymbol{F}_i$$

In the above equation $\boldsymbol{F}_i$ includes all the forces (internal and external) acting on m_i, but in the sum (over all masses of the body) the internal forces cancel out, because according to the third law these appear in pairs of forces equal in magnitude but in opposing directions. Therefore:

$$\frac{d}{dt}\boldsymbol{P} = \frac{d}{dt}(m\boldsymbol{V}) = \boldsymbol{F} \tag{6.4}$$

where, $\boldsymbol{F}$ is the (vector) sum of the *externally* applied forces on the body. We note that the above equation is identical with Eq. (6.1) except that now it applies not to a point particle but to the centre of mass of a body of finite dimensions, or a system of N interacting particles: $\boldsymbol{P}$ is the momentum (the sum of the momenta of its parts), m is the total mass of the body, and $\boldsymbol{V}$ is is the velocity of the centre of mass of the body.

6.2.2 Examples of Motion

In the examples we shall consider (only a few among the many that appear in Books 1 and 2 of the *Principia*) the body is assumed to be a solid of constant mass, in which case Eq. (6.4) becomes:

$$m\frac{d}{dt}\boldsymbol{V} = \boldsymbol{F} \qquad (6.5)$$

The motion of a projectile: A projectile of mass m goes off a gun with an initial (at time $t = 0$) velocity V_0 along a horizontal direction which we take as the x-axis, and we take the y-axis along the vertical direction pointing downwards (see Fig. 6.5). We chose the origin of coordinates so that at $t = 0$ the position of the projectile is given by $(X(0),Y(0)) = (0,0)$. We wish to know the position $(X(t),Y(t))$ of the body (by which we mean the position of its centre of mass) for $t > 0$, and the shape of its trajectory. We proceed as follows: We note that the only force acting on the body is that of its weight: $\boldsymbol{W} = mg\boldsymbol{j}$, where g denotes the acceleration (about 9.80 m/s^2, the same for all bodies) due to gravity. Therefore, resolving the projectile's motion into its components along the x and y axes, we obtain in place of Eq. (6.5) the following equations (note that the mass cancelled out of the equations):

$$\frac{d}{dt}V_x = 0 \quad \text{and} \quad \frac{d}{dt}V_y = g$$

The first equation tells us that V_x does not change. Therefore it preserves its initial value. Therefore $V_x = V_0$ throughout the motion. Therefore $X(t) = V_0 t + C$, where C denotes a constant. But $X(0) = 0$, so $C = 0$. Therefore $X(t) = V_0 t$. The second equation tells us that $V_y(t) = gt + C$. But $V_y(0) = 0$, so $C = 0$. Therefore $dY/dt = gt$, which integrated gives: $Y(t) = gt^2/2 + C$. But $Y(0) = 0$, therefore $C = 0$. Therefore $Y(t) = gt^2/2$.

In conclusion: $X = V_0 t$ and $Y = gt^2/2$. Substituting in the second of these equations the value of the time corresponding to a given X, i.e. putting $t = X/V_0$, we obtain: $Y = CX^2$, where $C = g/(2V_0^2)$. This equation describes the trajectory in the sense explained in Sect. 5.1, and it is indeed a parabola, as found originally by Galileo.

Note: we have assumed that the origin of coordinates in Fig. 6.5 is at some height (say L) from the ground. The equations we have derived are valid only until the moment the projectile hits the ground, i.e. for $t < \sqrt{(2L/g)}$.

A final note: The reader may wonder whether one is justified in using a coordinate system fixed to a place (A), on the surface of the earth, which is moving

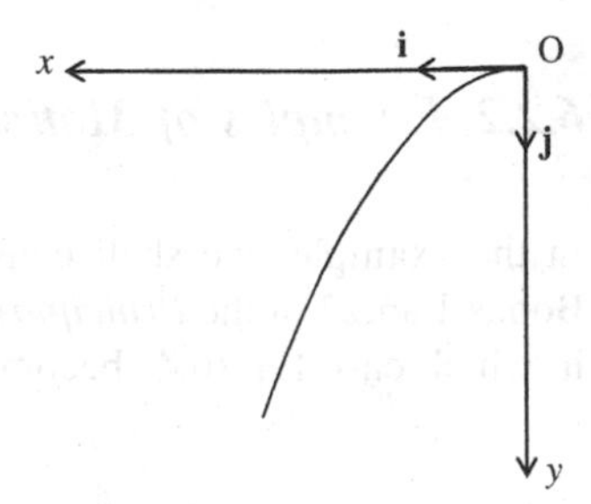

Fig. 6.5 The trajectory of a projectile

very fast about the centre of the earth (which is itself moving in its orbit about the sun). The fact is that for the *duration* of the experiment, A is, to a very good approximation, moving with a constant velocity $\boldsymbol{V}'$ (constant speed on a straight line) in Newton's absolute space. Now, the velocity of the projectile in this space is $\boldsymbol{V}' + \boldsymbol{V}$, where $\boldsymbol{V} = V_x\boldsymbol{i} + V_y\boldsymbol{j}$ is the velocity in the coordinate system attached to A (the one we have calculated). And because $\boldsymbol{V}'$ is constant for the duration of the experiment, the acceleration is the same in the two coordinate systems. And this is what matters. The acceleration determines the *change* in the velocity of the projectile and the trajectory of the projectile as described above (see also Sect. 6.2.4).

Harmonic motion: This is the motion $X(t)$ of a body of mass m moving on a straight line (we assume it to be the x-axis) about a centre (at the origin of coordinates) under the influence of a force $F = -kX$, where k is an appropriate constant. Such a motion will occur, for example, when a mass, hanging from an elastic spring along the vertical direction, is displaced from its equilibrium position ($X = 0$) by a small amount, so that at $t = 0$ is found not at the origin but at X_0. Assuming that the body is released from X_0 with zero speed, we wish to determine its subsequent motion, i.e. to find $X(t)$ for $t > 0$.

In this case $\boldsymbol{V}$ is reduced to one component $V_x = dX/dt$ and Eq. (6.5) becomes:

$$\frac{d}{dt}\left(\frac{dX}{dt}\right) = -\omega_0^2 X(t), \quad \text{where } \omega_0^2 = \text{k/m} \tag{6.6}$$

Now it is easy to verify [using Table 5.1 and Eq. (5.38)] that $X(t) = Acos(\omega_0 t) + Bsin(\omega_0 t)$ satisfies the above equation for any values of the constants A and B. The corresponding velocity is given by $V_x = dX/dt = -\omega_0 Asin(\omega_0 t) + \omega_0 Bcos(\omega_0 t)$. However, we demand that $V_x(0) = 0$; therefore (see Fig. 5.6) we must put $B = 0$. Therefore $X(t) = Acos(\omega_0 t)$, and because $X(0) = X_0$, we must put $A = X_0$. Therefore the motion of the body under consideration (often called a linear harmonic oscillator) is described by the function

$$X(t) = X_0\, cos(\omega_0 t) \tag{6.7}$$

The variation of $X(t)$ with the time is shown by the bottom diagram of Fig. 5.6 (replace x by $\omega_0 t$ and the '1' on the cosine axis by X_0). The body executes an oscillatory motion, between $-X_0$ and $+X_0$ (X_0 is called the amplitude of the oscillation) with a period $T = 2\pi/\omega_0$.

The above two examples show that: *if we know the position and the velocity of a body at some moment of time* (for convenience we measure time starting from this moment) *we can determine its position and velocity at any time* ($t > 0$), *if we know the (external) force (the sum of all such forces) acting on the body*. In some cases (as in the above two examples) we can do this analytically. In other cases we may have to do it numerically, as follows. Let us, for simplicity assume a linear motion (along the x-direction), and let us further assume that the force depends only on the position of the body. If we know the position $X(t)$ and the velocity $V(t)$ of the body at time t, we can obtain the same at time $t + \Delta t$ from the equations: $X(t + \Delta t) = X(t) + V_x(t)\Delta t$ and $V_x(t + \Delta t) = V_x(t) + [F(x(t))/m]\Delta t$, where the quantity in the square brackets is the acceleration of the body at time t, according to Newton's law. We can repeat the process to obtain the position and velocity of the body at time $t = 2\Delta t$, from those at time $t + \Delta t$, and so on indefinitely.

In 3-dimensional motion we apply the above procedure to all three components of the motion ($X(t)$, $Y(t)$, $Z(t)$) in parallel because in general the force will depend on all thee coordinates. And if the motion of a body is coupled (through some force) with the motion of another body (so that the force acting on one particle depends on the positions of both particles), then we apply the above procedure to the coordinates of both particles. And in principle we can apply the same method to a system of N bodies (involving a parallel calculation of $3N$ coordinates), as long as we know the force acting on each body. Some people extrapolating the above to the whole Universe did say that, everything that happens in the Universe is predetermined by the conditions that prevailed at the moment of its genesis!

Let me complete this section with a description of circular motion and the presentation of four theorems, proved in Book 1 of the *Principia,* which are necessary for an understanding of his theory of gravity. All four of them apply to *central* motions, which means that the force applied to the body (and it is the cause of its motion) is at all times directed to a fixed centre. Such a force is called central, or *centripetal* by Newton.

Circular motion: Consider the motion of a body moving with constant speed V (this being the magnitude of its velocity) on the circle (of radius R) shown in Fig. 6.6. Circular motion is obviously an accelerated motion since the direction of the body's velocity is not constant but changes continuously. At time t the body's velocity $\boldsymbol{V}(t)$ is normal to its position vector $\boldsymbol{R}(t)$, and similarly $\boldsymbol{V}(t + \Delta t)$ is normal to $\boldsymbol{R}(t + \Delta t)$. Therefore the angle between $\boldsymbol{V}(t + \Delta t)$ and $\boldsymbol{V}(t)$ is equal to the angle $\Delta\varphi$ between $\boldsymbol{R}(t + \Delta t)$ and $\boldsymbol{R}(t)$, as shown in Fig. 6.6. Now $\Delta\varphi$ is related to the segment $\Delta s = V\Delta t$, traversed by the body in the time Δt, by the equation $\Delta\varphi = \Delta s/R = (V\Delta t)/R$. At the same time we have (see the small diagram in Fig. 6.6) that $\Delta\varphi = |\Delta \boldsymbol{V}|/V$, where $|\Delta \boldsymbol{V}|$ is the magnitude of $(\boldsymbol{V}(t + \Delta t) - \boldsymbol{V}(t))$. Therefore $|\Delta \boldsymbol{V}|V = (V\Delta t)/R$ and therefore the magnitude of the acceleration of the body $|\Delta \boldsymbol{V}/\Delta t| = V^2/R$. We further note that the direction of $\Delta\, \boldsymbol{V}$, and therefore that of the acceleration $(\Delta \boldsymbol{V}/\Delta t)$ becomes normal to $\boldsymbol{V}(t)$ as Δt tends to zero. It points to the centre of the circle: in the opposite direction of $\boldsymbol{R}(t)$. We can summarize the above in the following formula[6]:

$$\frac{d}{dt}\boldsymbol{V}(t) = -\frac{V^2}{R}\boldsymbol{e}_R(t) \tag{6.8}$$

[6] The interested reader can derive this formula analytically as follows: We note that $\boldsymbol{R}(t) = R\boldsymbol{e}_R(t)$ and that $\boldsymbol{e}_R(t) = cos(\omega t)\boldsymbol{i} + \sin(\omega t)\boldsymbol{j}$, where $\boldsymbol{i}$ and $\boldsymbol{j}$ are unit vectors along the x and y axes of a Cartesian system of coordinates, and we have assumed that the centre of the circle is at the origin of the coordinates. We have also put $\omega = \Delta\varphi/\Delta t = V/R$. Now we obtain $\boldsymbol{V}(t) = R(d\boldsymbol{e}_R/dt) = R[-\omega sin(\omega t)\boldsymbol{i} + \omega cos(\omega t)\boldsymbol{j}]$, and therefore the acceleration $d\boldsymbol{V}/dt) = R[-\omega^2 cos(\omega t)\boldsymbol{i} minus; \omega^2 sin(\omega t)\boldsymbol{j}] = -\omega^2 R[cos(\omega t)\boldsymbol{i} + sin(\omega t)\boldsymbol{j}] = -(V^2/R)\ \boldsymbol{e}_R(t)$, in accordance with Eq. (6.8). It is clear that the analytical method is faster and simpler (after one gets used to it) than the synthetic (geometrical) method we used in the text.

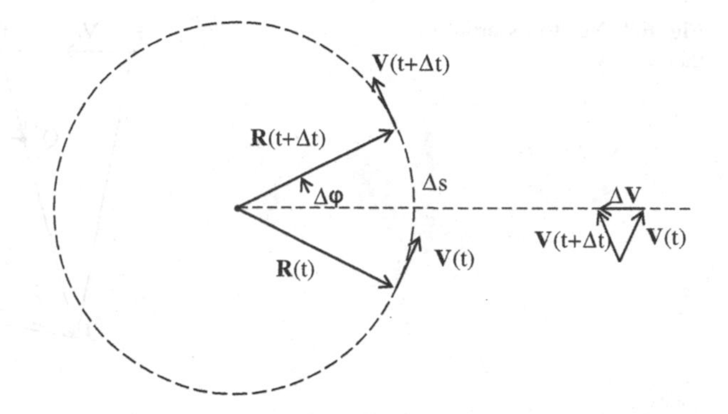

Fig. 6.6 Circular motion

where $e_R(t)$ is a *unit* vector in the direction of $R(t)$. The *minus* sign indicates that the acceleration vector points in the opposite direction to that of $R(t)$, i.e. towards the centre of the circle. The above formula was published by Huygens some years before the publication of the *Principia*. Apparently Newton knew the formula at the time of its publication but had not bothered to publish it.

Theorem A (In Newton's words): *The areas which bodies made to move in orbits described by radii drawn to an unmoving centre of force lie in unmoving planes and are proportional to the times.*

Newton tells us (see Fig. 6.7) that if at some moment of time (say at $t = 0$) a body is at some place P relative to an unmoving centre of force (which we can take as the origin of coordinates *O*) with a velocity V_0 and moves thereafter under the influence of a force F which (at all times) is directed from the body to the centre of force, then the body will stay on the plane defined by its initial position vector (*OP*) and its initial velocity V_0. And the area swept by the position vector of the body per unit time will be a constant (independent of time). And this will be so whatever the variation of the magnitude $F(R)$ of the force is with the distance (R) of the body from the centre of the force.[7]

Newton's proof of the theorem goes as follows. Look at Fig. (6.7). If no force is acting on the body, in a unit of time it will move from *P* to *Q*, and then from *Q* to *R*, and so on, traversing the same distance along the straight line (PQRS ...) per unit time. We note that the areas swept by the position vector per unit time are equal. The areas of the triangles *OPQ*, *OQR*, *ORS*, etc. are equal because the bases are equal ($PQ = QR = RS$ and so on) and the heights are the same (equal to the normal distance from *O* to the line *PQRS*). Now suppose that a thrust of central force is applied to the body when at *Q* as a result of which is found at Q', then, in consequence of its preexisting motion in the direction of *QR*, the body will be after

[7] The area swept by the position vector of a body per unit time is proportional to the so called 'angular momentum' of the body (see next chapter) and therefore the above theorem is often stated as follows: a central force (directed from the body to a constant centre) does *not* change the angular momentum of the body.

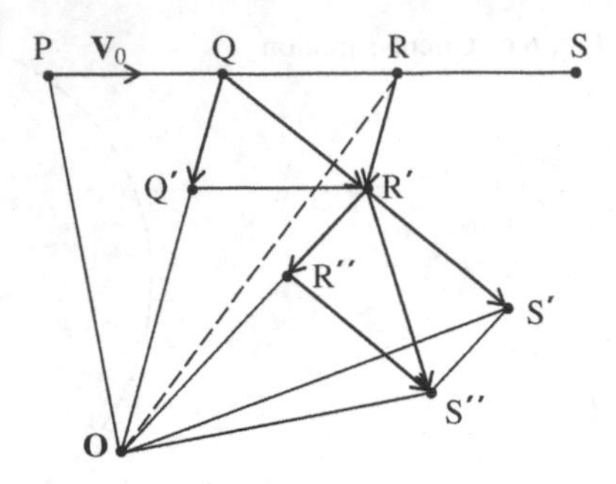

Fig. 6.7 Newton's proof of theorem A

a unit of time has elapsed at R'. One can prove that the area of the triangle OQR' is equal to the area of the triangle OQR (the two triangles have the same base OQ and the same height: the normal distance to OQ is the same from R′ as it is from R.) And now again a thrust of central force is applied to the body at R′ as a result of which it is found at R'', and after a unit of time has elapsed it is found at S''. And again one can show that the area of triangle OR′S″ is equal to the area $OR'S'$ which is equal to the area OQR'. By repeating the process we can show that the areas swept by the position vector in successive units of time are equal. Moreover by taking the unit of time smaller and smaller we get to a near continuity of thrusts which in the limit becomes a continuous force. We may therefore conclude, following Newton, that the theorem is valid generally.

Theorem B *If a body revolves in an ellipse,*[8] *the centripetal force acting on it tending towards a focus of the ellipse (the centre of force) has a magnitude $F(R)$ inversely proportional to the square of the distance R of the body from the centre of force: $F(R) = C/R^2$ where C is a constant.*

Theorem C *If a body departs from a place P along any straight line PR with any velocity whatever* and is at the same time acted upon by a centripetal force that is inversely proportional to the square of its distance from the centre of force, this body will move in some one of the conics*[9] *having a focus at the centre of the force; and conversely. (* Except when the body descends straight to the centre of force.)*

Theorem D *When a body is moving in an ellipse (as described in theorem B), the square of the period of its motion is proportional to the cube of the major axis of the ellipse.*

The proof of this theorem is straightforward in the case of the circular motion (the circle being a special case of the ellipse). In this case the magnitude of the

---

[8] A description of the ellipse is given in Sect. 5.1. See in particular Fig. 5.4.

[9] Conic is a figure formed by an intersection of a plane with a cone. If the plane is perpendicular to the axis of the cone the figure is a circle. If the intersecting plane is inclined to the axis at an angle in excess of half the apex angle of the cone the figure is an ellipse. If the plane is parallel to the sloping side of the cone the figure is a parabola.

centripetal force (which equals the mass m times the magnitude of the *acceleration* V^2/R) *is given by*: $F(R) = mV^2/R$. *But* $V = 2\pi R/T$, *where* T is the period of the motion. Therefore, the centripetal force is: $F(R) = 4\pi^2 mR/T^2$. On the other hand, according to theorem B, $F(R) = C/R^2$. Therefore:

$$T^2 = [4\pi^2 m/C]R^3 \tag{6.9}$$

in accordance to the above theorem. Newton proved that: *the periodic times in ellipses are the same as in circles whose diameters are equal to the major axes of the ellipses.* So for an elliptical orbit Eq. (6.9) becomes:

$$T^2 = [4\pi^2 m/C](R'/2)^3 \tag{6.10}$$

where R' is the major axis of the ellipse. We shall not reproduce the proofs of the above theorems here.

6.2.3 Gravity

A proper theory of gravity must account above all for the three laws of planetary motion discovered by Kepler (see Sect. 3.3) and confirmed by a number of other astronomers.

According to theorems A and B, stated above, the first two of Kepler's laws are accounted for, if one assumes that any one of the planets is attracted to the sun (the centre of the force at rest in absolute space) by a force whose magnitude is inversely proportional to the planet's distance from the sun: $F(R) = C/R^2$. We note that theorem D [Eq. (6.10)] does not necessarily account for Kepler's *third law*. For that to be the case, the value of $[4\pi^2 m/C]$ in Eq. (6.10) must be the same for *all* planets. And this is not obviously so, with m being the mass of the planet under consideration. One is therefore obliged to seek a formula for the force exerted by the sun on a planet of mass M_p in the form:

$$F(R) = C/R^2, \text{ where } M_p/C \text{ is the same for all planets} \tag{6.11}$$

Newton's great insight concerning the 'nature' of the centripetal force that keeps the planets in orbit about the sun came with his realization of the universal character of this force. He conceived of it as the cause not only of the motion of the planets about the sun and of the moon about the earth, but also of the falling of bodies towards the centre of the earth. In relation to the latter, the force (the weight of the body) would be given by $W = F(R_E + z) = C/(R_E + z)^2$, where C is a constant to be determined, and $(R_E + z)$ is the distance of the body from the centre of the earth. R_E is the radius of the earth and z is the position (height) of the body above the earth's surface, and because z is very much smaller than R_E (near the surface of the earth) we can disregard it. In which case we can write $W = C/(R_E)^2$. But of course the weight of the body is the cause of its falling towards the centre of the earth with a constant acceleration g (about 9.80 m/s^2, the same for all bodies), which implies that $W = mg$. And this, in turn, implies that the force attracting the body to the centre of the earth is proportional to the mass (m) of the body. But according to the third law of motion if body A attracts body B, then body B attracts body A with an equal (in magnitude) force. Therefore, he argued, if the law of gravity is to apply quite generally the attractive force between two bodies, of mass m_1 and m_2 respectively, must be as follows:

$$F(R) = G(m_1 m_2)/R^2 \tag{6.12}$$

where R is the distance between the centres of gravity of the two bodies, and G must be a universal constant to be determined once and for all. We have assumed, following Newton, that the bodies are spherical, and that the density of mass within each sphere may vary with the distance from the centre of the sphere but remains the same in every spherical shell at any distance from the centre.[10] When the bodies are a great distance away from each other (as is usually the case in astronomy) this is not a bad approximation.[11]

According to Eq. (6.11) the gravitational attraction between the sun (of mass M_S) and a planet (of mass M_P) is given by:

$$F(R) = G(M_S M_P)/R^2 \tag{6.12a}$$

And therefore Eq. (6.10) becomes:

$$T^2 = [4\pi^2/(M_S G)](R'/2)^3 \tag{6.13}$$

The above formula is in effect Kepler's third law (because the quantity in the square brackets is the same for all planets).

At this point the basics of the theory of gravity have been presented. We can, using the laws of motion and Eq. (6.12a), reproduce the observed orbit of a planet about the sun, if we know (observe) the position and the velocity of the planet at a point in time (chosen arbitrarily), and the same applies to the motion of a satellite about its planet (the attraction between these two is given by $F(R) = G(M_P M_{st})/R^2$ where M_{st} is the mass of the satellite and R its distance from the centre of the planet). One of course must also know the values of the constant quantities entering these equations: the mass of the sun or the planet under consideration, and of course

[10] The attraction between two bodies A and B (of any shape and any distribution of mass density) is in principle obtained as follows: we divide each body into a sufficiently large number of small bits, we calculate the attractive forces between all pairs of bits (the one bit in A and the other in B) using Eq. (6.12), and add them up. Newton proved that when this procedure is applied to spherical bodies (with mass density distributions which depend only on the distance from the centre of the sphere) the overall attraction between the bodies is given by Eq. (6.12).

[11] But in some instances a correction may be necessary. Because the shape of the earth is not exactly spherical, it bulges out at the equator, the total force exerted on it by the sun and the moon differs slightly from that obtained from Eq. (6.12), and it is this small deviation (pointed out by Newton) that causes the precession of the earth's axis described in Fig. 3.7.

the value of the gravitational constant (G). It is worth noting in this respect that the first accurate direct measurement of G was made by Henry Cavendish more than a hundred years after the publication of the *Principia*.[12] The accepted value of G today is: $G = 6.67259 \times 10^{-11}$ Nm^2/kg^2 (We rememember that the newton (N) is the unit of force: $1N = (1\,\text{kg})(1\,\text{m})/(1\,\text{s})^2$). Once G is known one can determine the mass of the sun (M_S) from Eq. (6.13) using the values of R' and T obtained from the observations of any one of the planets. Similarly one can obtain the mass of the earth (M_E) by applying the above formula (with M_S replaced by M_E) to the motion of the moon about the earth.

At this point we should say that Newton's theory of gravity had been very successful, its predictions in excellent agreement with the astronomical data: those available at the time of Newton and with those obtained thereafter. Of course approximations had been made in the construction of the theory we presented above, as a result of which there exist, in some instances, small but measurable discrepancies between the theoretical prediction and the measured value of certain quantities. For example: it has been assumed that the motion of any one of the planets is determined solely by the attractive force exerted on it by the sun. But in reality every one of the planets is also acted upon by the much weaker gravitational forces exerted on it by the other planets. The planets nearer to the sun (Mercury, Venus, Earth and Mars) act only slightly upon one another because their masses are small; and they are also little affected by the bigger planets because they lie far away from them. On the other hand, the action of Jupiter upon Saturn though small (its strength is about 200 times smaller than that of the sun) is not to be ignored according to Newton, who actually developed a perturbation method to evaluate the required correction to the orbit of this planet. We need not go into such details here. The fact is that when such a correction is necessary, the observed orbit does not obey *exactly* the laws of Kepler, for the simple reason that in this case the total force acting on the planet is not *exactly* centripetal (central). Another assumption made in the theory presented above is that the sun is at rest; when in reality it is the centre of mass of the solar system that is at rest, or moves ever so slowly with constant velocity in relation to the fixed stars. However, the centre of mass of the solar system never moves further than about the surface of the sun because the mass of the sun is so much larger than the mass in the planets, and because the different planets pull the sun in different directions. Therefore the said assumption is justified. Newton discusses these and other aspects of planetary motion and often provides estimates of the required corrections. Here we shall not be concerned with such matters. However, I should at least mention his theory of

[12] In Cavendish's experiment two small spheres of mass m are fixed to the ends of a light horizontal rod suspended by a very thin metal wire. Two large spheres of mass M are positioned near the small spheres, thereby causing the rod to rotate by a small angle which is measured by the deflection of a light beam reflected from a mirror attached to the vertical suspension. The deflection of light works in this case as an amplifier of the rotation. The experiment is repeated with different masses at different separations providing an experimental proof of the law of gravity and the value of the gravitational constant **G**. .

the tides, which he correctly attributed to the gravitational pull of the moon on the oceans.

The reader will have noted the absence from our exposition of any discussion about the rotation of the earth and of the moon and of the other planets about their axes. This is because Newton refers only occasionally to these motions in the *Principia*. He, at some point, refers to the important role that the gravitational attraction between different parts of a planet plays in the formation of the shape of the planet and in the continuity of its rotation about its axis, but offers no more in this respect.

The publication of the *Principia* stirred the scientific community of the time and was immediately recognized as a great breakthrough in the development of science. However some scientists, including Huygens, were reluctant to accept the main proposition of his theory of gravity [summarized by Eq. (6.12)], as it implied that the pull of one body by another occurs instantaneously and apparently without the need of a medium (an ether) for its propagation. Newton's view on this aspect of his theory was that his formulae did not require a propagating ether-like medium but did not necessarily exclude such a possibility. His formulae were correct in the only way that matters, he would say, which is to describe accurately the observed phenomena. The instantaneous propagation of the gravitational force, which remained a source of worry for scientists when ether was given up as a propagation medium, was finally settled by Einstein (see Chap. 12).

6.2.4 Galilean Relativity

Although Newton believed in absolute motion (relative to unmovable space) he recognizes (in the Introduction of the *Principia*) that: *When bodies are enclosed in a given space, their motions in relation to one another are the same whether the space is at rest* (in unmoving space) *or whether it is moving uniformly straight forward without circular motion. ... This is proved clearly by experience: on a ship, all the motions are the same with respect to one another whether the ship is at rest or is moving uniformly straight forward.*

In the years after Newton it was gradually realized, that the concept of an absolute motion, in relation to a system of coordinates $O(x,y,z)$ at rest in unmovable space, is not sustainable in physics. For there is no physical measurement we can make that will distinguish such a system from another $O'(x',y',z')$ which moves relative to it with a constant velocity V along a straight line (say along the x-direction, as in Fig. 6.8).

Referring to Newton's example we can say that an observer standing on the ship can not tell (by performing a measurement of some kind or other) that *he* is moving and not the quay. And the same applies to an observer standing on the quay. The following rule (known as the principle of Galilean Relativity) applies generally: The laws of motion apply equally well in all coordinate systems moving with uniform velocity with respect to each other. And because the laws of motion (and of physics in general) are expressed by mathematical equations, the principle

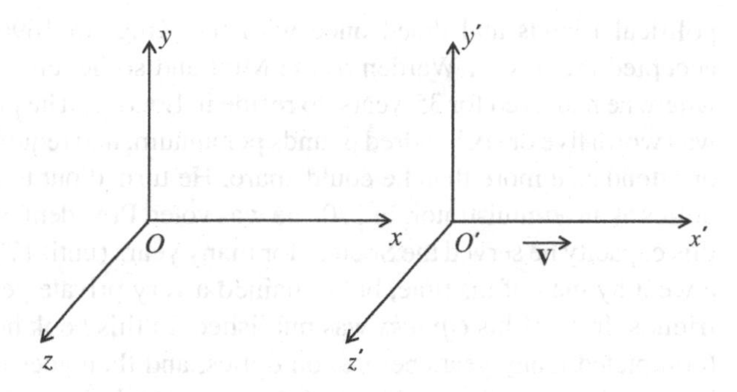

Fig. 6.8 The primed frame moves with respect to the unprimed coordinate frame. We assume that at time t = 0, the two systems coincide

of relativity means that these equations have the same form in all coordinate systems moving with uniform velocity relative to each other.

Let me demonstrate the above in relation to Newton's law of motion. The coordinates of a particle (a mass point) in the primed frame are obviously related to those in the unprimed frame by the equations:

$$x' = x - Vt,\ y' = y,\ z' = z \tag{6.14}$$

known as the Galilean transformation. Now suppose that we have a system of N such particles (i = 1, 2, 3, ..., N). The equations by which their motions are determined in the unprimed frame are:

$$m_i \frac{d}{dt}\left(\frac{dx_i}{dt}\right) = F_{ix}(\boldsymbol{r}_i - \boldsymbol{r}_j),\ m_i \frac{d}{dt}\left(\frac{dy_i}{dt}\right) = F_{iy}(\boldsymbol{r}_i - \boldsymbol{r}_j),\ m_i \frac{d}{dt}\left(\frac{dz_i}{dt}\right) = F_{iz}(\boldsymbol{r}_i - \boldsymbol{r}_j) \tag{6.15}$$

where a sum over j ($\neq i$) is implied on the right of each equation. And we have assumed that the force between any two particles is a function of the distance between the particles. Now it follows from Eq. (6.14) that $(\boldsymbol{r}_i - \boldsymbol{r}_j) = (\boldsymbol{r}'_i - \boldsymbol{r}'_j)$, i.e. the separations between particles are the same in the two frames. And because $dx'/dt = dx/dt - V$, where V is a constant, we obtain: $\frac{d}{dt}\left(\frac{dx'}{dt}\right) = \frac{d}{dt}\left(\frac{dx}{dt}\right)$. Therefore: Eq. (6.15) will still apply if we replace the unprimed coordinates with the corresponding primed ones.

And we expect the mathematical expression of any fundamental law of physics to have the same property: to be invariant (not change) under a Galilean transformation.

6.3 Newton's Life After the *Principia*

Newton's fame increased rapidly after the publication of the *Principia.* In 1688 he was offered the position of representing Cambridge University in Parliament. Newton was an MP (Member of Parliament) for about a year. He reported regularly to the Vice Chancellor but did nothing more, and spoke nothing. He made a number of political friends and dined once with the King. In 1696 he was offered and he accepted the post of Warden of the Mint and so he left academia and Cambridge, where he had lived for 35 years, to reside in London. The post of Warden of the Mint was worth five or six hundred pounds per annum, and required not too much business or attendance more than he could spare. He turned out to be very competent in his duties as an administrator. In 1703 he was voted President of the Royal Society and in this capacity he served the Society for many years (until 1725). He was now living as a wealthy man of his time, but remained a very private person and had no intimate friends. In 1704 his *Opticks* was published. In this book he describes his own work (completed many years before) on optics, and then goes on to discuss a variety of issues relating to the workings of the eye, metabolism, gravitation, and even such exotics as the Great Flood and the Creation. He died in 1727 at the age of 83.

Exercises

6.1. Given two vectors, $\boldsymbol{A} = 3\boldsymbol{i} - 2\boldsymbol{j}$ and $\boldsymbol{B} = -\boldsymbol{i} - 4\boldsymbol{j}$, obtain (a) $\boldsymbol{A} + \boldsymbol{B}$, (b) $\boldsymbol{A} - \boldsymbol{B}$, (c) the magnitudes, $|\boldsymbol{A} + \boldsymbol{B}|$ and $|\boldsymbol{A} - \boldsymbol{B}|$, of $\boldsymbol{A} + \boldsymbol{B}$ and $\boldsymbol{A} - \boldsymbol{B}$ respectively, (d) the directions of $\boldsymbol{A} + \boldsymbol{B}$ and $\boldsymbol{A} - \boldsymbol{B}$ as specified by the angle, θ, these vectors make with the x-direction (that of $\boldsymbol{i}$).

Answer: (a) $\boldsymbol{A} + \boldsymbol{B} = 2\boldsymbol{i} - 6\boldsymbol{j}$; (b) $\boldsymbol{A} - \boldsymbol{B} = 4\boldsymbol{i} + 2\boldsymbol{j}$; (c) $|\boldsymbol{A} + \boldsymbol{B}| = 6.32$, $|\boldsymbol{A} - \boldsymbol{B}| = 4.47$; (d) for $\boldsymbol{A} + \boldsymbol{B}$, $\theta = -71.6°$; for $\boldsymbol{A} - \boldsymbol{B}$, $\theta = 26.51°$.

6.2. A particle undergoes three consecutive displacements $\boldsymbol{A}$, $\boldsymbol{B}$ and $\boldsymbol{C}$ such that its total displacement is zero. If $\boldsymbol{A} = 2\boldsymbol{i} + 3\boldsymbol{j} - \boldsymbol{k}$ meters, and $\boldsymbol{B} = -3\boldsymbol{i} - \boldsymbol{j} + 2\boldsymbol{k}$ meters, what is the value of $\boldsymbol{C}$?

Answer: $\boldsymbol{C} = \boldsymbol{i} - 2\boldsymbol{j} - \boldsymbol{k}$ meters.

6.3. A projectile (a particle of mass m) is fired off at the origin of coordinates (at ground level) at an angle θ to the ground with velocity υ_0. What should the value of υ_0 be so that the projectile hits the ground a distance L away?

Answer: $\upsilon_0 = (\frac{gL}{2}\cos\theta\sin\theta)^{1/2}$, where g is the acceleration of gravity.

6.4. A spherical particle of mass m and volume V is falling in a viscous fluid of mass density ρ. What is the limiting (constant) speed, υ_f, of the particle if the viscosity force on it is given by $F_{visc} = -\kappa\upsilon$, where υ is the speed of the particle and κ a given constant.

Hint: When writing the equation of motion for the particle you must not forget Archimedes's principle.

Answer: $\upsilon_f = (m - V\rho)g/\kappa$, where g is the acceleration due to gravity.

6.5. A particle of mass m moves in a circular path of radius 0.4 m with constant speed, υ. If the particle makes five revolutions per second, determine (a) the value of υ; (b) the magnitude, a, of the acceleration.

Answer: (a) $\upsilon = 12.56$ m/s; (b) $a = 394.38$ m/s^2.

6.6. The position (in meters) of a particle as a function of time (measured in seconds) is given by: $\boldsymbol{r}(t) = 3\cos 2t\,\boldsymbol{i} + 3\sin 2t\,\boldsymbol{j}$. (a) Show that the path of the particle is a circle of a *3 m* radius centred on the origin of coordinates; (b) Calculate the velocity $\boldsymbol{\upsilon}(t)$ of the particle; (c) Calculate the acceleration $\boldsymbol{a}(t)$ and hence show that it points towards the origin.

Answer: (a) $|\boldsymbol{r}(t)| = 3$ m; (b) $\boldsymbol{\upsilon}(t) = -6\sin 2t\,\boldsymbol{i} + 6\cos 2t\,\boldsymbol{j}$; $\boldsymbol{a}(t) = -12\cos 2t\,\boldsymbol{i} - 12\sin 2t\,\boldsymbol{i} = -4\,\boldsymbol{r}(t)$.

6.7. The harmonic oscillator described by Eq. (6.6) is an ideal oscillator. In practice, a mass m attached to an elastic spring will not oscillate for ever; the amplitude of the oscillation will diminish with time, going to zero as $t \to \infty$. This is due to friction, through which the energy of the oscillator (see next section) is turned into heat. We can take friction into account through a 'force": $-\gamma m\upsilon$, where $\upsilon = dx/dt$ is the velocity of m, and γ is an appropriate constant. In which case Eq. (6.6) is replaced by the following equation of motion:

$$\frac{d^2x}{dt^2} + \gamma\frac{dx}{dt} + \omega_0^2 x = 0 \qquad \text{(A)}$$

(a) Show that: $x(t) = ae^{-\lambda t}\cos(\omega' t + \varphi)$, where a and φ are any constants, satisfies the above equation if the values of λ and ω' are chosen appropriately. Determine the appropriate values of λ and ω'.

Answer: $\lambda = \gamma/2, \omega' = \sqrt{\omega_0^2 - \gamma^2/4}$.

(b) The constants a and φ are determined by the position, x_0, and the velocity, υ_0, of m at $t = 0$. What should the values of a and φ be, if $x(t = 0) = x_0$ and $\upsilon(t = 0) = \upsilon_0$.

Answer: $a = \sqrt{x_0^2 + (\frac{\upsilon_0}{\omega'} + \frac{\gamma x_0}{2\omega'})^2}$, and φ is obtained from: $\cos\varphi = x_0/a$.

6.8. (a) Write the equation of motion for the harmonic oscillator of the previous exercise when a force $F = F_0\cos\omega t$ is acting on it.

$$\text{Answer}: \frac{d^2x}{dt^2} + \gamma\frac{dx}{dt} + \omega_0^2 x = f_0\cos\omega t, \text{ where } f_0 = F_0/m. \qquad \text{(B)}$$

(b) Show that: $x(t) = X_0\cos(\omega t + \Phi)$ is a solution of the above equation if X_0 and Φ are chosen appropriately. Determine the appropriate values of X_0 and Φ.

Answer: $X_0 = \frac{f_0}{(\omega_0^2-\omega^2)\cos\Phi + \gamma\omega\sin\Phi}$, and Φ is obtained from: $\frac{\sin\Phi}{\cos\Phi} = \frac{\gamma\omega}{\omega^2-\omega_0^2}$.

We note that X_0 obtains its maximum value when $\omega = \omega_0$; we then say that the oscillator is in resonance with the externally applied force. We further note that X_0 is greater the smaller the coefficient of friction (γ).

(c) Show that the general (so called) solution of Eq. (B) is obtained by adding to the solution obtained in (b) the (general) solution of Eq. (A) obtained in exercise 6.7. Therefore the (general) solution of Eq. (B) is:

$$x(t) = X_0\cos(\omega t + \Phi) + ae^{-\gamma t/2}\cos(\omega' t + \varphi), \text{ where } \omega' = \sqrt{\omega_0^2 - \gamma^2/4}$$

where the constants a and φ are determined by the position, x_0, and the velocity, υ_0, of m at $t = 0$. We note that for sufficiently large values of t, the second term in the above expression is very small and can be neglected.

6.9. Convince yourself that:
(a) The centre of mass of a cylindrical rod of uniformly distributed mass lies at the centre of the rod.
(b) The centre of mass of a sphere of uniformly distributed mass lies at the centre of the sphere.

6.10. The mass density of a cylindrical rod varies over its length, L, as follows: $\rho(x) = ax$, $0 \leq x \leq L$. Determine the position, X_{CM}, of its centre of mass.

Answer: $X_{CM} = 2L/3$.

6.11. Read footnote 10 of Chap. 6. Convince yourself that the gravitational force acting on a particle of mass m at point $\boldsymbol{r}$, due to a mass M uniformly distributed within a sphere of radius $R < r$, centered on the origin of coordinates, is given by:

$$\boldsymbol{F} = -\frac{GMm}{r^2}\boldsymbol{e}_r$$

where $\boldsymbol{e}_r$ is the unit vector in the radial direction (pointing away from the origin). In other words: the gravitational pull of the sphere is the same with that of a point particle of mass M at the centre of the sphere.

Hint: You must convince yourself that the spherical symmetry of the mass distribution within the sphere excludes any other possibility.

6.12. Substituting for $R'/2$ of Eq. (6.13) the mean distance of the moon from the centre of the earth (see Table 3.1) obtain an approximate value for the mass of the earth. Note: The radius of the earth is: $R_e \approx 6.37 \times 10^6$ m.

Answer: 6.2×10^{24} kg..

Chapter 7
Classical Mechanics

7.1 The Theory of Mechanics After Newton

In the years following the publication of the *Principia* scientists studied the motion of point particles and solid bodies on the basis of Newton's laws of motion using analytical methods exploiting the power of calculus. The most eminent of these scientists must be Leonhard Euler. He was born in 1707 in Basel (Switzerland), the son of a village minister. By the time he completed his studies, officially in theology but in practice he spent more time studying mathematics under the guidance of the mathematician Johan Bernouli, he had already acquired a reputation as a promising young scientist. He left Basel to join the newly created, by Peter the Great, Academy of Sciences in St. Petersburg, where he stayed until 1741, when he joined another new Academy, that of Frederick the Great in Berlin. In 1736 Euler published (in Latin) a two-volume book, *Mechanics, or the Science of Motion Described Analytically*, in which he presented Newtonian mechanics in analytical (algebraic) terms using calculus. By doing so, he could more easily solve not only the problems dealt with geometrically by Newton, but many others that would be practically unsolvable geometrically as well. This is of course how we do mechanics today (as we demonstrated already in our examples of Sect. 6.2.2). In 1765, a year before he left Berlin to return to St. Petersburg, Euler wrote another two-volume book, *Theory of the Motion of Solid Bodies,* which includes much of what students of physics learn today about the subject in a course of mechanics. In Sect. 7.2 we shall describe the simplest of the motions of a solid body: a rotation about a fixed axis.

Euler wrote in all more than 800 papers, three fifths of which were in mathematics with the rest in physics, engineering and astronomy. Though practically blind after 1771, he kept working on a variety of topics, assisted by a team of students and associates including his eldest son Johann Abrecht (one of 13 children), until his death (in 1783). Today, over 80 mathematical formulae, scientific concepts and theorems bear his name.

A. Modinos, *From Aristotle to Schrödinger*,
Undergraduate Lecture Notes in Physics, DOI: 10.1007/978-3-319-00750-2_7,

In this chapter I shall describe briefly some of the concepts introduced by Euler and other scientists and demonstrate their application to a small number of physical problems.

7.1.1 Angular Momentum

The *angular momentum* ($\boldsymbol{L}$) of a particle (of mass m) with respect to the centre (0) of a system of coordinates is defined by:

$$\boldsymbol{L} = \boldsymbol{r}(t) \times \boldsymbol{p}(t) \tag{7.1}$$

where $\boldsymbol{r}(t)$ is the position vector of the particle, $\boldsymbol{v}(t) = d\boldsymbol{r}/dt$ its velocity, and $\boldsymbol{p}(t) = m\boldsymbol{v}(t)$ its momentum (see Sect. 6.2). The *vector product* of Eq. (7.1) must be understood in accordance with the following general definition:

Given any two vectors $\boldsymbol{A}$ and $\boldsymbol{B}$, their vector product $\boldsymbol{A} \times \boldsymbol{B}$ is a vector normal to the plane defined by $\boldsymbol{A}$ and $\boldsymbol{B}$ pointing in the direction a right-turning screw advances when turning from $\boldsymbol{A}$ to $\boldsymbol{B}$, which implies that $\boldsymbol{B} \times \boldsymbol{A} = -\boldsymbol{A} \times \boldsymbol{B}$; the magnitude of $\boldsymbol{A} \times \boldsymbol{B}$ equals $ABsin(\delta)$, where A and B are respectively the magnitude of $\boldsymbol{A}$ and that of $\boldsymbol{B}$, and δ is the angle between them. We note (see Fig. 7.1) that $\boldsymbol{A} \times (\boldsymbol{B} + \boldsymbol{C}) = \boldsymbol{A} \times \boldsymbol{B} + \boldsymbol{A} \times \boldsymbol{C}$. Using this property and noting that the vector products (with one another) of the unit vectors $\boldsymbol{i}, \boldsymbol{j}, \boldsymbol{k}$ along the x, y, z axes of a Cartesian system of coordinates (see, e.g., Fig. 7.2) are: $\boldsymbol{i} \times \boldsymbol{i} = \boldsymbol{j} \times \boldsymbol{j} = \boldsymbol{k} \times \boldsymbol{k} = 0$; $\boldsymbol{i} \times \boldsymbol{j} = \boldsymbol{k}$, $\boldsymbol{j} \times \boldsymbol{k} = \boldsymbol{i}$, and $\boldsymbol{k} \times \boldsymbol{i} = \boldsymbol{j}$, we obtain for $\boldsymbol{A} \times \boldsymbol{B} = (A_x\boldsymbol{i} + A_y\boldsymbol{j} + A_z\boldsymbol{k}) \times (B_x\boldsymbol{i} + B_y\boldsymbol{j} + B_z\boldsymbol{k})$ the following expression:

$$\boldsymbol{A} \times \boldsymbol{B} = (A_yB_z - A_zB_y)\boldsymbol{i} + (A_zB_x - A_xB_z)\boldsymbol{j} + (A_xB_y - A_yB_x)\boldsymbol{k} \tag{7.2}$$

Applying the above definition of the vector product to Eq. (7.1), we find that the angular momentum ($\boldsymbol{L}$) of a particle moving in the xy-plane of a coordinate system (as in Fig. 7.2) is: $\boldsymbol{L} = L_z\boldsymbol{k} = mr(t)v(t)sin(\delta)\boldsymbol{k}$, where δ is the angle between $\boldsymbol{r}(t)$ and $\boldsymbol{v}(t)$; and we remember that a positive L_z means that the particle moves anticlockwise about the origin (0) of the coordinates. It turns out (read the caption of Fig. 7.2) that $L_z/(2\text{ m}) = dA/dt$, which is the area swept by the position vector $\boldsymbol{r}(t)$ of the particle per unit time.

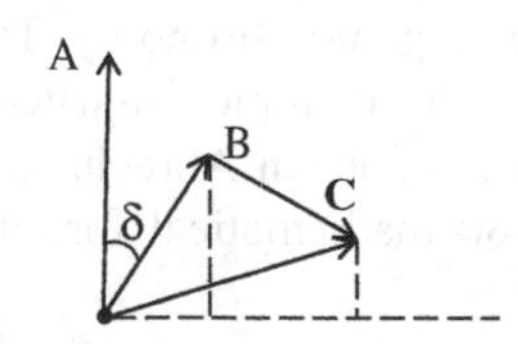

Fig. 7.1 We note that Bsin(δ) equals the projection of $\boldsymbol{B}$ on the normal to $\boldsymbol{A}$ (shown by the *broken line*). Now the projection of ($\boldsymbol{B} + \boldsymbol{C}$) on the normal to $\boldsymbol{A}$ is the algebraic sum of the projections of $\boldsymbol{B}$ and $\boldsymbol{C}$ on the normal to $\boldsymbol{A}$. Therefore $\boldsymbol{A} \times (\boldsymbol{B} + \boldsymbol{C}) = \boldsymbol{A} \times \boldsymbol{B} + \boldsymbol{A} \times \boldsymbol{C}$

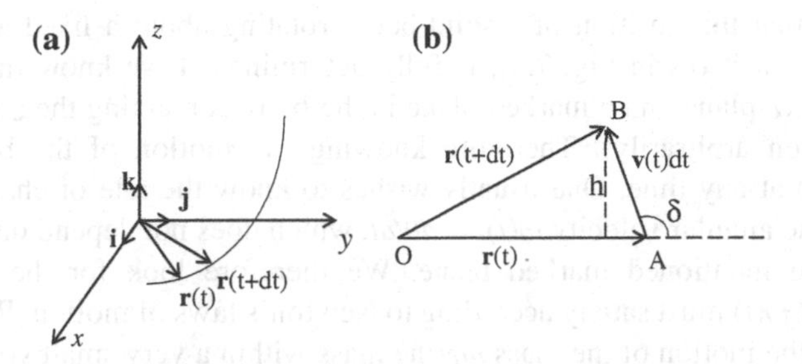

Fig. 7.2 **a** The area dA swept by $\boldsymbol{r}(t) = x(t)\boldsymbol{i} + y(t)\boldsymbol{j}$ in time dt is equal to the area of the triangle OAB, shown in **b**, which is equal to $(OA)h/2 = (OA) \times (AB)sin(\delta)/2 = r(t)(\upsilon(t)dt)sin(\delta)/2$. On the other hand, $\boldsymbol{L} = \boldsymbol{r}(t) \times \boldsymbol{p}(t) = L_z\boldsymbol{k} = mr(t)\upsilon(t)sin(\delta)\boldsymbol{k}$. Therefore $dA/dt = L_z/(2\,m)$

In Sect. 6.2.2 we presented Newton's geometrical proof of his Theorem A, which states that dA/dt is constant (a conserved quantity) if the force acting on the particle is a central force ($\boldsymbol{F} = f(r)\boldsymbol{r}(t)$). We now know that $dA/dt = L_z/2$ m, so that we can prove Theorem A by showing that the angular momentum ($\boldsymbol{L}$) of the particle does not change, i.e. $d\boldsymbol{L}/dt = 0$, when the force acting on it is a central one. The proof is straightforward. We have (from Eq. 7.1): $d\boldsymbol{L}/dt = d(\boldsymbol{r}(t) \times \boldsymbol{p}(t))/dt = \upsilon(t) \times \boldsymbol{p}(t) + \boldsymbol{r}(t) \times (d\boldsymbol{p}/dt)$. Now $\upsilon(t)$ and $\boldsymbol{p}(t)$ are in the same direction and therefore their vector product is zero (because $sin(\delta) = 0$); and $d\boldsymbol{p}/dt = \boldsymbol{F}$, by Newton's law (Eq. 6.1). Therefore:

$$d\boldsymbol{L}/dt = r(t) \times \boldsymbol{F} \tag{7.3}$$

The above equation is valid quite generally. It tells us, to begin with, that $d\boldsymbol{L}/dt = 0$ if $\boldsymbol{F} = 0$ (i.e. if there is no force acting on the particle). Therefore, the angular momentum with respect to the origin of coordinates of a particle moving with constant speed on a straight line is constant (it is conserved; it does not change with time).

But Eq. (7.3) tells us more: If $\boldsymbol{r}(t)$ and $\boldsymbol{F}$ are in the same direction, the angle δ between $\boldsymbol{r}$ and $\boldsymbol{F}$ vanishes, and therefore $sin(\delta) = 0$, and consequently the magnitude of $\boldsymbol{r}(t) \times \boldsymbol{F}$ vanishes. Therefore $d\boldsymbol{L}/dt = 0$, and so $\boldsymbol{L}$ is constant and therefore so is dA/dt, for a central force. The reader will agree, I hope, that this (analytical) proof is not only shorter but also more elegant than the geometrical proof of the Theorem A given in Sect. 6.2.2.

7.1.2 Rotation of a Solid Body About a Fixed Axis

We know that a door will close more easily if we exert a force on it at a point further away than nearer to its axis of rotation and perpendicularly to its surface rather than at a smaller angle to it. We should be able to account for this and other phenomena relating to the motion of solid bodies on the basis of Newton's laws of motion.

We note that the position of a solid body, rotating about a fixed axis (z-axis) passing through it (as in Fig. 7.3), is fully determined if we know the angle (θ) between the xz-plane and a marked plane in the body containing the z-axis (which can be chosen arbitrarily). Therefore knowing the motion of the body means knowing $\theta(t)$ at any time. One usually wishes to know the rate of change of $\theta(t)$, defined by the angular velocity $\omega(t) = d\theta/dt$, which does not depend on our choice of the above mentioned marked plane. We therefore look for the differential equation that $\omega(t)$ must satisfy according to Newton's laws of motion. We begin by considering the motion of the mass m_i (the mass within a very small volume of the body), which because of its smallness we can treat as a point particle. And because m_i is rotating about the z-axis, it is reasonable to begin by writing down the z-component of its angular momentum, which, according to Eqs. (7.1 and 7.2) is:

$$L_{iz}\boldsymbol{k} = \left(x_i p_{iy} - y_i p_{ix}\right)\boldsymbol{k} = \boldsymbol{r}_i' \times \boldsymbol{p}_i' = r_i' m_i \upsilon_i'(t) sin(\delta')\boldsymbol{k} \tag{7.4}$$

where $\boldsymbol{r}_i' = x_i\boldsymbol{i} + y_i\boldsymbol{j}$. We note that the magnitude (r_i') of this vector, which is equal to the radial distance of m_i from the z-axis (the axis of rotation), is constant (we assume that the solid does not expand or shrink during the rotation). Similarly, $\boldsymbol{p}_i' = m_i\boldsymbol{\upsilon}_i' = m_i\,(\upsilon_{ix}\boldsymbol{\imath} + \upsilon_{iy}\boldsymbol{j})$, and we note that $\boldsymbol{\upsilon}_i'$ is perpendicular to $\boldsymbol{r}_i'$; therefore, assuming an anticlockwise rotation, we put $sin(\delta') = 1$. Finally the magnitude of the velocity of m_i is given by: $\upsilon_i'(t) = \omega(t)\, r_i'$. Therefore, we can write Eq. (7.4) as follows:

$$\boldsymbol{r}_i' \times \boldsymbol{p}_i' = \omega(t)\, m_i r_i'^2 \boldsymbol{k} \tag{7.4a}$$

Taking the derivative with respect to time of both sides of the above equation, and using Newton's law (as in the derivation of Eq. 7.3), we obtain:

$$(d\omega/dt)\, m_i r_i'^2 \boldsymbol{k} = \boldsymbol{r}_i' \times \boldsymbol{F}_i' = N_i\boldsymbol{k} \tag{7.5}$$

where N_i is by definition the magnitude of $\boldsymbol{r}_i' \times \boldsymbol{F}_i'$. We note that $\boldsymbol{F}_i' = F_{ix}\boldsymbol{i} + F_{iy}\boldsymbol{j}$ is the force causing the change in the momentum of m_i along the y- and z-directions. The total force acting on m_i may have a z-component but this does not affect the rotation of the body about the z-axis. We refer to $N_i\boldsymbol{k}$ as the torque (due to $\boldsymbol{F}_i'$). We note (see Fig. 7.3b) that: $N_i = d_i F_i'$, where d_i is the normal

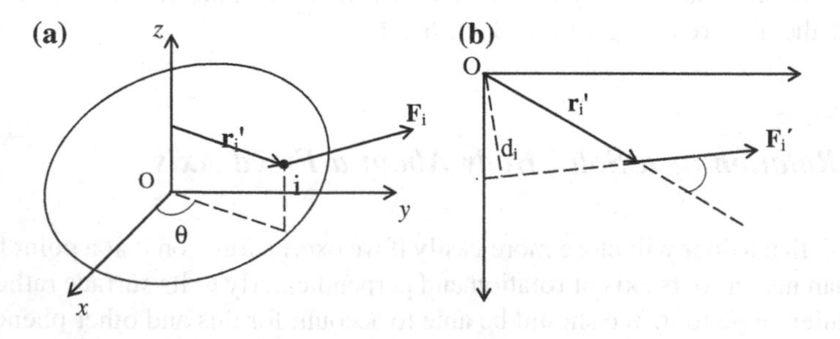

Fig. 7.3 **a** Rotation of a solid body about a fixed axis. **b** $N_i = d_i F_i'$

distance from the axis of rotation to the line of the force (defined by its direction). Therefore:

$$(m_i r_i'^2)(d\omega/dt) = N_i, \text{ where } N_i = d_i F_i' \tag{7.6}$$

We must emphasize that the torque $N_i\mathbf{k}$ is due to the *total* force $\boldsymbol{F}_i'$ acting on m_i, and as such includes internal forces (due to the other masses in the body). However, the internal forces always appear in pairs of forces, equal (in magnitude) but pushing in opposite directions along the same line, and therefore: summing up over *all* masses (m_i, $i = 1,2,3,\ldots N$) in the body, the contribution of the internal forces cancels out, and we obtain:

$$I\frac{d\omega}{dt} = N, \quad N = \sum d_i F_i' \tag{7.7}$$

where the sum for N includes all *externally* applied torques $N_i = d_i F_i'$; and we must remember that N_i is positive (negative) if on its own it produces an anti-clockwise (clockwise) rotation. The quantity (I) on the left of the equation, called the moment of inertia of the body with respect to the rotation axis (z-axis) is defined by:

$$I = \sum_i m_i r_i'^2 \tag{7.8}$$

where r_i' is the distance of m_i from the axis of rotation; and the sum includes all the masses in the body.

Finally, let me note that the kinetic energy (to be defined in Sect. 7.3) associated with the rotation of the body is given by:

$$K = \frac{1}{2}\sum_i m_i v_i^2 = \frac{1}{2}\sum_i m_i r_i'^2 \omega^2 = \frac{1}{2}I\omega^2 \tag{7.9}$$

It follows from Eq. (7.7) that, a solid body rotating about an axis will go on rotating with constant angular velocity (ω) in the absence of any external torque. It is also clear from Eq. (7.7) that the torque due to a force is greater when its normal distance (d) from the axis of rotation is greater. We can now understand why a door closes more easily if we push it at a point away from its axis.[1]

We have assumed, so far, that the moment of inertia of the solid body is constant. When the moment of inertia of the solid body changes during the rotation (the solid may expand or shrink), we must rewrite the equation of motion as

[1] The reader observing that the mass m_i of a rotating body executes a circular motion, will conclude that a force must be acting on it to keep it there. This is indeed so: the force is provided by the other masses (m_j, $j \neq i$) in the body. In the case of an ordinary solid body (such as a rotating door) this force arises through microscopic elastic deformations occurring in consequence of the induced rotation of the body. In the case of massive bodies (the earth and other celestial bodies rotating about their axes) gravitational forces contribute significantly as well.

follows. We note that Eq. (7.4a) remains valid, but now r_i' may change. If we take the time derivative of this equation and then take the sum over all masses (m_i) in the body, we obtain (instead of Eq. 7.7):

$$\frac{d}{dt}(I\omega(t)) = N \tag{7.10}$$

where $I\omega(t)$ is the angular momentum (L_z) of the body about the axis of rotation. And *it* will stay constant when the external torque (N) is zero. An example: when a person, sitting on a chair rotating freely with angular velocity ω, opens his arms (stretching them away from the axis of rotation), thereby increasing the moment of inertia about the axis, ω diminishes.

Finally we note that Eqs. (7.8 and 7.10) apply equally well to the rotation of a body about an axis passing through its centre of mass, whether it (the centre of mass) is fixed in space or moving. This is why the earth will go on rotating about its axis with constant angular velocity, if its moment of inertia with the respect to its rotation axis does not change.

7.1.3 The Simple Pendulum

The simple pendulum consists of a mass (M) at the end of a practically weightless rod of length (L) capable of rotation about an axis (0) as shown in Fig. 7.4a.

When displaced a little from its equilibrium position (the vertical position indicated by the broken line in the figure) and released with zero velocity, the pendulum oscillates about its equilibrium position for a long time (which would be infinite in the absence of friction). Let us see how this comes about. In the present case the moment of inertia of the pendulum with respect to the axis (0) is: $I = ML^2$, since all the mass is concentrated at the end of the rod. The only force acting on the pendulum (in the absence of friction) is its weight, which acts along the vertical direction; its magnitude equals Mg, and (see Fig. 7.4a) its normal distance from the axis of rotation is $L\sin(\theta)$. For *small* θ we have (see Eq. 5.7): $\sin(\theta) = \theta$. Therefore the corresponding torque is: $-MgL\theta$, where the minus sign takes into account the

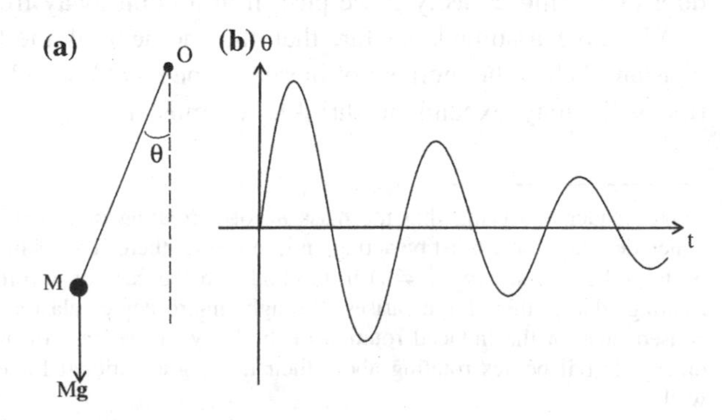

Fig. 7.4 **a** A simple pendulum. **b** A decaying oscillation

fact that the torque tends to bring the pendulum to its equilibrium position ($\theta = 0$). Therefore the equation of motion (7.7) becomes (we remember that $\omega = d\theta/dt$):

$$ML^2 \frac{d}{dt}\left(\frac{d\theta}{dt}\right) = -MgL\theta, \text{ which simplifies to: } \frac{d}{dt}\left(\frac{d\theta}{dt}\right) = -\frac{g}{L}\theta \quad (7.11)$$

Equation (7.11) is, apart from the difference in notation, the same as Eq. (6.6). Therefore its solution (if we assume that at $t = 0$ the pendulum is at an angle θ_0 with zero velocity) is (by analogy to Eq. 6.7):

$$\theta(t) = \theta_0 cos(\omega_0 t) \quad (7.12)$$

where $\omega_0 = \surd(g/L)$. Therefore the period of the oscillation is: $T = 2\pi/\omega_0 = 2\pi\surd(L/g)$ in good agreement with the observed motion of the pendulum.

In practice there is always a frictional (resistant) torque, however small, acting on the pendulum, given by: $-R(d\theta/dt)$. When this torque is taken into account the total torque on the pendulum is: $-MgL\theta - R(d\theta/dt)$, and consequently Eq. (7.11) is replaced by:

$$\frac{d}{dt}\left(\frac{d\theta}{dt}\right) = -\frac{g}{L}\theta - \frac{R}{ML^2}\left(\frac{d\theta}{dt}\right) \quad (7.13)$$

The solution of which (when the pendulum is at the angle θ_0 with $d\theta/dt = -\gamma\theta_0$ at $t = 0$) is: $\theta(t) = \theta_0 e^{-\gamma t}\cos(\omega_0' t)$, where $\gamma = R/(2ML^2)$ and $\omega_0' = \surd(\omega_0^2 - \gamma^2)$. We need not go into the details of the derivation of this formula here. The thing to note is that, in the presence of friction the amplitude of the oscillation ($\theta_0 e^{-\gamma t}$) diminishes exponentially with time (as shown schematically in Fig. 7.4b; for a description of the exponential function see Sect. 5.1); but as long as γT is much smaller than unity, it will not change much over the time of a few oscillations. Moreover, under the same condition, the period of the oscillation remains a constant (as noticed initially by Galileo) and not much different from that of the simple pendulum ($\omega_0' \approx \omega_0$).

The pendulum is of course a relatively simple body. In the case of an arbitrarily shaped body or one whose mass is not uniformly distributed, the evaluation of the moment of inertia of the body, as defined by Eq. (7.8), may not be as straightforward; but it can be done, if need be, with the help of a computer. And whatever the torque acting on the body is, we can always solve the equation of motion (Eq. 7.7), if need be, numerically, in the manner indicated in relation to translational motion (read the two paragraphs following Eq. 6.7).

When the axis of rotation, which passes through the body, is not fixed in space, we need more than one angle to fix the position of the body. When the axis goes through a fixed point, as it happens in a gyroscope, three angles suffice: two angles determine the direction of the axis (with respect to a system of coordinates whose origin coincides with the fixed point) and the third determines the rotation of the body about the axis. When we consider the rotation of a body about an axis that goes through its centre of mass, when the latter is moving, we need three angles to specify the position of the body relative to a system of coordinates attached to its centre of mass, and we need three more coordinates to specify the position of the centre of mass. But as a rule the motion of the centre of mass can be separated out from that of the motion of the body about its centre of mass. We shall not be concerned with such motions here.

7.2 The Vibrating String. Partial Differential Equations Fourier Series

A string of length L has its two ends fixed at $x = 0$ and $x = L$ on the x-axis. At $t = 0$ the string is given an initial displacement $y(x, t = 0)$ parallel to the y-axis, as shown, for example, in Fig. 7.5. In this case:

$$\begin{aligned} y(x,0) &= (2h/L)x, \quad && \textit{for} \quad 0<x<L/2 \\ &= 2h-(2h/L)x, && \quad L/2<x<L \end{aligned} \tag{7.14}$$

We wish to know $y(x,t)$ for $t > 0$.

The displacement $y(x,t)$ is obviously a function of two variables: x and t. We measure the rate of change of y with respect to x, at a given t, by the *partial derivative* $\partial y/\partial x$, which is defined as follows:

$$\frac{\partial y}{\partial x} = \lim(\Delta x \to 0)\frac{y(x+\Delta x,t)-y(x,t)}{\Delta x} \tag{7.15}$$

We note that in evaluating $\partial y/\partial x$, we treat the other variable (t) as a constant and then proceed as in the evaluation of an ordinary derivative (see Sect. 5.3). And in the same way we measure the rate of change of y with respect to t, at a given x, by the *partial derivative* $\partial y/\partial t$, which is defined as follows:

$$\frac{\partial y}{\partial t} = \lim(\Delta t \to 0)\frac{y(x,t+\Delta t)-y(x,t)}{\Delta t} \tag{7.16}$$

For example, if

$$y(x,t) = Asin(kx)cos(\omega t) \tag{7.17}$$

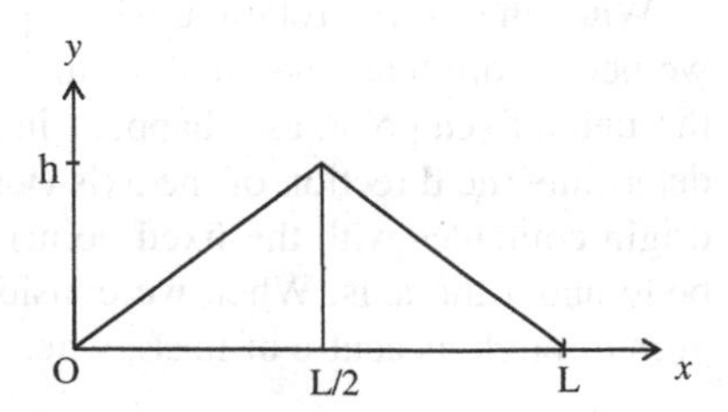

Fig. 7.5 A stretched elastic string (at $t = 0$) according to Eq. (7.14)

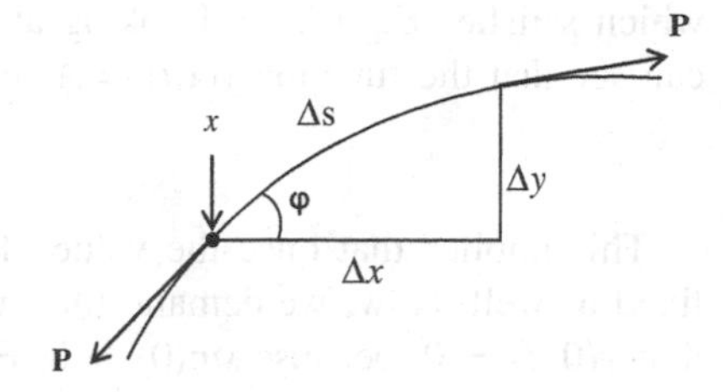

Fig. 7.6 An element of the stretched string between x and $x + dx$. We note that the y-component of the tension force (P) at x is: $-Psin(\varphi) = -P(\Delta y/\Delta s)$

Then:

$$\partial y/\partial x = kAcos(kx)cos(\omega t); \qquad \partial y/\partial t = -\omega Asin(kx)sin(\omega t)$$
$$\tfrac{\partial^2 y}{\partial x^2} = \tfrac{\partial}{\partial x}\left(\tfrac{\partial y}{\partial x}\right) = -k^2Asin(kx)cos(\omega t); \quad \tfrac{\partial^2 y}{\partial t^2} = \tfrac{\partial}{\partial t}\left(\tfrac{\partial y}{\partial t}\right) = -\omega^2Asin(kx)cos(\omega t)$$

We note, by the way, that most quantities in physics are functions of more than one variables and therefore often satisfy *partial differential equations*, which involve partial derivatives defined in the above manner.

Let us now look at an element of the stretched string as this might be at time t. This is shown schematically in Fig. 7.6. We assume that the mass of the string per unit length (δ) is the same along the entire string, which we assume also to be perfectly flexible, so that it can transmit tension but not bending or shearing forces. We shall also assume that the displacement parallel to the y-axis is generally small so that practically $\Delta s = \Delta x$, which implies that the length of each part of the string is practically unchanged and hence the tension is effectively constant along the string, and also (see Fig. 7.6) that the y-component of the tension force P at x is given by $-P(\Delta y/\Delta x)$, which (for Δx sufficiently small) becomes: $-P(\partial y/\partial x)$. Accordingly, the y-component of P at $x + dx$ is given by[2]:

$$P\frac{\partial y}{\partial x} + \frac{\partial}{\partial x}\left(P\frac{\partial y}{\partial x}\right)dx = P\frac{\partial y}{\partial x} + P\frac{\partial^2 y}{\partial x^2}dx$$

Therefore: the y-component of the net force on the element of the string between x and $x + dx$ is: $P(\partial^2y/\partial x^2)dx$. This force is equal (according to Newton's law of motion) to: the mass (δdx) of the element of the string between x and $x + dx$, times the acceleration (parallel to the y-axis) of this element given by $(\partial^2y/\partial t^2)$. Therefore:

$$\frac{\partial^2 y}{\partial t^2} = a^2\frac{\partial^2 y}{\partial x^2}, \quad a^2 = P/\delta \tag{7.18}$$

Therefore the displacement $y(x,t)$ of the string for $t > 0$ must satisfy the above *partial differential* equation, and it must also reduce to the expression of Eq. (7.14) for $t = 0$. Fortunately we already know what needs to be known to obtain $y(x,t)$

---

[2] We remember that $df/dx = (f(x+dx) - f(x))/dx$; so $f(x+dx) = f(x) + (df/dx)dx$. And of course the same applies to partial derivatives: $f(x+dx,t) = f(x,t) + (\partial f/\partial x)dx$.

which satisfies Eq. (7.18). Looking at $y(x,t)$ of Eq. (7.17) and its derivatives, we can see that the function $y(x,t) = Asin(kx)cos(\omega t)$ satisfies Eq. (7.18), provided:

$$\omega = a k. \tag{7.20}$$

This implies that once the value of k has been determined, the value of ω is fixed as well. Now, we demand that: $y(0,t) = y(L, t) = 0$ at any time t. We note that $y(0, t) = 0$, because $sin(0) = 0$. For $y(L,t)$ to be zero as well, we must have $sin(kL) = 0$. This means (see Fig. 5.6) that $kL = n\pi$, with $n = 1, 2, 3, 4, \ldots$. Therefore: any of the following (infinite many) functions $y_n(x,t)$, with A_n chosen arbitrarily, satisfies (7.18) and the boundary condition $y(0,t) = y(L, t) = 0$.

$$y_n(x,t) = A_n sin(n\pi x/L)cos(n\pi at/L), \quad n = 1, 2, 3, 4, \ldots \tag{7.21}$$

These functions are often called the normal modes of the vibrating string. We note that the sum of any number of these functions satisfies Eq. (7.18). This important property follows directly from the *linearity* of Eq. (7.18): meaning that there are *no* terms involving $(y(x,t))^l$ or $(\partial^2 y/\partial x^2)^l$ or any other derivative to the power of l (other than $l = 1$), or products of terms such as $y(x,t)(\partial y/\partial x)$ in Eq. (7.18). Consequently, the sum of two solutions of a linear equation is also a solution of the equation, and therefore the sum of three solutions is also a solution of the equation and so on. This is certainly not possible in the presence of non-linear terms. [It boils down to the following: if $y_1 = Df_1(x)$ and $y_2 = Df_2(x)$, where D stands for the derivative symbol d/dx, then $y_1 + y_2 = D(f_1(x) + f_2(x))$; but if, for example, $y_1 = (Df_1(x))^l$ and $y_2 = (Df_2(x))^l$, where l is any number other than 1, we can *not* write: $y_1 + y_2 = (D(f_1(x) + f_2(x))^l]$ In summary: because of the linearity of Eq. (7.18), the following sum:

$$y(x,t) = \sum_{n=1}^{\infty} A_n sin(n\pi x/L)cos(n\pi at/L) \tag{7.22}$$

satisfies Eq. (7.18), and moreover, $y(0,t) = y(L,t) = 0$, because every term in the above series vanishes at $x = 0, L$.

Now the following question arises: Can we choose the constants A_n so that:

$$y(x,0) = \sum_{n=1}^{\infty} A_n sin(n\pi x/L) = f(x) \tag{7.23}$$

where $f(x)$ is some given function, e.g., that of Eq. (7.14). Thanks to Fourier, we can answer this question positively.

This problem was initially considered by the great mathematician and physicist Joseph Louis Lagrange. He was born Giuseppe Lodovico (Luigi) Lagrangia in Turin in 1736, but lived part of his life in Prussia and part in France. In 1766 Lagrange succeeded Euler (on his recommendation) as the director of mathematics at the Prussian Academy of Sciences (in Berlin). In 1787, at the age of 51, he moved from Berlin to France and became a member of the French Academy. He survived the French Revolution and became the first professor of (mathematical) analysis at The Ecole Polytechnique when it opened in 1794. Napoleon made Lagrange a member of the Legion of Honour and a Count of the Empire (in 1808). He died in Paris in 1813 and is buried in the Pantheon. Lagrange was one of the creators (the other was Euler) of the *calculus of variations*, devising a method of maximizing and minimizing a *functional* (a function of a function, roughly speaking) similar to finding maxima and minima of ordinary functions. His formulation of Newtonian mechanics, known as Lagrangian mechanics, based on variational calculus has many merits especially in relation to constrained motion (e.g., when a ball is constrained to move in a curved tube); but it also proved extremely useful in relation to Quantum Mechanics. Among his early papers, some of them written with the aid of his students, there is one dealing with the vibrating string. In this paper he points out a lack of generality in the solutions obtained previously by Brook Taylor, D'Alebent, and Euler, and arrives at the conclusion that compound sounds (motions of the string) are superpositions of: $sin(mx)sin(nx)$.

Jean Baptiste Joseph Fourier was a doctoral student of Lagrange. He was born at Auxere (in the Yonne department of France) in 1768, the son of a tailor. He took a prominent part in his own district in promoting the French Revolution, and was rewarded by an appointment in 1795 at the Ecole Normale Superieure and subsequently by a chair at the Ecole Polytecnique. Fourier went with Napoleon on his Egyptian expedition in 1798 and was responsible for the organization of the workshops for the French army's munitions of war. And at the same time he contributed several mathematical papers to the Egyptian Institute which Napoleon founded at Cairo. It was after his return to France in 1801 that he began his work on the flow of heat, the results of which he published in his book *Theorie Analytique de la Chaler,* based on Newton's law of cooling, namely that the flow of heat between two adjacent molecules (elementary elements) of a medium is proportional to the extremely small difference of their temperatures. In this work he showed that: any function $f(x)$ defined over $0 < x < L$ can (even in the presence of discontinuities in the function or its derivative, as in the example of Fig. 7.5) be written as an infinite series of *sine* functions as in Eq. (7.23). These series (and the very similar *cosine* series) are now called Fourier series. The essentials of Fourier series can be summarized as follows (for the sake of clarity I restrict the presentation to the *sine* series):

1. The set of functions: $\sqrt{(2/L)}sin(n\pi x/L)$, $n = 1,2,3,\ldots$ are an orthogonal (and orthonormal) set, meaning that

$$\int_0^L \sqrt{(2/L)}sin(n\pi x/L)\sqrt{(2/L)}sin(m\pi x/L)dx = \delta_{nm} \tag{7.24}$$

where $\delta_{nm} = 1$ if $n = m$, and $\delta_{nm} = 0$, if $n \neq m$.

2. The above set of functions is a complete set, meaning that: any function $f(x)$ defined over the interval $0 \leq x \leq L$, is obtained (reproduced) by the following series:

$$f(x) = \sum_{n=1}^{\infty} A_n \sqrt{(2/L)}\, sin(n\pi x/L); \quad A_n = \int_0^L \sqrt{(2/L)} sin(n\pi x/L) f(x) dx \quad (7.25)$$

The above holds true even in the presence of discontinuities in $f(x)$ (as in: $f(x) = 1$ for $0 \le x \le L/2$; $f(x) = 2$ for $L/2 \le x \le L$) or in its derivative (as in the example of Fig. (7.5)). The equality of $f(x)$ with the Fourier series must be understood in the same way as Eq. (1.4) relating to the geometric series. Usually a few terms of the series are sufficient to represent reasonably accurately the given function.

Putting $f(x) = y(x,0)$, of Eq. (7.14), and assuming that $L = 2$ for the chosen unit of length, we obtain (I assure you that the integrals were evaluated correctly):

$$A_n = h \int_0^1 x sin(n\pi x/2) dx + h \int_1^2 (2 - x) x sin(n\pi x/2) dx = (8h/(\pi n)^2)\, sin(n\pi/2) \quad (7.26)$$

Therefore: $y(x, 0) = (8h/\pi^2) \sum_{n=1}^{\infty} (1/n^2) sin(n\pi/2) sin(n\pi x/2)$

And we note that $sin(n\pi/2) = 0$, and therefore $A_n = 0$, for $n = 2, 4, 6$, etc. In Fig. 7.7 we show graphically that a relatively small number of terms in the above series reproduces reasonably well the given function.

We conclude that:

$$y(x,t) = (8h/\pi^2) \sum_{n=1}^{\infty} (1/n^2) sin(n\pi/2) sin(n\pi x/2) cos(n\pi at/2)$$

is the required solution (for $L = 2$). It satisfies Eq. (7.18), the boundary conditions $y(0,t) = y(L,t) = 0$, and it reduces to the given $y(x,0)$ at $t = 0$.

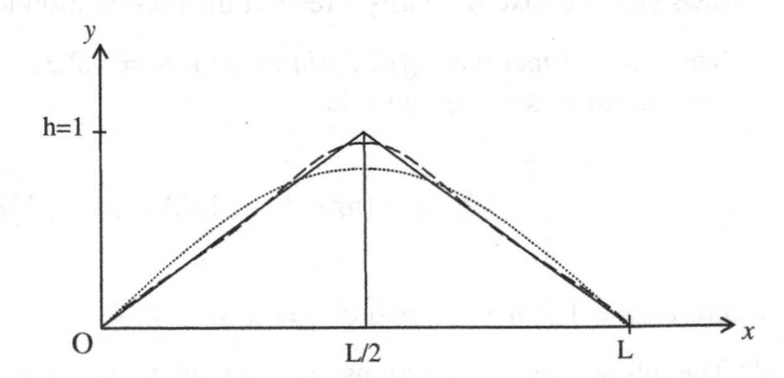

Fig. 7.7 The curves are calculated for $L = 2$. The dotted line corresponds to a single term ($n = 1$); the broken line is the sum of the first three non zero terms of the series (7.26). We see that the latter is reasonably close to the given curve (*solid line*)

7.3 Some Important Concepts of Analytical Mechanics

One important ingredient of analytical mechanics, which goes beyond the use of calculus, is the use of concepts which do not relate directly to observable quantities, these being the mass of a body, its position, etc. and quantities directly related to them such as the 'force' which we perceive as the cause of a body's acceleration (even if we do not know its exact nature). And it is true to say that some of these 'new' concepts are more 'abstract' (further away from the observed facts) than others. It is not possible to describe all the concepts that make up the vocabulary of analytical mechanics in the present book. But some of them are now widely used, and these we need to know. And because, it is not easy to establish who the author was of a particular concept, whether it was Newton himself, or Euler or Lagrange or somebody else, I shall introduce them without any reference to their authors.

We define the *work dw* done by a force $\boldsymbol{F}$ when it displaces a particle (of mass m) from $\boldsymbol{r}$ to $\boldsymbol{r} + \boldsymbol{dr}$ as follows[3]:

$$dw = \boldsymbol{F} \cdot \boldsymbol{dr} \tag{7.27}$$

or (which is the same thing): $dw = \boldsymbol{F} \cdot v(t)dt$

We say that dw is the *scalar product* of $\boldsymbol{F}$ with $\boldsymbol{dr}$.

The scalar product $\boldsymbol{A} \cdot \boldsymbol{B}$ of any two vectors $\boldsymbol{A}$ and $\boldsymbol{B}$ is a scalar quantity (a number with units attached to it when required) defined by:

$$\boldsymbol{A} \cdot \boldsymbol{B} = \boldsymbol{B} \cdot \boldsymbol{A} = AB\cos(\theta) \tag{7.28}$$

where θ is the angle between $\boldsymbol{A}$ and $\boldsymbol{B}$. We note that $\boldsymbol{A} \cdot \boldsymbol{B}$ is equal to the magnitude (A) of $\boldsymbol{A}$ times the projection ($Bcos(\theta)$) of $\boldsymbol{B}$ on $\boldsymbol{A}$. It follows (see Fig. 7.8) that: $(\boldsymbol{B} + \boldsymbol{C}) \cdot \boldsymbol{A} = \boldsymbol{B} \cdot \boldsymbol{A} + \boldsymbol{C} \cdot \boldsymbol{A}$ and therefore:

$$\boldsymbol{A} \cdot \boldsymbol{B} = \left(A_x\boldsymbol{i} + A_y\boldsymbol{j} + A_z\boldsymbol{k}\right) \cdot \left(B_x\boldsymbol{i} + B_y\boldsymbol{j} + B_z\boldsymbol{k}\right) = \left(A_xB_x + A_yB_y + A_zB_z\right) \tag{7.29}$$

because $\boldsymbol{i} \cdot \boldsymbol{i} = \boldsymbol{j} \cdot \boldsymbol{j} = \boldsymbol{k} \cdot \boldsymbol{k} = 1$ and $\boldsymbol{i} \cdot \boldsymbol{j} = \boldsymbol{i} \cdot \boldsymbol{k} = \boldsymbol{j} \cdot \boldsymbol{k} = 0$. It is worth noting that $\boldsymbol{A} \cdot \boldsymbol{B} > 0$ when $0 < \theta < \pi/2$, and $\boldsymbol{A} \cdot \boldsymbol{B} < 0$ *when* $\pi/2 < \theta < \pi$.

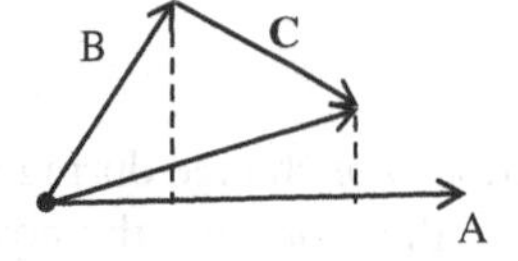

Fig. 7.8 $(B + C)\cdot A$ is equal to (the projection of $B + C$ on A) times A. Therefore: $(B + C) \cdot A = B \cdot A + C \cdot A$

[3] We say a particle, but the same formulae apply to a finite body. In that case dr is the displacement of the centre of mass of the body, $dr/dt = V$ is the velocity of the centre of mass, m is the total mass of the body, and F is the sum of the external forces acting on the body. And $K = mV^2/2$ is the kinetic energy associated with the *translational* motion of the body.

Why is the concept of *work done by a force* a useful one? You must wait. Let me introduce another concept: that of the *kinetic energy* of a particle of mass m moving with velocity $\boldsymbol{v} = d\boldsymbol{r}/dt$. Its kinetic energy, denoted by K, is defined by:

$$K = (m/2)v^2 = (m/2)\boldsymbol{v}\cdot\boldsymbol{v} \tag{7.30}$$

The time derivative of K is:

$$dK/dt = (m/2)(d\boldsymbol{v}/dt)\cdot\boldsymbol{v} + (m/2)\boldsymbol{v}\cdot(d\boldsymbol{v}/dt) = \boldsymbol{v}\cdot m(d\boldsymbol{v}/dt) = \boldsymbol{v}\cdot\boldsymbol{F}$$

Therefore: $dK = \boldsymbol{v}\cdot\boldsymbol{F}dt$, and according to Eq. (7.27) we obtain:

$$dK = dW = \boldsymbol{F}\cdot d\boldsymbol{r} \tag{7.31}$$

This tells us that the change in the kinetic energy of the particle is equal to the work done by the force acting on it. It is worth noting that a decrease in the kinetic energy of the particle ($dK < 0$) implies that, the displacement $d\boldsymbol{r}$ of the particle is at an angle $\theta > \pi/2$ to the direction of the force (the force opposes the motion of the particle).

The reader may rightly think that Eq. (7.31), though of some theoretical interest, it does not have an obvious practical value. In what way is it useful?

Let us look more carefully at some of the forces we know. And let us begin with the elastic force $F = -k\,x$, which underlies the harmonic motion of an oscillator (see Sect. 6.2.2). In this case the motion has only one component: $x(t)$; and Eq. (7.31) becomes: $dK = F(x)dx$. On the other hand, we observe (and this is the important thing) that:

$$F(x) = -dU/dx \tag{7.32}$$

if we put $U(x) = k\,x^2/2$. This tells us that: $dK = Fdx = -\,dU$. Therefore:

$$dK + dU = d(K + U) = 0 \text{ if Eq.(7.32) holds.} \tag{7.33}$$

When Eq. (7.32) holds, we say that the force acting on the particle is a *conservative* force, meaning that it is given by the spatial derivative of a scalar function U of the position of the particle (in the manner of Eq. (7.32) for one-dimensional motion). And we call U the *potential energy field* of the particle. According to Eq. (7.33), the (total) energy E of the particle, defined by

$$E = K + U \tag{7.34}$$

does *not* change during the motion of the particle. It is a *conserved* quantity; and it is, therefore, worth knowing. If a motion had an *Identity Card*, written on it would be the values of the conserved quantities of the motion (its unchangeable properties) in the same way that the identity card of a person notes the height of the person and the colour of his eyes which do not change with the time.[4]

[4] The unit of energy most commonly used is the joule (J): 1 J = 1 N m = 1 kg × m^2/s^2.

It is easy to see that the weight of a body is also a conservative force. If we take the z-axis in the vertical direction, the weight of the body can be written as follows:

$$-mg\boldsymbol{k} = -(dU/dz)\boldsymbol{k}, \quad \text{where } U(z) = mgz \tag{7.35}$$

A little problem: Assuming that a body (of mass m) falls to the ground from a height h, with zero initial velocity, find its speed υ_f the moment it hits the ground. Noting that the motion is one dimensional (along the z-axis) we write: $E = mgz + m\upsilon^2/2$, where $\upsilon = dz/dt$. Now, at $t = 0$: $\upsilon = 0$ and $z = h$. Therefore $E = mgh$ and does not change during the motion. At the moment the body hits the ground $z = 0$ and $\upsilon = \upsilon_f$ and so $E = m\upsilon_f^2/2$. Therefore $m\upsilon_f^2/2 = mgh$. Therefore: $\upsilon_f = \sqrt{(2gh)}$.

We must next consider how to generalize Eqs. (7.32 and 7.33) to three-dimensional motion.

Assume that a particle of mass m is attracted to a centre of force (at the origin of a Cartesian system of coordinates) by a force[5]:

$$\boldsymbol{F} = -(A/r^3)\boldsymbol{r} \tag{7.36}$$

And we remember that the magnitude r of $\boldsymbol{r} = x\boldsymbol{i} + y\boldsymbol{j} + z\boldsymbol{k}$ (the position vector of the particle) is: $r = \sqrt{(x^2 + y^2 + z^2)}$. Let us now introduce:

$$U(r) = -A/r \tag{7.37}$$

And let us evaluate the partial derivatives of U with respect to x, y and z remembering that, to obtain $\partial U/\partial x = (dU/dr)\partial r/\partial x$ we differentiate r with respect to x treating y and z as constants, and similarly in evaluating $\partial U/\partial y = (dU/dr)\partial r/\partial y$ we treat x and z as constants, and in the same way we evaluate $\partial U/\partial z = (dU/dr)\partial r/\partial z$. We find (see Eq. (5.41)):

$$\begin{aligned}(-\partial U/\partial x, -\partial U/\partial y, -\partial U/\partial z) &= (-(A/r^3)x, -(A/r^3)y, -(A/r^3)y) \\ &= -(A/r^3)\boldsymbol{r}\end{aligned}$$

Which is identical with the given force (of Eq. (7.36). We can write the left hand side of the above equation more neatly, using the concept of the gradient of $U(r)$.

The gradient ∇f of a scalar function $f(\boldsymbol{r})$, which is a function of position (but may depend also on time or other variables), is a vector defined at each point $\boldsymbol{r} = x\boldsymbol{i} + y\boldsymbol{j} + z\boldsymbol{k}$, as follows:

$$\nabla f = (\partial f/\partial x)\boldsymbol{i} + (\partial f/\partial y)\boldsymbol{j} + (\partial f/\partial z)\boldsymbol{k} \tag{7.38}$$

Using the above notation we can say that the force of Eq. (7.36) is the negative of the gradient of $U(r)$, of Eq. (7.37), and write it as follows:

$$\boldsymbol{F} = -\nabla U \tag{7.39}$$

[5] Putting the sun at the origin of coordinates, the force it exerts on a planet will be given by this formula with $A = GM_sM_p$ in accordance with Eq. (6.12a).

This is the generalization of Eq. (7.32) to three dimensions. When Eq. (7.39) holds, we say that the force acting on the particle is a *conservative* force, meaning that it is given by the gradient of a scalar function U of the position of the particle; and we call $U(x,y,z)$ the *potential energy* of the particle.

The change dK in the kinetic of a particle when displaced by $\boldsymbol{dr}$ under the action of a conservative force, is given by (see Eq. 7.27)[6]:

$$dK = \boldsymbol{F} \cdot \boldsymbol{dr} = -\nabla U \cdot \boldsymbol{dr} = -((\partial U/\partial x)dx + (\partial U/\partial y)dy + (\partial U/\partial z)dz) = -dU$$

where

$$dU = U(x + dx,\ y + dy,\ z + dz) - U(x,\ y,\ z)$$

Therefore:

$$d(K + U) = 0 \text{ if Eq. (7.39) holds.} \tag{7.40}$$

The above equation is the generalization of Eq. (7.33) to three dimensions. It tells that when the force acting on a particle is a conservative force, the total energy $E = K + U$, of the particle does not change, it is conserved.

Let me conclude this section with a consideration of a system of N particles kept together by mutual interactions. For example, the sun with its planets constitute such a system. And, as we shall later see, the atomic nucleus with a number of electrons moving about it, constitutes such a system. In each case we would like to know which quantities are conserved as the particles move about under the influence of each other, *if* no external forces are acting on the particles of the system.

We already know (see Eq. 6.4) that the total momentum $\boldsymbol{P}$, the sum of the momenta, $\boldsymbol{p}_i$, $i = 1, 2, 3, \ldots N$, of the particles that make up the system, is conserved in the absence of external forces ($\boldsymbol{F} = 0$ in Eq. (6.4) gives $d\boldsymbol{P}/dt = 0$). In consequence the velocity $\boldsymbol{V}$ of the centre of mass of the system remains constant. A most interesting application of the conservation of the total momentum of a system of many particles is to be found in the propulsion of rockets: by discharging (backward) a high speed gas, the main body of the rocket acquires a forward momentum (since the total momentum of the system must be conserved).

We can also easily establish that the total angular momentum $\boldsymbol{L}$ of the system with respect to the origin of coordinates is conserved, if there no external forces

---

[6] From the definition of the partial derivative (see Eq. (7.15) we obtain: $f(x + dx,y,z) = f(x,y,z) + (\partial f/\partial x)dx$. And similarly we can write:

$$f(x + dx, y + dy, z) = [f(x,y,z) + (\partial f/\partial x)dx] + \left(\frac{\partial}{\partial y}[f(x,y,z) + (\partial f/\partial x)dx]\right)dy$$
$$= f(x,y,z) + (\partial f/\partial x)dx + (\partial f/\partial y)dy,$$

leaving out the term proportional to $dxdy$ which is negligibly small for infinitesimal dx and dy. Doing the same for the displacement from z to $z + dz$, we finally obtain:

$$f(x + dx, y + dy, z + dz) = f(x,y,z) + (\partial f/\partial x)dx + (\partial f/\partial y)dy + (\partial f/\partial z)dz.$$

acting on the particles of the system. We have: $\boldsymbol{L} = \Sigma\, \boldsymbol{L}_i$, where the sum is over all particles ($i = 1,2, \ldots N$) of the system. Therefore: $d\boldsymbol{L}/dt = \Sigma\, d\boldsymbol{L}_i/dt = \Sigma\, \boldsymbol{r}_i \times \boldsymbol{F}_i$, according to Eq. (7.3). Here $\boldsymbol{F}_i$ is the force exerted on the the *i*th particle by the other particles in the system. $\Sigma\, \boldsymbol{r}_i \times \boldsymbol{F}_i = 0$, because all the forces involved appear in pairs of forces equal in magnitude but acting in opposing directions. Therefore: $d\boldsymbol{L}/dt = 0$, therefore $\boldsymbol{L}$ is constant.

Let us now assume (this is often the case) that the internal forces $\boldsymbol{F}_i\,(i = 1, 2, \ldots N)$ derive from a potential energy as follows:

$$\boldsymbol{F}_i = -((\partial U/\partial x_i)\boldsymbol{i} + (\partial U/\partial y_i)\boldsymbol{j} + (\partial U/\partial z_i)\boldsymbol{k});\; U = \sum_{i \neq j} u(|r_i - r_j|) \tag{7.41}$$

where $|r_i - r_j| = \sqrt{\left((x_i - x_j)^2 + (y_i - y_j)^2 + (z_i - z_j)^2\right)}$ is the distance separating particle i from particle j; and the sum includes all possible pairs of particles. Proceeding as in the case of a single particle (described above), we find that the total energy E of the system is conserved if there are no external forces acting on the particles of the system. We have:

$$E = K + U = constant \tag{7.42}$$

where U is given by Eq. (7.41), and $K = \Sigma\, m_i v_i^2/2$ is the total kinetic energy of the system (the sum of the kinetic energies of the N particles of the system).

7.4 Fluid Dynamics

The theory of fluid dynamics began most probably in 1741 when, according to some historians, Frederick the Great asked Euler to engineer a water fountain. Euler started by writing Newton's law for an incompressible fluid (of constant mass density ρ) as follows:

$$(\rho(\partial \boldsymbol{u}(\boldsymbol{r}, t)/\partial t) + \rho \boldsymbol{u} \cdot \nabla \boldsymbol{u}(\boldsymbol{r}, t) = -\nabla P(\boldsymbol{r}, t) + \boldsymbol{f}) \tag{7.43}$$

where, $\boldsymbol{u}(\boldsymbol{r},t) = u_x(\boldsymbol{r},t)\boldsymbol{i} + u_y(\boldsymbol{r},t)\boldsymbol{j} + u_z(\boldsymbol{r},t)\boldsymbol{k}$ is the velocity of the fluid at the position $\boldsymbol{r} = x\boldsymbol{i} + y\boldsymbol{j} + z\boldsymbol{k}$ (we assume as usual a Cartesian system of coordinates). It is important to remember that $\boldsymbol{u}$ *is different* from the velocity (usually denoted by $\boldsymbol{v}$) of an element of the fluid (of mass δm) which moves on with the flow. With the velocity so defined, the change in the momentum of the fluid (per unit volume), at the position $\boldsymbol{r}$, is given by the rather odd expression on the left of Eq. (7.43). The way to read $\boldsymbol{u}\cdot\nabla\boldsymbol{u}(\boldsymbol{r},t)$ is explained below. The physical origin of this term, known as convective acceleration, is explained schematically in Fig. 7.9. The expression on the right of Eq. (7.43) represents the force (per unit volume) at $\boldsymbol{r}$. The first term on the right is the negative of the gradient of the pressure (see Eq. 7.38): $\nabla P = (\partial P/\partial x)\boldsymbol{i} + (\partial P/\partial y)\boldsymbol{j} + (\partial P/\partial z)\boldsymbol{k}$; and $\boldsymbol{f} = f_x\,\boldsymbol{i} + f_y\boldsymbol{j} + f_z\,\boldsymbol{k}$ represents other forces (per unit volume) such as gravity.

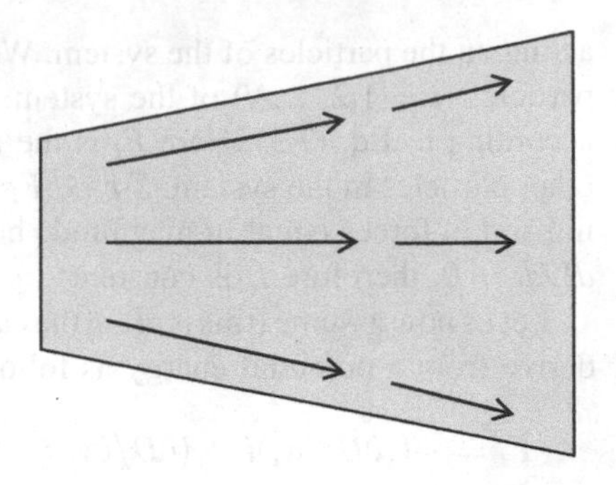

Fig. 7.9 An example of convection. Even for a steady flow ($\partial u/\partial t = 0$) the velocity of an incompressible fluid changes as it moves down a diverging duct. Therefore there is an acceleration happening over position

Equation (7.43) is a vector equation: it stands for three partial differential equations corresponding to the three components of $\boldsymbol{u}(\boldsymbol{r},t)$. By definition:

$$\boldsymbol{u}\cdot\nabla\boldsymbol{u}(\boldsymbol{r},t) = (\boldsymbol{u}\cdot\nabla u_x(\boldsymbol{r}))\boldsymbol{i} + (\boldsymbol{u}\cdot\nabla u_y(\boldsymbol{r}))\boldsymbol{j} + (\boldsymbol{u}\cdot\nabla u_z(\boldsymbol{r}))\boldsymbol{k}$$

and

$$\mathbf{u}\cdot\nabla u_x(\boldsymbol{r}) = u_x(\partial u_x/\partial x) + u_y(\partial u_x/\partial y) + u_z(\partial u_x/\partial z)$$

with the corresponding expressions for $\boldsymbol{u}\cdot\nabla u_y(\boldsymbol{r})$ and $\boldsymbol{u}\cdot\nabla u_z(\boldsymbol{r})$ obtained by replacing u_x in the parentheses by u_y and u_z respectively.

The three component-equations of (7.43) are, therefore, given by;

$$\begin{aligned} \rho(\partial u_x(\boldsymbol{r},t)/\partial t) + \rho\boldsymbol{u}\cdot\nabla u_x(\boldsymbol{r},t) &= -\,\partial P/\partial x + f_x \\ \rho(\partial u_y(\boldsymbol{r},t)/\partial t) + \rho\boldsymbol{u}\cdot\nabla u_y(\boldsymbol{r},t) &= -\,\partial P/\partial y + f_y \\ \rho(\partial u_z(\boldsymbol{r},t)/\partial t) + \rho\boldsymbol{u}\cdot\nabla u_z(\boldsymbol{r},t) &= -\,\partial P/\partial z + f_z \end{aligned} \tag{7.43a}$$

The above equations can not by themselves describe fully the fluid's flow. They are supplemented with an equation which expresses the conservation of mass, which for an incompressible fluid reduces to $\nabla\cdot\boldsymbol{u} = 0$, and there will be boundary conditions relating to the vessel (a tube or the banks of a river etc.) in which the fluid flows. We need not go into such details here. The important thing to note is, that the presence of the non linear term $\boldsymbol{u}\cdot\nabla\boldsymbol{u}$ makes solving the above equations and their complement a very difficult mathematical problem. And yet, some useful information can be obtained from the above equations relatively easily.

Daniel Bernoulli considered the *steady* flow ($\partial\boldsymbol{u}(\boldsymbol{r},t)/\partial t = 0$) *of* a fluid obeying Eq. (7.43), with $\boldsymbol{f}$ representing gravity: $\boldsymbol{f} = -\nabla U$ with $U = \rho g z$, where g is the acceleration of gravity (in the z-direction).[7] Bernoulli showed that this flow obeys the following conservation rule (now known as Bernoulli's theorem):

$$K + P + U = C \tag{7.44}$$

[7] Daniel Bernoulli, a Dutch-Swiss mathematician, born in Groningen in 1700, was a close friend of Euler and worked for some years (1724–1733) in St. Petersburg, before returning to Switzerland, where he stayed until his death (in 1782). He contributed significantly to mathematics (statistics) and physics (kinetic theory of gases and hydrodynamics).

where $K = \rho u^2(\boldsymbol{r})/2$ is the kinetic energy of the fluid (per unit volume) at $\boldsymbol{r}$; $U(\boldsymbol{r})$ is the potential energy (per unit volume) due to gravity at $\boldsymbol{r}$, and $P(\boldsymbol{r})$ is the pressure at $\boldsymbol{r}$. Their sum is a constant: C; it does *not* depend on the position $\boldsymbol{r}$ in the field of flow. One can think of the above equation as the expression of the conservation of energy in fluid motion.

The above theorem finds a useful application in aviation: Consider a plane in flight. Fix the system of coordinates on the plane. With respect to this system, the plane is stationary and the air flows. Now, by giving the plane the right shape, the flow of air below the plane slows down (its velocity decreases) relative to the air flowing above the plane. Therefore, in accordance with Eq. (7.44), the pressure on the plane from below is greater than the one from above, their difference producing an upward force which opposes the weight of the plane. The magnitude of this force depends on the angle the plane makes with the vertical and, therefore, the plane rises or comes down, or stays at the same height depending on this angle.

The reader will probably think that the application of Eq. (7.44), to the flow of air is not permissible. This equation is obtained on the basis of Eq. (7.43) which applies to an incompressible fluid (one with a constant density); and air is certainly compressible. Now, it turns out that, assuming a constant ρ is not a bad approximation for compressible fluids in those cases where compressibility does *not* play an active role (as it does, for example, in the propagation of sound).

When solutions of Eq. (7.43) were obtained for realistic situations (e.g. the non steady flow of a river down a slope) the predicted velocities were much greater than those observed. In 1827 Navier realized that the viscosity of the fluid, manifested by a sheer force between adjacent layers or small regions of the fluid, arising from the gradient of the velocity, plays an important role in the motion of a fluid (akin to that of friction in the motion of solid bodies) and must be taken into account in the formulation of the equation of motion of the fluid.[8] His work was extended by Stokes (in 1845) to give the Navier -Stokes equations: a set of five coupled equations, the two of them express the conservation of mass and energy and the remaining three the conservation of momentum.[9] For an incompressible fluid the momentum equations are obtained from the corresponding equations for *ideal* fluids (Eq. 7.43) by adding a viscosity term $\mu\ \nabla^2\boldsymbol{u}$ on the right of this equation as follows:

$$\rho(\partial\boldsymbol{u}(\boldsymbol{r},t)/\partial t) + \rho\boldsymbol{u}\cdot\nabla\boldsymbol{u}(\boldsymbol{r},t) = -\nabla P(\boldsymbol{r},t) + \mu\nabla^2\boldsymbol{u} + \boldsymbol{f} \tag{7.45}$$

---

[8] Claude Louis Navier was born in 1785 in Dijon (France). In 1824 he was elected as a Member of the French Academy of Sciences, and in 1830 became a professor at the Ecole Nationale des Ponts et Chaussees, and in the following year he succeeded Augustin Louis Cauchy as professor of calculus and mechanics at the Ecole Polytechnique. He died in 1836.

[9] George Gabriel Stokes was born in Ireland in 1819. After attending schools in Dublin, he studied at Cambridge University. In 1849 he was appointed to the Lucasian professorship of mathematics at the same University, a position he held until his death in 1903. Stokes made significant contributions, not only to hydrodynamics, but also to the theory of the polarization of light and related phenomena. He also contributed to mathematics (mostly in relation to numerical integration).

The symbol ∇^2, known as the *Laplace operator*, is defined as follows (*f*(***r***) is any function of ***r***):

$$\nabla^2 f = \partial^2 f/\partial x^2 + \partial^2 f/\partial y^2 + \partial^2 f/\partial z^2 \quad (7.46)$$

When ∇^2 appears in front of a vector, such as ***u***(***r***,*t*), its operation must be understood as follows:

$$\nabla^2 \boldsymbol{u} = (\nabla^2 u_x)\boldsymbol{i} + (\nabla^2 u_y)\boldsymbol{j} + (\nabla^2 u_z)\boldsymbol{k} \quad (7.47)$$

The coupled equations represented by Eq. (7.45) are therefore given by:

$$\rho(\partial u_x(\boldsymbol{r},t)/\partial t) + \rho \boldsymbol{u} \cdot \nabla u_x(\boldsymbol{r},t) = -\partial P/\partial x + \mu(\nabla^2 u_x) + f_x \quad (7.48)$$

and two similar equations for the *y* and *z* directions. The viscosity term in the above equations takes into account the conversion of kinetic energy of the fluid into heat during the flow.

[The viscosity coefficient of a fluid is obtained as follows: Let *F* be the force needed to maintain a velocity gradient *du/dz*, where *u* is normal to the *z*-direction, between two adjacent planes, at *z* and $z + dz$, of fluid area *A*. We write $F = \mu A(du/dz)$, where μ is the viscosity coefficient of the fluid. The kinematic coefficient of viscosity, η, is defined by $\eta = \mu/\rho$, where ρ is the density of the fluid. For water $\eta = 10^{-6}$ m^2/s].

Exact solutions of the Navier–Stokes Equations subject to realistic boundary conditions are practically impossible. And rough estimates are misleading. For example, one could estimate the magnitude of the velocity *u* in a big river as follows: Assuming that the water at the bottom of the river has practically zero velocity, then if the velocity at the surface of the river is *u*, the gradient *du/dz* will be about *u/L*, where *L* is the depth of the river. The force per unit volume required to maintain this gradient will be (see text in [...] above): $\mu u/L^2$]. Equating this force with $\rho g'$ (the effective gravity for the slope of the river) we obtain $u \approx \rho g' L^2/\mu = g' L^2/\eta$. Now for a big river dropping a few hundred meters over a distance of a thousand kilometers $g' \approx 10^{-3}$m/s^2 and $L \approx 10$ m, which gives $u \approx 10^5$ m/s, when the observed value of the velocity is about 1 m/s. The explanation of this discrepancy came about following a suggestion, made in 1894, by Reynolds,[10] that: the type of flow to be expected, in a given situation, depends on the relative strength of the nonlinear term in relation to the viscous term in Eq. (7.45), as measured by the following ratio (now known as Reynold's number):

$$Re = u\rho l/\mu \quad (7.49)$$

[10] Osborne Reynolds was born in Belfast in 1842, but he and his family soon afterwards moved to Essex (in England). He graduated from Cambridge University in 1867. In 1868 he was appointed professor of engineering at Owens College in Manchester. A professorship created and financed by a group of manufacturing industrialists in the Manchester area. He became a Member of The Royal society in 1877. Reynolds remained at Owens College to the end of his life. He died in 1912. Owens College became the University of Manchester in 1880.

where u is the magnitude of the velocity of the fluid, ρ its mass density, μ its coefficient of viscosity, and l is a characteristic length of the system under consideration, such as the diameter of a pipe. A smooth flow, dominated by the gradient of the pressure, is to be expected when $Re < 2000$. In this case, the non linear term can be dropped from Eq. (7.45) and analytical solutions become possible. In natural phenomena, however, one finds that $Re > 3000$ and consequently a *turbulent* motion is to be expected, and is found. In a turbulent motion large scale forces set up a cascade of energy transfers to smaller and smaller scales through nonlinearities of fluid motion until, eventually, at the smallest scales viscosity turns the motion into a random one (energy is dissipated into heat). In order to solve the Navier–Stokes Equations (to a reasonable approximation), one must some how average out this small scale randomness which is impossible to describe exactly. The problem is not straightforward and it still occupies the minds of many physicists and mathematicians.

Exercises

7.1. (a) Let $\boldsymbol{a} = 2\boldsymbol{i} + 3\boldsymbol{j} - \boldsymbol{k}$ and $\boldsymbol{b} = 3\boldsymbol{i} - \boldsymbol{j} + 2\,\boldsymbol{k}$. Evaluate $\boldsymbol{a} \cdot \boldsymbol{b}$ and $\boldsymbol{a} \times \boldsymbol{b}$.
Answer: $\boldsymbol{a} \cdot \boldsymbol{b} = 1$; $\boldsymbol{a} \times \boldsymbol{b} = 5\boldsymbol{i} - 7\boldsymbol{j} - 11\boldsymbol{k}$.
(b) Assume that $\boldsymbol{a}$ and $\boldsymbol{b}$ are functions of t: $\boldsymbol{a} = \boldsymbol{a}(t)$ and $\boldsymbol{b} = \boldsymbol{b}(t)$. Verify that: $d(\boldsymbol{a} \cdot \boldsymbol{b})/dt = (d\boldsymbol{a}/dt) \cdot \boldsymbol{b} + \boldsymbol{a} \cdot (d\boldsymbol{b}/dt)$ and $d(\boldsymbol{a} \times \boldsymbol{b})dt = (d\boldsymbol{a}/dt) \times \boldsymbol{b} + \boldsymbol{a} \times (d\boldsymbol{b}/dt)$.

7.2. Show that the angle $\boldsymbol{\theta}$ between two vectors $\boldsymbol{a}$ and $\boldsymbol{b}$ is given by:

$$\cos\theta = \frac{a_x b_x + a_y b_y + a_z b_z}{ab}$$

7.3. Use the formula of the previous exercise to evaluate the angle, θ, between $\boldsymbol{a}$ and $\boldsymbol{b}$ of exercise 7.1(a).
Answer: $\cos\theta = 0.071$.

7.4. Let O be a point in 3-dimensional space, and let A and B be two other such points. OA and OB can be taken as the adjacent edges of a parallelogram. OA defines a displacement vector $\boldsymbol{a}$, and OB a displacement vector $\boldsymbol{b}$. Show that the area, A, of the parallelogram is given by the magnitude of $\boldsymbol{a} \times \boldsymbol{b}$. We have: $A = |\boldsymbol{a} \times \boldsymbol{b}|$.

7.5. Consider the parallelogram of the previous exercise, and let C be a point in space not in the plane of the parallelogram. OC defines a displacement vector $\boldsymbol{c}$ not in the plane of $\boldsymbol{a}$ and $\boldsymbol{b}$. We note that OA, OB and OC can be taken as the adjacent edges of a parallelepiped. Show that the volume V of the parallelepiped is given by the absolute value of $(\boldsymbol{a} \times \boldsymbol{b})\cdot\boldsymbol{c}$, denoted by $|(\boldsymbol{a} \times \boldsymbol{b})\cdot\boldsymbol{c}|$. Note that the scalar product $(\boldsymbol{a} \times \boldsymbol{b})\cdot\boldsymbol{c}$ may be a positive or negative number.

7.6. Calculate the moment of inertia, I, of a very thin (denote its thickness by ΔR) hollow cylinder, of radius R, with respect to its axis. The mass density, ρ, of the cylinder is constant over its length (L).
Answer: $I = MR^2$, where $M = (2\pi R\Delta R)L\rho$ is the mass of the cylinder.

7.7. Calculate the moment of inertia, I, of a cylinder, of radius R and length L, with respect to its axis. The mass, M, of the cylinder is uniformly distributed over its volume.
Answer: $I = 2\pi L\rho \int_0^R r^3 dr = MR^2/2$.

7.8. A cylinder of radius R and mass M is rotating about its axis held in position by appropriate supports at its two ends. As a result of friction between the cylinder and the supports, the initial angular velocity, ω_0, diminishes with time according to the formula: $\omega(t) = \omega_0 e^{-\gamma t}$, if no other external torque is acting on the cylinder. Evaluate the torque that need be applied to the cylinder to maintain its rotation at a constant angular velocity, ω_0.
Answer: $N = MR^2\gamma\omega_0/2$.

7.9. Verify that $\theta(t)$, as given in the text following Eq. (7.13), is indeed a solution of that equation satisfying the given boundary conditions.

7.10. Calculate the minimum velocity, υ_{min}, a body (think of it as a particle of mass m) must have at the surface of the earth in order to escape from the gravitational field of the earth.
Solution: According to Eqs. (6.12), (7.36) and (7.37), the potential energy of the body in the gravitational field of the earth is given by: $U(r) = -GMm/r$, where M is the mass of the earth and r is the radial distance of the body from its centre. Noting that $U(r)$ is negative and tending to zero as $r \to \infty$, we deduce that the *conserved* total energy E of the body must be: $E \geq 0$, if the particle is to have a positive kinetic energy ($m\upsilon^2/2$) as $r \to \infty$, which must be so for the particle to escape. The conservation of the total energy, then requires that at the surface of the earth ($r = R_E$) we must have: $-GMm/R_E + m\upsilon^2/2 \geq 0$.
Answer: $\upsilon_{min} = \sqrt{(2GM/R_E)}$.

7.11. Calculate the escape velocity, υ_{min}, defined in the previous exercise, for a spacecraft with a mass of 6,000 kg. Estimate the energy (W) required to obtain this velocity.
Answer: $\upsilon_{min} = 1.14 \times 10^4$ m/s $\approx$ 41,000 km/h. $W = 3.90 \times 10^{11}$ J.

7.12. A cylinder of mass M and radius R is *rolling* down a plane inclined at an angle α to the ground.
(a) Write the equations which determine the motion of the centre of mass of the cylinder and the rotation of the cylinder about its axis.

Note: In rolling motion: the velocity, $\upsilon(t)$, of the centre of mass of the cylinder, parallel to the plane, relates to the angular velocity $\omega(t)$ of rotation about its axis according to: $\upsilon = \omega R$.
Answer:

$$M\frac{d\upsilon}{dt} = Mg\sin\alpha - T \tag{A}$$

$$I\frac{d\omega}{dt} = RT \quad \text{(B)}$$

Where: g is the acceleration due to gravity; I is the moment of inertia of the cylinder; and T is the friction-force between the plane and the cylinder (parallel to the plane and therefore tangential to the cylinder), whose value is to be determined. Assuming that rolling motion is possible in the given circumstances, we put $\omega(t) = \upsilon(t)/R$, in which case Eq. (B) becomes:

$$\frac{I}{R^2}\frac{d\upsilon}{dt} = T \quad \text{(B)}$$

In the event that T, as given by the above equation can not be maintained at the interface between the plane and the cylinder, rolling motion can not be sustained and the cylinder slides downward.
(b) Show that for a thin hollow cylinder Eqs. (A) and (B) are reduced to the following equation:

$$\frac{d\upsilon}{dt} = \frac{g}{2}\sin\alpha$$

(c) Assuming that the cylinder of (b) is released with zero velocity from a point on the plane a height h above the ground, how long will it take for it to reach the ground, and what will its velocity, υ_{gr}, be when it reaches the ground?
Answer: $t_{gr} = (2/sin\alpha)\sqrt{(h/g)}$; $\upsilon_{gr} = \sqrt{(gh)}$.
(d) Derive the above formula for υ_{gr}, using the constancy of the total (potential + kinetic) energy of the cylinder. Remember that in the present case the kinetic energy derives from a translational and a rotational motion.
Note on friction: The friction 'force', T, between two bodies acts parallel to the interface between the two bodies. It *opposes* motion of the one body relative to the other. When, as a result, the two bodies do not move relative to each other we have: $T \leq \mu_s N$, where N is the normal force between the bodies and μ_s is the static coefficient of friction. The equality sign applies when the one body is about to slide over the other. Typical values of μ_s vary from 0.05 for smooth surfaces to 1.5 for rough surfaces. When the two bodies move relative to each other, we write $T \leq \mu_k N$, where μ_k, the kinetic coefficient of friction may be a bit smaller than the static coefficient of friction and may also vary to some degree with the speed of the motion. In the present case, $N = Mgcos\alpha$ is due to the weight of the cylinder. It is evident that, in the case of a cylinder rolling down an inclined plane, if the angle α gets large enough, T will not be large enough for the rolling motion to be maintained and the cylinder will slide.

7.13. Friction plays an important role in many phenomena of every day life. Explain in as much detail as possible its role in helping us to walk on a road horizontal or otherwise, and its role in the motion of a car.
7.14. Consider two masses m_1 and m_2 on a straight line (say along the x-direction). Assume that m_2 lies in front of m_1. The two masses are connected by a string (a rope). A force F is acting (pulling) m_2 along the x-direction.

(a) Find the acceleration (a) of the masses, and estimate the tension (force) T in the string.
(b) How would you describe the physical origin of T?
Answer: (a): $a = F/(m_1 + m_2)$; $T = m_1 a$.
(b) It is an elastic force arising from the (uniform) stretching of the string, but the stretching required to produce the required force is so small that we do not see it.
7.15. (a) Verify Eq. (7.24). You can make use of published Tables of integrals.
(b) Verify Eq. (7.26).
(c) Write a short computer program to verify the result shown in Fig. 7.7.

7.16. Show that the set of functions: $\frac{1}{\sqrt{L}}, \sqrt{\frac{2}{L}}\cos\frac{n\pi x}{L}$ where $n = 1, 2, 3, \ldots$. are normalized and orthogonal to each other over the interval $0 \le x \le L$ (in the manner of Eq. 7.24).

7.17. Show that the set of functions: $sin(\pi x/L)$, $sin(2\pi x/L)$, $sin(3\pi x/L)$, …, 1, $cos(\pi x/L)$, $cos(2\pi x/L)$, $sin(3\pi x/L)$, … are orthogonal to each other over the interval: $-L \le x \le L$.
Normalize these functions.

7.18. The set of functions introduced in exercises (7.16) and (7.17) are complete sets: a function $f(x)$ defined over the interval $0 \le x \le L$ can be written as a sum (a Fourier series) of the functions given in exercise (7.16); and a function $f(x)$ defined over the interval $-L \le x \le L$ can be written as a sum (a Fourier series) of the normalized functions given in exercise (7.17). The coefficients in these sums are determined in exactly the same way as in Eq. (7.25). When the functions are defined over the interval $-L \le x \le L$ the integration extends of course over the same interval.
Write the Fourier series corresponding to the following functions. You can make use of published Tables of integrals.
(a) $f(x) = -x$ for $-L \le x \le 0$, $f(x) = x$ for $0 \le x \le L$

Answer:

$$\frac{L}{2} - \frac{4L}{\pi^2}\sum_{n=1}^{\infty}\frac{1}{(2n-1)^2}\cos\frac{(2n-1)\pi x}{L}$$

(b) $f(x) = x + x^2$, $-1 \le x \le 1$

Answer:

$$\frac{1}{3} + \frac{2}{\pi}\sum_{n=1}^{\infty}(-1)^n\left(\frac{2}{\pi n^2}\cos n\pi x - \frac{1}{n}\sin n\pi x\right)$$

7.19. A large cylindrical vessel, of cross section A, containing water is open to the atmosphere. The vessel is loosing water through a hole, of cross section $a \ll A$, at the bottom of the vessel. Using Bernoulli's theorem (Eq. 7.44) show that the velocity of the water at the hole is given by: $v_h = \sqrt{2g(z_s - z_h)}$, where g is the acceleration due to gravity, z_s is the height at the surface of the water and z_h the height at the hole.

Solution: Applying Bernoulli's theorem to a unit volume of water at the surface of the water in the vessel and at the hole we write:

$\frac{1}{2}\rho v_s^2 + P_s + \rho g z_s = \frac{1}{2}\rho v_h^2 + P_h + \rho g z_h$, where ρ is the mass density of water. We have $P_s = P_h =$ *atmospheric pressure.* Therefore the above equation reduces to:

$$\frac{1}{2}v_s^2 + gz_s = \frac{1}{2}v_h^2 + gz_h$$

Next we note that the conservation of mass demands that: $\rho v_s A = \rho v_h a$. Therefore: $v_s = v_h a/A$. Since $a \ll A$, we can put $v_s \approx 0$. Therefore: $v_h = \sqrt{2g(z_s - z_h)}$.

Chapter 8
The Beginnings of Chemistry

8.1 Chemistry's Relation to Alchemy

Chemistry as we know it began with Black, Cavendish, Priestley, Scheele and Lavoisier. But its roots lie in alchemy, which combined genuine technique in preparing and handling chemical substances with illusion and fraud. The idea, that all substances are made of a few basic 'elements', is of course ancient. Earth, water, air and fire were supposed to be these basic elements, or 'roots' of everything else (see Sect. 2.2). It was Boyle (whom we met in Sect. 5.2.1), an atomist by conviction, who pointed out that gold and silver had a better claim to be elementary substances than the above mentioned. And he said that on the basis of experimental evidence obtained by techniques which were the result of centuries of alchemical practice. Boyle pointed out that nobody ever extracted earth, water, air or fire or anything other than gold, from a piece of gold; and when gold is alloyed with other metals, it can always be recovered, its quantity and quality unchanged. Moreover, its properties are highly specific. Not only its mass density, and the corresponding specific weight (the weight per unit volume), is unique to it, as proclaimed by Archimedes, but its melting point and hardness and other physical properties are always the same and different from those of any other material. Therefore, there is good reason to believe that gold is a basic element and not a composite material. And the same is true for silver. A substance to be called a basic element must have properties like those of gold and silver, according to Boyle.

The oldest chemical technique is apparently that of distillation. In ancient times people knew that if they boiled sea water in a cauldron, drops of water would be formed on the under side of the lid, and that eventually, when all water had vaporized, there would be salt left in the cauldron. In the second century AD the alchemists had developed distillation sufficiently to volatilize mercury and arsenic from ores of these materials in order to alter the colour of metallic surfaces. Distillation, when mastered sufficiently, could be used to separate materials: removing the more volatile constituent of a mixture by vaporizing it out of the mixture and then liquefying it in a receiving vessel. Sublimation (transfer through

A. Modinos, *From Aristotle to Schrödinger*,
Undergraduate Lecture Notes in Physics, DOI: 10.1007/978-3-319-00750-2_8,

the vapour phase from solid to solid) was also known to the ancient alchemists. They did not, of course, know the 'physics' of the separation process. They were impressed by the efficacy of fire and the mystery of the vapour phase in which there was usually nothing to see. But this did not stop them applying the method to many substances beside sulphides of mercury and arsenic; and as a result they had hydrochloric acid, nitric acid, ammonia, a number of sulphur compounds and, by the Middle Ages, alcohol as well.

Another chemical technique which the alchemists discovered, perhaps by chance, is that of crystallization (the transfer from liquid to solid). Using the acids they obtained by distillation the alchemists prepared a number of salts in crystalline form. And they knew of course cupellation, a method used by the ancients (but still in use) to purify noble metals. One heats the alloyed metal in an open crucible with an absorbent lining until all the base metal has oxidized and been absorbed in the lining.[1]

8.2 The Pneumatic Chemists

These are the chemists who investigated gases in the 18th and 19th centuries. Pneumatic came to mean anything relating to air or wind; the word originates from the Greek πνεΰμα which means spirit (as in things spiritual or metaphysical). One may assume that the association of this word with the properties of gases has something to do with the vital importance of air to life, which was known but not understood at the time. The first pneumatic chemists were primarily interested in the nature of air.

Joseph Black: After the pioneering work of Robert Boyle (see Sect. 5.2) progress towards a fuller description of air was slow. That changed dramatically with the work of Joseph Black. Born in France, Black lived and worked practically all his life (1728–1799) in Scotland. Initially he studied medicine and wrote a doctorate thesis on the treatment of kidney stones with white magnesium (the salt magnesium carbonate). In the process he discovered that when heated strongly, white magnesium (a solid in powder form) gave off a gas and nothing else. Black surmised that all the gas came from white magnesium with a corresponding loss in weight of the powder. He then experimented with other caustic salts, such as limestone (calcium carbonate), and found that they all gave, when heated, the same gas. He called the gas *fixed air* (fixed to the salt before released from it by heating). Black also discovered that the same gas was produced when instead of heating the salts he added acid to them. He went on to study the properties of *fixed air*, which is of course what we now call carbon dioxide (CO_2), and he discovered that this

[1] This is how the ancients recovered pure silver from the metal obtained by smelting silver–lead ores and which looks like lead. Not knowing the mechanism by which this happened, the ancients believed that lead was turned into silver by the efficacy of fire. And because they never weighed the initial total mass and the final product, they were easily misled.

gas would extinguish the flame of a candle. He also discovered that quicklime (calcium oxide), the solid that is obtained from calcium carbonate when *fixed air* is removed from it by heat, is converted slowly back to limestone (calcium carbonate) simply by exposure to air. A discovery which showed that there was a small quantity of *fixed air* in the atmosphere; and that, therefore, air was not an element but a mixture of gases. Black established that gases were chemically the same as liquids and solids and, moreover, by meticulously weighing every reagent and product in the reactions he studied, he contributed considerably to the foundation of quantitative chemistry.

Two years after publishing his thesis *On Acid Humor Arising from Foods and on White Magnesia*, Joseph Black was Professor of Medicine at Glasgow University with a private medical practice on the side. He was a very good lecturer and often his students were the first to hear of his new findings such as, that *fixed air* was present in exhaled breadth and that it was also present in fermentation.

By his early thirties Black moved from chemistry to physics. In 1764 he articulated the difference between *quantity* and *intensity* in heat. He realized that while the *intensity of heat emanating from a* body is proportional to its temperature, the *quantity of heat in* the body is something different, depending on other factors as well. For he noticed that during boiling in spite of the supply of heat the temperature of the boiling substance did not change; and that the same applied to the melting of ice. Black termed the heat supplied to a melting or boiling substance *latent heat*. And he observed that during the reverse processes, steam condensing to water or water freezing to ice, latent heat was released. Such observations played an important role in the development of thermodynamics.

Henry Cavendish began experimenting with gases at about 1766, when he produced a paper on 'factitious airs', which is how gases produced by chemical treatment of solids or liquids were known at that time. The most interesting of these experiments involved dissolving metals in acids to produce hydrogen, which he called 'inflammable air'. Boyle had previously isolated hydrogen, but it was Cavendish who established that it was an element with specific properties and that it was highly inflammable.

At the time people believed that combustible materials contained a colourless, tasteless and weightless material, known as *phlogiston*, which was released when the material was burnt. The existence of such substance was proposed in 1669 by Johann Becher who called it *terra pinguis* (fat earth). According to Becher, wood consisted of ashes and *terra pinguis*; during burning the latter was released leaving behind the ashes. In the early 18th century Georg Stahl renamed the substance *phlogiston* (from the Greek word *φλόγα* which means flame), and extended the theory to include the calcination (oxidation) of metals. According to Stahl's theory, metals were composed of calx (today we would say an oxide in powder form) and phlogiston; when a metal was heated, the *phlogiston* was released and calx remained. Similarly, when calx was heated over charcoal (which was apparently rich in *phlogiston*), it absorbed the *phlogiston* released from the charcoal becoming metal again. The *phlogiston* theory was eventually demolished by Antoine Lavoisier some years later.

The extraordinary inflammability of hydrogen and its very small weight suggested to Cavendish that *it* was the phlogiston. At about the same time Joseph Priestley, who was a good friend of Cavendish, discovered another gas that he called *dephlogisticated air* which could not burn on its own but at the same time no material could burn without it being there. What he discovered was of course oxygen.[2] Cavendish constructed a special vessel (he called it an *endiometer*) which allowed him to burn precise amounts of gases and analyze their products. Using this instrument he found that on mixing his inflammable gas with Priestley's dephlogisticated air, the two combined explosively to make water. By weighing the substances involved he confirmed that the weight of water was equal (with very good accuracy) to the sum of the two reacting gases. This definitely proved that water was not an element (in the way that gold or silver is). He also discovered that the volume of hydrogen needed to make water was two times that of oxygen (measured at the same pressure and temperature). He then tried burning hydrogen in different amounts of common (atmospheric) air; he discovered that when all oxygen in the air had been used up what remained (which he called *phlogisticated air*, we call it nitrogen) would not support the combustion of hydrogen. After a series of careful experiments, he concluded that common air was approximately 79 % nitrogen, 21 % oxygen, and a remaining gas (less than 1 %) that he could not identify.[3]

Cavendish, who contributed much to science in other fields as well as to pneumatics (he was the first to measure the gravitational constant as we have already mentioned in Chap. 6) was in many ways a peculiar man. Born in 1731, an aristocrat and one of the richest men in 18th century Britain, he lived modestly and in relative isolation. He never completed his degree at Cambridge, partly because he objected to the religious requirements of his education but mainly because he was too shy to present himself for an oral examination by his teachers. He was so shy that he could not converse with more than one man at a time, and found it almost impossible to talk to any woman. He converted large parts of his house into laboratories for studying chemistry and physics and devoted his entire life to science. Another aspect of his life, which is even more difficult to understand, is the secrecy he kept about his findings. As a result many scientists were given credit for discoveries that Cavendish had made years earlier. He was recognized as a great scientist only many years after his death (in 1810), when James Clark Maxwell found in Cavendish's papers evidence that he had made discoveries in electricity that were fifty years ahead of his time. Ohm's law and Coulomb's law in

---

[2] The Swedish chemist Carl Scheele discovered oxygen, working independently, at about the same time as Priestley did. It appears that Scheele was first but failed to publish his discovery. Lavoisier produced oxygen later than Priestley, but it was he who gave the gas the name oxygen and the one who described his properties in more detail.

[3] One hundred years after Cavendish's experiment it was found that this gas (approximately 0.93 % of air) is the rare gas argon. There is also a small amount of carbon dioxide (0.04 % of air) in the atmosphere.

electricity and Charles' law of gases were apparently discovered by Cavendish, but he never announced these discoveries.[4]

Let me now tell you how **John Priestley**, who had already isolated and studied ten gases, including ammonia, sulphur dioxide, carbon monoxide, hydrogen chloride and nitrous oxide, discovered oxygen . At the time he was working as a political adviser to the second Earl Shelbum. One day he picked up some brick-red mercury oxide powder, that he had made earlier by heating mercury in air and, using a lens, focused on it the sun's rays. The sun's intense radiation turned the mercury oxide powder into shining globules of mercury driving off an intriguing gas in the process. This gas, Priestley discovered, had all the properties of atmospheric air but better. A flame burned more brightly and for longer in it. Mice were more active when they breathed it, and they survived much longer in a closed vessel containing this gas rather than air. Declaring the gas to be *five or six times better than common air for the purpose of respiration, inflammation and I believe in every other use of atmospheric air*, he reported that he had discovered *dephlogisticated air*. Priestley published the results of his gasses research in two volumes of *Experiments and Observations on Air*, of which the latter volume published in 1776 contains his discovery of oxygen.

John Priestley was in many ways a remarkable man. Born in 1733 to a family of weavers in West Yorkshire, he mastered 8 languages by the time he left school. He was intended for a career in the Church. He became a Rational Dissenter, a branch of Anglicanism that advocated rationality over dogma and religious mysticism. He was an enthusiastic teacher, and for some years, at the beginning of his career, he taught young boys and girls at a small school he organized at Nantwich (in Cheshire). He could teach practically any subject, including literature. He published a series of books relating to education, including one on grammar, several of which remained popular for many years in Britain and America. At some stage he became interested in electricity and made some worthwhile observations in that area. A 250,000-words history of electricity soon followed and a shorter version for the general public.

Three years after he discovered oxygen, Priestley fell out with Lord Shelbum and moved with his family to Birmingham. There he joined a group which called itself the Lunar Society, second only to the Royal Society (of which Priestley was a member) for organizing informal discussions on scientific and other matters, and not least politics. Priestley remained a non conformist, a sympathizer with slavery abolitionists and the French revolutionaries. When he attended a celebration on 14 July 1791 at a Birmingham hotel to mark the second anniversary of the fall of the Bastille, a rioting mob of political opponents attacked him and torched his house. He fled to London with his family, but with the French Republican government

[4] It is worth noting that it was another scientist, Daniel Rutherford, a former student of Joseph Black, who was given credit for the discovery of nitrogen (phlogisticated air). Because Cavendish did not publish his discovery, Rutherford who lived in Glasgow, did not know of it when he made his own discovery in 1772. And again, it was Lavoisier who described the properties of nitrogen in more detail.

declaring Priestley a French citizen, as it went to war against Britain, London proved no safer than Birmingham. His effigy was burnt for declaring the French Revolution was a harbinger of the second coming of Christ. In 1794, a week before Lavoisier was executed for treason by the Republican government in France, he emigrated to America. He stayed there for the rest of his life, devoting his time to writing books about education.

Antoine Lavoisier was born in 1743 to a wealthy family in Paris and inherited a large fortune at the age of five when his mother died. He had a good education, filled with the ideals of the French Enlightenment of that time. He received a law degree and was admitted to the bar, but never practiced as a lawyer. Influenced by Etienne Condillac, a prominent French scholar of the 18th century, he devoted himself to chemistry but, early in his career, he did also some research in thermodynamics in joint experiments with Laplace. At the age of 25 Lavoisier was elected a member of the French Academy of Sciences. At 26 he obtained a position as a tax collector in the *Ferme Generale*, where he attempted to introduce reforms in the French monetary and taxation system to help the peasants. In 1771, at the age of 28, he married the 13-year-old Marie-Ann Paulze, the daughter of a co-owner of the Ferme. Over time, Marie-Ann became a scientific assistant to her husband, translating English documents for him including the works of Black, Cavendish and Priestley.

Lavoisier did not discover new substances or fundamentally new equipment, but in elaborating on the work of others, mainly Priestley, Black and Cavendish, he made an important contribution to science. He is justly remembered for his articulation of the law of conservation of mass which states that, although matter can change its state in a chemical reaction, the total mass of matter is the same at the end as at the beginning of every chemical change. He demonstrated the above with a series of very careful experiments, concerning different reactions involving a variety of materials; in every case he was able to measure with sufficient accuracy the weights of all substances (reagents and products) involved in the reaction. And, of course, he is the one who demolished the *phlogiston* theory of combustion. After experimenting with Priestley's *dephlogistigated air* (which he named oxygen), Lavoisier demonstrated convincingly that combustion was an oxidation reaction, that the total mass was conserved in the process like in every other chemical reaction, and that therefore no such thing as the *phlogiston* is needed to understand the observed facts. And similarly he showed that a metal turns into calx when it is oxidized, i.e. by taking up oxygen, and not by releasing air (as claimed by the *phlogiston* theory).

Because of his prominent position in the deeply unpopular *Ferme Generale*, Lavoisier was branded a traitor during the Reign of Terror by the Revolutionists in 1794, although he was in fact one of the few liberals in his position. An appeal to save his life so that he could continue with his experiments failed. *The Republic needs neither scientists nor chemists*; *the course of justice can not be delayed*, the judge declared. Lagrange, the great mathematician, lamented the act with the words: *It took them only an instant to cut off his head, but France may not produce another such head in a century.*

One and a half years after his execution, Lavoisier was exonerated by the French government. When his personal belongings were returned to his widow, a note was attached to them which read: *To the widow of Lavoisier who was falsely convicted.*

8.3 John Dalton and the Atomic Theory of Matter

In the period between the publication of the *Sceptical Chymist* of Robert Boyle and the death of Lavoisier in 1794 the concept of the *chemical element* was introduced and clarified to mean: a substance that can not be resolved into other substances by the physical and chemical methods available to the chemist. The definition implied that a substance that appears to be a chemical element can turn out not to be one if and when progress in chemistry can show that it can be resolved into component substances. It was John Dalton, with his work at the beginning of the 19th century, who connected the concept of the chemical element (as defined above) with the 'atom' of Democritus.

John Dalton was born in Cumberland (England) into a Quaker family of weavers. At the age of 15 he joined his older brother Jonathan in running a quaker school in Kendal, and remained there until he moved to Manchester in 1793. In 1794 he was elected a member of the Manchester Literary and Philosophical Society. He learned a lot from John Gough, a blind philosopher and polymath, but he was mainly self educated. He was appointed teacher of mathematics and natural philosophy at the 'New College', a Dissenting academy in Manchester, until 1800, when financial reasons led him to resign his post. Subsequently he earned his living as a teacher and tutor of children of wealthy families in Manchester. Dalton's first publication was *Meteorological Observations and Essays* (in 1793), which contained the seeds of several of his later discoveries, but went unnoticed by the other scholars at the time. A second book of his, *Elements of English Grammar*, was published in 1801. Dalton died in 1844 in Manchester.[5]

Dalton was apparently a clumsy experimentalist, but he was very good at devising mechanical models to explain the observed facts. He believed that a gas consisted of atoms in contact with each other, as in the model (the static one) introduced by Boyle in relation to the gas law that he had discovered in 1660 (see Sect. 5.2.1). In the intervening time, that law was complemented by the discovery of a second law which states that, under constant pressure the change (ΔV) in the volume of a gas is proportional to the change (Δt) of the temperature (t):

$$\Delta V = (constant)\Delta t, \text{ at constant pressure.} \tag{8.1}$$

[5] The modern city of Manchester honours him by naming one of its colleges after him: The John Dalton College of Technology. For more on Dalton, Cavendish and and the other pneumatic chemists of that era, see: R. Uhlig, Genius of Britain, Collins, London, 2010.

This, the second law of gases, known also as Charles' law, resulted from experiments begun about 1787 by Jacques Charles.[6] The law was confirmed (in 1800), in a series of experiments with different gases by the French chemist Joseph Louis Gay-Lussac, and independently by John Dalton himself.

Joseph Louis Gay-Lussac was born at Saint-Leonard-de-Noblat in the department of Haute-Vienne. He became a professor of chemistry at the Ecole Polytechnique in 1802, and from 1808 to 1832 he was professor of physics at the Sorbonne. He died in 1850.

Gay-Lussac showed that:

$$V = V_0(1 + t/273), \quad \text{at constant pressure.} \tag{8.2}$$

In this equation, known also as Gay-Lussac's law, t denotes the temperature of the gas in degrees Celsius (°C), V is the volume of the gas at this temperature, and V_0 its volume at 0 °C.[7]

And then it was found that a similar equation (known as Charles' law of pressures) applies to the pressure of a gas:

$$P = P_0(1 + t/273), \quad \text{at constant volume.} \tag{8.3}$$

where, P denotes the pressure of the gas at temperature t and P_0 its pressure at 0 °C. And finally, John Dalton discovered the law of partial pressures (known as Dalton's law) which states that: if the gas is a mixture of gases (A, B, C, etc.), the (total) pressure (P) is given by:

$$P = P_A + P_B + P_C + \cdots \tag{8.4}$$

where P_A, P_B, P_C, ... are the pressures that A, B, C, etc. would have on their own in the given volume at the same temperature.

Boyle's static model of a gas could not provide a sure basis for a mechanical interpretation of the above laws, but many scientists including Dalton were not yet ready to accept the alternative model: of atoms moving randomly, independently of each other. It is worth noting in this respect that Dalton, knowing that common air consisted of nitrogen and oxygen (a well established fact by his time), was impressed by the fact that nitrogen, which was lighter than oxygen, was, at least at low heights, mixing more or less uniformly with it instead of spreading above it; a fact that he could not easily explain.[8] Had he accepted the alternative model of a

---

[6] Charles became a professor of physics at the Paris Conservatoire des Arts et Metiers, and the first man to make an ascent in a hydrogen balloon.

[7] The Celsius scale of temperature (originally known as the centigrade scale) increases uniformly from 0 °C (the temperature of ice in equilibrium with water at standard pressure) to 100 °C (the temperature of water in equilibrium with steam at standard pressure). It is named after the Swedish astronomer Anders Celsius (1701–1744) who devised the inverted form of this scale (ice point at 100°, steam point at zero degrees) in 1742.

[8] A volume of nitrogen weighs less than the same volume of oxygen under the same conditions of temperature and pressure.

gas consisting of unconnected atoms moving randomly in every direction he would be able to explain the observed fact in terms of diffusion, but he did not. He remained convinced that the atoms of a gas stuck to each other loosely, but connected to each other nevertheless, since the gas occupies all the space in the vessel it is in.

To be specific: Dalton believed that, *each atom consists of a hard core and a soft shell about it*. In a gas the soft shells of the atoms remain in contact with each other, and thereby interact with each other (there is no need for action at a distance). The diameter (including the outer shell) of the atoms that make up a substance, a chemical element (*A*), is *always* different from the diameter of the atoms that make up a different chemical element (*B*). According to Dalton, when *A* interacts chemically with *B* to produce substance *C* what actually happens is: an atom or more of *A* combine with an atom or more of *B* to produce a *composite atom C* (today we call it a molecule *C*). For example: relying on his own (not so accurate) experiments, Dalton concluded that, when nitrogen and oxygen combine to produce monoxide-nitrogen, the volume of oxygen participating in the reaction was about 80 % of the volume of nitrogen. He also determined, from the same experiments, that, under the same pressure and temperature, the oxide occupied about twice the volume of each of its constituents. From the above, and assuming (correctly in this case) that the carbon monoxide molecule consisted of an oxygen atom and a nitrogen atom bound together, he inferred, on the basis of his preferred model of a gas, that the volume of the oxygen atom must be about 80 % of the volume of the nitrogen atom, and that the volume of the monoxide-nitrogen molecule must be (to a good approximation) equal to the sum of the volumes of the constituent atoms. Similarly, observing that the production of (water) vapour from a mixture of hydrogen and oxygen (with the help of an electric spark) required 2 volumes of hydrogen and one volume of oxygen, he concluded, assuming that the water molecule consists of an oxygen and a hydrogen atom bound together, that the hydrogen atom is twice as large as that of oxygen. This is of course wrong. The water molecule consists of one atom of oxygen and two atoms of hydrogen; and the hydrogen atom is smaller than the oxygen atom. It is evident that Dalton's adoption of the static model of the gas led him to all kinds of wrong conclusions; but his idea that all substances are made of molecules which in turn are made of the atoms of a relatively few elementary substances proved correct.

One may summarize Dalton's ideas, contained in his book *A New System of Chemical Philosophy*, published in two parts in 1808 and 1810, as follows:

1. Elementary substances (the chemical elements) consist of identical atoms. These are indivisible unchangeable particles of matter too small to be seen in the microscope. The atoms of element *A* are different from the atoms of element *B*.
2. Atoms can neither be created nor destroyed. In chemical reactions atoms are rearranged but never change. As a result the total mass at the end of a chemical reaction is the same as at the beginning of the reaction, as proved experimentally by Lavoisier some years earlier.

3. Compounds (compound substances) are made of molecules. A pure substance is made of the same molecules. And different substances are made of different molecules. A certain molecule consists of certain atoms bound together. For example: the gas of carbon monoxide consists of carbon monoxide molecules; each carbon monoxide molecule is made of a carbon atom and an oxygen atom bound together. In the case of carbon dioxide, the corresponding molecule is made of a carbon and two oxygen atoms bound together; and similarly for other substances. Dalton assumed that atoms joined to form molecules in the simplest possible ratios: 1:1, as in the carbon monoxide; or 1:2, as in carbon dioxide (the molecules of the two carbon oxides had to be different); or some other simple ratio: 1:3, 2:3, etc., if that proved necessary.

We should note that the molecules of Dalton are nearly as spherical as the atoms: the hard cores of the atoms come very close to each other and are surrounded by the combination of their soft parts which come together to form the soft spherical shell of the molecule. And he introduced certain symbols to denote the atoms and the molecules made of them. For example, the oxygen atom was denoted by an empty circle, the hydrogen atom by a circle with a dot at its centre, the nitrogen atom by a circle with a diametrical bar in it, the carbon atom by a shaded circle, etc. Accordingly, he denoted the monoxide of nitrogen molecule by two touching circles: one empty for the oxygen atom, the other with a diametrical bar in it for the nitrogen atom; and he, similarly, denoted carbon dioxide by three touching circles in a row: a shaded one representing carbon, between two empty ones representing the oxygen atoms. The modern notation for atoms: H for hydrogen, O for oxygen, N for nitrogen, C for carbon, etc., and the corresponding notation for molecules: NO for the monoxide of nitrogen, H_2O for water, CO_2 for carbon dioxide etc. was introduced a few years later (in 1819) by the Swedish chemist Jöns Jacob Berzelius.

Dalton's atomic theory of matter, unlike that of Democritus, was based on experimental evidence. Experimental data available at the time showed that the ratio of the *weights* of the reacting gases was always the same (constant) in a given reaction. For example, the ratio of the weight of hydrogen to that of oxygen, when the two were reacting to produce water, was about 1:8. Dalton initially found this ratio to be about 1:6 which he later revised to 1:7. What is important in this case is not the actual number of this ratio but the fact that it is always the same (and always a simple ratio); which implies that the molecules of a substance are always the same: each molecule consisting of so many atoms A and so many atoms B, their numbers being always the same. It also implied, as Dalton pointed out, that the different atoms (A, B, etc.) involved in the formation of molecules, do so in simple ratios between them, such as 1:1, 1:2, 2:3, etc. And, of course, if one could establish these ratios (i.e. the correct chemical formula of a substance) one could thereby determine the ratios of the corresponding *atomic weights* of A, B, etc.

In 1808 Gay-Lussac, studied once more the explosive reaction of hydrogen with oxygen into (water) vapour, and he came out with the following observation: the ratio of the *volume* of hydrogen to that of oxygen, under the same temperature and pressure, was 2:1, within an accuracy of less than 0.1 %. And after a number of reactions involving other gases, he proposed that a simple ratio of this kind (1:1, 2:1, 3:2, etc.) was valid in every case. And he went further to suggest that, if the result of the reaction was another gas, then its volume had also a simple ratio to those of the reacting gases. For example: 2 volumes of hydrogen react with 1 volume of oxygen to produce (under the same temperature and pressure) 2 volumes of (water) vapour.

The above proposition implies that: A certain number (N) of molecules (or atoms) of substance A, may occupy exactly the same volume as N molecules (or atoms) of a different substance B. Such a proposition was evidently incompatible with Dalton's model of a gas. Dalton assumed that different molecules (like atoms) do have different volumes, but at the same time he believed (in accordance with his adopted model of a gas) that the molecules (or atoms) of a gas were in contact with each other and occupying the whole of the volume (V) available to the gas: $V = NV_M$ where V_M is the volume of the molecule. If V_M changes, so must V. Dalton would not easily accept Gay-Lussac's observations and the conclusions he drew from them, because he would not give up his cherished model of a gas.

8.4 Avogadro' Theory and the Law of Ideal Gases

Lorenzo Romano Amedeo Carlo Bernadette Avogadro, Count of Quaregna e Cereto, was born in 1776 in Turin (Italy) to a noble family of Piedmont. He graduated in ecclesiastical law at the age of 20, but soon after, he dedicated himself to science, and in 1809 became a teacher of mathematics and physics at a high school in Vercelli, where his family had property. In 1811 he published his *Essay on Determining the Relative Masses of the Elementary Molecules of Bodies and the Proportions by Which They Enter These Combinations*, which contains Avogadro's law (see below). In 1820 he became professor of physics at the University of Turin, where he taught until the end of his life (in 1856) except for the years 1823–1833 when he was expelled from the University for political reasons. He was apparently a religious man devoted to his work and his family (he married Felicita Mazzé and they had six children).

Amedeo Avogadro accepted Gay-Lussac's observations and in his 1811 essay he presented a theory to account for them. He proposed that:

1. The gas consists of atoms (or molecules) which occupy a small fraction of the volume of the gas, either because they mutually repel each other, or because they move very fast independently of each other (the question remains open).
2. The simple (common) gases need not be atoms; they could be molecules of these atoms, and there may be a variety of them (for example: HO, H_2O, H_3O, H_4O_2, etc.).

3. In equal volumes, at constant temperature and pressure, there is always the same number of molecules (free atoms, if such exist, count as molecules in this respect) in *any* gas, whether it consists of one kind or it is a mixture of gases. This is presently known as Avogadro's law.

Let me demonstrate Avogadro's law in relation to the chemical reaction of hydrogen with oxygen to produce (water) vapour: Suppose that in *two* volumes of molecular hydrogen (H_2) there are $2N$ molecules H_2 (where N is a certain number); then, according to Avogadro's law, in the *one* volume of oxygen (at the same temperature and pressure) there will be N molecules O_2. When the two gases interact we obtain a total of $2N$ molecules H_2O occupying two volumes of (water) vapour in accordance with Avogadro's law, and as found experimentally by Gay-Lussac. We should note that in our example we made use of the chemical formulae for the molecules of hydrogen, oxygen and water. By the same token, one can use Avogadro's law in conjunction with experimental data to help find the chemical formula of a gas.

The scientific community of the time paid little attention to Avogadro's work, so his theory was not immediately accepted, partly because of remaining ambiguities in relating to certain experimental measurements. It was finally accepted in 1860 (four years after Avogadro's death) when Stanislao Canizzaro showed that Avogadro's law was indeed valid quite generally.

I would like to finish this chapter with a restatement of the gas laws (Eqs. 5.16; 8.2 and 8.3) which takes into account Avogadro's law. For this purpose it is necessary to replace the Celsius scale of temperature (t) by the so-called absolute-temperature scale (T), which relates to it as follows:

$$T = t + T_0 \tag{8.5}$$

where $T_0 = 273.15°$. The absolute-temperature scale was introduced by Lord Kelvin, some years after the work of Avogadro, as the scale most appropriate to thermodynamics, but this need not concern us here (see Sect. 9.1.4). At this stage in the telling of our story we can look at Eq. (8.5) as a transformation of convenience. The thing to note is that a change of temperature by one Celsius degree (°C) is the same as a change of one degree on the absolute-temperature scale, denoted by K, and called degree Kelvin or, simply, kelvin. The one scale is simply obtained from the other by a displacement of 273.15 degrees. Consequently 0 °C is 273.15 K on the absolute-temperature scale; and −273.15 °C = 0 K. And it happens that, the thermodynamic (real) temperature can not go below 0 K.

Using Eq. (8.5) we write Eqs. (8.2) and (8.3) (with 273 replaced by 273.15) as follows: $V = (V_0/T_0)T = aT$ at constant P; and $P = (P_0/T_0)T = bT$ at constant V; and we remember [Eq. (5.16)] that $PV = c$ at constant T. Where a, b and c are constants for a *given* gas. These three equations tell us, respectively, that for a *given* gas: V is proportional to T when P is constant; similarly P is proportional to T when V is constant; and PV is constant when T is constant. One can easily verify that these statements (all three of them) can be put together into a single statement as follows: for a *given* gas PV is proportional to T, i.e. $PV = AT$, where A is a constant for a *given* gas. For, according to this equation, P is proportional to T when V is constant, and V is proportional to T when P is constant; and of course PV is constant when T does not change. So far we have not taken into account Avogadro's law, which states that the number (N) of all molecules in *any* gas of volume V is the same under the same pressure (P) and temperature (T). Which in turn implies that $A = PV/T$ must be a universal (the same for all gases) function of N. And if the law of partial pressures [Eq. (8.4)] holds, and it does, we must put[9]: $A = Nk$, where k is a universal constant. Therefore we write:

$$PV = NkT \tag{8.6}$$

This is known as the law of *ideal* gases. Experimentally it is found to be valid accurately for the noble (inert) gases, and in particular He, at low densities and not so low temperatures (only then Avogadro's assumption, that the number of molecules in the gas occupy a very small fraction of the volume of the gas, is valid).[10] For ordinary gases the law is valid only approximately (see Sect. 9.1.3).

We shall see (in the next chapter) that kT is a measure of the kinetic energy of an atom or molecule in the gas. The numerical value of the constant k, called the Boltzmann constant (sometimes written as k_B), as we know it today, is: $k = 1.380658 \times 10^{-23}$ J/K (or $\text{m}^2\ \text{kg s}^{-2}\ \text{K}^{-1}$). It depends, of course, on the chosen scale of temperature (on our choice of having 100° between the freezing and the boiling point of water). Finally it is worth noting that the number of molecules N in a gas of $V = 1\ \text{cm}^3$, at $T = 300$ K and $P = 1$ atm ($1.013 \times 10^5\ \text{N/m}^2$) is $N \approx 10^{20}$, a very large number indeed.

---

[9] Try anything else and the law of partial pressures will not hold. For example, if we put $P=N^2kT/V$, we obtain, for a mixture of two gases, $(N_1 + N_2)^2kT/V \neq N_1^2kT/V + N_2^2kT/V$, in contradiction to the law of partial pressures.

[10] Because Eq.(8.6) is accurately valid in this case, it can be used to define the absolute temperature as follows: Given PV for a gas at the temperature, T_B, at which water boils (at standard pressure) and at the temperature, T_F, at which it freezes (at standard pressure), we mark these two points on a diagram of PV versus T and draw a straight line through them as shown below.

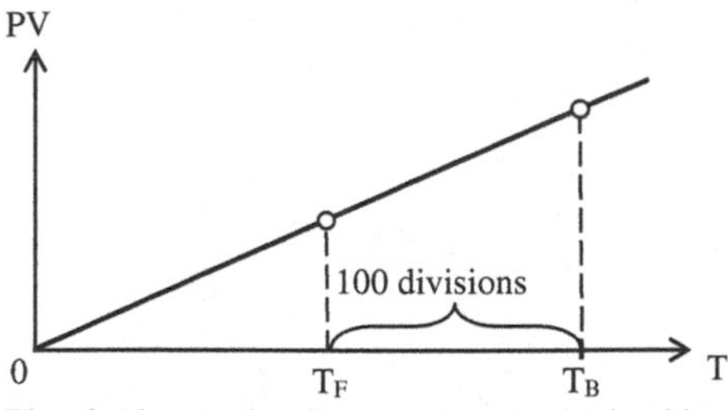

The absolute scale of temperature is obtained by dividing the interval between T_F and T_B into 100 equal steps (degrees Kelvin). The interception of this line with the T-axis defines the absolute zero temperature (T= 0). This way of defining the absolute temperature scale is equivalent to that of Lord Kevlin.

Exercises

8.1. A (gram)mole of a substance has a mass, expressed in grams, equal to its relative molecular mass. The relative molecular mass is the ratio of the mass of a molecule (it could be an atom) of a substance to 1/12 of the mass of a *carbon*-12 atom (about 12 times the mass of a *hydrogen* atom). It follows that the number of molecules in a mole of any substance is the same for all substances. It is known as the *Avogadro constant* or *Avogadro number*, and is denoted by N_A. It has the value:

$$N_A = 6.0221367(36) \times 10^{23}$$

Show that the law of ideal gases can be written as follows:

$$PV = nRT$$

where n is the number of moles in the gas, and $R = N_A k$, where k is Boltzmann's constant.

8.2 One mole of methane (CH_4) reacts with oxygen (O_2) to give carbon dioxide (CO_2) and water (H_2O). Assuming that all of methane has been used in the reaction, how much oxygen was required in the process?

Answer: Two moles of oxygen (O_2).

Chapter 9
Thermodynamics and Statistical Mechanics

9.1 Thermodynamics

9.1.1 Heat

In the present chapter we shall look at the concept of *energy* as it developed from empirical observations to become one of the two basic concepts (the other being *entropy*) of a branch of physics known as *thermodynamics*. The story begins with *heat*.

What is the nature of heat? People were aware, since ancient times, of the fact that heat flows from hot to cold bodies, and they knew that mixing a hot liquid with a cooler one would result in something between the two, and they knew of course that the sun was a source of heat, whatever that was. They came to believe that heat was a kind of fluid, Lavoïsier called it the *caloric*, that penetrated and diffused in all kinds of matter making them hotter in the process. Naturally, when a hot body comes into contact with one not so hot, some of this peculiar fluid flows from the first to the second body until some kind of equilibrium is established between the two. At the beginning they thought heat and the temperature of a body to be the same thing. It was Joseph Black, as we have already pointed out (in Sect. 8.2), that made the distinction between the two. The *caloric* theory of heat was not unconvincing at a qualitative level.

One scientist who did not accept the fluid model of heat was Count Rumford. He was born Benjamin Thompson in North Woburn of Massachusetts in 1753. He was made a count in 1798 by the ruler of Bavaria whom he served as a minister of war among other things. But in spite of his many official duties he found time for science. After a series of experiments, using the most reliable weighing machines of his time he concluded (in 1797) that: if heat is a fluid, it must be one without weight, for heating a body does not alter its weight. And because he thought such a fluid unlikely, he concluded that heat is associated with some kind of motion. However, without a model of this motion that could explain the observed phenomena, and in particular the transfer of heat from hot to cool bodies, the fluid model of heat remained the better model for most scientists.

A. Modinos, *From Aristotle to Schrödinger*,
Undergraduate Lecture Notes in Physics, DOI: 10.1007/978-3-319-00750-2_9,

Rumford pointed out that, there does not seem to be a limit in the amount of heat generated by friction between two bodies, which suggests that heat is not a fluid within the bodies, because there should be a limit to the amount of fluid a body can contain. A similar observation was made by Humphrey Davy (he was only 21-years-old at the time), who pointed out that the heat generated by rubbing two pieces of ice against each other was enough to melt the ice, when otherwise a considerable amount of heat would be required for the melting to occur. This meant, according to Davy, that heat was not a fluid, but something different, associated with vibrations within the body of some kind or other.

However, there was one phenomenon which appeared to have no explanation in terms of vibrations of ordinary matter: the flow of heat from the sun to the earth through empty space. This could more easily be understood in terms of the fluid model, and if that fluid had to be practically weightless, so be it, the supporters of the model would say.

During the first 20 or 30 years of the 19th century, experiments by William Hershel (in England), by Macedonio Melloni (in Italy) and by James Forbes (in Scotland) showed that thermal radiation reflected and refracted and in general behaved in much the same way as light. Therefore many scientists came to believe that heat in general was akin to light, not distinguishing between heat transfer between bodies in contact and radiated heat. At the same time, especially after the work of Thomas Young (in England) and Augustine Fresnel (in France), Newton's weightless particles of light were abandoned, and light was interpreted as vibrations of a weightless ether. Once ether was put in place to carry by its vibrations the heat from the sun, the last obstacle to Rumford's theory was removed, and his concept of heat as vibration, or rather as energy associated with vibration (mechanical motion of matter or the ether) was finally accepted.

During the first decades of the 19th century it also became apparent, by a variety of experiments that, some kind of equivalence existed between heat and other forms of energy. It is likely that the term *energy* was introduced at about 1810 to facilitate the description of this equivalence. The experiments related to electric and magnetic phenomena (see Chap. 10), to chemical reactions (heat is often required to initiate a reaction, and heat is often generated by a reaction), and there was of course the obvious transformation of heat into mechanical work effected by the steam engine, which was developed during the 18th century, although the way this transformation occurred, was not properly understood until about 1850. As to the sensation that heat produced in a human being, that was not more or less mysterious than the sensations produced in the same by light or sound. To this day, their explanation lies beyond the realm of physics.

9.1.2 Energy Conservation and the First Law of Thermodynamics

In 1842 Julius Robert Mayer, who was at the time a young doctor in the German town of Heilbronn, published a treatise where he proclaimed the equivalence of different *causes*. Of course, *cause* is not the right term (from a physicist's point of view) in this instance, but what Mayer meant by it, was what we now call *energy*. So in what follows I shall use this term rather than *cause* in describing Mayer's contribution. In his treatise he claimed that there are different forms of energy, that energy can change from one form into another but it can never disappear. He came to believe this for philosophical reasons, but he went on to argue that experimental results available at his time showed that this was indeed so in reality.

If two seemingly different quantities (as heat and mechanical work obviously are) are in reality different forms of *energy*, then there must be a definite relation between the units by which these quantities are measured. If such a relation can be established, and shown to be valid in all experiments where the one form changes into the other, then (and only then) is the equivalence of the two forms of energy established. Mayer established the equivalent of the unit of heat, the *calorie*, in units of mechanical work as defined by Eq. (7.27). The unit of the latter, according to Eq. (7.27), is: (1 Newton) × (1 m) = 1 Nm, [We remember that 1 N = 1 kg m/s^2. Today the unit 1 Nm is called a *joule* (J)]. On the other hand the *calorie* is, by definition, the amount of heat required to raise the temperature of 1 g of water by 1 °C. [Strictly speaking from 15 to 16 °C, but this quantity does not vary significantly with the temperature.]

To better describe Mayer's results I need to tell you what the specific heat c of a material is. It is the amount of heat required to raise the temperature of 1 g of this material by 1 °C. By definition, $c_{water} = 1$ (cal/°C) g^{-1}. For lead one obtains $c_{lead} = 0.03$ (cal/°C)g^{-1}. The specific heats of most solids and liquids are somewhere between these two values.[1] For gases things are a bit different. The specific heat c_P of a gas, when its temperature is raised under constant pressure (in which case its volume increases), is greater than its specific heat c_V when the temperature of the gas is raised under constant volume.

It was known at the time of Mayer that 1 g of air requires 0.17 cal to increase its temperature by 1 °C under constant volume, but 0.24 cal when it is done under constant pressure of 1 atm (equal to 1.013×10^5 N/m^2). Mayer argued that the heat required to raise the temperature of the gas was exactly the same in both instances, but when that occurred with an accompanying expansion of the gas, additional heat (0.07 cal in the present case) was required to compensate for the work done by the gas during the expansion. His claim, that the amount of heat required to raise the temperature of the gas was the same in both instances, was supported by Gay–Lussac's finding (in 1807) that, no energy is required when a

[1] The large value of c_{water} means that water does not cool easily; this is why it is used in radiators and hot water bottles.

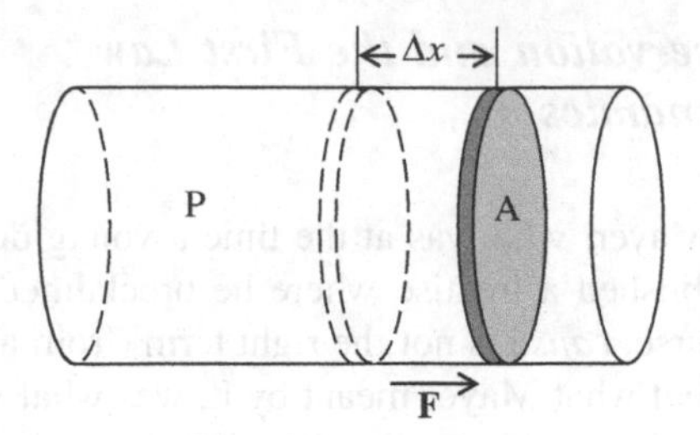

Fig. 9.1 At equilibrium, the pressure on either side of the cap on the right of the cylinder (of area A) is $P = 1$ atm. Therefore the work done by the gas when the cap is displaced by Δx is: $\Delta W = F\,\Delta x$, where $F = P\,A$ is the force that the gas in the cylinder exerts on the cap. Therefore $\Delta W = P\,A\,\Delta x = P\,\Delta V$

gas expands freely (into an empty space). This result was subsequently (in 1845) confirmed by Joule. [We shall later see that the energy of an ideal gas, and therefore that of a dilute gas depends only on its temperature, in full agreement with the mentioned experimental results.] Applying the above to the case under consideration meant that 0.07 cal was transformed into the work ΔW done by the gas in expanding its volume by the observed amount, $\Delta V = 2.83 \times 10^{-6}$ m^3, under constant pressure $P = 1.013 \times 10^5$ N/m^2. The work done is given by (see Fig. 9.1): $\Delta W = P\Delta V = 0.286$ N m). Therefore: 0.07 cal = 0.286 J. Therefore: 1 cal = 4.1 J, approximately (in view of the experimental errors involved in the measurements).[2]

The work of Mayer did not attract much attention in the years that followed its publication, though many scientists were asking questions similar to the ones he dealt with in his treatise, and in spite of the fact that he continued to publish extensions of his original idea concerning the conservation of energy in chemistry and biology. Disappointed by the lack of interest in his work, Mayer had a mental breakdown, but fortunately he recovered from it and towards the end of his life he received the acknowledgement that was due to him for his pioneering ideas. He died in 1878.

The acceptance of Mayer's ideas was made easier by the work of James Prescott Joule. Joule was born in Manchester in 1818, the son of a beer merchant, whose business he inherited. But at the same time he was devoted to science. When 17-years-old he became a student of John Dalton, and a few years later, at the age of 22, he began a series of experiments determined to prove that, whenever mechanical work [as defined by Eq. (7.27)] turns into heat or the other way round, the ratio of the two is always the same number: in each case 1 cal of heat is equivalent to the same number of *joules* (which of course he called *newton-meters*) of mechanical work, which meant that the two are different forms of energy, and that energy is conserved in every transformation. This is of course what Mayer

[2] Today's value is: 1 cal = 4.186 J.

claimed. Joule did know, though not completely Mayer's work, when he presented his first paper in 1843. In that paper he showed that the ratio of mechanical work needed to operate a generator of electricity, to the heat produced by the electric current obtained is equal to about 4.51 J/cal. The difference between this value and that of Mayer (4.1 J/cal) lies easily within the margin of the experimental error of the corresponding measurements. Later in the same year Joule measured the heat generated by water flowing in a narrow tube due to friction between the two, and the work done by the force preserving the flow and found that the ratio between the two amounted to 4.1 J/cal. He went on to perform a variety of other experiments, with increasing accuracy, until it became apparent that 1 cal = 4.20 J, to a very good approximation (see Footnote 2).

When, in 1847, Joule presented his results to a meeting of scientists, the chairman asked Joule to shorten his presentation because, he said, not many were interested in what he had to say. However, things turned out very different from what the chairman expected, because of the curiosity and the intelligent questions and remarks of a young man in the audience. His name was William Thomson, later Lord Kevlin, who was to become one of the renowned physicists of his time.

By 1850 Joule's work was more widely known, and his ideas and those of Rumford and Mayer were gaining ground. Within ten years or so every one agreed that heat was a form of energy associated with the motion of matter or the ether; that energy appears in different forms, and that includes mechanical work; and though energy can change form, the total energy is always conserved. The conservation of energy was expressed mathematically as follows:

$$\Delta Q = \Delta E + \Delta W \tag{9.1}$$

The equation applies to any process and any physical system: ΔQ denotes the amount of heat supplied to the system ($\Delta Q > 0$) or taken away from the system ($\Delta Q < 0$); ΔE is the accompanying change in the (internal) energy E of the system (the difference between the final state energy and the initial state energy of the system) and, of course, ΔE can be positive or negative; and ΔW stands for work done, it is positive when work is done *by* the system (as, for example, when the gas expands in the system of Fig. 9.1) and negative when work is done *on* the system (as when the gas is compressed in the system of Fig. 9.1).

The conservation of energy, expressed by Eq. (9.1), became the first law of thermodynamics.

9.1.3 Thermodynamic Functions, the Equation of State and Phase Diagrams

One may ask at this point: what does E, the energy of the physical system, consist of? Assuming that the system consists of interacting particles, we can think of E as the sum of the kinetic and potential energies of the particles (in the manner of Sect. 7.3).

However, in thermodynamics we are *not* concerned with the details of the internal (microscopic) structure of the system under consideration. What we want to know is E as a function of macroscopic quantities such as the (absolute) temperature T, the pressure P or the volume V, which characterize the macroscopic state of the system at equilibrium (when it does not change with time).[3] And, of course, there are other quantities beside E which we would like to know as functions of the same variables. Let me give you an example. The specific heat c_V of a substance is by definition the heat ΔQ required to raise the temperature of a gram of the substance by 1 K under constant volume. In this case, Eq. (9.1) tells us that: $\Delta Q = \Delta E$ because $\Delta W = P\Delta V = 0$. It follows that:

$$c_V = \left(\frac{\partial E}{\partial T}\right)_V \tag{9.2}$$

where we have assumed that, for a substance of given mass, the energy can be written as a function of V and T. The thing to note is that the above equation expresses c_V in terms of quantities which are properties of the given physical system (while ΔQ, which appears in the definition of c_V, is not a property of the system). In order to express c_P (the specific heat of the substance under constant pressure) in the same way, we introduce a new quantity H, the *enthalpy* of the system, which we define by: $H = E + PV$, so that: if P stays constant, $dH = dE + PdV$, which is equal, according to Eq. (9.1), to the heat supplied to the substance when together with the energy the volume also increases under constant pressure. It follows that:

$$c_P = \left(\frac{\partial H}{\partial T}\right)_P \tag{9.3}$$

where we have assumed (see below) that H can be written as a function of P and T.

There is a number of thermodynamic quantities beside E and H, which are used to describe a physical system at equilibrium, and we shall introduce some of them in what follows. They belong to two different categories. They are either *extensive*, meaning that the quantity is proportional to the mass of the system (V, E, H are examples of such quantities), or *intensive*, meaning that the quantity does *not* depend on the amount of mass in the system (P and T are examples of such quantities).

The *equation of state* of a system is of particular importance. When P, V, and T are used to describe a system in equilibrium (we assume that the mass of the system is constant, unless otherwise explicitly stated), the equation of state is an equation involving these three variables and, therefore, it reduces the independent variables from three to two. In most cases we do not have an analytical expression of the equation of state, but only a graphical one, obtained empirically, which

[3] Sometimes a macroscopic variable (a physical quantity) is called a thermodynamic variable; and similarly a macroscopic state is called a thermodynamic state.

allows us to represent the equilibrium state of the system as a point in a 3-dimensional P-V-T space, so that a continuous variation of the system is represented by a continuous line in this space. I shall give you an example of this in relation to phase diagrams (see below). But to begin with, let me consider the equation of state of an ideal gas, which we know already; it is given by Eq. (8.6). For a gas of given mass (which means that N is constant), knowing the values of any two of the three variables P, V, T, determines the value of every other quantity of the gas. Let us take as the two independent variables, the quantities V and T. Then every other property of the gas (in equilibrium) can be written as a function of V and T. Obviously $P = NkT/V$, but the important thing is that, every other property of the gas can also be written as a function of V and T. For example, we shall see later in this chapter that, the energy E of the gas is given by[4]:

$$E = 3N\,k\,T/2 \tag{9.4}$$

And therefore, according to Eq. (9.2): $c_V = 3Nk/2$, where N is the number of molecules in 1 g of the gas. And for the enthalpy of the gas we obtain:

$$H = E + PV = 3NkT/2 + NkT = 5NkT/2 \tag{9.5}$$

And therefore, according to Eq. (9.3): $c_P = 5Nk/2$. So that: $c_P/c_V = 1.66$. [The difference between this value and the value for air (1.42), used by Mayer in his treatise (of 1842), is mainly due to the fact that air consists mostly of diatomic molecules see footnote 4].

It is evident from the above that the way one chooses the independent variables is a matter of convenience; but we must remember which two we have chosen. It is for this reason that we write: $c_P = (\partial H/\partial T)_P$ and not, simply, $(\partial H/\partial T)$, as we would have done if the independent variables were unchangeable. The sub-index is there to remind us of the other variable. And, finally, there may be situations where it is more convenient to use as independent parameters, other quantities such as the energy E, the enthalpy H, or some other parameter, in place of P, V or T; and there may be systems, when more than two independent variables are required, but this need not concern us here.

We have already mentioned that the equation of state of a real gas at not so low densities will differ from Eq. (8.6). In the 1870s Johannes van der Waals proposed a modification of Eq. (8.6) arguing as follows: The effective volume V_{eff} available to a molecule of the gas,[5] when the presence of the other molecules is taken into account, is smaller than the total volume V of the gas, because colliding molecules

[4] This is true only for a gas of monoatomic molecules (e.g., Ar or He). The energy given by Eq. (9.4) derives from the translational motion of the molecules. For a gas of diatomic molecules at ordinary temperatures there is an additional contribution to the energy, equal to NkT, associated with the rotational motion of the molecules. Therefore the total energy $E = 5NkT/2$. In which case $c_P/c_V \approx 1.4$.

[5] By this time practically every one agreed that a gas consisted of molecules moving randomly in the space available to them.

are forbidden to come very close together by the repulsive nature of the interaction between them at small separation, as shown schematically in Fig. 9.2. Accordingly $V_{eff} = (V - b)$ where b is a constant characteristic of the gas under consideration (it will depend on the diameter of the molecules and the number of them in the gas). At the same time at a larger separation between two molecules an attraction exists between them according to the potential energy curve of Fig. 9.2. As a result of this attraction, molecules are to some degree held back from hitting the walls of the vessel by other molecules present in the vicinity of the walls. The number of such molecules should be proportional to $(N/V)^2$; therefore the pressure P exerted on the walls of the vessel is reduced relative to P_{kin}, (the would-be pressure if the above attraction between molecules did not exist) as follows: $P = P_{kin} - a/V^2$, where a is a constant depending on the gas under consideration. But, on the other hand we have that: $V_{eff} P_{kin} = NkT$; therefore we obtain:

$$(V - b)(P + a/V^2) = NkT \tag{9.6}$$

This is van der Waals' equation of state of a real gas; and it is a remarkably good one (provided the constants a and b which characterize the gas are chosen appropriately).

Analytic expressions of the equations of state of liquids and solids are difficult to obtain, and one is satisfied with graphical representations of the functional relation between P, V and T in these cases. Of particular interest are the so-called phase diagrams of a substance, which tell us the regions of P-V-T space over which a substance is respectively a gas, a liquid or a solid. A schematic phase diagram for a typical substance is shown in Fig. 9.3a. We note that the diagrams of Fig. 9.3 are not drawn to a uniform scale; for example, the region around the critical point in Fig. 9.3a has been magnified relative to other regions to make the picture clearer. The P-T diagram (Fig. 9.3a) tells that at low temperatures (below that of the *triple* point) the substance is a gas for P, T on the right of the line (S-line) from 0 to the

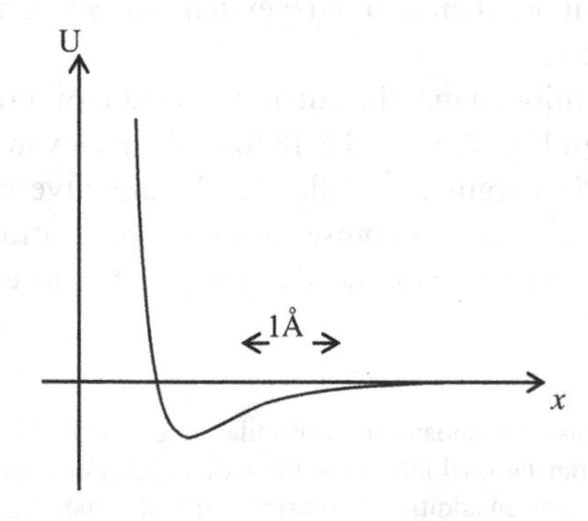

Fig. 9.2 Schematic representation of the interaction between two molecules according to van der Waals. $F = -\, dU/dx > 0$ (repulsive) at small separation (x); and $F = -\, dU/dx < 0$ (attractive) for larger x. Later calculations (using quantum mechanics) confirmed the essential correctness of this interaction. $1\text{Å} = 10^{-8}$ cm

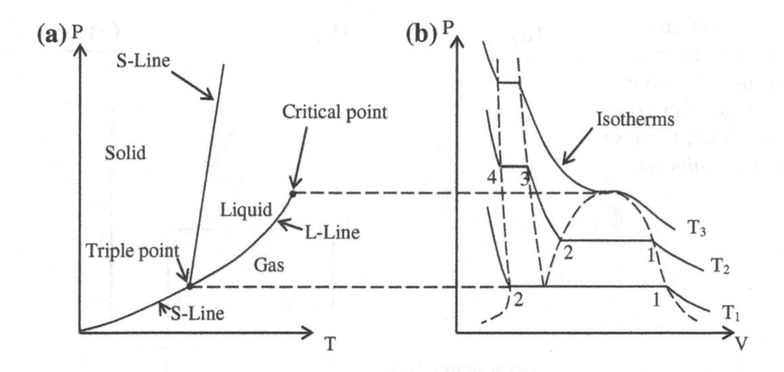

Fig. 9.3 *P-T* and *P-V* diagrams for a typical substance (not to scale)

triple point, a solid on the left of this line, and a mixture of the solid and gas phases for the *P, T* points along this line. This implies that along the *S*-line from 0 to the *triple* point, the given substance transforms from the gas phase to the solid phase, or the other way round, directly (without the intermediary of the liquid phase). For *P* above the triple point and below the critical point, the substance is a liquid for *P,T* points between the *S*-line and the *L*-line, and a gas on the right of the *L*-line. And it is of course a mixture of the two phases along the line that separates these two phases. At the triple point the substance is a mixture of all three phases. Finally, above the critical point the liquid and gas phases merge into a single phase.

The *P-V* diagram of Fig. 9.3b supplements the phase diagram of Fig. 9.3a. It shows what happens during a phase transition. The different curves in this diagram are called *isotherms* because the temperature (T) is the same along a given *P-V* line. We note that $T_1 < T_2 < T_3$. Let us look carefully at the T_2.-isotherm moving from right (large volume) to left (small volume). At the larger volume the substance is a gas occupying a volume V (as shown in Fig. 9.4c): as V gets smaller, P increases; the slope of the isotherm (when drawn to a uniform scale) determines the *isothermal compressibility* κ_T of the gas at the given temperature.[6] When by compressing the gas we reach point 1 on the isotherm (which is a point on the *L*-line of Fig. 9.4a), further reduction of volume does not change the pressure, but turns some of the gas into liquid (as shown in Fig. 9.4b). Since the total mass is conserved, the above tells us that *the (mass) density changes* when the substance changes from the gas to the liquid phase. A transition of this kind is called a first-order transition. As V gets smaller, more gas is turned into liquid, until we reach point 2 on the isotherm: at this point all gas has been transformed into liquid (as

---

[6] The isothermal compressibility of a substance (solid, liquid, or gas) is defined by:

$$\kappa_T = -\frac{1}{V}\left(\frac{\partial V}{\partial P}\right)_T$$

.

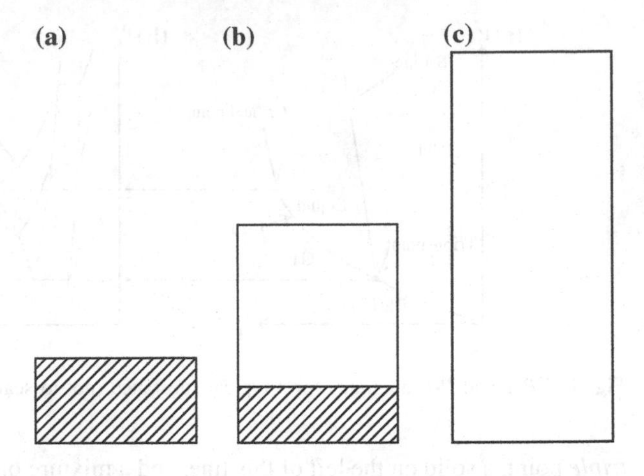

Fig. 9.4 In a first-order phase transition the total volume changes as more material changes from one phase to the other, because the phases have *different* densities

shown in Fig. 9.4a). The part of the isotherm from 2 to 3 tells us how the volume V of the liquid is reduced with increasing pressure at constant temperature ($T = T_2$); the slope of the curve (when drawn to a uniform scale) determines the *isothermal compressibility* κ_T of the liquid at the given temperature (T_2). When point 3 is reached (which is a point on the S-line of Fig. 9.3a) further reduction of volume does not increase the pressure, but turns a part of the liquid into solid, until point 4 is reached. At this point all liquid has been turned into solid. Beyond that point a larger pressure will of course reduce the volume of the solid: the slope of the isotherm now determines the isothermal compressibility of the solid at the given temperature, which is obviously smaller than that of the liquid, which in turn is much smaller than that of the gas.

9.1.4 The Second Law of Thermodynamics: Entropy

Thermodynamic relations apply to systems in equilibrium. However, when a thermodynamic system changes *sufficiently slowly* with time, we can assume that equilibrium is established *in effect instantaneously* at every step of the process. This implies that: the internal (microscopic) processes by which a system returns to equilibrium, when it is put off equilibrium by an (externally induced) change in one of its parameters (for example its volume), are much faster than the rate of change of this parameter. For example, in Fig. 9.1: when the cap is displaced by Δx in time Δt, increasing the volume of the gas by ΔV in the same time, P will adjust to the changing value of V so quickly that P will be given by its equilibrium value at every step in the process of expansion. [This would not be valid if $\Delta x/\Delta t$ was greater, say, than the speed of sound in the gas.] In what follows we assume that the processes we are considering are sufficiently slow in the above sense.

We must next distinguish between *reversible processes* and *irreversible processes*. A process that changes a (thermodynamic) state A of a system to a state

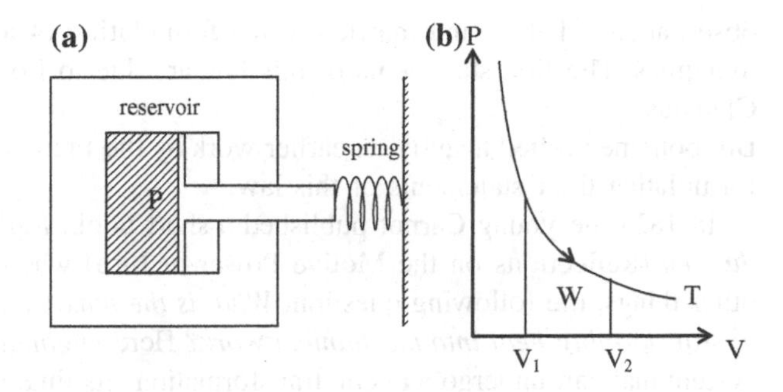

Fig. 9.5 A reversible process

B is reversible, if the system can return from B to A following in reverse order all the steps that led from A to B, in such a way that nothing in the system or its environment (the rest of the world) is different from what it was at the beginning. A process which is not reversible is de facto irreversible. And we note that *sufficiently slow* processes are not necessarily reversible.

Let us, for example, consider the system shown in Fig. 9.5: an ideal gas in a cylinder with a movable cap (piston) connected to an elastic spring. And we assume that no other pressure is exerted on the cap other than the one due to the spring. The system is in thermal equilibrium with a reservoir of temperature T and therefore while it expands, its temperature and its energy E (which for an ideal gas depends only on the temperature) do not change. Therefore the work (W) the gas does in expanding from V_1 *to* V_2, which is equal to the area under the curve between V_1 and V_2 in Fig. 9.5b,[7] is equal to the heat taken from the reservoir during the expansion. And W, in turn, is transformed into potential energy of the elastic spring which is compressed by the expanding gas. Now the process we have described *is* reversible, because when the spring relaxes, it pushes the cap backward, doing work ($-W$) *on* the gas. And because the contraction is done isothermally this work is retuned as heat to the reservoir; and therefore the system and the reservoir return to their respective initial states.

We next disconnect the spring and allow the gas to expand freely from from V_1 *to* V_2. In this case the gas does no work and therefore does not receive or give heat to the reservoir. The latter does not change during the expansion. But the gas can *not* return to its initial state (compressed to V_1) unless work is done on it by the rest of the world (which will accordingly change). Therefore free expansion is an irreversible process. We expected that much. Everyone knows that if we release an amount of gas at a corner of a vessel, the gas will quickly occupy the whole of the vessel. But the opposite (the gas crowding into one corner of the vessel) never happens.

And, of course, everyone knows that heat will by itself flow from a hot body to a cooler one, but never will by itself flow from a cool to a hot body. It is

---

[7] In the present case $P = NkT/V$ and with T being constant, the corresponding integral is easily evaluated (see Table 5.1, Chap. 5): $W = \int_{V_1}^{V_2} PdV = \int_{V_1}^{V_2} \frac{NkT}{V} dV = NkT\,log(V_2/V_1)$.

observations of this kind that led to the formulation of a second law of thermodynamics. The first statements of this law are due to Lord Kevlin and to Rudolf Clausius.
But both men relied heavily on earlier work by the French engineer Sadi Carnot in formulating their statements of this law.

In 1824 the young Carnot published a short book: *Réflexions sur la Puissance du Feu* (Reflections on the Motive Power of Fire) where he considered, among other things, the following question: *What is the maximum efficiency of an engine in transforming heat into mechanical work?* Here *engine* means a thermodynamic system that can undergo a cyclic transformation (its final state is exactly the same as its initial state) and in the process does the following things and nothing else:

1. It absorbs an amount of heat Q_2 from a reservoir of temperature T_2.
2. It gives an amount of heat $Q_1 > 0$ to a reservoir of temperature $T_1 < T_2$.
3. It does an amount of work $W > 0$.

The efficiency of this engine is defined by: $\eta = W/Q_2$. And because in a cyclic transformation $\Delta E = 0$, we have $W = Q_2 - Q_1$ and therefore:

$$\eta = 1 - Q_1/Q_2 \tag{9.7}$$

Carnot proved the following theorem: No engine operating between two given temperatures is more efficient than the one defined as follows (now we call such an engine a Carnot engine): It consists of any system (substance) that goes through the *reversible* cyclic transformation described in Fig. 9.6.

Lord Kelvin realized that an absolute scale of temperature can be established on the basis of Carnot's theorem,[8] if one accepts that such exists, i.e. that there is an absolute zero of temperature (a well defined lowest limit of temperature, even if it can not be reached in practice). When the temperature is measuerd on any absolute scale, the efficiency of a Carnot engine will be given by:

$$\eta = 1 - T_1/T_2 \tag{9.8}$$

In other words $Q_1/Q_2 = T_1/T_2$ for a Carnot engine. Therefore if we fix (arbitrarily) the value of T_2, a Carnot engine can be run between this fixed temperature and any unknown temperature T_1, whose value we can determine by measuring Q_1 and Q_2. We should say, however, that in practice it is very difficult to construct a Carnot engine and to measure accurately Q_1 and Q_2; and that methods relying on the law of gases to fix the absolute scale of temperature are preferred (see Footnote 10 of Chap. 8). Needless to say, that the absolute scales obtained by the two methods are equivalent.

On the basis of Carnot's theorem, and noting, in relation to Eq. (9.8), that in practice T_1 can never be exactly zero, Kelvin asserted that: *There exists no*

[8] It is worth noting that, though we used the symbol T to denote the temperature in our definition of the Carnot engine, such is not necessary and was not the case in Carnot's statement of his theorem.

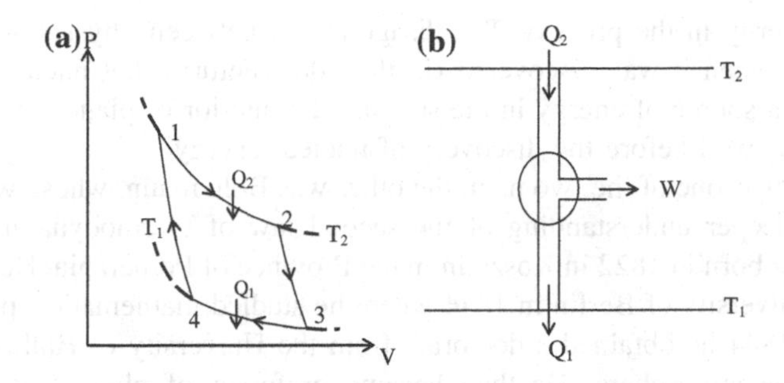

Fig. 9.6 **a** The Carnot cycle: From *1* to *2* the process is isothermal ($T = T_2$) and during it, the system does work which is equal to the heat the substance of the engine (e.g. vapour) absorbs from a reservoir of temperature T_2. [We note that the work done *by* the engine from *1* to *2* is given by $\int_1^2 PdV$ which is equal to the area between the isothermal curve and the *V*-axis between *1* and *2*. From *2* to *3* the process is adiabatic (meaning that the system does not receive or give heat between these two points). The work done by the engine between *2* and *3* (equal to the area between the adiabatic curve and the *V*-axis between *2* and *3*) is done at the expense of the internal energy of the substance of the engine. From *3* to *4* the process is isothermal and during it the engine rejects heat to a reservoir of temperature $T_1 < T_2$, equal to the work done *on* the engine (equal to the *negative* of the area between the isotherm and the *V*-axis between *3* and *4*). Finally from *4* to *1* the process is adiabatic, during which work is done *on* the system (equal to the *negative* of the area between the adiabatic curve and the *V*-axis between *4* and *1*) increasing its internal energy to the value it had at the beginning of the cycle. **b** Schematic description of the Carnot engine

thermodynamic transformation whose sole effect is to extract a quantity of heat from a given heat reservoir and to convert it entirely into work. This is Kelvin's statement of the second law of thermodynamics.

Lord Kelvin was born William Thomson in Belfast in 1824. He became a professor of natural philosophy at Glasgow University in 1846. He did important experimental work in electromagnetism, contributing to the development of telegraphy, beside his work relating to the second law of thermodynamics and the absolute scale of temperature based on it. He was, among other things, very much interested in the geological problem concerning the cooling of the earth. Following Fourier (the one who invented Fourier series) who also studied heat conduction in earth, Kelvin estimated that it took the earth between a 100 and 200 million years to reach its present state, assuming that initially it had a temperature of 4,000–6,000 °C. His estimate was far too short compared with the independent estimates of geologists and evolutionists (Darwin's *Origin of Species* was published in 1869) for whom a much longer period was a prerequisite for evolution. And he also underestimated the time it would take for the earth to cool below temperatures that could sustain life, and the time it would take the sun to extinguish itself. It is worth noting that in Kelvin's calculations the major source of energy (to compensate for the heat lost by radiation to the cool universe) was gravity, as a result of which great masses within the sun and the planets contracted releasing energy in the process. The disagreement between physics and geology disappeared when it was discovered (in the 20th century) that nuclear reactions was the main source of energy in the sun and the interior of planets. Lord Kevlin died in 1907, well before the discovery of nuclear energy.

Clausius was one of the two men, the other was Boltzmann, whose work, made possible a deeper understanding of the second law of thermodynamics. Rudolf Clausius was born in 1822 in Koszalin in the Province of Pomerania. He graduated from the University of Berlin in 1844 where he studied mathematics, physics and history. In 1844 he obtained a doctorate from the University of Halle on optical effects in the atmosphere. He then became professor of physics at the Royal Artillery and Engineering School in Berlin, and in 1855 he became professor at the ETH Zürich, the Swiss Federal Institute of Technology in Zürich, where he stayed until 1867, when he moved to Würzburg and, two years later to Bonn. During the Franco–Prussian war (in 1870) he organized an ambulance service, was wounded in battle and was left with a lasting disability. After his wife died in childbirth in 1875, he raised their six children, and that left him little time for research. In 1886 he remarried and had another child. He died in Bonn in 1888.

Clausius's work relating to the second law of thermodynamics began with his own statement of the law as follows: *There exists no thermodynamic transformation whose sole effect is to extract heat from a colder reservoir and deliver it to a hotter one.*

Clausius's statement of the second law is equivalent to that of Kelvin. One can prove that if one of them is false, then the other is also false. Let us assume that Kelvin is wrong. Then we can extract an amount of heat Q from a reservoir of temperature T_1 and convert *all* of it to work ($W = Q$), which we can then deliver as heat into a reservoir of temperature $T_2 > T_1$ in *contradiction* to Clausius's statement of the law. And by similar arguments one is convinced that if Clausius is wrong, then Kelvin is also wrong.

Clausius major breakthrough relies on the following theorem, which he was able to prove using Carnot's theorem.

Clausius's theorem: Consider a *cyclic* transformation consisting of N infinitesimal steps ($i = 1, 2, 3, \ldots N$). At each step the system receives an amount of heat Q_i from a reservoir at temperature T_i. The following is true:

$$\sum_{i=1}^{N} \left(\frac{Q_i}{T_i}\right) \leq 0 \text{ or } \oint \frac{dQ}{dT} \leq 0. \tag{9.9}$$

where the equality holds only for a reversible cycle. And we understand that the integral is only a convenient expression for the sum: for N sufficiently large (which we assume to be the case) the two are identical. The *circle* on the integral sign reminds us that we are dealing with a cyclic transformation.

The following is an important corollary of the above theorem: *for a reversible transformation, the integral* $\int_A^B \frac{dQ}{T}$ [understood as a sum of infinitesimal steps as in Eq. (9.9)] is independent of the path and depends only on the initial state (A) and the final state (B) of the system undergoing the transformation. This allows us to define a new thermodynamic function, the *entropy* S of the system under consideration, by choosing arbitrarily a *fixed* initial state (to be denoted by 0), and evaluating the integral below along any *reversible* transformation between 0 and A (we assume that there exists at least on such transformation):

$$S(A) = \int_0^A \frac{dQ}{T} \tag{9.10}$$

According to the above definition, the entropy of the system (in state A) is known only up to an arbitrary additive constant.[9] However the difference ΔS between two states A and B of the system is completely determined (a reversible transformation between A and B is assumed):

$$\Delta S = S(B) - S(A) = \int_A^B \frac{dQ}{T} \tag{9.11}$$

We note the following properties of entropy (which follow from Clausius's theorem):

1. For an *arbitrary* transformation

$$\int_A^B \frac{dQ}{T} \leq S(B) - S(A) \tag{9.12}$$

where the equality holds only for reversible transformations.

2. The entropy of a thermally isolated system *never* decreases.

This follows from Eq. (9.12). A thermally isolated system can not receive or give heat to its environment, therefore $dQ = 0$, and therefore $S(B) - S(A) \geq 0$.

The above two properties can be taken as a statement of the second law of thermodynamics.

One or two examples will clarify the physics underlying the above formulae. Consider the transfer of heat from a reservoir of temperature T_2 to a reservoir of temperature $T_1 < T_2$. The transfer occurs by conduction along a metallic bar which connects the two reservoirs. The total system (the two reservoirs and the metallic bar) is insulated from the rest of the world and is therefore thermally isolated. In a unit of time the T_2-reservoir rejects an amount of heat ΔQ, therefore its entropy changes by $\Delta S_2 = -\Delta Q/T_2$; the conducting bar allows the heat to go through it

[9] We simplify the description if we postulate a *third law* of thermodynamics (suggested by Walther Nernst in 1906) which states that: The entropy of a system at the absolute zero of temperature is a universal constant, which we can put equal to zero.

(it does not absorb or generate heat so its entropy does not change); the T_1-reservoir absorbs ΔQ, therefore its entropy changes by $\Delta S_1 = \Delta Q/T_1$; therefore the change in entropy of the total system is $\Delta S = \Delta Q/T_1 - \Delta Q/T_2 > 0$, which is what we expect, for the process under consideration is an irreversible one (heat will never flow across the bar from the cooler T_1-reservoir to the hotter T_2-reservoir.

For a second example, let us look again at the reversible expansion of a gas described in Fig. 9.5. The expanding gas (in the cylinder), the heat reservoir and the (frictionless) spring taken together constitute the total system which is practically a thermally isolated system; its entropy (the sum of the entropies of its constituent parts) can never decrease. A frictionless spring does not receive or give heat, so its entropy does not change; therefore we need only consider the changes in the entropy of the gas and of the reservoir associated with the expansion of the gas (and the compression of the spring). The expansion is isothermal (at temperature T), therefore the energy E of the (ideal) gas does not change during the expansion, therefore the heat ΔQ absorbed by the gas is equal to the work W done by the expanding gas on the spring. We have: (see footnote 7) $W = NkT\, log(V_2/V_1)$. Therefore the associated change in the entropy of the gas is given by $(\Delta S)_{gas} = \Delta Q/T = W/T = Nklog(V_2/V_1)$. [We note that this depends only on the final and initial states of the gas in accordance with Eq. (9.11).] The heat ΔQ is given to the gas by the reservoir at temperature T, so the change in the entropy of the latter is $(\Delta S)_{reservoir} = -\ \Delta Q/T$; therefore the change in the entropy of the total system: $(\Delta S)_{gas} + (\Delta S)_{reservoir} = 0$ during the expansion, as expected for a reversible transformation. It is also worth noting that the transformation of ΔQ into W is not the *sole* effect of the transformation (the volume of the gas changed as well) so Kelvin's statement of the second law is not violated. By similar arguments we can show that the change in entropy of the total system associated with the reverse process, by which the potential energy of the compressed spring returns as heat to the reservoir, is also zero.

Let us next look at the free expansion of the gas (we disconnect the expanding gas from the spring) at temperature T. In this case the gas does no work in expanding from V_1 to V_2, and receives no heat from the reservoir, therefore $(\Delta S)_{reservoir} = 0$ in free expansion; but the entropy of the gas is a property of the state of the system, and this has changed in the same way as before: $(\Delta S)_{gas} = Nklog(V_2/V_1)$. And therefore the change in the entropy of the total system: $(\Delta S)_{gas} + (\Delta S)_{reservoir} > 0$ during the expansion, as expected for an irreversible transformation. One may observe that by allowing the gas to expand freely we lost potentially useful energy. This is true in general: an increase in the entropy S of a system of energy E reduces the amount of energy that can be transformed into work. This is how the concept of the *free energy* (F) arose. This is a thermodynamic quantity of a system (at equilibrium) defined by:

$$F = E - TS \tag{9.13}$$

It is often called the *Helmholtz free energy,* in honour of the German physicist Hermann von Helmholtz who introduced the concept and who contributed significantly to the development of thermodynamics. The physical significance of F can be seen from the following: According to Eq. (9.1) we have: $\Delta W = \Delta Q - \Delta E$, but Eq. (9.12) tells us that for an arbitrary transformation $\Delta Q \leq T\Delta S$, and therefore for an arbitrary transformation: $\Delta W \leq T\Delta S - \Delta E$. And we remember that the equality holds for a reversible transformation. On the other hand, from Eq. (9.13) we obtain: $\Delta F = \Delta E - S\Delta T - T\Delta S$, which for an isothermal transformation ($\Delta T = 0$) becomes $\Delta F = \Delta E - T\Delta S$. Therefore for an isothermal (but otherwise arbitrary transformation): $\Delta W \leq -\Delta F$. Which tells us that: the maximum work obtainable from a system under constant temperature is equal to the *reduction* in its free energy ($-\Delta F = F_{initial} - F_{final}$). The maximum work is, of course, obtained for a reversible transformation.

Another very useful thermodynamic quantity is the Gibbs' energy (or Gibbs' potential) defined by:

$$G = F + PV \tag{9.14}$$

It has the following property: in a transformation where both the temperature T and the pressure P stay constant,

$$\Delta G \leq 0 \tag{9.14a}$$

And the equality holds for a reversible transformation. An interesting application of the above formula relates to the phase diagram of Fig. 9.3a. We note to begin with that V, E, S, F, and G are extensive quantities (proportional to the mass m of the system (substance) under consideration), whereas P and T are intensive quantities independent of m. This implies that, when P and T are used as independent variables, G takes the form:

$$G = m\mu(P, T) \tag{9.14b}$$

The quantity $\mu(P,T)$, a function of P and T but not of m, is known as the chemical potential of the substance. We next observe that when a small quantity of mass Δm transforms from the liquid to the gas phase or the other way round, at constant P,T (on the L-line separating the two phases in Fig. 9.3a), it does so reversibly; which means, according to Eq. (9.14a), that the potential G for the total system (liquid and gas) does not change: $\Delta G = \Delta G_{liquid} + \Delta G_{gas} = 0$; which in turn implies that:

$$\mu_{gas}(P, T) = \mu_{liquid}(P, T) \tag{9.15}$$

along the $P-T$ line separating the two phases. The above equation, demonstrated schematically in Fig. 9.7, defines in fact the P–T line.

We observe that the slope of the curve in Fig. 9.7 changes abruptly (discontinuously) at the transition point:

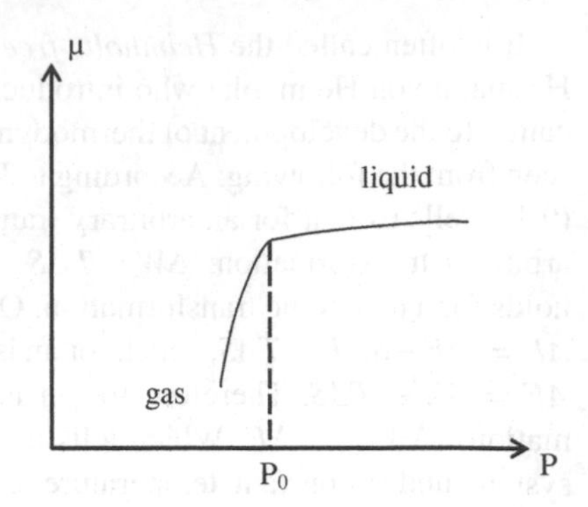

Fig. 9.7 Variation of $\mu_{gas}(P,T)$ and $\mu_{liquid}(P,T)$ with P at constant temperature, $T = T_0$ The phase transition occurs at (P_0, T_0)

$$\left(\frac{\partial \mu_{gas}}{\partial P}\right)_T > \left(\frac{\partial \mu_{liquid}}{\partial P}\right)_T \tag{9.16}$$

Now, it can be shown that $(\partial\mu/\partial P)_T = \upsilon$, where υ denotes the volume per unit mass of the substance. Therefore what the above inequality tells us is that the volume per unit mass is greater in the gas phase than in the liquid phase, which is of course what we see in reality (see Fig. 9.4).

After Clausius, a variety of formulae were derived, by a number of scientists, on the basis of the two laws of thermodynamics. These formulae establish useful relations between observable quantities and facilitate greatly our description of natural phenomena.

A final note concerning the second law of thermodynamics: The following question is often asked: Does the law apply to the Universe as a whole? The answer is not straightforward. The second law of thermodynamics is an empirical law, and as we shall see in the next section it relates to the microscopic motion of the particles that make up a macroscopic system. As such it does not necessarily apply to the entire universe. It is also worth noting that the second law defines a *direction in time*, that of increasing entropy. This is certainly not the case in Newton's mechanics, or in quantum mechanics. In both mechanics, the following is true: if a motion leading from A to B is possible, then the reverse process is (equally) possible, whatever the force involved (gravitational, electromagnetic or nuclear).[10]

One may then ask: does not the second law of thermodynamics invalidate the reversibility of the laws of mechanics? The answer is definitely *not*. The motions of the individual particles that make up a substance (say a gas) are certainly reversible; what is not likely to happen is that an enormous number of particles will by themselves reverse their velocities at the same moment of time.

[10] We note, in this respect, that friction, though often represented by a sigle force is *not* one; friction involves a number of random collisions between the surface atoms of two bodies. Each collision by itself is a reversible process but the entire process is not.

9.2 Statistical Mechanics

9.2.1 The Kinetic Theory of Gases

By the middle of the 19th century the kinetic theory of gases was accepted by most scientists. In his paper of 1857: *In relation to the nature of the motion which we call heat,* Clausius summarizes his own view as follows. He agrees with other scientists (John Herapath, James Joule, and August Krönig who had published a short paper on the subject in 1856) that a gas consists of molecules moving randomly in the space available to them colliding with each other and with the walls of the vessel containing the gas. He attributes (like Krönig) the pressure that the gas exerts on the walls of the vessel to these collisions. And he asserts (for the first time) that the molecules of the gas are likely to also vibrate (the way a system of two masses joined by an elastic spring vibrates) and rotate (the way two balls often rotate after a collision) and that, therefore, the total energy of the gas consists in part of vibrational and rotational energy.

A quantitative theory relating the translational motion of the molecules of a gas to the pressure it exerts on the walls of its container is easily established within the context of the kinetic theory. It goes like this: Let there be N molecules (each of mass m) in a cubic vessel of volume $V = L^3$. We assume the gas to be at equilibrium at temperature T. When a molecule with a speed υ_x along the x-direction is reflected by the wall normal to this direction, the x-component of its momentum changes by an amount $\Delta p_x = 2\, m\upsilon_x$. In time Δt the molecule will hit the wall $\Delta t/\tau$ times, where $\tau = 2L/\upsilon_x$ is the time it takes the molecule to complete a journey from one wall to its opposite one and back again. So in time Δt the molecule suffers a change of momentum equal to $(\Delta p_x)_{total} = (2\, m\upsilon_x)\Delta t/\tau = (m\upsilon_x^2/L)\Delta t$. This implies that the force exerted by the wall on the molecule is $F = (\Delta p_x)_{total}/\Delta t = m\upsilon_x^2/L$. And by Newton's third law a force of the same magnitude is exerted by the molecule on the wall. If there are N molecules in the vessel, the combined force on the wall is $F = (\mathrm{N/L})m\overline{v_x{}^2}$, where the bar indicates that the quantity under it is an average over the N molecules of the gas. And therefore the pressure on the wall (whose area equals L^2) is given by: $P = F/L^2 = (N/V)m\overline{v_x{}^2}$, since $L^3 = V$. And because $\overline{\upsilon_x^2} = \overline{\upsilon_y^2} = \overline{\upsilon_z^2} = (\overline{\upsilon_x^2} + \overline{\upsilon_y^2} + \overline{\upsilon_z^2})/3 = \overline{\upsilon^2}/3$, we can write:

$$(3/2)PV = N\,(m\overline{\upsilon^2}/2) \tag{9.17}$$

We note that the quantity on the right of Eq. (9.17) represents the kinetic energy of the gas associated with the *translational* motion of all the molecules in the gas. We note, in passing, that according to the above equation: $\overline{\upsilon^2} = 3P/\rho$, where $\rho = Nm/V$ is the (mass) density of the gas. For air (at $P = 1$ atm and at a temperature of 20 °C), $\rho = 1.29$ kg/m^3; this gives $\upsilon_{r.m.s} = \sqrt{\overline{\upsilon^2}} = 500$ m/s which is about two times the speed of sound (320 m/s). [$\upsilon_{r.m.s} = \sqrt{\overline{\upsilon^2}}$ is called the *root mean square* (*r.m.s*) of the velocity.] Finally: if we *assume* that the kinetic theory of gases is valid, then Eq. (8.6) and Eq. (9.17) taken together, tell us that the kinetic energy associated with the translational motion of all the molecules in the gas is given by:

$$E = N(m\overline{v^2}/2) = (3/2)NkT \tag{9.18}$$

A result we stated in advance in Eq. (9.4). It follows from the above, that the (translational) energy of a molecule is on average equal to $3(kT/2)$, i.e. $kT/2$ for each of the three component motions (along the x, y and z directions) of the molecule. For a present-day physicist this is an undisputed fact, but that was not so at the time of Clausius. For it is true that, what we presented so far is an argument in favour of the kinetic theory (if it were valid it would explain very nicely how the pressure the gas exerts on the walls of its container comes about) but it does not prove that this is what nature does. It would certainly be easier to prove its reality experimentally (by some method or other) if one knew not just the average velocity of the molecules, but the way these velocities are distributed.

9.2.2 The Maxwell–Boltzmann Distribution

It is clear that the molecules of a gas (at equilibrium at temperature T) can not all have the same speed; for even when two molecules have the same speed before they collide, they are likely to have different speeds after the collision. And after many such collisions it is likely that there will be some considerable spread in the distribution of their speeds.

Thanks to the work of Laplace, Gauss, and Poisson and other mathematicians, the physicists of the 19th century were familiar with statistical concepts. These could be applied in the analysis of scientific data, but also to other data relating to economics and social phenomena. We note in this respect that Laplace's book, *Philosophical Essays on Probabilities*, was published in 1814. The most commonly used statistical distribution at the time of Clausius had to do with the way the results of measurements of a given quantity (under the same conditions) are distributed about a mean value due to unavoidable random errors, or with the way some property, such as the 'height' of an individual, is distributed about a mean height in a given country at a given time. In these cases the data are usually distributed symmetrically about a mean value, in accordance with the bell-shaped curve shown in Fig. 9.8. It is known as the Gaussian distribution, and corresponds to the following mathematical function:

$$P(x) = A\,exp\left\{-\left(\frac{x-\bar{x}}{\delta}\right)^2\right\} \tag{9.19}$$

where $\bar{x}$ is the *mean* value of the measured quantity; the parameter δ determines the spread of the distribution: a larger value of δ means that the measured values of

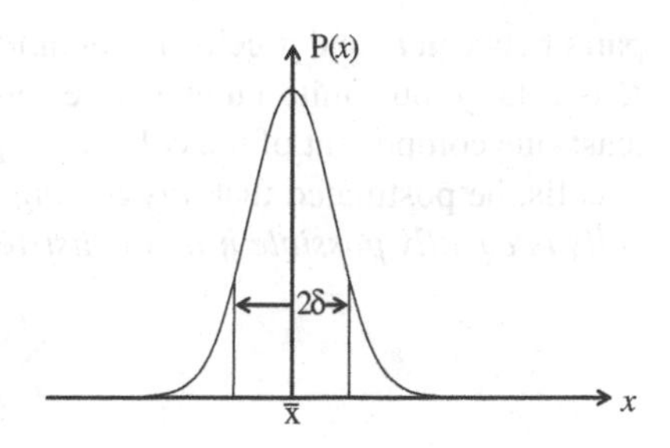

Fig. 9.8 The Gaussian distribution (schematic)

x spread more widely about their mean value; and A is a normalization constant, fixed by the requirement that[11]:

$$\int_{-\infty}^{\infty} P(x)dx = 1 \tag{9.20}$$

We can say that $P(x)dx$ (the area below the Gaussian curve between x and $x + dx$) is the *fraction* of a large number of measurements resulting in a value of the measured quantity between x and $x + dx$. Sometimes we say this differently: We say that, $P(x)dx$ is the *probability* that a measurement of the given quantity will give a value between x and $x + dx$. If, for example, the said fraction is 0.1, we say that the probability of finding a value between x and $x + dx$ is 10 %, by which we understand that we expect this to happen once in ten measurements, or ten times in a hundred measurements.

James Clerk Maxwell suggested (in 1860) that the velocities of the molecules of a gas at equilibrium would be distributed in some way about a mean velocity, but not symmetrically as in the Gaussian distribution, because there is obviously a lower limit (zero) to the speed of a molecule but apparently no upper limit to it. And if there was an upper limit, it would be so much larger than the mean velocity of the molecules, so that an asymmetric distribution would be obtained. Maxwell went on to propose an appropriate distribution for the velocities of the molecules of a gas as follows. He divided the space (**r**-space) within the volume V of the gas into a large (but finite) number of space-cells (**r**-cells) of small (but not infinitesimal) volume $\Delta V = \Delta x \Delta y \Delta z$; and, similarly, he divided momentum space (**p**-space) into a large (but finite) number of small (but not infinitesimal) momentum cells (**p**-cells) of 'volume' $\Delta p = \Delta p_x \Delta p_y \Delta p_z$; to be consistent with his assumption of a finite number of **p**-cells he assumed that the translational energy of a molecule ($p^2/(2m)$) never exceeds a large (but finite value). He combined the **r**-cells and the **p**-cells into **r**-**p**-cells, of 'volume' $\Delta V \Delta p$, consisting of all possible

[11] The limits of the integration are to be understood as follows: we consider the integral $I(\alpha) = \int_{-\alpha}^{+\alpha} P(x)dx$. By choosing α sufficiently large, we obtain: $|I(\alpha) - 1| < \varepsilon$, however small ε is. In other words: I is the limit (in the Weierstrass sense explained in Sect. 1.1) of $I(\alpha)$ as $\alpha \to \infty$.

pairs between **r**- and **p**-cells. He enumerated the **r**-**p**-cells: $i = 1, 2, 3, \ldots, K$. Where K is a large but finite number (we remember that two **r**-**p**-cells are different if at least one component of the cells, **r**- or **p**, is different. With **r**-**p**-space separated into K cells, he postulated that *any arrangement of the N molecules of the gas in* ***r - p*** *cells is equally possible if it is consistent with the given macroscopic conditions*[12]:

$$\sum_{i=1}^{K} n_i = N \tag{9.21a}$$

$$\sum_{i=1}^{K} n_i \varepsilon_i = E \tag{9.21b}$$

where n_i is the number of molecules in the ith cell; $\varepsilon_i = p_i^2/(2m) = m\,\upsilon_i^2/2$ is the (translational) energy of a molecule in the ith cell; N is the number of molecules in the gas and E the energy of the gas (the sum of the translational energies of the N molecules). It is worth remembering that K is much larger than N and therefore many **r**-**p**-cells remain empty, and more so when a number of molecules find themselves in the same cell. We must also remember that a given distribution $\{n_i\}$ is obtained by a number of different arrangements of the molecules. For example, in a given distribution, moving molecule '1' from the ith cell to the jth cell and at the same time molecule '34' from the jth cell to the ith cell produces a different arrangement of the molecules but does not change the distribution $\{n_i\}$, which is determined by *how many* molecules are in a particular cell and *not by which these molecules are*. Since any arrangement of molecules that satisfies Eqs. (9.21a, 9.21b) is equally possible, it is reasonable to claim (and this is what Maxwell did) that the probability of a distribution $\{n_i\}$ being actualized is proportional to the number of different arrangements of the molecules compatible with $\{n_i\}$. We call this number the *statistical weight* of $\{n_i\}$, and denote it by $\Omega\{n_i\}$. A formula for $\Omega\{n_i\}$ can be established as follows: Imagine we know the N 'positions' the N atoms occupy; some cells will have a number of occupied positions (we enumerate them all as if they were different), and most cells will have no such positions at all. We calculate the number of different ways we can arrange the molecules on these positions as follows: the first of the molecules can be placed in any of the N positions (i.e. in N different ways), the second in any of the remaining $(N - 1)$ positions (i.e. in $N - 1$ different ways), the third in any of the remaining $(N - 2)$ positions (i.e. in $N - 2$ different ways), and so on. So the number of different ways the N molecules can be arranged on N positions is: $N \times (N - 1) \times (N - 2) \times (N - 3) \times \ldots \times 3 \times 2 \times 1 = N!$ (according to the definition of $N!$ [see Eq. (5.9)]. In deriving the above we have considered the different positions in the *same* cell as different, but they are not.

[12] The molecules in the gas are of the same type (e.g., oxygen molecules) but we assume that we can distinguish between one oxygen molecule and another as if each one had a number attached to it. We note also that if the gas consisted of N_1 molecules of one kind and N_2 molecules of another kind the distribution we shall obtain applies to them separately. This corresponds to the *law of partial pressures* stated in Sect. 8.3.

We can easily take this into account by noting that the number of different sequences of n objects is $n!$. We finally obtain the following formula for $\Omega\{n_i\}$:

$$\Omega\{n_i\} = \frac{N!}{n_1!\, n_2! \ldots n_K!} \tag{9.22}$$

And we remember that for an empty cell, $n = 0$ and $0! = 1$.

Maxwell went on to determine the distribution which maximizes $\Omega\{n_i\}$, subject to the conditions imposed by Eqs. (9.21a, 9.21b). We need not go into the details of his mathematical derivation. He found that this is given by

$$n_i = C\, e^{-\beta\varepsilon_i} \tag{9.23}$$

where, C and β are constants to be determined from Eq. (9.21a, 9.21b). We remember that the ith cell is a small 'volume' $\Delta V \Delta p = \Delta x \Delta y \Delta z\ \Delta p_x \Delta p_y \Delta p_z$ about a point $\boldsymbol{r}$ in ordinary space and a point $\boldsymbol{p}$ in momentum space, and that the sums in Eqs. (9.21a, 9.21b) extend over all $\boldsymbol{r}$-space and all $\boldsymbol{p}$-space. Their evaluation involves turning them into appropriate integrals, which again is a technical thing that need not concern us here. At the end of this procedure one obtains [in place of Eq. (9.23)] the following distribution:

$$f(r, \upsilon) d^3 r\, d^3 \upsilon = n \left(\frac{\beta m}{2\pi}\right)^{3/2} e^{-\frac{\beta m \upsilon^2}{2}} d^3 r\, d^3 \upsilon \tag{9.24}$$

$$3/(2\beta) = \mathrm{E}/N \tag{9.24a}$$

We note that according to Eq. (9.18): $E/N = 3kT/2$. Therefore:

$$\beta = 1/kT \tag{9.24b}$$

$f(r,\upsilon)d^3rd^3\upsilon$ is the number of molecules in $\Delta V \Delta p = m\ d^3 r d^3 \upsilon$. We put $p = m\upsilon$, where m is the mass of the molecule and υ its velocity; and we replaced $\Delta V \Delta p$ by $md^3rd^3\upsilon = mdxdydzd\upsilon_x d\upsilon_y d\upsilon_z$. [Because ΔV is so small compared to the volume of the gas, and Δp is so small compared to the variation of $f(\boldsymbol{r},\boldsymbol{p})$ with $\boldsymbol{p}$, it is permissible to treat them as infinitesimals when turning the sums into integrals.] And $n = N/V$ is the number of molecules per unit volume and, as we expected, is independent of $\boldsymbol{r}$ (the molecules are distributed uniformly in the volume V).[13] Therefore the above distribution is reduced to the following velocity distribution:

$$f(\upsilon) d^3 \upsilon = n \left(\frac{\beta m}{2\pi}\right)^{3/2} e^{-\frac{\beta m \upsilon^2}{2}} d^3 \upsilon \tag{9.25}$$

It is worth noting that what has been shown so far is that the distribution of Eq. (9.25) is the one (among all possible distributions) with the maximum statistical

[13] That would not be the case if ε_i depended on $\boldsymbol{r}$, as it happens when the gas exists in a potential field $U(\boldsymbol{r})$ which varies significantly over the region occupied by the gas. Disregarding the potential energy of the molecules due to the gravitational field ($U = mgz$) in the derivation of Eq. (9.25) implies that $mgL \ll kT$, so that $exp(-mgz/kT) \approx 1$, for $0 < z < L = V^{1/3}$.

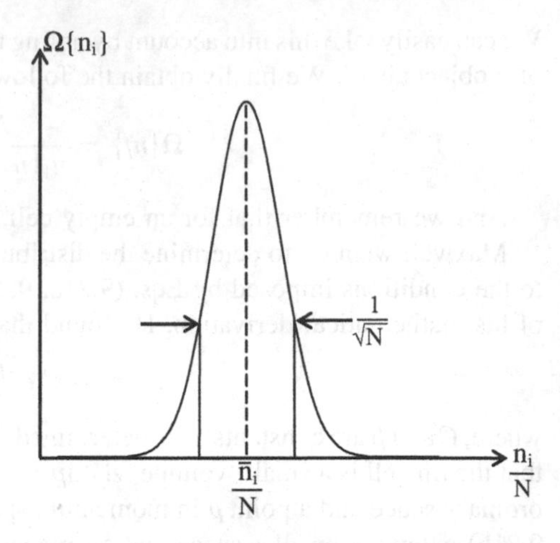

Fig. 9.9 Statistical weight of $\{n_i\}$. The most probable distribution $\{\bar{n}_i\}$ corresponds to the Maxwell distribution [Eq. (9.25)]

weight. Before we accept it as a realistic representation of the actual distribution, we must be convinced that the statistical weight of all other distributions taken together is negligibly small compared to the statistical weight $\Omega\{f\}$. It turns out that this is indeed the case if the number N of molecules in the gas is large. This is shown schematically in Fig. 9.9. Nevertheless, it is worth remembering that, however small their statistical weight, these other distributions do exist and in some instances are manifested as fluctuations about the (observed) mean distibution (see also Sect. 9.2.5). Finally we note that $f(\upsilon)$ depends *only* on the magnitude υ of the velocity, which allows us to write the number of molecules with speed between υ and $\upsilon + d\upsilon$, per unit volume, as follows:

$$f(\upsilon)4\pi\upsilon^2 d\upsilon = n\left(\frac{\beta m}{2\pi}\right)^{3/2} e^{-\frac{\beta m\upsilon^2}{2}} 4\pi\upsilon^2 d\upsilon \qquad (9.26)$$

In Fig. 9.10 we show the percentage of Nitrogen molecules in intervals of $\Delta\upsilon = 10$ m/s about selected values of υ, at 0 °C. In the same figure we show the velocity $\upsilon_{max} = \sqrt{(2kT/m)}$ at which the distribution $f(\upsilon)4\pi\upsilon^2$ acquires its maximum value, and $\upsilon_{r.m.s} = \sqrt{\overline{\upsilon^2}}$, where $\overline{\upsilon^2}$ is the average value of υ^2 given by:

$$\overline{\upsilon^2} = \frac{1}{n}\int_0^\infty \upsilon^2 f(\upsilon)4\pi\upsilon^2 d\upsilon = 3kT/m$$

We have: $\overline{\upsilon^2} = \overline{\upsilon_x^2} + \overline{\upsilon_y^2} + \overline{\upsilon_z^2}$, and obviously $\overline{\upsilon_x^2} = \overline{\upsilon_y^2} = \overline{\upsilon_z^2}$, therefore:

$$\frac{m\overline{\upsilon_x{}^2}}{2} = \frac{m\overline{\upsilon_y{}^2}}{2} = \frac{m\overline{\upsilon_z{}^2}}{2} = \frac{kT}{2} \qquad (9.27)$$

We can say that the motions along the x, y, and z directions are independent motions, each contributing $kT/2$ to the mean kinetic energy associated with the

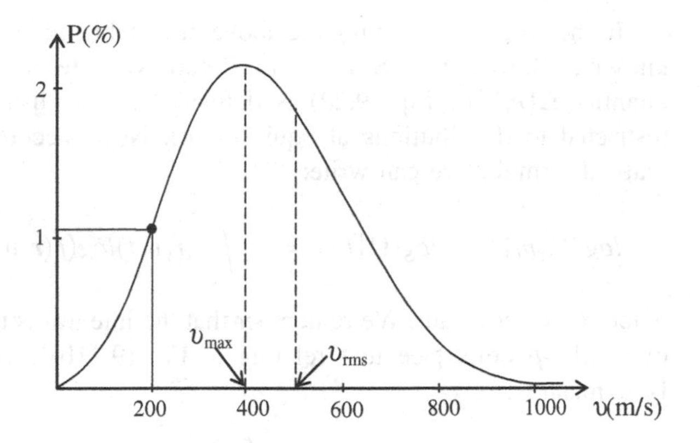

Fig. 9.10 Percentage of Nitrogen molecules in intervals of $\Delta\upsilon = 10$ m/s about selected values of υ, at 0 °C. (From G. Holton and S.G. Brush, *Introduction to Concepts and Theories in Physical Science*, 2nd Edition, Princeton University Press, 1985)

translational motion of the molecule. The above result is in agreement with the equipartition theorem to be discussed in Sect. 9.2.4.

In 1872 Ludwig Boltzmann developed an equation that now bears his name which solves the following problem. He assumed that at time $t = 0$, the distribution of the molecules of a gas is *not* the equilibrium one, given by $f(\upsilon)$ in the manner we have described, but an arbitrary distribution $f_A(\boldsymbol{r},\upsilon)$. And he asked: how will that distribution change with time?

It was known (see Fig. 9.2) that when two molecules collide (come close to each other) they interact, and after the collision their velocities do change. Boltzmann constructed an equation which determines how the distribution function $f(\boldsymbol{r},\upsilon,t)$ changes as a result of these collisions. The equation takes into account the nature of the intermolecular interaction (Boltzmann's assumptions in relation to this interaction were sufficiently general) and also how often such collisions happen.[14] Boltzmann's equation is a *nonlinear integro-differenial* equation involving the partial derivatives of $f(\boldsymbol{r},\upsilon,t)$ with respect to the time t and with respect to x,y,z and $\upsilon_x, \upsilon_y, \upsilon_z$ and also integrations over the υ-space. It is a monster equation and we shall not write it down here. Boltzmann showed that the distribution $f(\upsilon)$ of Eq. (9.25) satisfies his equation when $\partial f/\partial t = 0$, *i.e.* at equilibrium, whatever the specific nature of the intermolecular interaction. And he also established that whatever the initial distribution $f(\boldsymbol{r},\upsilon,t = 0) = f_A(r,\upsilon)$ is, finally (as $t \to \infty$)[15]: $f(\boldsymbol{r},\upsilon,t) \to f(\upsilon)$ of Eq. (9.25).

[14] A useful concept in this respect is the *mean free path λ of a molecule*. It is the average distance a molecule travels before it collides with another molecule. The average duration τ of a free path is called the collision time: $\tau = \lambda/\sqrt{(2kT/m)}$. For example: for a H_2 gas at its critical point (see Fig. 9.3a) $\lambda \approx 10^{-7}$ cm and $\tau \approx 10^{-11}$ s.

[15] In practice, e.g., for a H_2 gas under normal conditions, any non uniformity in density or pressure or temperature, over distances of the order of 10^{-7} cm will be smoothed out in about 10^{-11} s. Variations over macroscopic distances persist over longer times. This is how, for example, the propagation of sound in air becomes possible. The establishment of local equilibrium is a prerequisite of the macroscopic variation which constitutes the sound wave.

In the process of proving the above, he established an important theorem, now known as Boltzmann's *H*-Theorem. Before we state the theorem, we note that the quantity $\Omega\{n_i\}$ of Eq. (9.22) is defined for any distribution $f(\mathbf{r},\upsilon,t)$ (it is not restricted to distributions at equilibrium). Now, according to a known mathematical formula, we can write:

$$log\,(\Omega\{n_i\}) = log(\Omega\{f\}) = -\int f(\mathbf{r},\upsilon,t)log(f(\mathbf{r},\upsilon,t))d^3rd^3\upsilon + C \qquad (9.28)$$

where *C* is a constant. We remember that the integral is to be understood as a sum over all ***r-p*** cells [see text relating to Eq. (9.21b)]. The quantity called *H* by Boltzmann is:

$$H\{f(\mathbf{r},\upsilon,t)\} = \int f(\mathbf{r},\upsilon,t)log(f(\mathbf{r},\upsilon,t))d^3rd^3\upsilon \qquad (9.29)$$

And we note that *H,* like $\Omega\{f\}$, is defined for any distribution (non-equilibrium ones as well). Boltzmann's theorem states that: *if the collisions between the molecules of the gas happen at random, then*

$$\frac{dH}{dt} \leq 0 \qquad (9.30)$$

And the equality holds only for $f(\upsilon)$ of Eq. (9.25), which from now on we shall call the *Maxwell– Boltzmann* distribution function. Boltzmann also showed that the *entropy S* of the gas, as defined by Clausius Eq. (9.10), is given within an additive constant by:

$$S = klog(\Omega\{f\}) \qquad (9.31)$$

In words: the entropy of the gas is proportional to the logarithm of the number of microstates of the gas (the different arrangements of the molecules of the gas in ***r-p*** space) compatible with the macroscopic state of the gas. And by an extension of this definition to non equilibrium states in accordance with Eqs. (9.28)–(9.29), we obtain: $S = -kH\{f(r,\upsilon,t)\}$; which implies, according to Eq. (9.30), that at each step on the way to equilibrium the entropy of the gas increases. This is the nearest one ever got to a 'derivation' of the second law of thermodynamics.

Finally, as we shall see in Sect. 9.2.3, Eq. (9.31) can be generalized so as to apply not only to a gas, but to *any* physical system, and as such it provides a definition of entropy which clarifies its physical meaning in a most satisfactory way.

Ludwig Eduard Boltzmann was born in 1844 in Vienna, the first son a taxation officer Ludwi Gottfried and Maria Pauernfeind, the daughter of a wealthy merchant. He had a private primary education before joining a gymnasium at Linz (where his family moved because of his father's work). He was the best student in his class, enthusiastic about mathematics and science, but interested in other subjects as well. He also took piano lessons from the composer Anton Bruckner, and was throughout his life a keen pianist. His father died of tuberculosis, when Boltzmann was only 15, and this loss affected Boltzmann throughout his life. After gaining a Ph.D. in physics in 1866 from the University of Vienna, he became an assistant professor there, and in 1869 he took up the chair of mathematics at the University of Graz. This is when he began to study the connection between microscopic and macroscopic physics, which culminated in his great, 100-page paper containing among other things the integro-differential transport equation that now bears his name. His paper, published in the *Proceedings of the Imperial Academy of Sciences of Vienna* in 1872 was not widely read at the time, partly because it was too long and difficult to read. Even the great Maxwell (about whom I shall write in the next chapter) opined about it (in a letter to his colleague Peter Tait): *By the study of Boltzmann I have been unable to understand him. He could not understand me on account of my shortness, and his length was and is an equal stumbling-block to me. Hence I am very much inclined to join the glorious supplanters and to put the whole business in about six lines.* In 1873 Boltzmann accepted a chair in Vienna, where he remained for three years. In 1876 he returned to the University of Graz and stayed there for 14 years, which were the happiest of his life. He married Henriette von Aigentler and they had two sons and two daughters. In 1877 he wrote another long paper on the relation between microphysics and macrophysics, which contained among other things his famous equation relating *entropy* to the density of microstates, Eq. (9.31).

In the late 1880s Boltzmann began to develop psychological problems, partly at least, due to the tragic loss of his son at the age of 11 from an appendicitis. In 1890 he went to the University of Munich, four years later he returned to Vienna; in 1900 he went to Leipsig, and in 1902 he returned to Vienna. In his late years, Boltzmann became more interested in matters of philosophy, relating mostly to science but also more generally. His views attracted the attention of members of the Vienna circle and related thinkers such as Wittgenstein. Considered a champion of materialism, his views were quoted with approval by a number of people including Lenin. Boltzmann's views appear to be similar to those held by most scientists today. Towards the end of his life his sight deteriorated dramatically and his periods of depression worsened. On 5 September 1906 he committed suicide by hanging himself.

Boltzmann's understanding of the pursuit of knowledge is summarized in the following extract from Schiller which he quoted in one of his essays: *Effort that never tires, that always creates but never destroys, that to the building of eternity brings but grains of sand and yet from the heavy debt of time sweeps minutes, days and years.*

Experimental verification of the Maxwell–Boltzmann distribution did not come quickly. It eventually happened in the 1920s (by this time the atomic theory of matter had been firmly establishrd), when Otto Stern and his co-workers managed to measure the velocity distribution of atoms evaporated from a hot wire (of silver at temperatures a few hundred degrees below the melting point).[16] In the Stern apparatus the hot wire is along the axis of a cylinder with a slit-diaphragm which can open momentarily (like the diaphragm of a camera) to allow a bunch of emitted atoms to pass through it on their way to a screen on the inside surface of an outer cylinder which rotates with constant speed about its axis (the hot wire). Now, if the screen was not moving the atoms would arrive at the same spot on the screen, get adsorbed there forming a linear image of the slit. But because the screen is moving the slower atoms hit it at a place behind that of the faster ones and therefore one obtains a continuous tape-like layer of adsorbed atoms. A strip between x and $x + \Delta x$ on the screen corresponds to speeds of the atoms (which can be calculated from the known geometry and the known speed of the screen) between υ and $\upsilon + \Delta\upsilon$; and the density of the adsorbed atoms between x and Δx determines the distribution of their velocities.

9.2.3 Gibbs' Statistical Physics

The last important step in the development of statistical mechanics was made by Josiah Willard Gibbs. He was born in 1839 in New Heaven (Connecticut, U.S.A.). His father was a professor of sacred literature at the Yale Divinity School. He graduated from Yale in 1858 with a degree in mathematics and Latin. In 1863 he was awarded the first Ph.D. degree in engineering in the U.S.A. from the Sheffield Scientific School at Yale for a thesis *On the Form of the Teeth of Wheels in Spur Gearing*. In 1866 he went to Europe to study, and stayed there for three years: one in Paris, the other two in Berlin and Heidelberg, where he was influenced by Gustav Kirchhoff and Hermann von Helmholtz. In 1871 he became Professor of Mathematical Physics at Yale (the first such professorship in the U.S.A.), a post he held to the end of his life. His monograph *On the Equilibrium of Heterogeneous Substances,* based largely on his own work in the 1870s, is considered by many as one of the foundations of chemical thermodynamics and physical chemistry and one of the greatest scientific achievements of the 19th century. In the 1880s he wrote on vector analysis and on optics. After 1889 he worked on statistical mechanics laying the foundation not only of this field but also of the statistical interpretation of the still to come quantum mechanics. In 1901 Gibbs was awarded

[16] Otto Stern was born in 1888 in Sorau (in Silesia, Germany). He obtained his Degree in Physical Chemistry from the University of Breslau in 1912. He worked at the University of Frankfurt am Main, and Rostock, where he became a professor of Physical Chemistry in 1923. In 1933 he moved to the United States, being appointed Research Professor of Physics at the Carnegie Institute of Technology in Pittsburgh. Apart from the work mentioned here, he did important work, in cooperation with Gerlach, on the deflection of atoms by the action of magnetic fields on the magnetic moments of the atoms. He also demonstrated the wave nature of atoms by observing interference effects in rays of hydrogen and helium atoms. And he also measured the magnetic moments of sub-atomic particles including the proton. Otto Stern was awarded the Nobel Prize for Physics for the year 1943. He died in 1969.

the Copley Medal of the Royal Society of London, the highest honour for a scientist possible at his time. He died in 1903.

Let us imagine, following Gibbs, an ensemble (a collection) of N physical systems (N being a very large number) all of which are macroscopically identical with a given system: of the same mass m, the same volume V and the same other parameters (such as the value of an externally applied electric or magnetic field) if such are required to specify the macroscopic state of the given system. And we assume that the N members of the ensemble have the same temperature: we assume them to be in the same heat bath of temperature T. And then we denote by $i = 1,2,3,\ldots,K$ the micro-states of the given system. By a micro-state we understand a complete description of the motion of all the particles that make up the given system according to the laws of mechanics and the nature of the forces operating between these particles as well as the externally applied forces on the system. In the present case the index i stands for a set of parameters that specify the state (the positions and velocities) of all particles in the system [in the way that i describes the state of one molecule in Eqs. (9.21a, b)].[17] Let the energy of the system in state i be E_i. It consists of the kinetic energies of the particles in the system (in state i), of the potential energy of their mutual interactions and the potential energy of the particles in the external field if such exists. Then the question arises: in equilibrium, what *fraction* of the N systems in the ensemble are in state i, if the total energy (the sum of the energies of the N systems), $E_{ensemble} = N\overline{E}$, is a given constant. Here $\overline{E}$ denotes the average energy per system, which is eventually determined by the temperature. The reader will recognize that the problem we have to solve is very much the same with the one solved by Maxwell in relation to the velocity distribution of N *molecules* when the sum of their energies is constant (determined eventually by the temperature). Therefore, proceeding in the same manner, we obtain [by analogy to Eq. (9.23)] that: the fraction $f(i)$ of the N systems of the Gibbs' ensemble in state i is given by:

$$f(i) = Ce^{-E_i/kT} \tag{9.32}$$

where $C = \left(\sum f(i)\right)^{-1}$ is a normalization constant so that $\sum_i f(i) = 1$. Replacing β (of Eq. (9.23) with $1/kT$ in Eq. (9.32) is obviously the only way to connect the imaginary assembly of the N systems with reality. It is through collisions with the molecules of the heat bath (it could be a large quantity of some fluid at temperature T) that the equilibrium distribution is established. Now, in practice we do not have N systems to play with, but it is reasonable to assume that within a reasonable time interval: one or two orders of magnitude larger than the time $\tau \approx 10^{-11}$ s between the above mentioned collisions, an individual member of the ensemble (and therefore the actual system under consideration) will be in state i for a fraction of the time given by Eq. (9.32).

---

[17] From now onwards we shall, when no ambiguity arises, refer to the micro-states of the system simply as the states of the system.

The value of any physical quantity A associated with the given system is in principle known (can be calculated) if the state i the system is in, is known. We denote this value by A_i. Therefore the measured ensemble-averaged value $\overline{A}$ of this quantity is given by:

$$\bar{A} = \sum_i f(i)A_i \tag{9.33}$$

where $f(i)$ is given by Eq. (9.32).[18]

In practice there is a large number of states i of a macroscopic system with energy between E and $E + \Delta E$, and it is not practical to enumerate them one by one, as implied in the above equation. We talk instead about the density of states $\rho(E)$ of a given system which is defined as follows: $\rho(E)\Delta E$ is the number of states of a given system between E and $E + \Delta E$. Let me demonstrate the usefulness of this concept in relation to the average energy $\overline{E}$ of a system (of given mass and given volume) in equilibrium at temperature T. Using Eq. (9.33) together with Eq. (9.32), and turning the sums into integrals, we obtain:

$$\bar{E} = \frac{1}{C}\int E\rho(E)e^{-E/kT}dE \tag{9.34}$$

where $C = \int \rho(E)e^{-\frac{E}{kT}}dE$. We note that for an ordinary system, consisting of a *large* number of molecules, the fluctuations of the energy about its mean value are inversely proportional to the number of molecules in the system and are, therefore, negligible; therefore we refer to the mean energy $\overline{E}$ of the system as the energy of the system and denote it simply by E. It is of course a function of the temperature.

Finally, identifying the entropy S of the given physical system with the logarithm of the number of (micro) states of the system underlying its macroscopic state, we can write (by analogy to Eq. (9.31)):

$$S = klog(\rho(E)\Delta E) = klog(\rho(E)) \tag{9.35}$$

---

[18] Gibbs derived the same result in a somewhat different manner by considering, to begin with, a microcanonical ensemble: all members of which have the same mass, the same volume V and the *same energy* (between E and $E + \Delta E$). Assuming that all microstates i of the given system occur with equal probability $f(i) = 1$ if E_i lies between E and $E + \Delta E$, and $f(i) = 0$ otherwise, one can evaluate the average value $\overline{A}$ of any thermodynamic quantity A, in the manner of Eq. (9.33), as follows: $\overline{A} = \frac{1}{C}\sum_i f(i)A_i$, where $C = \sum_i f(i)$. One obtains the same average value $\overline{A}$ as the one obtained with the ensemble we have considered in the text, known as the canonical ensemble. This is because the fluctuations about the average value of the energy, obtained in the canonical ensemble (where T rather than E is the common property of the members of the ensemble), are very small (negligible for ordinary systems of large mass). The canonical ensemble and its properties derive from those of the microcanonical ensemble by straight forward arguments which we need not present here. Finally, it is worth noting that Boltzmann considered such ensembles (in relation to substances other than gases) in a paper he wrote in 1884 but he did not elaborate on them in the way that Gibbs did some years later.

where we put $\Delta E = 1$; the unit of energy being a minute quantity compared to the energy E of the system. If we could calculate E and S in terms of the microstructure of the system we would then be able, using established thermodynamic relations, to evaluate many other macroscopic properties of the system. This is very rarely possible in practice, but it is often possible to construct *approximate* models that help us explain the observed phenomena.

9.2.4 The Equipartition Theorem

The theorem applies to physical systems whose energy can be written as a sum of terms of the form: $A\xi^2$, where A is a constant and ξ is a continuous variable. The case of a free particle is a good example: the translational energy of the particle

$$\frac{mv_x^2}{2} + \frac{mv_y^2}{2} + \frac{mv_z^2}{2} \tag{9.36}$$

is the sum of three $A\xi^2$ terms ($-\infty < \xi < \infty$), with $A = m/2$ and $\xi = v_x, v_y, v_z$ respectively. Each one of these terms contributes to the equilibrium energy $\overline{E}$ of the particle (at temperature T) an amount given, according to Eq. (9.33), by

$$\overline{A\xi^2} = \frac{\int_{-\infty}^{+\infty} A\xi^2 \exp(-A\xi^2/kT)d\xi}{\int_{-\infty}^{+\infty} \exp(-A\xi^2/kT)d\xi} = \frac{kT}{2} \tag{9.37}$$

It follows from the above that the (average) translational energy of a free particle at temperature T equals $3kT/2$, as we expect from Eq. (9.27).

The equipartition theorem applies as well to a gas of linear harmonic oscillators. We remember that the energy of an oscillator is

$$E_v = \frac{m}{2}v^2 + \frac{k}{2}x^2 \tag{9.38}$$

where the velocity v and the position x may have any value between $-\infty$ and $+\infty$. The two terms in Eq. (9.38) have the form: $A\xi^2$, and therefore the (average) vibrational energy of an oscillator at temperature T equals $2(kT/2) = kT$.

Equation (9.37) applies also to the (kinetic) energy associated with the rotation of a molecule about an axis, which is given [see Eq. (7.9)] by:

$$E_{rot} = \frac{I}{2}\omega^2 \tag{9.39}$$

where I is the moment of inertia of the molecule about the given axis and ω the angular velocity about this axis, which again may take any value. Therefore, in a

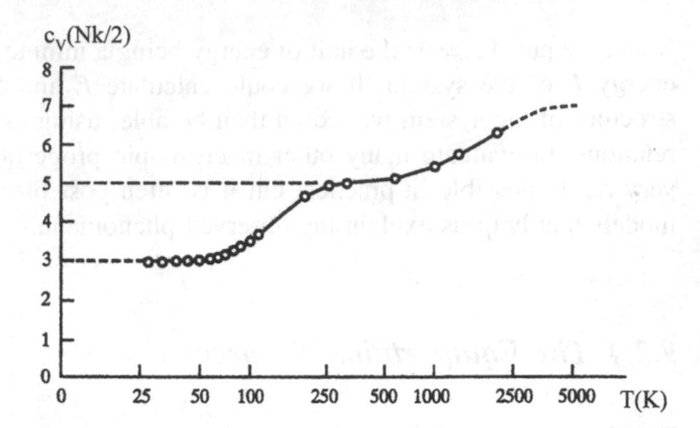

Fig. 9.11 Variation of the specific heat (in units of $Nk/2$) of a hydrogen gas with the temperature T. Note that the scale for the temperature is a logarithmic one. (From G. Holton and S.G. Brush, *loc.cit*)

gas of molecules at temperature T, the (average) energy associated with an independent rotation of a molecule equals $kT/2$.

Let us apply the above to a gas of N diatomic molecules, say hydrogen molecules, at temperature T. We note that in the case of a diatomic molecule, the moment of inertia about the axis of the molecule (the line joining the two atoms) is practically zero; therefore we have only two independent rotations, about the two axes normal to the axis of the molecule. Therefore the rotational contribution to the energy $\overline{E}$ of the gas is: $2(kT/2)N$. To this we must add the vibrational contribution: $2(kT/2)N$, and the translational energy: $3(kT/2)N$. Therefore we expect the (total) energy of the gas to be: $\overline{E} = 7(kT/2)N$. Accordingly we expect the specific heat c_V [defined by Eq. (9.2)] of the gas to be given by $c_V = (7/2)kN$. It turns out that this is indeed the case at sufficiently high temperatures but not so at low temperatures (see Fig. 9.11). It appears that at low temperatures the rotational and the vibrational mechanisms are frozen and that they are activated only at higher temperatures. The explanation of why this is so could not be obtained in classical physics; the quantum mechanical explanation of it will be given in Sect. 14.6.3.

9.2.5 Fluctuations

We noted (in Sects. 9.2.2 and 9.2.3) that at thermal equilibrium the fluctuations about the mean values of thermodynamic quantities are, for large systems containing many molecules, very small, and for most purposes they can be neglected. But they are there and in some instances are manifested in observable phenomena. Let me give you two examples.

The first one relates to the colour of the sky. The number $\overline{\Delta N}$ of molecules in volume ΔV within a much larger volume V of a gas at thermal equilibrium is practically constant throughout the gas, and any deviations from $\overline{\Delta N}$ will be negligible when ΔV is large enough. Now consider the passage of light through matter. The critical volume ΔV in this case is given by $\Delta V = \lambda^3$, where λ is the wavelength of light, which for visible light is (see Sect. 10.4.3) about 10^{-6} m. ΔV is therefore about a thousand times larger than the molecular dimensions and large enough to be treated as macroscopic. Now if the fluctuations about the mean value $\overline{\Delta N}$ of the number of molecules in ΔV were negligible, as it is the case when light passes through glass or some other material, light would be refracted by it but *not* scattered by it. But as it is, the number of molecules in $\Delta V \approx \lambda^3$ of the gas is relatively small and the fluctuations in the density of the gas not entirely without effect. They are in fact responsible for the scattering of sunlight as it passes through the gases of the atmosphere, and this is the reason the sky appears not black but blue.

A second example of the effect of fluctuations is provided by the *Brownian motion*. When small particles (with dimensions of about 10^{-6} m) are introduced into a liquid, they move about in an irregular fashion, as shown in Fig. 9.12. This irregular motion bears the name of Robert Brown, a Scottish botanist who first observed the phenomenon in 1827. Brown studied, using an ordinary microscope, the behaviour of pollen grains in water. The pollen grains were particles about 1/5,000 of an inch obtained from various plants. *These motions*, Brown wrote, *were such as to satisfy me, after frequently repeated observation, that they arose neither from currents in the fluid, nor from its gradual evaporation, but belong to the particle itself.* The phenomenon was subsequently observed with a variety of particles in water and other liquids by Brown and other scientists. In 1865 a team of investigators showed that the motion continued unabated for an entire year.

Half a century after Brown's discovery several scientists offered a qualitative explanation of the phenomenon which turned out to be correct. They argued that the Brownian motion could be explained if one was prepared to accept that the liquid was composed of molecules in thermal agitation, in which case a particle suspended in it will be continually bombarded from all sides in all directions by the molecules of the liquid. It must be remembered that at that time the atomic/molecular theory had not been completely accepted by the scientific community. Though many

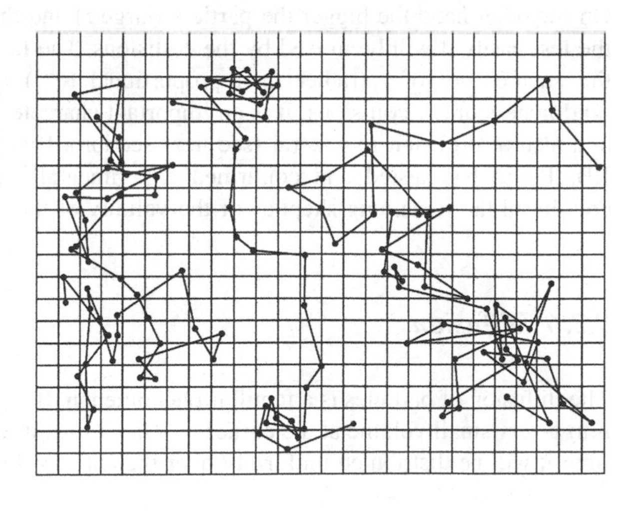

Fig. 9.12 Brownian motion of a particle in a liquid drop as seen through a microscope (projected on a *horizontal plane*). The straight lines join successive positions of the particles at 30-s intervals

scientists favoured the atomic point of view, this was still no more than a promising hypothesis. Its connection with the Brownian motion was still vague when Einstein wrote his paper on the subject in 1905. The paper, published in *Annalen der Physik* was entitled: *On the Movement of Small Particles Suspended in a Stationary Liquid Demanded by the Molecular Kinetic Theory of Heat.* In his paper Einstein considers the problem defined in the above title quite generally, and refers to the Brownian motion incidentally. *It is possible*, he writes at the beginning of his paper, *that the movements to be discussed here are identical with the so-called Brownian molecular motion; however the information available to me regarding the latter is so lacking in precision, that I can form no judgement in the matter.*

Because the particle suspended in the liquid is small, the number of molecules with which it collides per unit time is relatively small and therefore the *effective* force on the particle because of these collisions *fluctuates* considerably from the average of the force over a relatively long period, or the force that a larger body would experience, which would of course be zero for a stationary liquid in thermal equilibrium. Einstein assumed that the effective force acting on a suspended particle was a random one; that it was equally likely that the suspended particles got kicked in any possible direction. To simplify the calculation Einstein represented the suspended particles by tiny spheres with radii of about 10^{-3} cm, while the molecules of the liquid have radii of about 10^{-8} cm. He derived the following formula for the average displacement Δ in time t:

$$\Delta = k\sqrt{\frac{Tt}{r\eta}} \tag{9.40}$$

where, k is a constant (the same for all liquids), T is the temperature of the liquid, r is the radius of the suspended (spherical) particle, and η is a number which measures the viscosity of the liquid. The significance of these quantities is clear: the greater the temperature the more agitated is the thermal motion of the liquid and hence more violent the collisions of its molecules with the suspended particle. On the other hand the bigger the particle (large r) and the more viscous the fluid, the less easily it will be moved by the collisions. The fact that Δ is propotional to the square root of t (rather than proportional to t) is a consequence of the randomness of the collisions. It is an important characteristic of Brownian motion and also of diffusion in general (see next section). By 1908 the French physicist J.B. Perrin had tested and confirmed experimentally Einstein's formula. This convinced the remaining skeptics of the validity of the atomic hypothesis.

9.2.6 *Diffusion*

The diffusion of particles is a familiar phenomenon. If we release a quantity of gas into a very small volume at the corner of a large box, it will diffuse and after some time it will be distributed uniformly over the entire volume of the box, if there is no external force acting on the particles (molecules) of the gas. I shall assume, for the sake of simplicity, that the box is a long rectangular one (extending from $-\infty$ to $+\infty$) with a cross section area denoted by A; and that the density $n(x,t)$ of the particles depends only on the x-direction. In this case, $n(x,t)Adx$ gives the number of particles in the system between x and $x + dx$ at time t. Because the total number of particles in the system is conserved (it does not change wth the time) we have:

$$\frac{\partial n(x,t)}{\partial t} A dx = (J_{dif}(x,t) - J_{dif}(x+dx))A = -\frac{\partial J_{dif}}{\partial x} A dx \tag{9.41}$$

where $J_{dif}(x,t)$ is the diffusion current density: the number of particles crossing a unit area (at x) per unit time (at time t). We expect J_{dif} to be proportional to the gradient of the particle density:

$$J_{dif} = -D\frac{\partial n}{\partial x} \tag{9.42}$$

where D is known as the diffusion constant: a number, with dimensions of m^2/s, which depends on the properties of the diffusing particles, those of the medium in which they move, and on the temperature. Substituting Eq. (9.42) in Eq. (9.41) we obtain:

$$\frac{\partial n(x,t)}{\partial t} = D\frac{\partial^2 n}{\partial x^2} \tag{9.43}$$

This is known as the diffusion equation (in one dimension).[19] It must be solved subject to the initial condition:

$$n(x, t = 0) = \textit{a given distribution} \tag{9.44}$$

The reader will verify that the following distribution satisfies Eq. (9.43):

$$n(x,t) = n_0 P(x,t);\ P(x,t) = \frac{1}{2\sqrt{\pi D t}} \exp\left(-\frac{x^2}{4Dt}\right) \tag{9.45}$$

And we note that: $\int_{-\infty}^{+\infty} P(x,t)dx = 1$ at any time, which tells us that the total number of particles does not change with the time, as we expect. In Fig. 9.13 we show $P(x,t)$ at two moments of time: t_1 and $t_2 > t_1$. As $t \to 0$, the distribution becomes a very narrow peak about $x = 0$, which implies that at $t = 0$ all particles are within an infinitesimally small region about $x = 0$. On the other hand, as $t \to \infty$, the distribution becomes uniformly small everywhere.

---

[19] The diffusion equation, derived here on an intuitive basis, can be derived, at least for a dilute gas, from Boltzmann's transport equation, which would show the dependence of D on the temperature and the properties of the particles involved in the collisions that drive the phenomenon. We note that the diffusion equation applies not only to a dilute gas, but to a variety of systems as well. In most cases the dependence of D on the properties of the diffusing particles, those of the medium in which they move and on the temperature is determined experimentally.

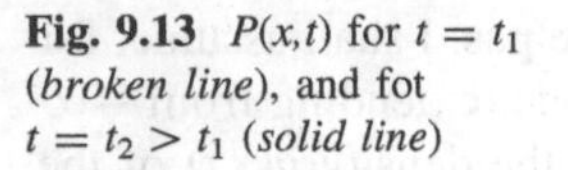
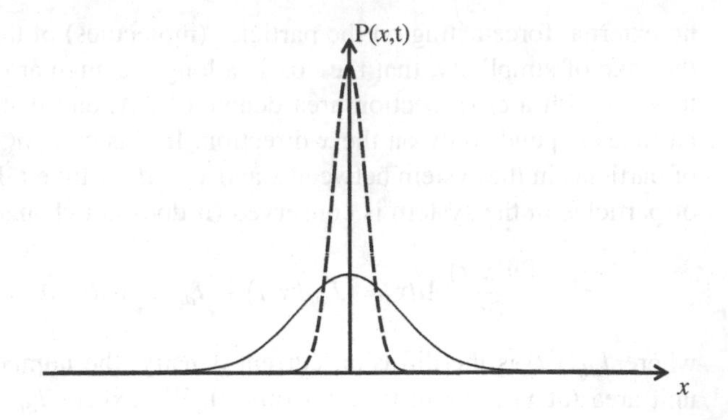

Fig. 9.13 $P(x,t)$ for $t = t_1$ (*broken line*), and fot $t = t_2 > t_1$ (*solid line*)

We can interprete $P(x,t)dx$ as the probability that a particle, initially (at $t = 0$) at $x = 0$, will be found at time t, between x and $x + dx$. In which case the mean square distance traveled by the particle in time t is:

$$\overline{x^2} = \int_{-\infty}^{+\infty} x^2 P(x,t)dx = 2Dt \tag{9.46}$$

So that:

$$\Delta \equiv \sqrt{\overline{x^2}} = \sqrt{2Dt} \tag{9.47}$$

Which varies with the time in the same way as Δ of the Brownian motion Eq. (9.40). And this is so, because diffusion, like the motion of Brownian particles, is driven by random collisions. We can not, of course, tell how much D depends on the properties of the diffusing particles, the medium in which they move and on the temperature, unlike the case of Eq. (9.40), which was derived by consideration of the microscopic motion (collisions) of the particles.

Exercises

9.1. Calculate the amount of heat absorbed by an ideal gas in equilibrium with a heat reservoir at temperature T, when it expands from volume V_0 to volume V.

Answer: $Q = NkT\log(V/V_0)$, where N is the number of atoms/molecules in the gas and k is Boltzmann's constant.

9.2. Calculate the work W done by an ideal gas when it expands adiabatically ($dQ = 0$) from volume V_0 at temperature T_0 to volume V at temperature T (to be determined).

Hint: For an ideal gas: $dE = c_V dT$, therefore: $-c_V dT = dW = PdV = NkT(dV/V)$, which leads to: $T = T_0(V/V_0)^{1-\gamma}$, where $\gamma = Nk/c_V + 1 = c_P/c_V$. Where N is the number of atoms/molecules in the gas and k is Boltzmann's constant.

Answer: $W = \frac{P_0 V_0^\gamma}{(1-\gamma)}\left[\frac{1}{V^{\gamma-1}} - \frac{1}{V_0^{\gamma-1}}\right]$, where we used $NkT_0 = P_0V_0$.

9.3. (a) When a liquid of mass M vaporizes at constant pressure P (and constant temperature; see Fig. 9.3b) its volume increases from V_l (in the liquid state) to V_v (in the vapour state). Find the work done (W) in the process and the change ΔE in the energy of the system. The heat required to vaporize all of the liquid is $Q = ML_v$, where L_v is by definition the *latent heat* of the vaporization of the liquid.

Answer: $W = P(V_v - V_l)$; $\Delta E = Q - W = ML_v - P(V_v - V_l)$;

(b) One gram of water has a volume of 1 cm^3 at atmospheric pressure, and when it boils the steam has a volume of 1,671 cm^3. Estimate the change ΔE in the energy of the system. We note that L_v (for water) = 570 cal/g.

Answer: $\Delta E = 2{,}090$ J.

9.4. Each of two containers A and B contains N molecules of the same ideal gas, at the same pressure P, but at different temperatures T_A and T_B respectively. The two containers are brought into thermal contact at pressure (P). Evaluate the change ΔS in the entropy of the system after equilibrium is established and show that $\Delta S \geq 0$.

Solution: The temperature (T_F) after equilibrium is established is $T_F = (T_A + T_B)/2$. Assuming that the change in the temperature occurs in reversible infinitesimal steps we can write (m is the mass of each molecule):

$$\Delta S = mNc_P \int_{T_A}^{T_F} \frac{dT}{T} + mNc_P \int_{T_A}^{T_F} \frac{dT}{T} = mNc_P \log \frac{(T_A + T_B)^2}{4T_AT_B} \geq 0.$$

9.5. Using the equations: $F = E - TS$; $dS = dQ/T$; and $dQ = dE + PdV$, show that:

$$P = -\left(\frac{\partial F}{\partial V}\right)_T \text{ and } S = -\left(\frac{\partial F}{\partial T}\right)_V$$

Which are known as the first and second, respectively, of Maxwell's eight equations relating various thermodynamic quantities.

9.6. A diaphragm separates the volume V of a cylinder in two parts of volume V_1 and V_2, where the pressure is P_1 and P_2 respectively. The diaphragm is allowed to find its equilibrium position at constant temperature T. Find the condition which determines its equilibrium position.

Answer: At equilibrium, at temperature T, the free energy of the system: $F = F_1(V_1,T) + F_2(V_2,T)$, where $V_2 = V - V_1$, is minimum. Therefore at equilibrium: $(\frac{\partial F}{\partial V_1})_T = (\frac{\partial F_1}{\partial V_1})_T + (\frac{\partial F_2}{\partial V_1})_T = 0$, which we may write as: $(\frac{\partial F_1}{\partial V_1})_T = (\frac{\partial F_2}{\partial V_2})_T$ which, using the first of Maxwell's equations (see exercise 9.5), we can write as: $P_1 = P_2$. This tells us that at equilibrium the pressure is the same in V_1 and V_2, as we expect.

9.7. Prove Eq. (9.14a).

Proof: It follows from Eq. (9.14) that: $-\Delta G = -\Delta F \;- P\,\Delta V \;- V\Delta P$. We know [see text after Eq. (9.13)] that when $T =$ constant, we have: $-\Delta F \geq P\Delta V$. So if P is also constant, $\Delta P = 0$, and we obtain: $-\Delta G \geq 0$.

9.8. Using $\Delta S = \Delta Q/T$, write the first law of thermodynamics as follows: $\Delta E = T\Delta S - P\Delta V$. Hence show that: $T = (\frac{\partial E}{\partial S})_V$, another of Maxwell's eight equations.

9.9. Beginning with the definition of enthalpy show that: $T = (\frac{\partial H}{\partial S})_P$, which is another of Maxwell's equations, and one more way of looking at the absolute temperature.

9.10. A material has the following properties:

1. On expansion under constant temperature T_0 from volume V_0 to volume V, it does work: $W = RT_0 \log(V/V_0)$.
2. Its entropy is given by: $S = R\frac{V_0}{V}(\frac{T}{T_0})^\alpha$, where R, V_0, T_0, α are given constants.

Calculate: (a) The free energy F of the material.

(b) The equation of state of the material.

(c) The work done when the expansion from V_0 to V happens under an arbitrary (but constant) temperature T.

Solution: (a) Using (2) and the second equation of Maxwell (see exercise 9.5) we find:

$$F(T,V) = -\frac{RV_0T^{\alpha+1}}{VT_0^\alpha(\alpha+1)} + f(V) \tag{A}$$

where $f(V)$ is a function of V which we can determine as follows. Using the property of the free energy: $-\Delta F = \Delta W$ at constant T, and property (1) of the material we find:

$$F(T_0,V) - F(T_0,V_0) = -RT_0 \log\left(\frac{V}{V_0}\right) \tag{B}$$

Substituting (A) in (B) we can determine $f(V)$. Therefore we obtain:

$$F(T,V) = -\frac{RV_0T_0}{V(\alpha+1)}\left(\frac{T}{T_0}\right)^{\alpha+1} + \frac{RT_0}{(\alpha+1)}\left(1-\frac{V_0}{V}\right) - RT_0\log\left(\frac{V}{V_0}\right) + C$$

where C is a constant.

(b) Knowing $F(T,V)$ we can obtain the equation of state of the material using the first equation of Maxwell (see exercise 9.5). We obtain:

$$P = \frac{RV_0T_0}{V^2}\left(\frac{V}{V_0} - \frac{1}{(\alpha+1)}\left(\frac{T}{T_0}\right)^{\alpha+1} - \frac{1}{\alpha+1}\right)$$

(c) $W = \int_{V_0}^{V} P(V,T)dV = RT_0\log(\frac{V}{V_0}) - \frac{(V-V_0)RT_0}{(\alpha+1)V}[(\frac{T}{T_0})^{\alpha+1} - 1]$

9.11. What is the most probable value, υ_{max}, of the velocity of an atom/molecule of an ideal gas in thermal equilibrium at temperature T?

Hint: It is that value of υ for which $4\pi\upsilon^2 f(\upsilon)$, where f is the Maxwell–Boltzmann distribution, is maximum. There $d(4\pi\upsilon^2 f(\upsilon))/d\upsilon = 0$.

Answer: $\upsilon_{max} = (2kT/m)^{1/2}$, where m is the mass of the atom/molecule.

9.12. Prove that [see text preceding Eq. (9.27)] that: $\upsilon_{r.m.s.} = (3kT/m)^{1/2}$.

9.13. You should consider this and the following exercise after familiarizing yourself with the concept of a *magnetic dipole* (exercise 10.7) and that of an electric dipole (see Sect. 10.4.3), and the spherical coordinates presented in Figs. 14.1 and 14.3.

Consider a 'gas' of non-interacting magnetic dipoles, each of which has a magnetic dipole moment $\boldsymbol{\mu}$, in a uniform magnetic field $\boldsymbol{B}$, in the z-direction. In the absence of the field the magnetic dipoles will point to any direction with equal probability and the magnetization of the gas (magnetic moment per unit volume of the gas) will be zero. In the presence of the magnetic field the number of magnetic dipoles pointing in the direction defined by the angles θ, φ within a solid angle $d\Omega = sin\theta d\theta d\varphi$, as defined in Fig. 14.3, is given according to Eq. (9.32) by

$$dN(\theta, \varphi) = Ce^{-U(\theta)/kT} \sin\theta d\theta d\varphi \qquad \text{(A)}$$

where: $U(\theta) = -\mu B cos\theta$ is the potential energy of the dipole in the magnetic field (see exercise 10.7) and C is a normalization constant. Accordingly the magnetization $\boldsymbol{M}$ of the gas (in the direction of the magnetic field) is given by

$$M = N\mu\overline{\cos\theta}$$

where: N is the number of dipoles per unit volume of the gas and $\overline{\cos\theta}$ is the mean value of the cosine of the angle between the magnetic field and the dipoles, obtained according to Eq. (9.33).

(a) Show that $\overline{\cos\theta} = -\frac{1}{\alpha} + \frac{e^{\alpha}+e^{-\alpha}}{e^{\alpha}-e^{-\alpha}}$, where $\alpha = \frac{\mu B}{kT}$
(b) Show that as $T \to 0$, $\overline{\cos\theta} \to 1$: all dipoles are oriented parallel to the field.
(c) Show that at high temperatures ($T \to \infty$), $\overline{\cos\theta} \to 0$: the dipoles are oriented entirely randomly.

Note: $\int xe^{ax}dx = \frac{e^{ax}}{a^2}(ax - 1)$

9.14. Consider a 'gas' of non-interacting electric dipoles, each of which has a dipole moment $\boldsymbol{p}$, in a uniform electric field $\boldsymbol{E}$, in the z-direction. In the absence of the field the dipoles will point to any direction with equal probability and the polarization of the gas (dipole moment per unit volume of the gas) will be zero. In the presence of the field the number of dipoles pointing in the direction defined by θ, φ within a solid angle $d\Omega = sin\theta d\theta d\varphi$ is given by Eq. (A) of the previous exercise, but in the present case: $U(\theta) = -pEcos\theta$ is the potential energy of the dipole in the electric field (see exercise 10.8) and C is a normalization constant. Accordingly the polarization $\boldsymbol{P}$ of the gas (in the direction of the electric field) is given by

$$P = Np\overline{\cos\theta}$$

where N is the number of dipoles per unit volume of the gas and $\overline{\cos\theta}$ is the mean value of the cosine of the angle between the electric field and the dipoles.

Show that the results (a), (b) and (c) of the previous exercise are also valid in the present case except that now $\alpha = pE/kT$.

9.15. Verify that $n(x,t)$ of Eq. (9.45) satisfies the diffusion Eq. (9.43).

9.16. Consider a slab of thickness L. We assume that the temperature varies with the position along the direction (x) normal to the slab. Then the flow of heat (the amount of heat passing through a unit area normal to the x-direction per unit time) is given by

$$F(x,t) = -K\frac{\partial u}{\partial x} \tag{A}$$

where: $u(x,t)$ denotes the temperature of the slab at the position x at time t. And we remember that $0 \le x \le L$. The above is the mathematical expression of Newton's law of cooling mentioned in Sect. 7.2. The constant K is known as the coefficient of thermal conductivity of the material. Now, if more heat enters the element (volume) of the slab between x and $x + dx$, the temperature of this element will change (per unit time) according to

$$(Sdx)\rho c\frac{\partial u}{\partial t} = (F(x,t) - F(x + dx,t))S$$

where: S is the area of the slab, Sdx is the volume of the slab between x and $x + dx$, ρ is the mass density of the slab, and c is its specific heat. Since $F(x + dx,t) = F(x,t) + (\partial F/\partial x)dx$, the above expression becomes:

$$\rho c\frac{\partial u}{\partial t} = -\frac{\partial F}{\partial x} = K\frac{\partial^2 u}{\partial x^2}$$

where we used Eq. (A) in the second of the above equations. Finally, putting $\kappa = K/\rho c$, known as the diffusivity of the material, we write the above equation in a more compact form as follows:

$$\frac{\partial u}{\partial t} = \kappa\frac{\partial^2 u}{\partial x^2} \tag{B}$$

We note that the above, which is known as the heat equation has the same mathematical form as the diffusion Eq. (9.43) of Sect. 7.2.

Problem: The lateral surface of a slab is insulated against the flow of heat. At $t = 0$ the temperature equals zero throughout the slab. The temperature at the end of the slab at $x = 0$ is zero at all times, while the other end is maintained at temperature T_0 at all times. Find the distribution of the temperature $u(x)$ in the slab after a steady (not depending on t) state has been established and evaluate the corresponding steady flow of heat.

Answer: $u = (T_0/L)x$; $F = -KT_0/L$.

9.17. What does $n(x,t)$ of Eq. (9.45) tell us about the diffusion of heat injected at $t = 0$ in the middle of a very long rod (whose cylindrical surface is thermally insulated)?

9.18. Special relativity tells us (see Chap. 12) that no particle can have a velocity, υ, greater than that of light, c. Convince yourself that in a gas, in thermal equilibrium at temperatures as high as 3,000 K, the fraction of the atoms/molecules with velocities greater than c, according to the Maxwell–Boltzmann distribution, is negligibly small and that, therefore, this distribution remains valid at such temperatures.

9.19. You may prefer to consider this exercise after reading Chaps. 13 and 14.

(a) We have seen [Eqs. (9.34) and (9.35)] how the energy (E) and the entropy (S) of a macroscopic system relate to the *density of the microstates* ($\rho(E)$) of the system. Are these equations invalidated by quantum mechanics?

Answer: Definitely not, because their derivation does not depend on the exact nature of the microstates. But $\rho(E)dE$ must be understood as the number of microstates as defined in quantum mechanics: these are the energy-eigenstates between E and $E + dE$ of the Hamiltonian corresponding to the given physical system.

(b) Is the counting of the microstates more difficult in quantum mechanics?

Answer: The counting of the microstates is, at least in principle, easier in quantum mechanics. The quantum numbers (however many) that define an energy-eigenstate of the system are well defined, whereas the division of the $\boldsymbol{r}$-space and the $\boldsymbol{p}$-space into cells [as in Eqs. (9.21a, b)], which is necessary for the description in classical terms of the motion of the particles that make up the system, is to some degree arbitrary.

Chapter 10
Electromagnetism

10.1 Early History

Magnetic phenomena were known to the ancient Greeks since about the 9th century BC. They noticed that a kind of stone (what is presently *known* as magnetite: $Fe_3 O_4$) attracts pieces of iron. The word *magnetic* derives from *Magnesia*, the name of a Greek region where magnetite was found. And they also knew that a piece of amber, the Greek name of which is *ήλεκτρον* (*electron*), when rubbed could attract such things as straw and feather. In the same way that sometimes a comb gets electrified when run though somebody's hair and then can attract small pieces of paper. But not much was done in the way of investigating these phenomena.

In 1269 Pierre de Maricourt mapped out the directions taken by a needle at various points on the surface of a spherical natural . He noted that these directions were tangential to lines (which were later named *magnetic field lines*) which come off the spherical magnet at one point (which was subsequently named the *north pole* of the magnet) and, turning round circular-like, enter it again at a diametrically opposite point (which was named the *south pole* of the magnet). It was subsequently found that every magnet has a *north* (*N*) pole and a south (*S*) pole and that the field lines from north to south pole follow, more or less, the same pattern (see Fig. 10.1). And it was noted that like-poles repel each other while unlike-poles attract each other. In 1750 John Mitchell was able to determine experimentally, using a torsion balance (as in Cavendish's measurement of the gravitational constant), that this force is inversely proportional to the square of the distance between the poles.

Another important discovery was made in 1600 by William Gilbert who found that a magnetic (compass) needle orients itself in a particular direction at each place on earth, suggesting that the earth was a permanent spherical magnet (giant in size but relatively weak) with *N* and *S* magnetic poles, where the geographical poles are.

A. Modinos, *From Aristotle to Schrödinger*,
Undergraduate Lecture Notes in Physics, DOI: 10.1007/978-3-319-00750-2_10,

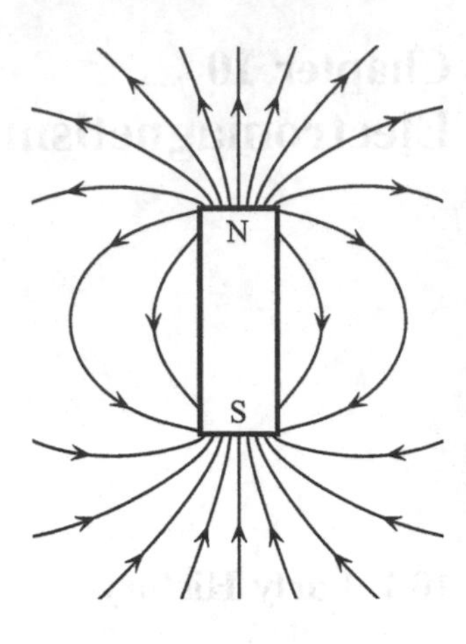

Fig. 10.1 Magnetic field lines of a bar magnet

10.2 Electrostatics, Steady Currents and Magnetism

10.2.1 Electrostatics

The first steps towards a scientific description of the attraction between amber and straw and similar phenomena of electricity were made by Benjamin Franklin. Franklin, the first distinguished American physicist, was born in 1706 and excelled not only as a scientist but also as a politician and author. Among other things he organized a Philadelphia discussion group which later became the American Philosophical Society, and helped in the creation of an academy which in due course became the University of Pennsylvania. Many people will know of him in relation to his discovery of the thunder conductor, a discovery not unconnected to his more basic observations that will concern us here.[1] He died in 1790.

An example of Franklin's experiments is shown in Fig. (10.2). In this example, a rod of glass which has been rubbed with silk attracts the suspended (from a non metallic thread) rod of hard rubber that has been rubbed with fur, when the two rods come near to each other. On the other hand two rubber rods that have been rubbed with fur repel each other; and two glass rods that have been rubbed with silk also repel each other. The above observations are easily explained, as Franklin suggested, if we accept that *there are two types of electric charge*, and that charges of the same type repel each other, while charges of different type attract each other. He named (arbitrarily) the charge on the glass rod *positive* and the charge on the rubber rod *negative*. Accordingly, the charge on any object would be positive if it were attracted to the negatively charged rubber rod (or any other negatively charged object); or repelled by the positively charged glass rod (or any other

---

[1] We now know that a thunderbolt occurs when an electrically charged cloud discharges via a spark its charge to earth. The thunder conductor by being a good conductor of electricity picks up this charge and guides it safely on to earth.

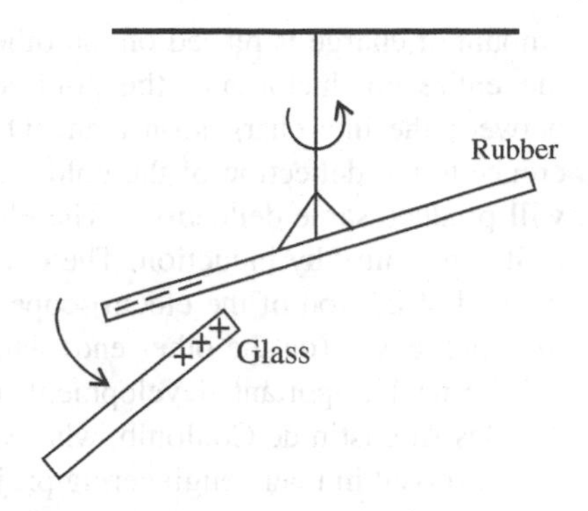

Fig. 10.2 Schematic description of Franklin's experiments

positively charged object). Franklin suggested that the electrified state of an object arises when electric charge (what ever the carrier of charge is) is *transferred* from one body to the other. When a glass rod is rubbed with silk, the silk obtains a negative charge which is equal in magnitude to the positive charge on the glass rod.[2] Charge is never created or destroyed.

It was soon established that while some materials, such as rubber, silk and glass, hold the electric charge in one place (where it was generated initially by friction), others, such as metallic rods or wires or wet surfaces (including earth) disperse the electric charge. So that when charge is generated on a metallic rod or sphere it spreads uniformly on the entire surface of the rod or the sphere. The first category (glass, etc.) became known as insulators; the second category (metals, etc.) as conductors (of electricity). One way of generating electric charge on a conductor, let it be a metallic rod suspended from an insulating thread or held by an insulating handle, that had been used widely is the following. One produces (by friction) an amount of charge on one end of an insulating rod. When the charged end of this rod is brought near an end of the metallic rod, the metallic rod polarizes: an amount of charge equal in magnitude but of opposite type to that on the insulating rod accumulates on the end of it near the charged insulator; an equal amount of charge of the same type as that on the insulator is repelled to the other end of the conducting rod. One can get rid of this charge by connecting this end of the metallic rod momentarily (and while the other end is kept charged by its proximity to the charged end of the insulating rod) to earth via a wire. When the metallic rod is eventually taken away from the insulating road the charge on it spreads uniformly on the surface of the rod. And by repeating the process we can add more charge to the conducting rod. And the same thing will happen if we use a metallic sphere instead of a rod.

Starting from the above observations an instrument was devised to detect electric charge. In the *gold-leaf electroscope* two rectangular gold leaves are attached to the end of a conducting rod held in an insulating frame. When an

[2] We now know that negative charge (electrons) is transferred from the glass to the silk during the rubbing process.

amount of charge is placed on the other end of the conducting rod it spreads over the entire conductor and the gold leaves move apart owing to the repulsion between the like charges on them. One can compare amounts of charge by reference to the deflection of the gold leaves they produce: same amounts of charge will produce same deflections. The electroscope could also detect electric charge in its proximity by induction. The external charge induces an opposite charge on the end of the rod of the electroscope near it, pushing charge of the same type to the gold leaves (on the other end) which move apart in consequence.

The next important development arose from the work of a French scientist, Charles Augustin de Coulomb, who was born in 1736. He trained as an engineer, was involved in many engineering projects and in 1773 wrote a treatise on friction and cohesion in statics relating to architecture. However, his fame rests on the work he did on electricity and magnetism between 1785 and 1791. He died in Paris in 1806.

In his famous experiments two metallic spheres are suspended in vacuum at a distance from each other. The diameters of the spheres are very small compared to the distance separating them so that they can be taken as points in the analysis of the data. The spheres are charged and the force between them is measured using a torsion balance similar to the one used by Cavendish to measure the gravitational constant. He established that the magnitude F of the force between two point charges Q_1 and Q_2 is inversely proportional to the square of the distance r between them. His findings are summarized in the following equation, which is presently known as Coulomb's law:

$$F = k_E \frac{Q_1 Q_2}{r^2} \tag{10.1}$$

And we remember that the force is repulsive between like charges and attractive between unlike ones. The constant k_E in the above equation depends on the units one uses. In the originally used units, known as the electrostatic units, $k_E = 1$, and the unit of charge is taken to be that amount which produces a unit of force on the same unit charge at a unit of distance away.

Coulomb went on to confirm Mitchell's results concerning the force between poles of different magnets, i.e. that this force (repulsive between like-poles and attractive between unlike-ones) is also inversely proportional to the square of the distance between them.

It will be useful to us, in what follows, to write Coulomb's law in the following form. Assume a Cartesian system of coordinates, and assume that there is a charge Q *fixed* at the origin of the coordinates. Then the force $\boldsymbol{F}$ exerted by Q on a charge q at the position $\boldsymbol{r}$ is given by:

$$\boldsymbol{E}(\boldsymbol{r}) = (k_E Q/r^3)\boldsymbol{r} \tag{10.2}$$

$$\boldsymbol{F}(\boldsymbol{r}) = q\boldsymbol{E}(\boldsymbol{r}) \tag{10.3}$$

We read the above equations as follows: The charge Q generates a (static) electric field $\boldsymbol{E}(\boldsymbol{r})$ in the space about it given by Eq. (10.2); a charge q at position $\boldsymbol{r}$ in this field has a force acting on it given by Eq. (10.3). $\boldsymbol{E}$ has the direction of $\boldsymbol{r}$ (pointing away from the origin) if Q is positive, and has the direction of $-\boldsymbol{r}$ (pointing towards the origin) if Q is negative. $\boldsymbol{F}$ has the direction of $\boldsymbol{E}$ for positive q and the opposite direction for negative q.

We note that the above force has the form of Eq. (7.36) and can therefore be written as the gradient of a potential energy $U(r)$, as in Eq. (7.39). By analogy to Eq. (7.37), and having in mind Eqs. (10.2) and (10.3) we write:

$$\boldsymbol{F} = -\nabla U \tag{10.4}$$

$$U(\boldsymbol{r}) = qV(\boldsymbol{r}); V(\boldsymbol{r}) = k_E Q/r \tag{10.5}$$

We say that $U(\boldsymbol{r})$ is the potential energy of q in the electrostatic field of Q. We call $V(\boldsymbol{r})$ the potential field due to Q; it gives the potential energy of a unit positive charge in this field. It follows [see Eq. (7.40)] that the work done by this field in displacing q from $\boldsymbol{r}$ to $\boldsymbol{r} + d\boldsymbol{r}$ is given by:

$$\boldsymbol{F} \cdot d\boldsymbol{r} = -dU = U(\boldsymbol{r}) - U(\boldsymbol{r} + d\boldsymbol{r}) = q(V(\boldsymbol{r}) - V(\boldsymbol{r} + d\boldsymbol{r})) \tag{10.6}$$

[The above way of describing the force between two charges (through an active field) avoids the need for *action at a distance*. The interaction between the charges occurs via a *field* which has a real existence in space. As we shall see the same concept applies to magnetic interaction. Its full significance becomes apparent when the electric and magnetic fields vary with the time. Electromagnetic interaction is not instantaneous, it propagates with the speed of light. For static charges the interaction appears of course instantaneous. Similar arguments apply to gravity, but these were not obvious at the time of Newton (see section on general relativity).]

10.2.2 The Discovery of the Battery

The next important step in the science of electricity happened with the discovery of the battery. The story began like this. In 1786 the Italian Luigi Galvani noticed that when a hook (made of copper) was inserted into a frog's spinal cord hung from an iron railing, the leg muscles of the frog contracted. He tried other metals in place of the hook and railing and the same thing happened. He thought that the phenomenon was due to electric charge from the muscles or nerves of the frog and called it 'animal electricity'.

Alessandro Volta was born in 1745 in Como (Italy) and taught in the public schools there, becoming a professor of physics at the Royal School in Como in 1774. He became interested in Galvani's animal electricity around 1791. He soon realized that the electricity arose not from the frog's leg and that it could arise when an appropriate pair of metals (we shall call them *electrodes*) were placed in the right inorganic substance (we shall call it the *electrolyte*). This led him to the construction (in 1800) of the first *electrochemical cell* (the first battery). It consisted of two electrodes, a rod of zinc (Zn) and the other a rod of copper (Cu), partly immersed in (diluted) sulphuric acid (SO_4H_2). An underlying electrochemical process results in the Zn-electrode becoming negatively charged and the Cu-electrode positively charged.[3] An electrostatic field is therefore created in the region between the two poles of the battery. The force on a charged particle (with charge q) in this region will be given by

$$\boldsymbol{F} = -\nabla U(\boldsymbol{r}) = -q\nabla V(\boldsymbol{r}) \tag{10.7}$$

where $U(\boldsymbol{r}) = qV(\boldsymbol{r})$ is the sum of terms, like those of Eq. (10.5), due to all charges on the poles of the two electrodes. It is worth remembering the important property of a force that is the gradient of a potential energy field $U(\boldsymbol{r})$, which is: the work done by the force in moving a charge q from a point $\boldsymbol{r}_A$ to a point $\boldsymbol{r}_B$ is the same whatever the path from $\boldsymbol{r}_A$ to $\boldsymbol{r}_B$, and it is given by: $U(\boldsymbol{r}_A) - U(\boldsymbol{r}_B)$. In most applications we need only know the work done by the electric field (by the battery) when charge q is moved from the positive pole to the negative pole of the battery. And because this is proportional to q, we characterize the battery by noting the work done when a unit charge ($q = 1$) is moved from the positive to the negative pole of the battery, which is given, according to Eq. (10.6), by the *potential difference*: *V*(*at the positive electrode*) −*V*(*at the negative electrode*). This quantity is often called the *voltage* of the battery and is denoted by V. The name *voltage* derives from a unit of this quantity, the *Volt*, named so in honour of Vollta (we shall introduce this unit in Sect. 10.2.5).

Volta went on to construct a more efficient battery which consisted of alternate disks of silver and zinc, separated by cloth soaked in dilute acid. This layered structure provided a steady potential difference between the two ends (poles of the battery), with excess positive charge at the silver end and an equal amount of negative charge at the zinc end. And although the potential difference V obtained in this way was small it could provide a steady current in a metallic wire connecting the two poles of the battery. We remember that before the discovery of the battery, charge transfer could only occur discontinuously by friction and induction,

---

[3] The free energy of the acid solution (electrolyte) is minimized when a certain fraction of the acid exists in the form: $2H^+$ and SO_4^{2-}. Some of the SO_4^{2-} react with the Zn electrode transferring their excess charge (two electrons) to it; and the positive hydrogen ions capture electrons from the Cu electrode forming bubbles of hydrogen gas ($2H^+ + 2e \rightarrow H_2$). The process stops when an electrochemical equilibrium is established. During the process 'chemical' energy is transformed into potential energy associated with the electrostatic field in the region between the poles of the battery (the ends of the electrodes outside the electrolyte). When the two poles are connected with a conducting wire charge flows between them, which transforms the potential energy into other forms (e.g. heat); at the same time the charge on the poles is maintained by the above mentioned internal electrochemical process until the battery is exhausted.

and by a spark between two highly charged bodies when brought sufficiently near each other.

Volta was made a count by Napoleon in 1810 in recognition of his important discovery. He died in 1827.

10.2.3 Electric Currents and Magnetism

At this stage in the development of electromagnetism, electric and magnetic phenomena seemed not to relate to each other. Static electric charges had some features in common with magnetic poles (the repulsion/attraction between like/unlike charges was similar to the repulsion/attraction between like/unlike magnetic poles) but there was also an obvious difference between them arising from the fact that magnetic poles always appeared in pairs (the pair consisting of the N and the S pole of a given magnet) whereas positive and negative charges could exist in isolation. More importantly (static) electric charges did *not* interact with magnetic poles. There was no evidence to suggest that there was an intimate connection between electricity and magnetism. The availability of steady (constant in time) electric currents, following the discovery of the battery, led eventually to the discovery that electricity and magnetism were in fact related to each other. This happened in 1820 when the Danish physicist Hans Oersted made the following observation.[4] He placed a magnetic needle (naturally pointing from South to North) below a long horizontally stretched wire along the same direction (South to North). When the (conducting) wire was connected to the poles of a battery the magnetic needle changed its direction orienting itself in the direction West to East.[5] *The electric current was obviously generating a magnetic field about it which was very much stronger than the magnetic field of the earth.* In this respect, electric charge *in motion* (that is what an electric current is) behaves very differently from static charge.

The next important development came from the French Andre-Marie Ampere. He was born in Lyons in 1775. A child prodigy, had it seems no formal education, but was tutored by his father.[6] In 1809 he became professor of mathematical

---

[4] Hans Christian Oersted was born in 1777 in Rudkøbing in Denmark. He studied at the University of Copenhagen, and earned a doctorate in 1799 for a dissertation based on the works of Kant entitled: The Architectonicks of Natural Metaphysics. He became a professor at the University of Copenhagen in 1806. His research was mainly on electric currents and acoustics. He also contributed to chemistry; he was apparently the first to discover aluminum in 1825. And he was also a published poet. His poetry collection Luftskibet (The Airship) was inspired by the balloon flights of fellow physicist and magician Etienne-Gaspard Robert. He died in 1851.

[5] The same thing was observed in 1802 by the Italian jurist Gian Dominico Romognosi but it went unnoticed, perhaps because it was published in a newspaper, the Gazetta de Trentino, instead of a scholarly journal.

[6] A talented but unlucky man, Ampere's father was beheaded by the Jacobins in 1793.

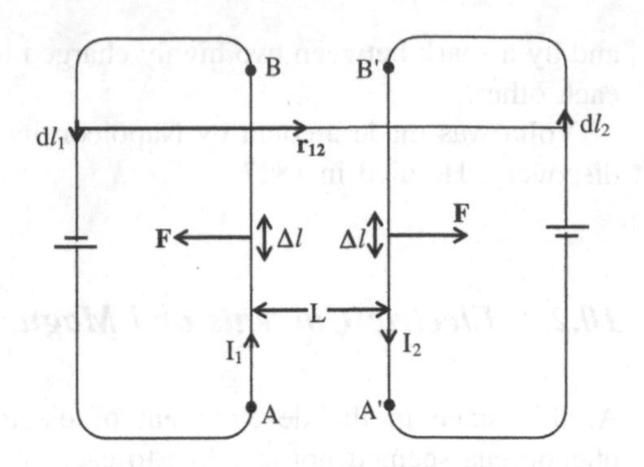

Fig. 10.3 The force between two current-carrying straight-line wires. The currents in the closed circuits are maintained by batteries (denoted by | 1)

analysis at the Ecole Polytechnique, and in 1814 was elected to the Academy of the Sciences of the Institute de France, when he defeated his former student Cauchy who was the other candidate for the post. In 1824 Ampere became also professor at the College de France where he taught electrodynamics. Ampere was productive in mathematics and wrote among other things on variational calculus and Taylor series. In physics, he apparently rediscovered (in 1814) Avogadro's law (see Sect. 8.4) but his work on this subject passed unnoticed like that of Avogadro. He also did work on the refraction of light. However, his fame rests on his work (summarized in what follows) relating electric currents to magnetism.

In 1820, after hearing of Oersted's discovery, he determined experimentally that two parallel straight-line wires carrying electric currents I_1 and I_2 respectively attract each other if the currents flow in the same direction and repel each other if the currents flow in opposite directions. One measures the current I flowing in a wire by the amount of charge crossing a cross section of the wire per unit time. The direction of the current is that in which positive charge flows; so if the carriers of charge are negatively charged particles, the direction of the current is opposite to the velocity of the particles. Ampere established that the magnitude of the force that the one wire exerts *on a segment* Δl of the other wire is given by:

$$F = 2k_M \frac{I_1 I_2}{L} \Delta l \tag{10.8}$$

where L is the distance separating the wires (see Fig. 10.3) and k_M is a constant which depends on the units of measurement (the 2 factor is there to facilitate our subsequent discussion of these units). The above formula is very good when Δl is somewhere in the middle of the straight-line segments (as shown in Fig. 10.3) and better the longer these segments are.

Assuming that it remains reasonably good over most of the length l (from A to B) of these segments, the total force between these segments is N times that of Eq. (10.8), where $N = l/\Delta l$. It is worth remembering that the electric current I_1 (the same argument applies to I_2), can only exist in a closed circuit maintained by a battery (as shown in Fig. 10.3). Therefore what one needs, eventually, is a formula which gives the force which circuit *1* exerts on circuit 2 *and* vice versa.[7]

In 1819 Felix Savart[8] a young doctor who spent more time studying physics than treating patients left Metz for Paris, where he met Jean-Baptiste Biot,[9] a professor of physics at the College de France since about 1800, wanting to discuss with him problems relating to the acoustics of musical instruments. At the time Biot was lecturing on acoustics and welcomed Savart. However, when they heard of Oersted's discovery, they decided to work together on this new topic. Using the deflection of a magnetic dipole (needle) to determine the strength of the magnetic field about a current carrying wire, they were able to propose a formula for the evaluation of the magnetic field $\boldsymbol{B}$ at any point in the space about any circuit of steady current. Their formula, now known as the Biot and Savart law, was presented to the academy of Sciences in 1820. It is conveniently written as follows:

$$\boldsymbol{B}_2 = \mathrm{k}_M I_1 \oint_1 d\boldsymbol{l}_1 \times \boldsymbol{r}_{12}/(r_{12})^3 \qquad (10.9)$$

We read the above formula in conjunction with Fig. 10.4 as follows. We divide circuit 1 into (say) N elements $d\boldsymbol{l}_1$. Each of these elements contributes to the magnetic field at position 2 (the position of $d\boldsymbol{l}_2$ in the figure) a term proportional to $d\boldsymbol{l}_1 \times \boldsymbol{r}_{12}/(r_{12})^3$, where the vector product is evaluated according to its definition (see Sect. 7.1.1). The integral sign in the above equation tells us that the total field at position 2 due to circuit 1 is the (vector) sum of the N terms corresponding to the N elements $d\boldsymbol{l}_1$ of circuit 1.

---

[7] The force $\boldsymbol{F}_2$ exerted on circuit 2 by circuit 1 is given by the following formula: $\boldsymbol{F}_2 = -k_M \oint_1 \oint_2 [d\boldsymbol{l}_1 \cdot d\boldsymbol{l}_2 \boldsymbol{r}_{12}/(r_{12})^3] I_1 I_2$. The peculiar double integral in the square brackets is to be understood as follows: we divide circuit 1 into a large number of elements $d\boldsymbol{l}_1$ pointing in the direction of the flow of I_1 and we divide circuit 2 into a large number of elements $d\boldsymbol{l}_2$ pointing in the direction of I_2. For *every pair* of elements $d\boldsymbol{l}_1$ and $d\boldsymbol{l}_2$ we construct the quantity in the square brackets which is a vector in the direction of $\boldsymbol{r}_{12}$, the displacement vector from $d\boldsymbol{l}_1$ to $d\boldsymbol{l}_2$ (see Fig. 10.3); r_{12} is the magnitude of $\boldsymbol{r}_{12}$ and $d\boldsymbol{l}_1 \cdot d\boldsymbol{l}_2$ is the scalar product of $d\boldsymbol{l}_1$ and $d\boldsymbol{l}_2$ to be evaluated according to the definition of Eq. (7.28). The integral signs tell us that we must take the (vector) sum of all the vectors obtained in this way. [If there are N $d\boldsymbol{l}_1$—elements in circuit *1* and M $d\boldsymbol{l}_2$—elements in circuit *2* there will be NM pairs and therefore NM vectors to add up.] Finally we note that the force $\boldsymbol{F}_1$ exerted on circuit *1* by circuit *2* is obtained by the above formula by replacing $\boldsymbol{r}_{12}$ by $\boldsymbol{r}_{21} = -\boldsymbol{r}_{12}$ which leads to $\boldsymbol{F}_1 = = -\boldsymbol{F}_2$, in agreement with Newton's third law.

[8] Felix Savart was born in 1791 in Mezieres (France). After training at Metz Hospital he served in Napoleon's army, from 1812 to 1814, as a regimental surgeon. In 1816 he graduated with a medical degree from Strasburg University. He set up a medical practice in Metz in 1817, but in 1817 he gave it up to pursue his interests in physics. In 1827 he was elected to the physics section of the Academy of Sciences (in Paris) to replace Fresnel who had died earlier that year. Savart died in 1841.

[9] Jean-Baptiste Biot was born in Paris in 1744. He was appointed professor of mathematics at Beauvais in 1797, and became a professor of physics at the Collége de France in 1800, and three years later he was elected as a member of the Academy of Sciences. He did research in electromagnetism, optics, and astronomy. He died in1862.

Fig. 10.4 The law of Biot and Savart

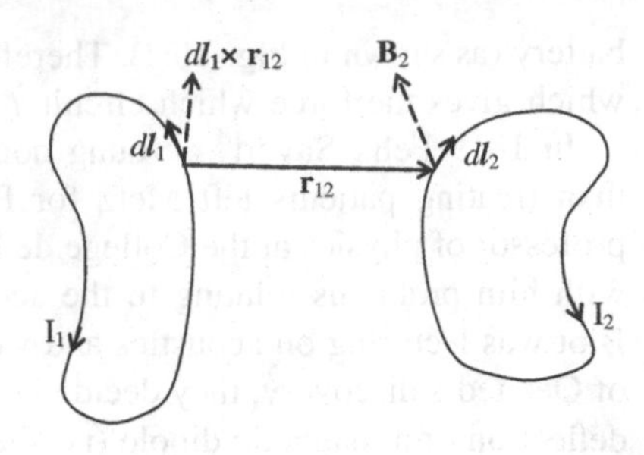

Knowing the magnetic field in the space about a current carrying circuit (1) we can calculate the force on an element dl_2 of a circuit 2 (see Fig. 10.4) if we have a formula for the force acting on it due to the magnetic field at its position. Obviously this formula must at the end give the same force between two straight wires as that found by Ampere. It turns out that the required formula is a simple one. The force dF on dl_2 due to a magnetic field B_2 at its position is given by the vector product of dl_2 with B_2 multiplied by the current I_2 that flows through it. We have:

$$dF_2 = I_2\, dl_2 \times B_2 \tag{10.10a}$$

and we remember that dl_2 points in the direction of the current. The force F_2 exerted on the whole of circuit 2 by circuit 1 is, accordingly, given by the (vector) sum of the forces acting on all the elements of this circuit (we assume as usual that circuit 2 is divided into a large but finite number of elements of length dl_2). We write this sum as follows:

$$F_2 = I_2 \oint_2 dl_2 \times B_2 \tag{10.10b}$$

Substituting Eq. (10.9) in Eq. (10.10b) we obtain the formula for the force between the two circuits (see Footnote 7).

The magnetic field B about a long straight-line wire (say along the z-axis of a coordinate system and in theory extending to infinity on either direction) calculated from Eq. (10.9) is shown schematically in Fig. 10.5. The figure shows a magnetic field line (B-line) associated with this field: the magnetic field lines are circles about the wire in planes normal to the wire. The magnitude B of the field (tangential to the B-lines) is proportional to the current I flowing in the wire, and inversely proportional to the radius L. We have: $B = 2k_M I/L$ which implies that in this case the following is true:

$$\oint B \cdot dl = (2k_M I/L) 2\pi L = 4\pi k_M I \tag{10.11}$$

where the integral on the left is taken over a circle of radius L [Because B is tangential to the circle, $B \,.\, dl = Bdl = (2k_M\ I/L)dl$. Taking the integral of this quantity over the circle gives $(2k_M\ I/L)2\pi L$].

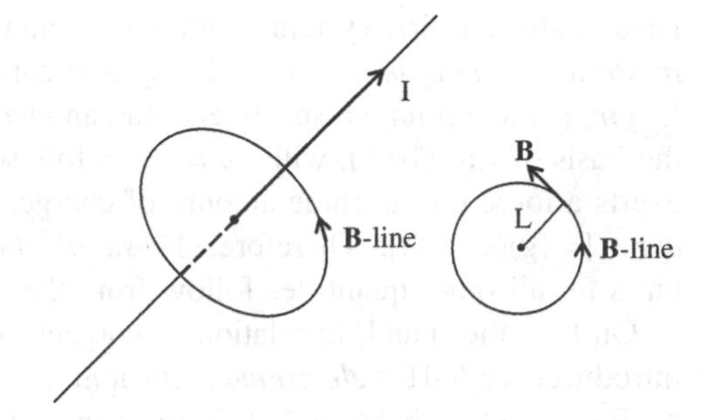

Fig. 10.5 **a** The magnetic field of a *long straight wire*. **b** The wire is normal to the page, the current flowing out of the page

In 1825, after much experimental and theoretical research, Ampere proposed that

$$\oint_C \boldsymbol{B} \cdot d\boldsymbol{l} = 4\pi k_M I \tag{10.12}$$

is true for *any path* C encircling a conductor carrying a steady current (I). The encircling (closed) path can have any shape: circle, ellipse, rectangle, any shape; it need not even be on the same plane, it can be for example a closed path on the surface of a sphere as long as the current I flows through it. We can choose the positive direction of the current flow arbitrarily, but having chosen this direction, the integration in Eq. (10.12) must proceed anticlockwise about it (as in Fig. 10.5b). The above equation, known as Ampere's law, was verified experimentally and played a crucial role in the development of electromagnetism. We note that Ampere's law remains true if not one but a number of conducting wires go through C, but in this case I is the algebraic sum of the currents in the wires.

In 1827 Ampere published a long memoir summarizing his work on electric and magnetic phenomena, of the preceding seven years, both theoretical and experimental. He was apparently the first to construct a galvanometer, an instrument to measure currents. The basic instrument consists of a rectangular coil mounted vertically on a spring in the magnetic field between the N and the S pole of an appropriately shaped permanent magnet. The greater the current through the coil the greater the angle it turns before the torque exerted by the spring equals the torque due to the magnetic field. The reading of the meter is shown by a pointer, attached to the coil, which moves over a scale.

In the years after the publication of his memoir Ampere lost interest in physics. He died in 1836.

10.2.4 Electromagnetic Units

I have so far avoided any reference to the units used in the measurements of electric and magnetic quantities. I should, at this point say something about them.

Historically the first system of units to include electrical quantities was the *electrostatic system of units (esu)*. This system consists of the *cgs* units: *cm, g* and *s* for *length, mass* and *time* respectively, and an *electrostatic unit of charge* obtained on the basis of Eq. (10.1), with $k_E = 1$, as follows. By definition, one *esu of charge* exerts a force on the same amount of charge, at a distance of 1 cm, equal to one *dyne* [= (g)(cm/s^2)]. Therefore: 1 *esu of charge* = $[(\text{g})(\text{cm/sec}^2)(\text{cm})^2]^{1/2}$. The units for all other quantities follow from the above.

On the other hand, in relation to magnetostatics, a unit of electric current was introduced, called the *electromagnetic unit (emu) of current* (or the *abampere*), based on Eq. (10.8) with $k_M = 1$. It is the constant current $I = I_1 = I_2$ in two straight parallel infinitely long wires *1 cm* apart that would produce a force of 2 *dyne/cm* between the two. Therefore according to Eq. (10.8): (1 *emu of current*)^2 = 2 *dyne.* This unit of current taken together with the *cgs* units makes up the *electromagnetic system of units.*

The ratio between *esu* and *emu* units can be obtained as follows. Let *I* be the current flowing through a conductor with a cross section of area *A*. The current density $\boldsymbol{j}$ is a vector in the direction of the current with magnitude $j = I/A$. The current density is related to the charge density ρ (the charge per unit volume) in the conductor as follows: $\boldsymbol{j} = \rho \boldsymbol{u}$, where $\boldsymbol{u}$ is the velocity of the charge carriers. The charge density expressed in electrostatic units (ρ_{esu}) is related to the current density expressed in electromagnetic units ($\boldsymbol{j}_{emu}$) as follows:

$$\boldsymbol{j}_{emu} = \rho_{esu}\, \boldsymbol{u}/c \tag{10.13}$$

where $\boldsymbol{u}$ is the velocity (measured in cm/sec in both systems of units) of the charge carriers (we now know that in metallic conductors these are electrons), and *c* is the ratio of the units. This ratio was first determined (at about 1860) by Weber and Kohlrausch by measuring the discharge of a capacitor whose electrostatic capacity was known.[10] They found that *c* was (within the accuracy of the measurements) numerically equal to the speed of light. This was an important discovery for it suggested (at least Maxwell thought so) that *light was intrinsically connected with electromagnetism.* Another (more obvious) conclusion to be drawn from the above equation relates to the comparative strength of the electric force described by Eq. (10.1) and the magnetic force between currents described by Eq. (10.8) (or the more general formula given in footnote 7). The magnetic force is weaker by a factor of c^2!

Weber in collaboration with Gauss developed a unified system of units based on electrostatic and electromagnetic units, known as the Gaussian units, but we need not describe it here. The system of units presently used (and throughout in this book) is known as the MKS system of units devised by E. Giorgi in 1901. It is based on the meter (m), kilogram (kg) and second (s) and on the ampere (A) as the unit of electric current. The latter was initially taken as the 1/10 of the *emu* of

---

[10] A capacitor is an arrangement of conductors (for example, two metallic plates as shown in Fig. 10.15, or spheres) separated by an insulator, used for storing charge.

current, but its more modern definition is: 1 A *is the constant current that, maintained in two straight parallel infinite conductors of negligible cross section placed one meter apart in vacuum, would produce a force between the conductors of* 2×10^{-7} *Newton/meter* (N/m). The MKS unit of charge, called *Coulomb* (*C*), is accordingly defined as *the charge transferred by a current of one ampere in one second.* The units of all other quantities are expressed in terms of the above units. Finally in the MKS system of units the constants k_E [appearing in Eq. (10.1)) and k_M (appearing in Eqs. (10.8), (10.11) etc. of the previous section] are given by:

$$k_E = 1/(4\pi\varepsilon_0); \varepsilon_0 = 10^7/(4\pi c^2) = 8.85434 \times 10^{-12}\,\mathrm{F/m} \qquad (10.14)$$

where c is the speed of light. F/m stands for *farad/meter*; $1\ F = C^2 s^2\ \mathrm{kg}^{-1}\ \mathrm{m}^{-2}$. The constant ε_0 is known as the permittivity of vacuum.

$$k_M = \mu_0/(4\pi); \mu_0 = 4\pi \times 10^{-7}\,\mathrm{H/m} \qquad (10.15)$$

H/m stands for *henry/meter*; 1\,H = kg m^2 C^{-2}. The constant μ_0 is known as the (magnetic) permeability of vacuum. We note that:

$$\varepsilon_0 \mu_0 = 1/c^2 \qquad (10.16)$$

10.2.5 Ohm's Law

The availability of constant electric currents following the discovery of the battery allowed scientists to measure the current in wires made of different materials. One of the first physicists to do so was the German Georg Simon Ohm. Born in 1787, Ohm taught at the universities of Cologne, Berlin, Nuremberg and finally (from 1849 until his death in 1854) at Munich. He formulated the law [Eq. (10.17)] that bears his name in 1827.

Imagine a wire connected to a battery (one end of the wire connecting to the positive electrode and the other to the negative electrode of the battery). The electric current (positive charge flowing from the positive to the negative electrode or, equivalently, negative charge flowing in the opposite direction) can be measured using a galvanometer. Ohm determined that the current I in the wire is proportional to the potential difference V (between the poles) of the battery whatever the material the wire is made of. The relation between I and V is a simple one:

$$I = V/R \qquad (10.17)$$

R is a constant, known as the resistance of the conductor. For a wire of length L and a uniform cross section (of area A), it is given by:

$$R = \frac{1}{\sigma}(L/A) \tag{10.18}$$

where σ, a constant which characterizes the material the wire is made of, is known as the conductivity of the material (its inverse, $1/\sigma$ is called the resistivity of the material).

The physical picture behind Ohm's law is made clearer if we put $V = EL$ where E is the magnitude of the electric field in the wire, and $I = jA$ where j is the current density. With these substitutions and using Eq. (10.18) we obtain:

$$j = \sigma \mathrm{E} \tag{10.19}$$

Moreover, remembering that $j = \rho u$, where ρ is the charge per unit volume of the wire and u the (average) velocity of the charge carriers, we may conclude that this velocity is constant: $u = (\sigma/\rho)E$, when the electric field $E = V/L$ is constant. This means that after *a very short period of time* following the application of the electric field (connection to the battery), over which the current grows from zero to its steady value [of Eq. (10.17)], a friction-like force develops, which opposes the force exerted on the charge carriers by the electric field, so that their velocity stays constant thereafter. As a result the work done by the electric field (by the battery) in maintaining the current is dissipated as heat. This is how electric heaters work.

Finally, remembering that the current density and the electric field are vector quantities (in the case of a wire conductor their direction is that of the wire pointing in the direction of the current flow) we write:

$$\boldsymbol{j} = \sigma \boldsymbol{E} \tag{10.20}$$

The above formula is valid quite generally, whatever the geometrical shape of the conductor and the variation of the electric field with position.

A word on the MKS units of the above quantities: Potential difference is measured in *volts*: 1 volt (V) is the potential difference between two points on a conductor carrying a constant current of one ampere when the power dissipated between the points is one *watt* (one watt is one joule per second). The resistance R is measured in *ohms* (named so in honour of Ohm): 1 Ω (ohm) = (1 V)/(1 A).

It turns out that the resistivity ($1/\sigma$) of metals (generally good conductors) is of the order of 10^{-6} Ωm and does not very much from metal to metal; it increases linearly with the temperature at a different rate for different metals but never by too much. In semiconducting materials (such as silicon and germanium) the resistivity may vary from 10^{-4} to 10^{7} Ωm depending on the material, on its purity and on the temperature. In some insulating materials the resistivity can be even larger.

A final note on Ohm's law: In Eq. (10.17), V is the potential difference between the poles of the battery when there is a current flowing in the closed circuit which includes the battery and the (external) resistance R. The potential difference between the poles of the battery in an open circuit ($I = 0$), known as the *electromotive force* (*e.m.f.*) of the battery, is a little different from V. A more accurate form of Ohm's law would be:

$$I = (e.m.f.)/(R + r) \quad (10.21)$$

where r is the contribution to the resistance from within the battery (an internal resistance). However, r is usually much smaller than R, and can be neglected: $(e.m.f.) = IR + Ir \approx IR = V$. The *e.m.f.* is equal to the work done by the battery on a unit of charge moved over any closed path from the positive pole of the battery to the negative pole (along a wire outside the battery) and then from the negative pole to the positive pole (inside the battery); which translates into the following formula:

$$e.m.f. = \oint \boldsymbol{E} \cdot dl \quad (10.22)$$

where $\boldsymbol{E}$ denotes the electric field, and the integral is to be taken over the said path. The above formula shows that the *e.m.f.* generating the current is *not* electrostatic in nature. The electric field maintaining the current is not conservative (it can not be the gradient of a potential field), because such a field can not produce any work over a closed path. The *e.m.f.* of the battery derives from the electrochemical processes going on inside the battery. In the absence of a current these processes lead to the accumulation of positive and negative charge respectively on the two electrodes of the battery, which give rise to the external electrostatic field as described in Sect. 10.2.2.

10.2.6 Electricity and Chemistry

Humphrey Davy was born in 1778 in Penzance (England). When his father died in 1794, he apprenticed himself with an apothecary surgeon, in order to provide for his mother and four younger siblings, and set about teaching himself all kinds of subjects from mechanics and physics, to anatomy. After learning French from a visiting French priest, he read the original text of Lavoisier's *Fraite Elementiare de Chemie* and became fascinated with chemistry. When twenty-years old he was offered a job at the Pneumatic Institute in Bristol, which had been recently created to promote the investigation of the therapeutic properties of gases. It was there that he discovered nitrous oxide, known as the laughing gas, and experienced its effect by inhaling it (he found that it lowered inhibitions and led to emotions enthusiastic and sublime). Some say he became addicted to it. Davy was certainly a colourful character, a bohemian of his time who wrote poems (and published them) and counted among his friends the poet Samuel Coleridge and the great William Wordsworth. But he was above all a gifted scientist. In 1800 Davy became assistant lecturer in chemistry at the Royal Institution (in London) and within a year he was promoted to professor and director of the chemical laboratory of the Institution. On hearing of Volta's discovery of the battery, Davy did his own experiments confirming Volta's results and went further, arguing that if electricity could arise from a chemical reaction, the reverse should also be possible:

electricity could engineer a chemical reaction as well. Davy pioneered *electrolysis*: the production of a chemical reaction by passing an electric current through a solution of certain substances (the process being the reverse of that underlying the operation of a battery). He used electrolysis to isolate a number of previously unknown metals, such as potassium from potash, sodium from molten sodium hydroxide, calcium from lime, as well as magnesium, boron, barium, strontium and aluminium. In 1810 he discovered chlorine and investigated hydrochloric acid; and he disproved Lavoisier's contention that all acids contained oxygen, and insisted (and he was right) that all acids contain hydrogen instead.

In 1812, while experimenting on the first high explosive, nitrogen trichloride, which had recently been discovered by Pierre Louis Dulong, had an accident which damaged his eyesight.[11] As a result he hired an assistant, a young man who approached him after one of his public lectures. The name of the young man was Michael Faraday. Within the year Davy resigned from the Royal Institution, married a wealthy widow and set out on a tour of Europe, taking Faraday with him as a personal assistant.

When he returned from Europe (in 1815) he was consulted by the Society for the Prevention of Accidents in Mines in relation to explosions in mines which caused many deaths every year. The explosions were attributed to firedamp which at the time was thought to be hydrogen. After a few weeks of research Davy established that the firedamp was a mixture of methane with air and that it would explode only at high temperatures. Miners carried with them canaries in cages to detect the gas, but by the time the bird fell off its perch it was too late to prevent the explosion. Davy's miner's lamp was fitted with a two-layer metal gauze chimney surrounding the flame so that: oxygen could still reach the flame, but most of the heat generated by it was dissipated by the metal chimney and prevented from reaching the firedamp, which therefore stayed below its detonation temperature. It is for this discovery, which saved many lives, that Davy is remembered by most people today. And it is worth noting that he refused to patent his invention. When he was urged to do so by a friend he replied: *My good friend I never thought of such a thing; my sole object was to serve the cause of humanity… I am amply rewarded in the gratifying reflection of having done so.* In 1820 Davy became president of the Royal Society. He died in 1829.

10.3 Faraday's Electromagnetic Induction

Michael Faraday was born in 1791 in London, one of ten children of a poor blacksmith. When fourteen-years old he was apprenticed with a newsagent and bookbinder called George Riebau. At the time he had no knowledge of

[11] In that he was more lucky than Dulong, who lost one eye and two fingers during one of his experiments.

mathematics but he could read. And he could not resist peeking inside the covers. Riebeau did not mind; he encouraged him to read as much as possible. One day, going through a copy of the Encyclopedia Britanica, which a customer brought in for repairs, he read for the first time about electricity. And he wanted to know more about it. He read as many books on science as he could find, including Lavoisier's famous textbook on chemistry. In 1810 he started attending lectures at the City Philosophical society. By 1812, when his apprenticeship was coming to an end, he decided to become a scientist, and this is when he approached Davy for a job and was given one. Davy was no doubt flattered by Faraday's notes of his last four lectures (386 pages with coloured illustrations beautifully bound). Faraday moved into rooms on the top floor of the Royal Institution quarters, accepting a salary of a guinea a week, less than what he had previously earned as a bookbinder. He worked as a bottle washer and at the same time he learned under the guidance of Davy. Six months after joining the Royal Institution, Faraday resigned, to accompany Davy and his wife on an eighteen-month tour of Europe, officially as Davy's valet but in effect as his experimental assistant. In Europe he met many of the outstanding scientists of the time including Volta and Ampere. Faraday was quietly becoming an expert on his way to outshining his master and Davy resented this development. He would later unjustifiably accuse Faraday of stealing some of his own ideas. On his return from Europe (in 1815) Faraday was recognized as a most able scientist on his own merit and was re-employed at the Royal Institution, as a laboratory assistant to the new professor of chemistry, on a higher salary. As a chemist he investigated electrolysis, and made a number of other discoveries: liquefying gases under pressure, isolating benzene (of great importance to organic chemistry), using platinum as a catalyst, among other things. But his fame rests on his discovery of electromagnetic induction,

In 1821 Faraday, a devout Christian, made his confession of faith in the Sandemanian Church, was appointed Superintendent of the House of the Royal Institution, and married Sarah Barnard. In the same year he constructed the first electromagnetic rotor exploiting the force between a magnet and an electric current (see Fig. 10.6). The physics underlying the operation of the rotor is essentially that of Eq. (10.10), assuming that the magnetic field of a permanent magnet is due to some internal current. The two experiments of Fig. 10.6 demonstrate that a current exerts a force on a magnet as much as a magnet on a current, as one would expect from Newton's third law.

In the years following his construction of the rotor, Faraday was looking for ways of generating electric current from a magnetic source. If current gives rise to a magnetic field, the contrary must be true: a magnetic field must somehow give rise to an electric current, he and others argued. But how would this happen? He had his first clue sometime in 1831. He noticed that a current was induced in a circuit (without a battery) when sufficiently near another circuit in which a current was flowing, but only momentarily: when the battery was switched on, and when it was switched off. This suggested to Faraday that the induced current was connected with the changing (varying with time) current in the inducing circuit. He investigated further using an arrangement of two circuits as shown in Fig. 10.7.

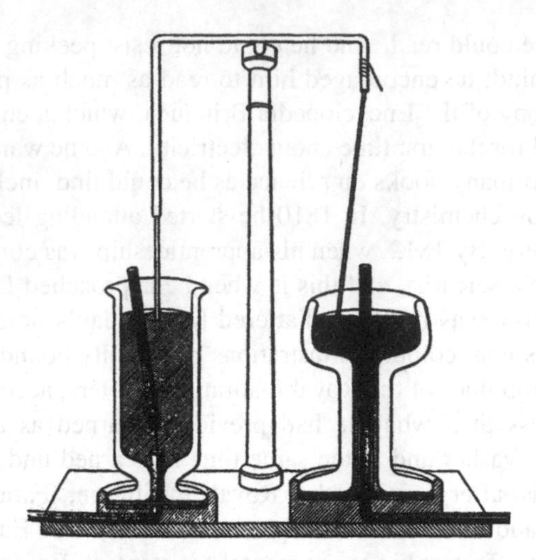

Fig. 10.6 Faraday's rotor. (Reproduced from: G. Holton and S. G. Brush, *loc. cit.*) The vessels in both figures contain (*liquid*) mercury which is a good conductor of electricity. In both cases a closed electric circuit is formed by a wire connecting the base of the vessel to the thin metallic rod hanging on top of it. On the left, this rod is fixed in space, but the magnet (the rod in the *liquid* mercury) is free to rotate about its end-point (its *S*-pole) which is fixed to the base of the vessel. When a current flows through the circuit, generated by a battery (not shown), the magnetic field about the metallic rod exerts a force on the *N*-pole of the magnet forcing it into a rotation about the metallic rod. On the *right*, the magnet is fixed in space (in the *liquid mercury*), but the metallic rod is free to rotate about its fixed *top end*. In this case when a current flows through the circuit, the current-carrying metallic rod is forced into a rotation by the field of the magnet

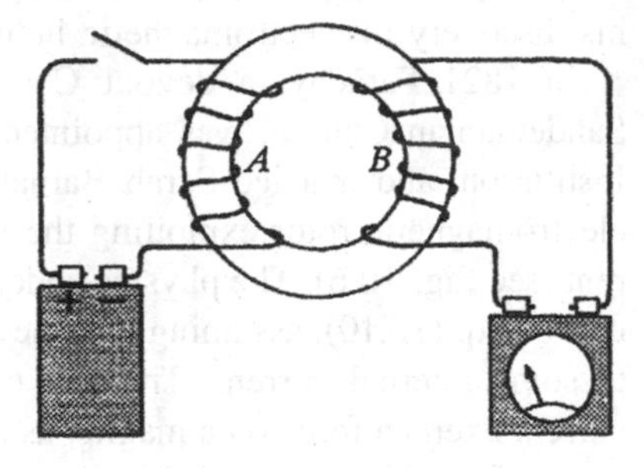

Fig. 10.7 Faraday's electromagnetic induction experiment (Reproduced from: G. Holton and S. G. Brush, *loc. cit.*)

The primary circuit *A* consists of a wire (a few turns of it) about one part of a soft iron ring,[12] and the secondary circuit *B* consists of another wire about another part of the iron ring (which is of course insulated from the wires). After the battery in *A* is switched on and during the short period it takes for the current to increase

---

[12] A soft iron core with a coil of insulated wire wound round it constitutes an *electromagnet*: When a current flows through the wire the core becomes a magnet; when the current stops the core looses its magnetization. Evidently Faraday used the iron core in his experiment of Fig. 10.7 to *magnify* the weaker magnetic field that would exist in its absence.

from zero to its steady value, the current in A varies with time; and similarly the current varies with time (diminishing from its steady value to zero) when the battery is switched off. Faraday observed a current (an induced current) in B during these periods. Faraday assumed (a tentative hypothesis to begin with) that *the current in B is induced by the variation (with time) of the magnetic flux (which he perceived as a bunch of magnetic field lines generated by the current in A) passing through B.* He discovered electromagnetic induction.[13] Faraday repeated the above experiment without an iron ring; he again observed an induced current in B but weaker in agreement with his hypothesis. And he went on to show that a current is induced in a closed circuit if there is a magnetic flux through it that changes with time, *whatever the origin of this flux.* It could be, for example, due to a magnetic rod passing through it. He eventually summarized his results in the following statement (which is known as Faraday's law of induction): *Whenever there is a change in the magnetic flux linked with a circuit an electromotive force (e.m.f.) is induced, the magnitude of which is proportional to the rate of change of the flux through the circuit.*[14]

A second law of electromagnetic induction, known as Lenz law, was formulated by Heinrich Lenz in 1835.[15] It states that: *The direction of the induced current is always such as to oppose the change producing it.*

The two laws (taken together) are summarized in the following formula:

$$\textit{Induced e.m.f.} = \oint_C \boldsymbol{E} \cdot d\boldsymbol{l} = -d\Phi/dt \tag{10.23}$$

where the integral is over a closed line C, for example that of a closed circuit, and Φ is the magnetic flux through it. The induced current I in the circuit is of course proportional to the induced *e.m.f.*. Let me clarify the above equation by reference to Fig. 10.8. The flux $d\Phi$ of $\boldsymbol{B}(\boldsymbol{r},t)$ through the element of $d\boldsymbol{S}$ at $\boldsymbol{r}$ at time t is given by the scalar product $d\Phi = \boldsymbol{B}(\boldsymbol{r},t) \cdot d\boldsymbol{S}$. This can be positive, negative or zero

---

[13] Apparently Faraday was not the first to discover electromagnetic induction. Joseph Henry, an American physicist born in 1797, discovered the phenomenon a few months earlier but did not publish it until a year later. He made his discovery while he was a teacher at an Academy in New York. In 1832 he became a professor at Princeton University. He discovered among other things self-induction. The unit *henry* (*H*) for measuring self-inductance is named after him. He died in 1878.

[14] By definition, the *e.m.f.* is equal to the work done on a unit of charge in one complete circle over a closed line encircling the (changing) magnetic flux. See Eq. (10.22).

[15] Heinrich Friedrich Emil Lenz was born in 1804 in Dorpat (now Tartu, in Livonia), in the Russian Empire at that time. He studied physics and chemistry at the University of Dorpat. During a three-year (1823–1826) expedition around the world with the navigator Otto von Kotzebue he studied climatic conditions and the physical properties of sea water. On his return (in 1826), he joined the University of St. Petersburg where he later served as the Dean of Mathematics and Physics, from 1843 to 1863 and Rector from 1863 until his death (in 1865).

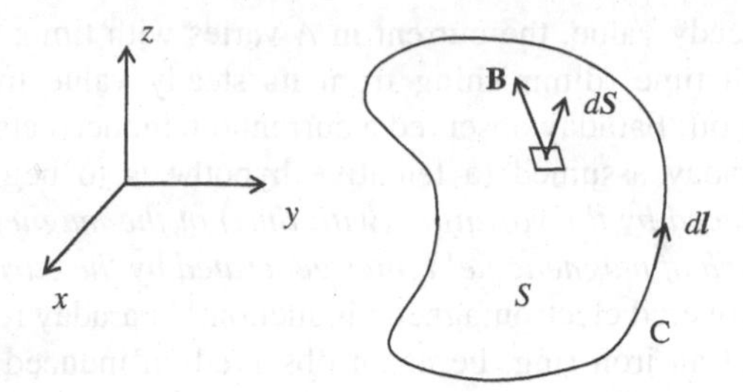

Fig. 10.8 In this diagram a surface S (it need not be a *plane surface*) is enclosed by a close line C (to be identified with the circuit). An element of the surface is described by a vector $d\boldsymbol{S}$ *normal* to the surface with magnitude dS equal to the area of this element. For a closed surface (as in a *sphere*) $d\boldsymbol{S}$ points outwards. In the case of an open surface (as in the diagram) the direction of $d\boldsymbol{S}$ is decided arbitrarily; the direction of the surface element $d\boldsymbol{S}$ (chosen arbitrarily) determines the positive direction in the closed line C (that of $d\boldsymbol{l}$). It goes anticlockwise about $d\boldsymbol{S}$

depending on the angle between the two vectors.[16] The magnetic flux Φ through C is obtained by summing up the contributions from all the elements $d\boldsymbol{S}$ that make up the surface S. We denote it by an integral as follows:

$$\Phi(t) = \oint_S \boldsymbol{B}(\boldsymbol{r}, t) \cdot d\boldsymbol{S} \tag{10.24}$$

Therefore:

$$\frac{d\Phi}{dt} = \oint_S (\partial \boldsymbol{B}/\partial t) \cdot d\boldsymbol{S} \tag{10.25}$$

Please note that with the direction of $d\boldsymbol{S}$ known (by convention), the sign of $d\Phi/dt$ is unambiguously determined. For example: Let it be *negative*: the flux through C *diminishes with time* (this will be the case if, for example, $\boldsymbol{B}$ diminishes in magnitude but does not change its direction). Then $(-d\Phi/dt) > 0$, which implies [according to Eq. (10.23)] that the direction of the induced electric field $\boldsymbol{E}$, and of the corresponding induced current, will be that of $d\boldsymbol{l}$ (see Fig. 10.8). The reader can verify (on the basis of Eq. (10.9)) that the contribution to the magnetic flux through C due to this (induced) current is *positive* (*tending to increase* Φ *through* C) in agreement with Lenz law.

The discovery of electromagnetic induction influenced dramatically our technological development. As much as the discovery of the steam engine. But whereas thermodynamics was developed to a large degree in response to the steam

[16] The physical significance of this can be seen more easily in the case of the flux $\boldsymbol{J} = \rho \boldsymbol{u}(\boldsymbol{r})$ of a fluid of constant density ρ, whose velocity varies with position. Clearly the flux through an element of surface $d\boldsymbol{S}$ (the amount of mass passing through this element per unit time) will be maximum if $\boldsymbol{u}(\boldsymbol{r})$ is normal to the surface element (i.e. in the direction of $d\boldsymbol{S}$) and it will be zero if $\boldsymbol{u}(\boldsymbol{r})$ is parallel to the surface element (i.e. normal to $d\boldsymbol{S}$).

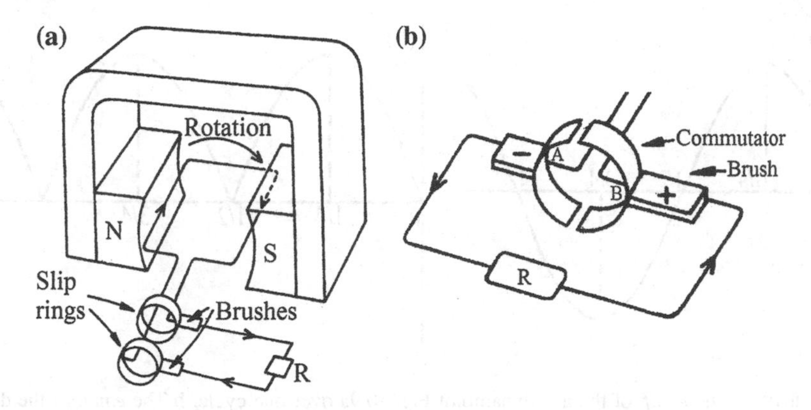

Fig. 10.9 **a** A simple a.c. dynamo. **b** The commutator of a simple d.c. dynamo

engine, electrical technology developed out of science. Electromagnetic induction is the mechanism by which a *dynamo* (a generator of electric current) operates. How this is done is demonstrated schematically in Fig. 10.9 for a simple a.c. (alternate current) dynamo.

The rectangular coil is made to turn in the field of a magnet at a constant frequency of so many *hertz* (Hz); 1 Hz = 1 cycle/s. In the U.K., for example, this frequency is 50 H. The ends of the coil are connected to two slip rings mounted on the coil spindle. Through them the ends of the coil are connected to two brushes: each end to its own (always the same) brush. The brushes are connected to the ends of the external part of the circuit. The *e.m.f.* obtained from this a.c. dynamo over one cycle is shown in Fig. 10.10a. We can understand the shape of it as follows: The magnetic flux through the coil is: $\Phi(t) = BAcos(2\pi t/T)$; where B is the magnitude of the permanent magnetic field, A is the area of the rectangle of the coil, and $(2\pi t/T)$ is the angle, at time t, between the normal to the rectangle and the direction of the field, and T is the duration of one cycle. We have: $e.m.f. = - d\Phi/dt = (2\pi/T)BAsin(2\pi t/T)$, as shown schematically in Fig. 10.10a.

In a d.c. (direct current) dynamo the two ends of the coil are solded to the two halves of a copper split ring or commutator as shown in Fig. 10.9b. Each brush connects through the commutator to a different end of the coil every half cycle, so that the current in the external circuit flows in the same direction. The corresponding *e.m.f.* is shown in Fig. 10.10b.

I should add that the systems described in Fig. 10.9 can be used in reverse mode as motors: when an electric current flows in the coil a force is exerted on it which makes it turn.

We need not describe here the various other applications of electromagnetic induction in the production and transmission of electrical energy.

Faraday went on to study other aspects of electromagnetism. He investigated, among other things the electrostatic field associated with charged conductors of certain shapes, discovering in the process the *Faraday cage*: an earthed screen of metallic wire that surrounds an electric device so as to shield it from external

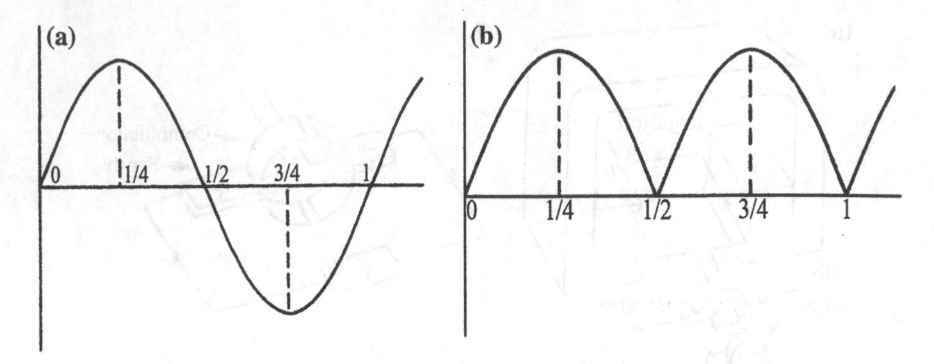

Fig. 10.10 **a** The *e.m.f.* of the a.c. dynamo of Fig. 10.9a over one cycle. **b** The *e.m.f.* of the d.c. dynamo over one cycle

electrical fields. And in his endeavours to establish the connection between electromagnetism and light he discovered the *Faraday effect*: the rotation of the plane of polarization of light when it passes through a region exposed to a magnetic field.

Finally, Faraday made much effort to make sense of electromagnetic phenomena in terms of models that he constructed in terms of *field lines,* a term which he introduced to physics. Today when we use *field lines* to describe a field, say the magnetic field due to two parallel straight wires [shown in Fig. 10.11) or the electrostatic field due to two static charges of equal magnitude (shown in Fig. 10.12)], we do so to facilitate a description of the field: the actual field is tangential to the field line, points in the direction of the line, and by having more lines crossing a unit area at a given point we indicate an increased magnitude of the field there leading to a correspondingly larger flux. Faraday believed these lines to have real existence, he thought of them as invisible mechanical transmission lines in an ethereal medium, along which electric and magnetic forces were propagated by the mutual influences of contiguous regions, like elastic strains and stresses in an elastic medium. And in the case of time-varying fields, he imagined rings of $\boldsymbol{E}$ interlocking with rings of $\boldsymbol{B}$ so that when the one ring expands the other contracts and vice versa.

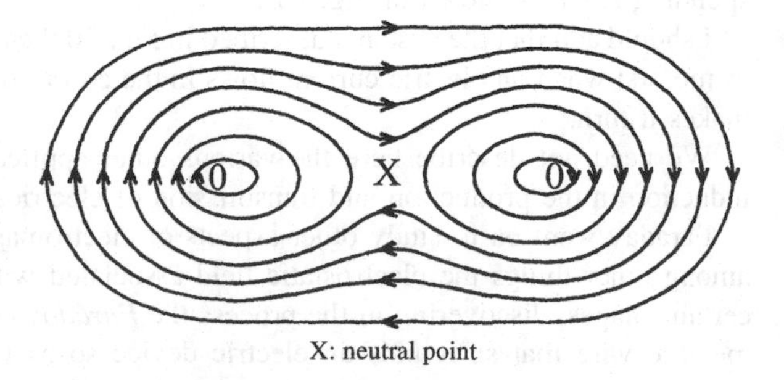

Fig. 10.11 The magnetic field of two *parallel straight wires* (normal to the page at the points **0**) carrying a current *I* in the same direction

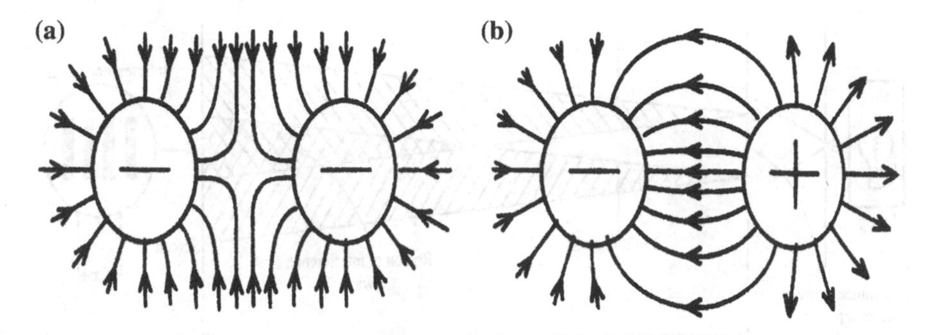

Fig. 10.12 The electrostatic field of two point charges in the plane of the page

Apparently, Faraday could not accept action between bodies at a distance from each other (through vacuum), especially if the interaction were to be transmitted as a wave with a finite velocity (which he thought impossible without a medium). Other scientists, especially those of the French school (Coulomb, Ampere, Cauchy and others) following Newton, were more pragmatic in their theoretical description of the observed phenomena. They did not insist on a specific physical model behind their mathematical formulae; they were content if the formulae described correctly the observed facts.

Faraday was appointed first Fullerian Professor of Chemistry at the Royal Institution in 1833, he was offered (but refused) the Presidency of the Royal Society in 1858 (he had been a fellow of the Society since 1824). He continued to lecture at the Royal Institution until 5 years before he died (in 1867).

10.4 Electromagnetic Waves

10.4.1 Confirmation of the Wave Nature of Light

Though Huygens' theory of light provided an elegant explanation of Snell's law of refraction (see Sect. 5.2.2), it did not prove beyond any doubt that light was a wave. This came about with the work of Young and Fresnel. Thomas Young was born in Somerset (England) in 1773. A child prodigy, especially in languages (he could speak many languages before he was 19), he eventually chose medicine. He studied at the Universities of London, Göttingen, Edinburgh and Cambridge before he began a medical practice in London in 1800. Young was a polymath who made important discoveries in physics, engineering, physiology and philology. Many engineers and physicists will know *Young's modulus*, a measure of elasticity defined as the ratio of the stress acting on a body to the strain produced. To physiologists he will be known as the man who explained how the eye can focus on objects at varying distances; and to the historian of ancient languages for his

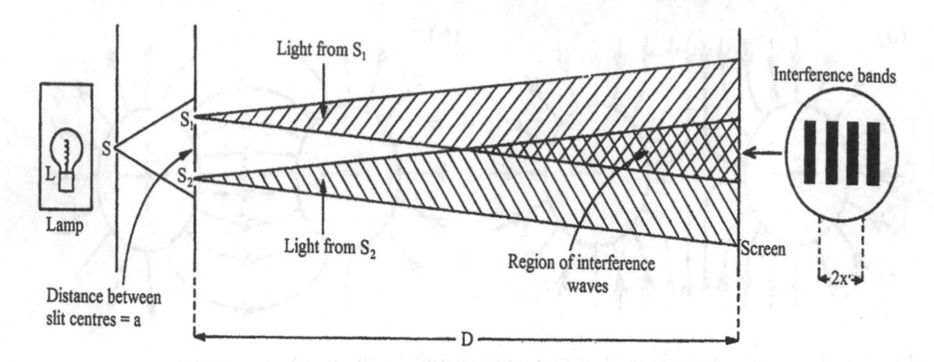

Fig. 10.13 Demonstration of optical interference—the double slit experiment

work towards deciphering the Rosetta stone and the hieroglyphic script, which led to Jean Champolion's breakthrough (in 1882) in understanding the hieroglyphics of Egypt. His extensive knowledge of many subjects was reflected in his Royal Institution lectures (he became a member of the Royal Society at the age of twenty-one) published in 1807 as *A course of lectures on Natural Philosophy and the Mechanical Arts.*[17] However, Young, who died in 1829, is best known for his discovery (about 1804) of optical interference, which proved beyond doubt that light is a wave.

A modern version of his famous double-slit experiment is described in Fig. 10.13. In Young's original experiment two pin-holes were used instead of two slits. The experiment goes like this: Light is passed from a lamp through a slit S in a screen and used to illuminate two adjacent slits (S_1 and S_2) separated by a distance a on a second screen. When the light from S_1 and S_2 falls on a third screen, a series of parallel interference bands is formed. If the lamp used produces monochromatic light, of a single wavelength λ, we obtain alternate bands of brightness and darkness as shown in the figure. Following Young we explain the bands as follows: We imagine light to be sinusoidal-like oscillations (waves) in time and space (like sea-waves or sound waves) with (positive) maxima and (negative) minima about a zero mean. Where maxima of the two waves from the slits coincide a bright band occurs (constructive interference), and where maxima of one wave coincide with minima of the other wave dark bands are produced (destructive interference). The distances shown in the figure are connected by the formula: $x' = \lambda D/a$, where x' is the fringe separation, and D is the distance between the plane of the slits and the screen.

If a lamp is used in the above experiment which produces white light, which is a mixture of colours (wavelengths), the constructive and destructive interference bands for the different wavelengths, will occur at slightly different positions. The central band will be white but colours will be seen after two or three fringes on either side of it.

[17] Young's lectures were reprinted as recently as 2002.

The interference phenomenon observed by Young implies that light is a wave, but does not tell us whether this wave is *longitudinal* (the oscillation of the thing that oscillates occurs in the direction of propagation of the wave) or *transverse* (the oscillation is normal to the direction of propagation). If one assumed (as Young and most scientists of that time did) that light was a vibration of the *ether* (a weightless fluid), it was reasonable to assume that light was a longitudinal wave (like sound in air).

The phenomenon that held the key to a proper answer to the above question came to be known as the *polarization,* of light. The first systematic observations relating to this phenomenon were made in 1669 by the Danish mathematician Erasmus Bertholinus, who noted a peculiar property of Icelandic Spar (optical quality calcite, $CaCO_3$). This naturally occurring transparent crystal separates an image into two displaced images when looked through in certain directions. Bertholinus described his observations in a 60-pages essay, but could not offer any explanation of the phenomenon. In 1808 a young Frenchman, Etienne Malus, noticed that the intensity of light reflected from a glass, looked through a crystal of Icelandic Spar, varied when he rotated the crystal. Intrigued by this observation he made more experiments the results of which could be explained as follows: light was not a longitudinal but a transverse vibration; when light is reflected from glass (and many other materials transparent or opaque, but not polished metals) a percentage of it becomes polarized: the transverse vibration that is light occurs in a *specific* plane; finally he assumed that certain crystals (such as Icelandic Spar) allow light to go through them only if the light is polarized in certain planes in relation to the geometry of the crystal. Malus communicated his results to Young pointing out that his (Malu's) observations proved Young's assumption of longitudinal waves was wrong. He wrote back to Malus: *Your experiments show the insufficiency of a theory which I have adopted, but they do not prove it false.* A few years later Young came to the conclusion (independently of Fresnel) that light waves are at least partly transverse waves.

Augustin-Jean Fresnel was born in 1788 in Broglie (France), the son of an architect. As a child he learned slowly (he could not read until he was eight). But at seventeen he entered the Ecole Polytechnique, where he acquitted himself with distinction. He served as an engineer at various places in France, and in Paris from about 1816 until his early death in 1827. Building on the experimental work of Thomas Young, extended the wave theory of light and by 1821 was able to show that the polarization of light could be explained only if light waves were entirely transverse. Among other things he was the first to construct a special type of lens, now called a Fresnel lens, as a substitute of mirrors in lighthouses. In 1823 he became a member of the French Academy and in 1825 was elected to the Royal Society of London. However, his work did not at the time receive the recognition it deserved and some of his papers were published many years after his death. He did not seem to mind. *All the compliments,* he wrote to Young, *that I have received from Arago, Laplace and Biot never gave me so much pleasure as the discovery of a theoretical truth, or the confirmation of a calculation by an experiment.*

10.4.2 Two Important Theorems Concerning Vector Fields

I shall begin with Gauss's theorem. But let me first say a little about the man, whose name we have already mentioned in relation to the Gaussian distribution (in Sect. 9.2.2) and in relation to the Gaussian units of electromagnetism (in Sect. 10.2.4). Johann Carl Friedrich Gauss made fundamental discoveries in algebra and the theory of numbers and is recognized as one of the greatest mathematicians of all times. He made also important contributions to statistics, differential geometry, electromagnetism, astronomy and optics. Gauss was born in Lower Saxony (Germany) in 1777, the son of poor working-class parents. He was a child prodigy and made his first ground-breaking mathematical discoveries while still a young student at the University of Göttingen from 1792 to 1795, and at the age of 21 he completed his *Disquisitiones Arithmeticae,* a fundamental work in the theory of numbers. During these early years Gauss was supported financially by the Duke of Braunschweig. In 1807 he was appointed professor of astronomy and director of the astronomical observatory in Göttingen, a post he held to the end of his life. His personal life was marked by the early death of his first wife Johanna Osthoff, in 1809, followed soon after by the death of a son. Gauss fell into depression from which he never fully recovered. He married again, to Johanna's best friend, Friederica Waldeck. After his second wife died in 1831, his daughter, Therese, one of his six children, took over the household and cared for him until his death in 1855.

In 1831 Gauss began a fruitful collaboration with the professor of physics, Wilhelm Weber, on problems of electromagnetism, which led to, among other things, the construction (in 1833) of the first electro-mechanical telegraph (which connected the astronomical observatory with the institute of physics in Göttingen). However, Gauss's most important contribution to electromagnetism must be his formulation of the law (known as Gauss's law) for electric fields, which states that: *the electric flux through a closed surface is proportional to the algebraic sum of the charges in the volume enclosed by the surface.*

Consider a volume V enclosed by a surface S as shown in Fig. 10.14. We divide the surface into a large but finite number of sufficiently small surface elements. Each surface element is represented by a vector $\boldsymbol{dS}$ normal to the surface pointing outward with a magnitude equal to the area of the surface element. The electric flux $d\Phi$ of the electric field $\boldsymbol{E}(\boldsymbol{r})$ through the element of $\boldsymbol{dS}$ at $\boldsymbol{r}$ is given by the scalar product $d\Phi = \boldsymbol{E}(\boldsymbol{r}) \cdot \boldsymbol{dS}$. It is positive when $\boldsymbol{E}(\boldsymbol{r})$ points outward (the angle between $\boldsymbol{E}(\boldsymbol{r})$ and $\boldsymbol{dS}$ is less than 90°), it is negative when $\boldsymbol{E}(\boldsymbol{r})$ points inward (the angle between $\boldsymbol{E}(\boldsymbol{r})$ and $\boldsymbol{dS}$ exceeds 90°), and equals zero when $\boldsymbol{E}(\boldsymbol{r})$ is normal to $\boldsymbol{dS}$ (the field is parallel to the surface at the given point). The electric flux Φ through S is obtained by summing up the contributions from all the elements $\boldsymbol{dS}$ that make up the surface S. We denote it by an integral as follows: $\Phi = \oint_S \boldsymbol{E}(\boldsymbol{r}) \cdot \boldsymbol{dS}$. Gauss's law tells us that this quantity is proportional to the total charge (the algebraic sum of the positive and negative charges) in the volume V enclosed by S. We convince ourselves that the law is true as follows: The law is obviously satisfied by the flux, through a spherical surface of radius r, of an electric field due to a single charge Q at the origin of coordinates. According to Eq. (10.2) the field is normal to the spherical surface at every point and has a magnitude (in MKS units) of $E = (1/4\pi\varepsilon_0)Q/r^2$. Therefore the flux through the spherical surface (of area $4\pi r^2$) is: $\Phi = E(r)4\pi r^2 = Q/\varepsilon_0$. We note that if $Q > 0$ then $\Phi > 0$ (outward flux) and if $Q < 0$ then $\Phi < 0$. The reader will convince himself (herself) that because of the way $d\Phi$ has been defined ($d\Phi = \boldsymbol{E}(\boldsymbol{r},t) \cdot d\boldsymbol{S}$) only the component of $d\boldsymbol{S}$ in the direction of $\boldsymbol{E}$ contributes to $d\Phi$, which in turn means that Φ through *any* closed surface (not just a spherical one) about the charge Q will give the same result: $\Phi = Q/\varepsilon_0$. And finally because the electric field due to a number of charges is the (vector) sum of the fields due to the individual charges, we may conclude that Gauss's law is valid quite generally:

$$\oint_S \boldsymbol{E}(\boldsymbol{r}) \cdot \mathrm{d}\boldsymbol{S} = Q/\varepsilon_0. \tag{10.26}$$

where Q is the total charge in the volume V enclosed by S. We can write Q as a *volume integral* of charge distribution: we divide the volume V into a large but finite number of volume elements $dV = dxdydz$; the charge dQ in dV at the position $\boldsymbol{r}$ is given by $dQ = \rho(\boldsymbol{r})dV$, where $\rho(\boldsymbol{r})$ is the charge density (the charge per unit volume) at $\boldsymbol{r}$. The total charge in V is the sum of the charges from all the volume elements that V has been divided into; we denote it by an integral: $Q = \oint_V \rho(\boldsymbol{r})dV$. We can then write Gauss's law (Eq. (10.26)) as follows:

$$\oint_S \boldsymbol{E}(\boldsymbol{r}) \cdot d\boldsymbol{S} = \frac{1}{\varepsilon_0}\oint_V \rho(\boldsymbol{r})dV \tag{10.27}$$

The advantage of writing Gauss's law in the above form derives from the *divergence theorem*, known also as Gauss's theorem, or as Ostrogradsky's

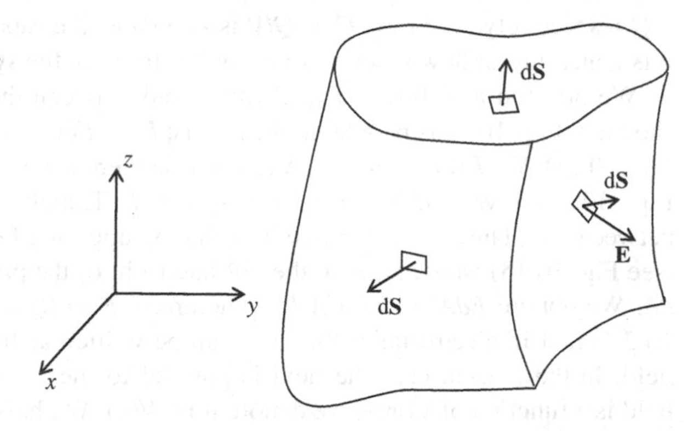

Fig. 10.14 Gauss's theorem and Gauss's law are valid for any volume V and for any shape of the *enclosing* surface S

theorem, named so after Michel Ostrogradsky who stated it independely of Gauss (in 1831).

Consider a vector field (we assume as usual a Cartesian system of coordinates): $\boldsymbol{A}(\boldsymbol{r}) = A_x(\boldsymbol{r})\,\boldsymbol{i} + A_y(\boldsymbol{r})\boldsymbol{j} + A_z(\boldsymbol{r})\boldsymbol{k}$, where $\boldsymbol{i}, \boldsymbol{j}, \boldsymbol{k}$ denote, as usual, unit vectors along the x,y,z axes. The *divergence* of $\boldsymbol{A}(\boldsymbol{r})$, denoted by $\nabla \cdot \boldsymbol{A}$ or $div\boldsymbol{A}$, is a scalar quantity (a number with appropriate units attached to it where necessary) defined at any position $\boldsymbol{r} = x\,\boldsymbol{i} + y\boldsymbol{j} + z\boldsymbol{k}$, as follows:

$$\nabla \cdot \boldsymbol{A} = \partial A_x/\partial x + \partial A_y/\partial y + \partial A_z/\partial z \qquad (10.28)$$

We note that $\boldsymbol{A}$ may depend on the time t as well as on the position coordinates $x,y,z,$ but only the variation with position enters the divergence theorem. The *divergence theorem* (we shall not prove it here) states that:

$$\oint_S \boldsymbol{A}(\boldsymbol{r}) \cdot d\boldsymbol{S} = \oint_V \nabla \cdot \boldsymbol{A} dV \qquad (10.29)$$

where $\boldsymbol{A}(\boldsymbol{r})$ is *any* vector field, S is *any* closed surface and V is the volume enclosed by it.

Applying the above theorem to the left-hand-side of Eq. (10.27), we obtain:

$$\oint_V \nabla \cdot \boldsymbol{E}(\boldsymbol{r}) dV = \frac{1}{\varepsilon_0} \oint_V \rho(\boldsymbol{r}) dV \qquad (10.30)$$

And because the above applies whatever the volume V, we must conclude that:

$$\nabla \cdot \boldsymbol{E}(\boldsymbol{r}) = \rho(\boldsymbol{r})/\varepsilon_0 \qquad (10.31)$$

The above equation is the basic equation of electrostatics. Let me demonstrate its usefulness by an example: Consider two parallel rectangular conducting plates of area A at a distance d from each other. The two plates are uniformly charged as shown in Fig. 10.15. Such a system is called a capacitor. We wish to know the potential difference V between the plates, when the charge on them is $+Q$ and $-Q$ respectively. The ratio $C = Q/V$ is known as the capacitance of the capacitor: it is a measure of how much charge can be stored in the system for a given voltage V. We proceed as follows: The electric field between the plates of the capacitor shown in Fig. 10.15 is parallel to the x-axis: $\boldsymbol{E} = E(x)\boldsymbol{i}$, and therefore, according to Eq. (10.28), $\nabla \cdot \boldsymbol{E}(\boldsymbol{r}) = dE/dx$. And because there is no charge between the plates, Eq. (10.31) gives: $dE/dx = 0$ for $0 < x < d$. Therefore $E(x) = E$ (a constant) between the plates; we determine its value by applying Eq. (10.27) to a small disc (see Fig. 10.15) whose sides to the left and right of the plate at $x = 0$ have an area dA. We obtain: $EdA = Q(dA/A)/\varepsilon_0$. Therefore: $E = (Q/A)/\varepsilon_0$. We know (see Sect. 10.2.1) that in electrostatics the field can be written as the gradient of a potential field. In the present case the field is parallel to the x-direction and the potential field is a function of x only; we denote it by $V(x)$. We have $-dV/dx = E$; therefore $V(x) = -Ex + (a\ \ constant)$; therefore the potential difference $V = V(0)$

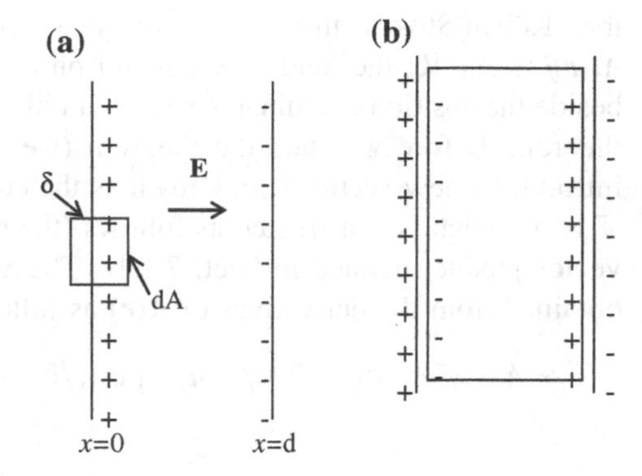

Fig. 10.15 **a** A parallel plate capacitor. **b** The same with a polarized insulating material between the plates

$-V(d) = Ed = (Q/A)d/\varepsilon_0$. Therefore the capacitance of the parallel plate capacitor of Fig. 10.15 is: $C = Q/V = \varepsilon_0 A/d$. When a polarizable material with permittivity $\varepsilon > \varepsilon_0$ is inserted between the plates the capacitance is increased to: $C = \varepsilon A/d$ (see Sect. 10.4.4).

We shall return to Eq. (10.31) in our discussion of Maxwell's equations (next section). At this point I would like to present a pictorial interpretation of Gauss's theorem, based on the field lines associated with an electric field (see, e.g., Fig. 10.12). We note that the number of field lines crossing a unit of surface determine the flux through it. And we note that the electric field lines originate on charges and terminate on charges (where $\rho(\boldsymbol{r})$ and therefore $\nabla \cdot \boldsymbol{E}(\boldsymbol{r})$ are *not* zero). Therefore Gauss's law tells us that the electric flux through a closed surface will be different from zero [it can be positive (outward) or negative (inward)] only if at least some field lines originate or terminate within the surface. Assuming that the same picture applies to the magnetic field, we conclude that the magnetic flux through a closed surface S is always zero, because the magnetic field lines are always closed lines (they do not originate and do not terminate any where). This is certainly true for the magnetic fields induced by electric currents, and it is true for permanent magnets if we further assume that the lines shown in Fig. 10.1 continue through the magnet. The fact that the poles of a magnet do not exist in isolation suggests strongly that this is so. We may therefore conclude that:

$$\oint_S \boldsymbol{B}(\boldsymbol{r}) \cdot d\boldsymbol{S} = 0 \tag{10.32}$$

for any closed surface S. And from the above and Gauss's theorem [Eq. (10.29)] we further obtain that:

$$\nabla \cdot \boldsymbol{B}(\boldsymbol{r}) = 0 \tag{10.33}$$

The above equation is another important result leading to Maxwell's equations.

A second and equally important theorem of vector analysis was proved by Stokes (known to us from his work on fluid dynamics; see Sect. 7.4) following a suggestion of Lord Kelvin. The theorem, which is known as Stokes's theorem (or the Kelvin-Stokes theorem) applies to any vector field $\boldsymbol{A}(\boldsymbol{r}) = A_x(\boldsymbol{r})\,\boldsymbol{i} + A_y(\boldsymbol{r})\boldsymbol{j} + A_z(\boldsymbol{r})\boldsymbol{k}$; the field may depend on other variables (for example the time) beside the position coordinates x,y,z, but only the variation with position enters the theorem. Before we state the theorem (we shall not prove it here) we need to introduce a new vector field, known as the curl of $\boldsymbol{A}$, and denoted by ***curl*** $\boldsymbol{A}(\boldsymbol{r})$ or $\nabla \times \boldsymbol{A}$, which is constructed as follows (the reader will note the analogy with the vector product defined in Sect. 7.1.1). The x, y and z components of $\nabla \times \boldsymbol{A}$ are obtained from the derivatives of $\boldsymbol{A}(\boldsymbol{r})$ as follows:

$$\nabla \times \boldsymbol{A} = (\partial A_z/\partial y - \partial A_y/\partial z)\boldsymbol{i} + (\partial A_x/\partial z - \partial A_z/\partial x)\boldsymbol{j} + (\partial A_y/\partial x - \partial A_x/\partial y)\boldsymbol{k} \tag{10.34}$$

Stokes theorem states that, for any vector field $\boldsymbol{A}(\boldsymbol{r})$ we obtain:

$$\oint_S (\nabla \times \boldsymbol{A}(\boldsymbol{r})) \cdot d\boldsymbol{S} = \oint_C \boldsymbol{A}(\boldsymbol{r}) \cdot d\boldsymbol{l} \tag{10.35}$$

where C is any closed line on any surface S ($d\boldsymbol{S}$ and $d\boldsymbol{l}$ are defined as in Fig. 10.8).

Let us now go back to Ampere's law as stated in Eq. (10.12). We remember that the law applies to the total current I passing through any surface S bounded by C. The current dI passing through an element dS of the surface is given by $dI = \boldsymbol{j} \bullet d\boldsymbol{S}$, where $\boldsymbol{j}$ is the current density at the position of $d\boldsymbol{S}$ and need not be in the direction of $d\boldsymbol{S}$ (normal to the surface). The total current is the sum (to be denoted by an integral as usual) from all the surface elements that make up S. Therefore, the right-hand-side of Eq. (10.12) becomes [in MKS units; see Eq. (10.15)]: $\mu_0 \oint_S \boldsymbol{j} \cdot d\boldsymbol{S}$. And using Stokes' theorem [Eq. (10.35)] we can replace the line integral over C on the left-hand-side of Eq. (10.12) by the surface integral: $\oint_S (\nabla \times \boldsymbol{B}(\boldsymbol{r})) \cdot d\boldsymbol{S}$. Therefore Ampere's law relating to *stationary* currents and their respective magnetic fields becomes:

$$\oint_S (\nabla \times \boldsymbol{B}(\boldsymbol{r})) \cdot d\boldsymbol{S} = \mu_0 \oint_S \boldsymbol{j} \cdot d\boldsymbol{S} \tag{10.36}$$

And because the above is valid for any S, we may conclude that:

$$\nabla \times \boldsymbol{B}(\boldsymbol{r}) = \mu_0 \boldsymbol{j}(\boldsymbol{r}) \tag{10.37}$$

Finally, substituting for $d\Phi/dt$ on the left of Eq. (10.25) the expression given by Eq. (10.23) we obtain:

$$\oint_C \boldsymbol{E} \cdot d\boldsymbol{l} = -\oint_S (\partial \boldsymbol{B}/\partial t) \cdot d\boldsymbol{S} \tag{10.38}$$

which is one way of writing Faraday's law of electromagnetic induction. We remember that in Faraday's demonstration of the above law, C coincides with a current carrying wire. However, it is reasonable to assume, following Faraday and Maxwell, *that the above holds generally*, whether there is a current carrying wire along C or not. We can then, using Stokes' theorem, write the above as follows:

$$\oint_S (\nabla \times \boldsymbol{E}) \cdot d\boldsymbol{S} = -\oint_S (\partial \boldsymbol{B}/\partial t) \cdot d\boldsymbol{S} \tag{10.39}$$

And because the above is valid for any S, we finally obtain:

$$\nabla \times \boldsymbol{E}(\boldsymbol{r},t) = -\partial \boldsymbol{B}(\boldsymbol{r},t)/\partial t \tag{10.40}$$

Before concluding this section let me write down an important property of $\nabla \times \boldsymbol{A}$:

$$\nabla \cdot (\nabla \times \boldsymbol{A}) = 0 \tag{10.41}$$

In words: The divergence of the ***curl*** of *any* vector field equals zero.

10.4.3 Maxwell and His Equations

James Clerk Maxwell was born in 1831. His family was one of the most distinguished and most wealthy families in Edinburgh and both his parents were deeply cultured. After they got married they began developing their estate in Glenlair, and because there was no school there the boy Maxwell was taught by his mother. And her death when he was only eight affected him deeply. After two unhappy years with a private tutor he was sent to the Edinburgh Academy. His first scientific paper (he generalized the definition of an ellipse and succeeded in producing true ovals identical to those studied earlier by Descartes) appeared when he was fourteen, but his interests were much wider. In fact, a poem of his was published in the *Edinburgh Courant* six months before his scientific paper. And he continued to read and write poetry throughout his life. He studied at Edinburgh University from 1847 to 1850 acquiring a broad education based on philosophy, and in 1850 he moved to Cambridge University to take the Mathematical Tripos, his studies there lasting 3 years and a term. In 1854 he just missed the position of senior wrangler in the mathematics examination, coming second to E.J. Routh. 2 years later he was made a Fellow of Trinity College (Cambridge), before returning to Scotland in 1856 as professor of natural philosophy at the Marischal College in Aberdeen. It was there that he met and married Katherine Dewar, a daughter of the principal of the College. In 1860 Aberdeen's two colleges (Marischal and King's) merged, and Maxwell was one of the professors who lost their job. He did not suffer financially as a result as he had a private income (from his Glenlair estate) far in excess of a professor's salary. He moved to King's College (London) and in 1861 was elected to the Royal Society. In 1865 he resigned his chair at King's College and went back to Scotland to enlarge Glenlair House and to write his famous *Treatise on*

Electricity and Magnetism. As the proprietor of an 1,800 acre estate, Maxwell had all the qualities of the better kind of Victorian country gentleman: cultivated, considerate of his tenants, active in local affairs, and an expert swimmer and horseman. In 1871 he returned to Cambridge as the first professor of experimental physics. With funding from the seventh Duke of Devonshire he created the Cavendish Laboratory, which opened in 1874 and in due course became one of the great research centres of the world. He died in Cambridge in 1879 at the age of 48 from abdominal cancer. His mother had died at the same age and from the same illness.

Maxwell made significant contributions to a number of fields. The longest paper he wrote (he spent 4 years, from 1856 to 1860, working on it) concerns the nature of Saturn's rings. He showed that the planet's rings are not solid, liquid or gas, but consisted of a vast number of independent particles. But his fame surely rests on the contributions he made to thermodynamics and statistical mechanics (see Sect. 9.2.2), and above all on his formulation of the basic equations of electromagnetism that bear his name. In 1865, while at King's College, he ended a letter to his cousin Charles Cay with the following remark: *I have also a paper afloat containing an electromagnetic theory of light, which till I am convinced to the contrary, I hold to be great guns.* He was right!

As we have seen in the preceding sections of this chapter, a lot of progress had been made in electromagnetism and optics. But there was not a consistent all-embracing theory of the observed phenomena. In particular, though Fresnel's formulae provided a correct description of reflection, refraction and of the polarization of light in terms of vibrations of an elastic ether, the fundamental relation between electromagnetism and light had not been discovered. As we shall see Maxwell was able by extending appropriately the existing formulae, to combine them into a consistent set of equations (Maxwell's equations) which could explain light as one aspect of electromagnetism, and to predict that light was only a part of a much wider spectrum of *electromagnetic waves*, which includes waves of much lower and much higher frequencies. Maxwell developed his theory in a number of papers he wrote in the 1860s and in his *Treatise on Electricity and Magnetism* which was published in 1873.

We remember that Faraday believed that field lines were not just convenient geometrical lines (to help us visualize a vector field) but actual physical lines rather like elastic bands with an extra sideways repulsion. Electric and magnetic field lines interlocking with each other were responsible for the observed electric and magnetic forces. Maxwell thought Faraday's *lines* far too simple a model to account for the complexity of the observed phenomena but, like Faraday, he thought it was of central importance both to elucidate the laws and to reveal the underlying fundamental mechanisms even if that meant postulating the existence of unobservable ethereal models. Whether this insistence on an underlying mechanical system derived from the need to avoid action at a distance or from an almost religious conviction about the nature of the physical world we can not tell. Maxwell, like Faraday and Kelvin, was certainly deeply religious. In any case, in a paper he wrote in 1861, Maxwell devised an *ether* full of tiny rotating *molecular*

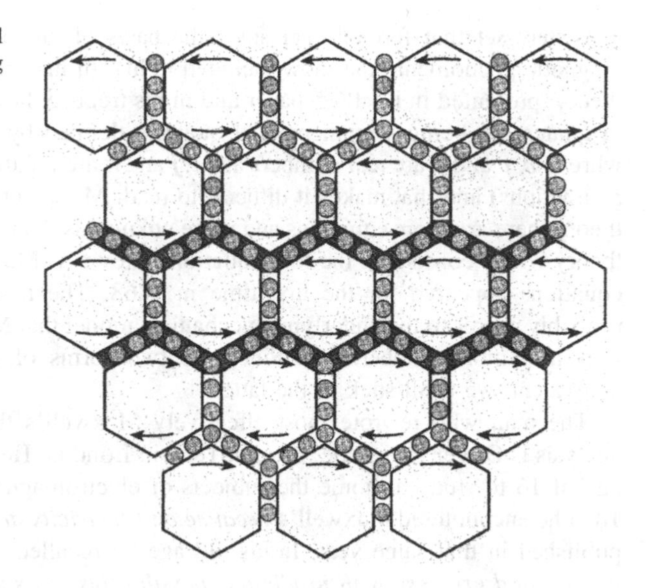

Fig. 10.16 Maxwell's model of ether, consisting of rotating molecular vortices meshed with smaller gearwheel particles (Redrawn from F. Everitt: *J. C. Maxwell*, Physics World, December 2006.)

vortices meshed with smaller *gearwheel particles* (as shown schematically in Fig. 10.16) and reasoned that: like tiny spinning Earths, each vortex shrinks axially and expands sideways, giving rise to the stress patterns suggested by Faraday.

Maxwell emphasized that the above model was speculative and not the real thing, but he maintained that it was a useful way to understand electromagnetism: In a wire, the particles flow freely giving rise to an electric current. In space they are counter-rotating idle wheels between vortices to make successive ones turn in the same direction, explaining in this fashion the magnetic force in Faraday-like terms. At the same time the model, by identifying the energy associated with the magnetic field with the kinetic energy of the vortices suggested that this energy is in the space surrounding the wire. Maxwell considered the electric force after submitting two papers on the magnetic force. He assumed that the ether is elastic and that the electric force arises from the potential energy needed to distort this elastic ether. Arguing that an elastic ether ought to transmit waves, he proceeded to calculate the speed of propagation of these waves in terms of the interlocking electric and magnetic fields in the ether. Using the ratio of magnetic to electric force, which had been determined experimentally by Weber (see Sect. 10.2.4), he discovered that the speed of the waves in the ether equaled the speed of light, which was known at the time with an accuracy of 1 %. *We can scarcely avoid the inference,* he wrote, *that light consists in the transverse undulations of the same medium which is the cause of electric and magnetic phenomena.*

By the mid 1860s Maxwell had assembled a number of equations which related electric and magnetic phenomena to each other and to electromagnetic waves. The latter could have frequencies, according to the equations, extending from near zero to visible light (about 10^{14} cycles per second) and to much higher frequencies (with no apparent upper limit). Maxwell was able to collect these equations into a consistent set that formed ever since the basis of our understanding of electromagnetic phenomena (that now included light). In his original formulation of the theory, presented in his 1865 paper and in his treatise, he made use of quaternions (a quaternion being a mathematical quantity defined by $Q = a + b\boldsymbol{i} + c\boldsymbol{j} + d\boldsymbol{k}$, whre a, b, c and d are real numbers and $\boldsymbol{i}, \boldsymbol{j}$ and $\boldsymbol{k}$ are the unit vectors in the x, y and z directions) and that makes it difficult to read. Moreover, in this first form of the theory there are more equations and more unknowns than in the second form of the theory which consists of the four equations known as Maxwell's equations. These equations appeared for the first time in 1868. The first form of the theory is probably nearer to the underlying mechanical model that Maxwell had devised, but it is also more cumbersome. Since the two forms of the theory are *in effect* equivalent, we shall stick to the latter.

The man who rewrote most effectively Maxwell's theory using only vector analysis is Oliver Heaviside. Born in 1850 in London, Heaviside left school at the age of 16 to study at home the subjects of electromagnetism and telegraphy. In 1873 he encountered Maxwell's *Treatise on Electricity and Magnetism* which was published in that same year. In his old age he recalled: *I saw that it was great, greater and greatest, with prodigious possibilities in its power. I was determined to master the book and set to work. I was very ignorant. I had no knowledge of mathematical analysis. It took me several years before I could understand as much as I possibly could. Then I set Maxwell aside and followed my own course. And I progressed much more quickly …It will be understood that I preach the gospel according to my interpretation of Maxwell.* In the 1880s it became gradually apparent that Heaviside's reformulation of Maxwell's theory (based entirely on vector analysis) was the more effective and economical and was in due course universally accepted. In 1891 the Royal Society recognized Heaviside's contribution to the mathematical description of electromagnetic phenomena by accepting him as a Fellow. Heaviseside went on to make other significant contributions to mathematics (among other things, he introduced the *step function* that bears his name) and to the theory of electromagnetism. He died in 1925.

Let me now introduce Maxwell's equations (in the manner of Heaviside) starting from the equations given in Sect. 10.4.2.

Maxwell assumed Eqs. (10.31) and (10.33) to be valid not only for static (time-independent) fields but generally, and this includes fields which vary with time, which derive from charge densities $\rho(\boldsymbol{r},t)$ and current densities $\boldsymbol{j}(\boldsymbol{r},t)$ which vary with time. And as we have already noted, he assumed Faraday's law of electromagnetic induction in its generalized form of Eq. (10.40). The generalization of Ampere's law [Eq. (10.37)] to time-dependent fields is more difficult for the following reason: taking the divergence of both sides of Eq. (10.37) we obtain: $\nabla \cdot (\nabla \times \boldsymbol{B}(\boldsymbol{r})) = \mu_0 \nabla \cdot \boldsymbol{j}(\boldsymbol{r})$. Now, $\nabla \cdot (\nabla \times \boldsymbol{B}) = 0$ whether $\boldsymbol{B}$ depends on the time or not, because of Eq. (10.41); but the same is not true for $\nabla \cdot \boldsymbol{j}$. Before explaining why this is so, let me introduce an equation which is useful not only in relation to the present argument but in many other situations as well.

Consider a volume V enclosed by a surface S. The total charge $Q(t)$ in V at time t is given by: $Q(t) = \oint_V \rho(\boldsymbol{r}, t) dV$. We have:

$$\frac{dQ}{dt} = \oint_V \frac{\partial \rho}{\partial t} dV = -\oint_S \boldsymbol{j}(\boldsymbol{r}, t) \cdot d\boldsymbol{S} \tag{10.42}$$

where $\boldsymbol{j}$ is the current density (the charge crossing a unit area per unit time in the direction the current flows), and $\boldsymbol{j}(\boldsymbol{r},t) \cdot d\boldsymbol{S}$ is the charge flowing *outward* through the surface element $d\boldsymbol{S}$ at point $\boldsymbol{r}$ on the surface (we remember that $d\boldsymbol{S}$ is normal to the surface pointing outward), so the *negative* of this quantity represents inward flow of charge, and therefore the expression on the right of Eq. (10.42) gives dQ/dt (the charge flowing into V per unit time). But, according to the divergence theorem [Eq. (10.29)]: $\oint_S \boldsymbol{j}(\boldsymbol{r}, t) \cdot d\boldsymbol{S} = \oint_V \nabla \cdot \boldsymbol{j}(\boldsymbol{r}, t) dV$. Therefore:

$$\oint_V \frac{\partial \rho}{\partial t} dV = -\oint_V \nabla \cdot \boldsymbol{j}(\boldsymbol{r}, t) dV$$

And because this equation is valid for any V, we conclude that:

$$\frac{\partial}{\partial t} \rho(\boldsymbol{r}, t) + \nabla \cdot \boldsymbol{j}(\boldsymbol{r}, t) = 0 \tag{10.43}$$

The above is known as the *continuity equation.* It is the mathematical way of saying that electric charge is conserved: it is never lost or created from nothing. The same equation applies in relation to *mass conservation* when as a consequence of mass flow, the mass density changes with time. In that case, $\rho(\boldsymbol{r},t)$ and $\boldsymbol{j}(\boldsymbol{r},t)$ will denote mass density and flow of mass density respectively. It is also worth noting that: $\boldsymbol{j}(\boldsymbol{r},t) = \rho(\boldsymbol{r},t)\boldsymbol{u}(\boldsymbol{r},t)$, where $\boldsymbol{u}(\boldsymbol{r},t)$ is the velocity of the mass carriers (atoms or molecules) or charge carriers (electrons or ions).

We can now return to the problem relating to the generalisation of Ampere's law to a.c. currents and time-dependent fields. In the case of an a.c. current in a circuit of metallic wire the current density is given by $j(t) = I(t)/A$ [the current $I(t)$ divided by the cross section A of the wire]; its magnitude and direction (along the wire) changes with the time, but at any given moment it has the same value along the wire and therefore $\nabla \cdot \boldsymbol{j}(\boldsymbol{r}, t) = 0$. When the circuit contains a capacitor (as in Fig. 10.17) an a.c. current will still flow through the circuit, but in this case there will be charge on the plates of the capacitor. On each plate, the charge changes from positive (during half of the cycle) to negative (during the other half of the cycle). Therefore: $\partial \rho / \partial t \neq 0$ on the plates and consequently $\nabla \cdot \boldsymbol{j} \neq 0$ there. It is then clear that, Ampere's law [Eq. (10.37)] must be modified in order to apply generally. How to modify it, is not at all obvious. Maxwell's ingenious solution to this problem was to add another term to the current which does *not* involve actual motion of charge. We remember that in his model of the ether, shown schematically in Fig. 10.16, the gearwheel particles exist outside the wire, as well as within it, in order to sustain the magnetic force. This additional current,

Fig. 10.17 When the e.m.f. varies with time (for example, $\varepsilon(t) = \varepsilon_0 cos(\omega t)$) an a.c. current will flow in the circuit but not in phase with the e.m.f.

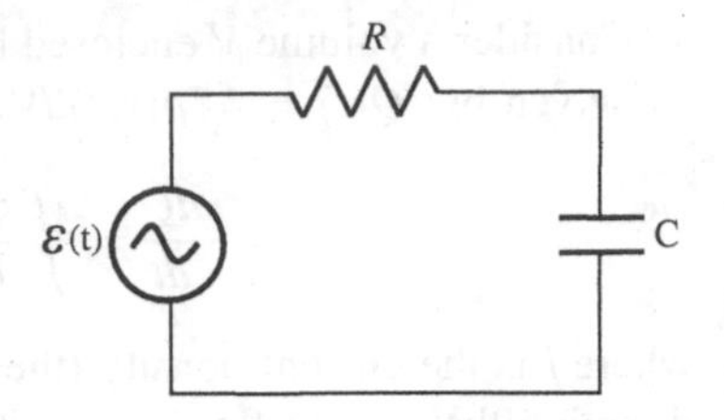

known as the *displacement current*, is given by $\varepsilon_0 \partial \boldsymbol{E}/\partial t$ according to Maxwell. Therefore Ampere's law becomes:

$$\nabla \times \boldsymbol{B}(\boldsymbol{r},t) = \mu_0(\boldsymbol{j}(\boldsymbol{r},t) + \varepsilon_0 \partial \boldsymbol{E}(\boldsymbol{r},t)/\partial t) \tag{10.44}$$

Let us verify that the divergence of the vector on the right of the above equation is indeed zero. We have:

$$\nabla \cdot (\boldsymbol{j} + \varepsilon_0 \partial \boldsymbol{E}/\partial t) = \nabla \cdot \boldsymbol{j} + \varepsilon_0 \partial \nabla \cdot \boldsymbol{E}/\partial t = \nabla \cdot \boldsymbol{j}(\boldsymbol{r},t) + \partial \rho(\boldsymbol{r},t)/\partial t = 0$$

Where we made use, first of Eq. (10.31), and then of Eq. (10.43).

In the example of Fig. (10.17), the displacement current density, $\varepsilon_0 \partial \boldsymbol{E}/\partial t$, is significant in the space between the plates of the capacitor (where $\boldsymbol{j} = 0$). The field $\boldsymbol{E}(t)$ there is of course related to the charge on the plates of the capacitor (see Fig. 10.15). Finally, the validity of Eq. (10.44) was confirmed experimentally by observing the magnetic field generated about the above displacement current. In a metallic wire the displacement current is not zero but it is much smaller than $\boldsymbol{j}$ even at the highest possible (technologically) frequencies, and can for practical purposes be neglected.

We have now all of Maxwell's equations; let me put them together:

$$\nabla \cdot \boldsymbol{E} = \rho/\varepsilon_0 \tag{10.31/10.45a}$$

$$\nabla \cdot \boldsymbol{B} = 0 \tag{10.33/10.45b}$$

$$\nabla \times \boldsymbol{E} = -\partial \boldsymbol{B}/\partial t \tag{10.40/10.45c}$$

$$\nabla \times \boldsymbol{B} = \mu_0(\boldsymbol{j} + \varepsilon_0 \partial \boldsymbol{E}/\partial t) \tag{10.44/10.45d}$$

These are known as Maxwell's equations for vacuum. All quantities are in general (or can be) functions of $\boldsymbol{r}$ and t. *If* we know $\rho(\boldsymbol{r},t)$, a charge distribution in vacuum, and $\boldsymbol{j}(\boldsymbol{r},t)$, a current density distribution in vacuum, the above equations allow us to determine the corresponding fields $\boldsymbol{E}(\boldsymbol{r},t)$ and $\boldsymbol{B}(\boldsymbol{r},t)$ in vacuum. I may add that in most cases *air* is for practical purposes as good as vacuum, and currents in metallic wires are in effect currents in vacuum (as we have indeed assumed in our exposition so far). It is worth noting, to begin with, that the formulae we have given in the previous sections relating to electrostatics and to stationary currents are obtained from the above equations by putting $\partial \boldsymbol{B}/\partial t = \partial \boldsymbol{E}/\partial t = 0$. And electromagnetic induction is of course described by Eq. 10.40/10.45c. Maxwell

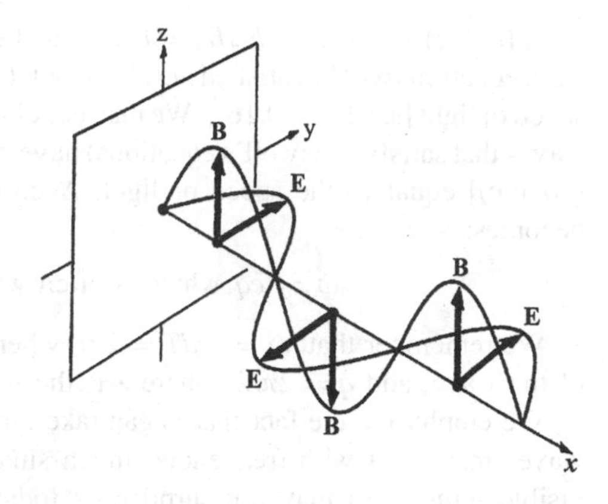

Fig. 10.18 An electromagnetic wave propagating in the x-direction

showed that the above equations describe light as well. Visible light can be identified with *electromagnetic waves* within a limited region of frequencies; electromagnetic waves exist (the above equations tell us) at frequencies which extend way above and way below the visible region! Let me show you how this comes about.

The electromagnetic (EM) wave shown in Fig. 10.18 has an electric component $\boldsymbol{E}$ in the y-direction and a magnetic component $\boldsymbol{B}$ in the z-direction, and it propagates in the x-direction in empty space. We write:

$$\boldsymbol{E} = E_y(x,t)\boldsymbol{j} = E_0 cos(qx - \omega t)\boldsymbol{j} \tag{10.46a}$$

$$\boldsymbol{B} = B_z(x,t)\boldsymbol{k} = B_0 cos(qx - \omega t)\boldsymbol{k} \tag{10.46b}$$

The wavefront (an imaginary surface of constant phase; see also Sect. 5.2.2) of the above wave is a plane normal to the x-direction. Its speed of propagation is determined by the displacement of, say, the maxima of the fields, which occur at $qx - \omega t = 0$. Which implies that the wave propagates (advances) with a speed:

$$\upsilon = \omega / q \tag{10.46c}$$

In calling the wave described by Eq. (10.46) an EM wave, we *assumed* that the $\boldsymbol{E}$ and $\boldsymbol{B}$ satisfy Eq. (10.45) in *empty* space. Knowing the facts, after Maxwell's discovery, it is indeed easy to show that the $\boldsymbol{E}$ and $\boldsymbol{B}$ given above satisfy the said equations, provided we put: $\upsilon = c$ (the speed of light). Let us do so. In empty space, $\rho = 0$ and $\boldsymbol{j} = 0$ and Eq. (10.45) simplify accordingly. Eq. (10.45a) becomes: $\nabla \cdot \boldsymbol{E} = 0$; from the definition [Eq. (10.28)] of the divergence, and the fact that $\boldsymbol{E}$ has only a y-component which depends only on x, we obtain: $\nabla \cdot \boldsymbol{E} = \partial E_y / \partial y = 0$; therefore Eq. (10.45a) is satisfied. Similarly, $\nabla \cdot \boldsymbol{B} = \partial B_z / \partial z = 0$ and therefore Eq. (10.45b) is also satisfied. In relation to the remaining two equations we note that: according to the definition [Eq. (10.34)] of the ***curl*** of a vector, $\nabla \times \boldsymbol{E} = (\partial E_y / \partial x)\boldsymbol{k} = -qE_0 sin(qx - \omega t)\boldsymbol{k}$, when $\boldsymbol{E}$ is given by Eq. (10.46a); and similarly when $\boldsymbol{B}$ is given by Eq. (10.46b), $\nabla \times \boldsymbol{B} = -(\partial B_z / \partial x)\boldsymbol{j} = qB_0 sin(qx - \omega t)\boldsymbol{j}$. On the other hand: $\partial \boldsymbol{B} / \partial t = \omega B_0 sin(qx - \omega t)\boldsymbol{k}$, and $\partial \boldsymbol{E} / \partial t = \omega E_0 sin(qx - \omega t)\boldsymbol{j}$. Therefore

Eq. (10.45c) reduces to $E_0/B_0 = \omega/q$; and Eq. (10.45d) to $E_0/B_0 = (\varepsilon_0\mu_0)^{-1}q/\omega$. Both equations will be satisfied *only* if we put: $(\omega/q)^2 = (\varepsilon_0\mu_0)^{-1} = c^2$, where c *is* the speed of light [see Eq. (10.16)]. We may conclude that EM waves (by which we mean waves that satisfy Maxwell's equations) have a propagation velocity [defined by Eq. (10.46c)] equal to the speed of light. Accordingly, for EM waves Eq. (10.46c) becomes:

$$\omega = cq, \text{which is often written as} : \lambda f = c \tag{10.47}$$

We remember that: $\omega = 2\pi/T = 2\pi f$, where T is the period and f the frequency of the wave; and $q = 2\pi/\lambda$, where λ is the wavelength of the wave.

We emphasize the fact that ω can take any value: $0 < \omega < \infty$. Therefore EM waves may exist with frequencies much smaller and much greater than those of visible light. This may not surprise us today, but it was not so at the time of Maxwell's discovery. In conclusion: Eq. (10.46a) represent an EM wave propagating in the x-direction with a frequency, ω, which can have any positive value; the corresponding q is determined by Eq. (10.47); the value of E_0 can be chosen arbitrarily but B_0 must be determined accordingly: $B_0 = c^{-1}E_0$.

I give below the frequency f in Hz (cycles per second) and the wavelength λ in meters, for different regions of the EM spectrum:

a.c. currents $f < 10^2$ Hz; $\lambda > 10^6$ m
radio waves $10^4 < f < 10^8$ Hz; $10^4 > \lambda > 1$ m
microwaves $10^8 < f < 10^{13}$ Hz; $1 > \lambda > 10^{-4}$ m
visible light f about 10^{14} Hz; λ about 3×10^{-6} m
X-rays $10^{16} < f < 10^{19}$ Hz; $10^{-8} > \lambda > 10^{-11}$ m
gamma-rays $f > 10^{19}$ Hz; $\lambda < 10^{-11}$ m

Finally, the EM waves described by Eqs. (10.46) propagate in the x-direction, but obviously an EM wave can propagate in *any* direction. The direction of propagation of an EM wave is that of its *wave vector*: $\boldsymbol{q}$. In Eqs. (10.46) this is: $\boldsymbol{q} = q\boldsymbol{i}$. In general: $\boldsymbol{q} = q_x\boldsymbol{i} + q_y\boldsymbol{j} + q_z\boldsymbol{k}$ but *its magnitude* q *is always given by* Eq. (10.47); the corresponding wave (its electric component) is given by:

$$\boldsymbol{E}(\boldsymbol{r}, t) = \boldsymbol{E}_0(\boldsymbol{q})cos(\boldsymbol{q} \cdot \boldsymbol{r} - \omega t) \tag{10.48a}$$

Where $\boldsymbol{E}_0(\boldsymbol{q})$ is a vector normal to $\boldsymbol{q}$, as shown schematically in Fig. (10.19), but otherwise arbitrary; we note that for each $\boldsymbol{q}$ there are two independent plane waves corresponding to the two independent directions (to be denoted by $e = 1, 2$) of $\boldsymbol{E}_0(\boldsymbol{q})$ normal to $\boldsymbol{q}$. In each case, the associated magnetic field is given by:

$$\boldsymbol{B}(\boldsymbol{r}, t) = \boldsymbol{B}_0(\boldsymbol{q})cos(\boldsymbol{q} \cdot \boldsymbol{r} - \omega t), \text{with } \boldsymbol{B}_0(\boldsymbol{q}) = \omega^{-1}\boldsymbol{q} \times \boldsymbol{E}_0(\boldsymbol{q}) \tag{10.48b}$$

And we remember that $\omega = cq$, and that q may have any positive value. We note that that $\boldsymbol{E}$ and $\boldsymbol{B}$ are normal to each other and to the wave vector $\boldsymbol{q}$ as one expects for a *transverse* wave. And we further note that the wavefront is a plane normal to $\boldsymbol{q}$: for any point $\boldsymbol{r}$ on this plane: $\boldsymbol{q} \cdot \boldsymbol{r} = d$, where d is the normal distance to the plane, shown by the broken line in Fig. (10.19). And therefore at any time

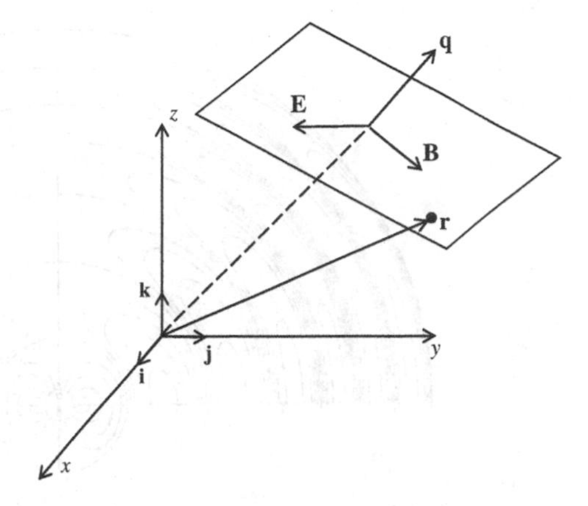

Fig. 10.19 The wavefront of a plane wave is a plane normal to $\boldsymbol{q}$

t all points on this plane will have the same phase. For this reason, the waves described by Eq. (10.48a) are called *plane waves.*

It is worth noting that a sum of plane waves [given by Eq. (10.48a)], i.e.

$$\boldsymbol{E}(\boldsymbol{r},t) = \sum_{q,e} \boldsymbol{E}_{0e}(\boldsymbol{q})cos(\boldsymbol{q}\cdot\boldsymbol{r}-\omega t) \tag{10.49}$$

where e takes the values 1, 2, for given $\boldsymbol{q}$, and $\boldsymbol{q}$ may take any value, together with the associated with it magnetic field, is also a solution of Maxwell's equations in free space (this follows from the linearity of these equations; see Sect. 7.2). Which means that EM waves can transmit signals of practically any shape across space.

We have, so far, said nothing about the generation of EM waves. According to Maxwell's equations, charge and current density distributions [$\rho(\boldsymbol{r},t)$ and $\boldsymbol{j}(\boldsymbol{r},t)$ in Eqs. (10.45)] which vary with time generate about them EM waves propagating with the speed of light.[18] We need not go into the mathematical details which prove the above. Figure (10.20) shows a snapshot of the electric field lines of the EM wave field about an oscillating dipole. The simplest dipole consists of two point charges $-q$ and $+q$ at a distance l from each other; the dipole moment $\boldsymbol{p}$ is defined by: $\boldsymbol{p} = q\boldsymbol{l}$, where $\boldsymbol{l}$ points from the negative charge to the positive one. In an oscillating dipole the one charge oscillates about the other so that $\boldsymbol{p} = \boldsymbol{p}_0 cos(\omega t)$. The wave field generated by this dipole has of course the same frequency.

After Maxwell's prediction that EM waves can be generated in the above manner, many scientists tried to produce them in the laboratory. Of course they could not produce visible light (it was not possible then and it is not possible now to produce by mechanical means currents oscillating at such high frequencies

[18] In the case of an a.c. current of low frequency (of say $f = 50$ Hz) generated by an ordinary a.c. dynamo, the wave field (of the same frequency) has a wavelength so long (many kilometers) that it is not observable. At these low frequencies the magnetic field due to the current can be obtained using the same formulae as for stationary currents.

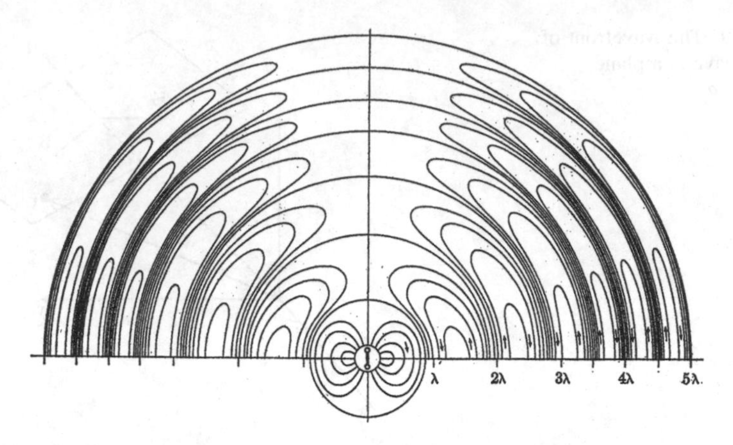

Fig. 10.20 Electric field *lines* produced by an oscillating *dipole*. The figure shows a cross section of the field in a plane containing the *dipole* (at the centre of the figure). The wave field is symmetric about the *horizontal axis*

(comparable to those of light). However, it is possible to generate EM waves at lower frequencies. The first to do so was the German physicist Heinrich Hertz.

Heinrich Hertz was born in Hamburgh in 1857, the son of a prosperous and cultured family. He showed at an early age an aptitude for science and for languages (he learned Arabic and Sanskrit). He studied physics and engineering at Dresden, Munich, and Berlin where he obtained his Ph.D. in 1880. He remained there working with Hermann von Helmholtz until 1883, when he took a post as a lecturer in theoretical physics at the University of Kiel. In 1885 Hertz became a professor at the University of Karlsruhe, and it was there that he generated, the first man-made EM waves. His (radio wave) transmitter consisted of a high-voltage induction coil and a capacitor in a circuit which included a small spark gap (between two spheres of 2 cm radius) to cause a spark discharge between them oscillating at a frequency determined by the values of the capacitor and the induction coil. [A circuit consisting of a high-voltage coil (one with many turns) and a capacitor can maintain an a.c current once the capacitor is charged: as the capacitor discharges through the coil, its electric (potential) energy becomes magnetic (kinetic) energy associated with the current through the coil; which in turn becomes electric energy as the capacitor is charged again in reverse fashion. The oscillating current eventually stops when the energy is dissipated into heat due to the resistance of the wire and partly due to the radiation loss (energy carried away by the generated EM waves). The wave field generated about the spark gap in Hertz's circuit is much greater than it would be in the absence of it.] The wave field produced by Hertz using the above transmitter had a frequency of about 5×10^8 Hz with a corresponding wavelength of about 60 cm, and it could be detected by a receiver several meters away from the transmitter. The receiver used by Hertz was a piece of copper wire 1 mm thick, bent into a circle with a diameter of 7.5 cm, with a small gap (a few hundredths of a *mm*) at one end of which was a small metallic sphere opposite a pointed wire. The presence of induced current in the receiver was signaled by sparks across the gap between the pointed wire and the sphere.

In subsequent, more advanced, experiments, Hertz measured the speed of the generated electromagnetic radiation and confirmed that it was the same with that of light, and he also showed that it reflected and refracted in the same way as light (see next section). Finally, Hertz devised his own mathematical methods to calculate the radiation field in accordance, always, to Maxwell's equations. But he did not realize, at the time, the practical importance of his discovery. He stated that: *It is of no use whatsoever,... this is just an experiment that proves Maestro Maxwell was right. We just have these mysterious electromagnetic waves that we cannot see with the naked eye. But they are there.* Hertz died in 1894, at the age of 36, in Bonn. The unit of frequency, the *Hertz* (*Hz*), is named after him.

The practical importance of the Hertzian waves was demonstrated by Guglielmo Marconi, an Italian engineer born in 1874, who made the first radio transmission in the year of Hertz's death. Moving to London in 1896, he worked for a few years on improving the range and reliability of his equipment; and in 1901 he was able to transmit Morse signals across the Atlantic Ocean, establishing radiotelegraphy and the use of radio waves as a communication medium. In 1909 Marconi shared the Nobel Prize for physics with Karl Braun. He died in 1937.

I would like to conclude this section with a brief discussion relating to the energy associated with an electromagnetic field. For this purpose allow me to write down an equation which derives from Maxwell's equations (but which we need not prove here). The equation is:

$$\frac{\partial}{\partial t}\int_{v}\frac{1}{2}(B^2/\mu_0+\varepsilon_0 E^2)dV = \int_{v}\frac{1}{2}(\boldsymbol{E}'\cdot\boldsymbol{j}-j^2/\sigma)dV - \int_{S}(\boldsymbol{E}\times\boldsymbol{B}/\mu_0)\cdot d\boldsymbol{S} \tag{10.50}$$

The volume—integral on the left of the equation represents the energy of the EM field in the volume *V*, and its time derivative tells us how much it changes per unit time. The volume integral involving ($\boldsymbol{E}'\cdot\boldsymbol{j}$) is the work done per unit time by the e.m.f. which maintains the current density $\boldsymbol{j}$. [We remember that $\boldsymbol{j}(\boldsymbol{r},t)$ and $\rho(\boldsymbol{r},t)$, in Maxwell's equations are given quantities which generate $\boldsymbol{E}(\boldsymbol{r},t)$ and $\boldsymbol{B}(\boldsymbol{r},t)$.] The volume integral involving j^2/σ, where σ is the conductivity of current-carrying wires within the volume *V*, represents the energy dissipated as heat in these wires per unit time (see Sect. 10.2.5). In the absence of the last term in the above equation, the EM energy in *V* will increase ($\frac{\partial}{\partial t}\int_{v}\frac{1}{2}(B/\mu_0+\varepsilon_0\mathrm{E})dV>0$) if the work done by the e.m.f. exceeds that which is dissipated as heat per unit time. The last term on the right of the above equation (taken with the *minus* sign) represents the flow of electromagnetic energy into the volume *V* through the surface *S* which encloses *V*.

The vector

$$N = (1/\mu_0)(E(r,t) \times B(r,t)) \qquad (10.51)$$

is known as the Poynting vector. According to our analysis of Eq. (10.50), N can be taken to represent the electromagnetic field energy flow per unit area (normal to N) per unit time. In the same way we can think of

$$U(r,t) = \frac{1}{2}(B^2/\mu_0 + \varepsilon_0 E^2) \qquad (10.52)$$

as the 'energy density' of the EM field. We remember that B and E, and therefore their magnitudes are functions of r and t.

Applying Eqs. (10.51) and (10.52) to the field of a plane wave, described by Eqs. (10.48), we obtain:

$$U(r,t) = \frac{1}{2}(B_0^2/\mu_0 + \varepsilon_0 E_0^2)(cos(q \cdot r - \omega t))^2 \qquad (10.53)$$

We are interested in the time-average over a period of this quantity. [We can equivalently take the average over a volume equal to λ^3, where $\lambda = 2\pi/q$.] The average over a period of the square of the cosine is ½; therefore the average energy density $U(q)$ associated with a plane wave (of wave vector q) is given by

$$U(q) = (B_0^2/\mu_0 + \varepsilon_0 E_0^2)/4 \qquad (10.54)$$

But, according to Eq. (10.48b), $B_0(q) = E_0(q)/c$; and because $\varepsilon_0\mu_0 = c^{-2}$, we have $B_0{}^2/\mu_0 = \varepsilon_0 E_0{}^2$. Therefore:

$$U(q) = \varepsilon_0 E_0^2(q)/2 \qquad (10.55)$$

Finally, the time-averaged Poynting vector of the plane wave described by Eqs. (10.48) is given by:

$$N(q) = U(q)c\,n \qquad (10.56)$$

where n is a unit vector in the direction of propagation ($n = q/q$).

The Poynting vector [Eq. (10.51)] is particularly useful in deciding whether energy is radiated (flowing out) from a system enclosed by a surface S. We note, however, that using it at individual points in space could be misleading. For example, in static superposed electric and magnetic fields there may be points in space where N is different from zero, but its integral over *any* closed surface S [as in Eq. (10.50)] *will be zero*.

Often the charge density $\rho(r,t)$ and the current distribution $j(r,t)$ which give rise to a radiation field (electromagnetic waves) are continuous functions of r. But this is not always the case. We know that a particle of charge e at rest creates an electrostatic field about it. A particle moving with *constant* velocity u generates an electric and magnetic field about it which, however, does *not* radiate energy. We can see why this is so, as follows: the generation of energy (radiation) can not depend on our choice between two inertial systems the one moving with constant velocity with respect to the other; the charge being stationary in one, and moving

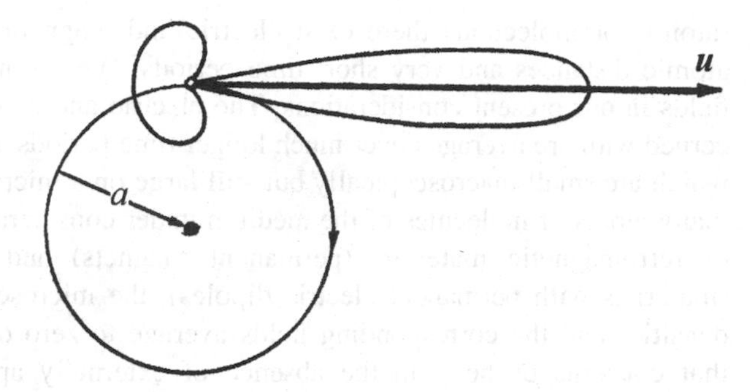

Fig. 10.21 Radiation field (in the plane of motion) of a charge e, whose acceleration $\gamma = du/dt$ is at *right angles* to its velocity $\boldsymbol{u}$. ($u = a\omega$, where a is the radius of the circle, and ω is the angular velocity of e. The magnitude of the acceleration is, of course, $\gamma = a\omega^2$.) For large velocities the radiation is more intense in the direction of $\boldsymbol{u}$ (as shown)

in the other. But Maxwell's equations tell us that an accelerated charge e, whatever the acceleration, whether e moves on a straight line or on a circle (as in Fig. 10.21) or any other path, generates propagating EM waves about it, i.e., it radiates energy. In the example of Fig. (10.21), the energy radiated per unit time is given by:

$$\frac{dW}{dt} = \frac{e^2 a^2 \omega^4}{6\pi\varepsilon_0 c^3}(1 - u^2/c^2)^2 \tag{10.57}$$

An exposition of Maxwell's equations is not complete without the formula which gives the force exerted on a particle of charge q moving with velocity, $\boldsymbol{u}$, in a field of $\boldsymbol{E}$ and $\boldsymbol{B}$. This force which is known as the Lorentz force (although it appeared in Maxwell's papers of the 1860's) is given by:

$$\boldsymbol{F} = q(\boldsymbol{E} + \boldsymbol{u} \times \boldsymbol{B}) \tag{10.58}$$

The reader will note that the component of the force due to $\boldsymbol{B}$ is an obvious extension of the force exerted on an element of current given by Eq. (10.10a).

For an example, let us consider the force exerted on a charge q by the $\boldsymbol{E}$ and $\boldsymbol{B}$ components of a plane wave [Eq. (10.48)]. The electric component of this force is proportional to E_0; the magnetic component of the force is proportional to $E_0(u/c)$, where c is the speed of light; therefore the magnetic component of the force is usually (when u is much smaller than c) negligible compared to the electric component of the force.

10.4.4 Electromagnetic Fields in Material Media

Matter consists of atoms and molecules which in turn are made up of positively charged nuclei and negatively charged electrons. Therefore on a microscopic scale (atomic or molecular) there exist electric and magnetic fields which vary over atomic distances and very short time periods. We are not concerned with these fields in our present considerations. The electric and magnetic fields we are concerned with are averages over much longer time periods and over regions of space which are small macroscopically but still large on a microscopic scale, containing many atoms or molecules of the medium under consideration. With the exception of ferromagnetic materials (permanent magnets) and ferro-electric materials (materials with permanent electric dipoles), the microscopic charge and current densities and the corresponding fields average to zero on the macroscopic scale that concerns us here, in the absence of externally applied fields. But in the presence of external electric and/or magnetic fields, induced charges and currents do appear which are macroscopically relevant and must be taken into account in the evaluation of the effective electric and magnetic fields in a material medium. For example, when a dielectric (non-conducting) medium (such as paper or glass, etc.) is placed between the charged plates of a capacitor (shown in Fig. 10.15b), the molecules in the dielectric become polarized to some degree: each molecule acquires a dipole moment pointing in the direction of the electric field between the charged plates of the capacitor. Inside the dielectric the induced charges on the dipoles cancel on averaging, but the induced charges on the surfaces of the dielectic remain and must be taken into account when evaluating the (macroscopic) electric field $\boldsymbol{E}$ in and about the dielectric.

Following Maxwell we proceed, in general, as follows: We introduce a polarization charge density ρ_P (arising from the polarization of the medium), to be distinguished from the (true) charge density ρ (on the plates of the capacitor in our example). The first of Maxwell's equations [Eq. (10.45)] becomes:

$$\varepsilon_0 \nabla \cdot \boldsymbol{E} = \rho + \rho_P \tag{10.59}$$

We next introduce a polarisation field $\boldsymbol{P}$ as follows:

$$\nabla \cdot \boldsymbol{P} = -\rho_P \tag{10.60}$$

Adding the above two equations we obtain:

$$\nabla \cdot (\varepsilon_0 \boldsymbol{E} + \boldsymbol{P}) = \rho \tag{10.61}$$

We define the *electric displacement* vector $\boldsymbol{D}$ to be:

$$\boldsymbol{D} = \varepsilon_0 \boldsymbol{E} + \boldsymbol{P} \tag{10.62}$$

In terms of which, Eq. (10.59) becomes:

$$\nabla \cdot \boldsymbol{D} = \rho \tag{10.63}$$

We note that $\boldsymbol{D}(\boldsymbol{r},t)$ depends only on the (true) charge density $\rho(\boldsymbol{r},t)$; but in order to obtain from it the electric field $\boldsymbol{E}(\boldsymbol{r},t)$, we must know $\boldsymbol{P}$, which can not be determined from Maxwell's equations alone, as it depends on the properties (the molecular structure) of the medium as well.

In many cases we can write:

$$\boldsymbol{P} = \chi \varepsilon_0 \boldsymbol{E} \tag{10.64}$$

where χ is a positive number, called the electric susceptibility of the medium. In such cases, Eq. (10.62) becomes:

$$\boldsymbol{D} = \varepsilon_0 \boldsymbol{E} + \boldsymbol{P} = \varepsilon_0 (1 + \chi) \boldsymbol{E} = \varepsilon \boldsymbol{E} \tag{10.65}$$

where $\varepsilon = \varepsilon_0\ (1 + \chi)$ is called the permittivity of the medium. Therefore, in a medium described by Eq. (10.64), the first of Maxwell's equations remains valid, except that the permittivity of vacuum ε_0 must be replaced by the permittivity of the medium ε. We have:

$$\nabla \cdot \boldsymbol{E} = \rho / \varepsilon \tag{10.66}$$

We must next consider how the last of Maxwell's equations is to be modified in a material medium. In a material medium there are *two* induced current densities to be added on the right of Eq. (10.45d). The first derives from the time variation of $\boldsymbol{P}$, and it is given by: $\partial \boldsymbol{P}/\partial t$. The second derives from induced (magnetization) currents in the medium and Maxwell chose to represent it by $\boldsymbol{j}_M = \nabla \times \boldsymbol{M}$, where $\boldsymbol{M}$ is known as the magnetisation of the medium. Taking into account these terms, Eq. (10.45d) becomes:

$$\nabla \times \boldsymbol{B} = \mu_0 (\boldsymbol{j} + \nabla \times \boldsymbol{M} + \varepsilon_0 (\partial \boldsymbol{E}/\partial t) + \partial \boldsymbol{P}/\partial t) \tag{10.67}$$

We note that the choice made by Maxwell for $\boldsymbol{j}_M$ secures that the divergence of the expression on the right equals zero as it must. Next, following Maxwell, we introduce a magnetic field $\boldsymbol{H}$, as follows:

$$\boldsymbol{H} = \boldsymbol{B}/\mu_0 - \boldsymbol{M} \tag{10.68}$$

It became customary to call $\boldsymbol{H}$ the magnetic field, and to call $\boldsymbol{B}$ the magnetic field intensity, but here (to avoid confusion) we shall call them: the $\boldsymbol{H}$ field, and the $\boldsymbol{B}$ field respectively. Using Eqs. (10.62) and (10.68), we can write Eq. (10.67) as follows:

$$\nabla \times \boldsymbol{H} = \boldsymbol{j} + \partial \boldsymbol{D}/\partial t \tag{10.69}$$

which tells us that $\boldsymbol{H}$ depends only on the (true) current density $\boldsymbol{j}$ and (through the second term) the (true charge density). But, again, in order to know $\boldsymbol{B}$, we must know $\boldsymbol{M}$, which can not be determined from Maxwell's equations alone, as it depends on the properties (the molecular structure) of the medium as well.

In many cases we can write:

$$\boldsymbol{M} = \chi_m \boldsymbol{H} \tag{10.70}$$

where χ_m is a positive number, called the magnetic susceptibility of the medium. In such cases, Eq. (10.68) becomes:

$$B = \mu_0(1 + \chi_m)H = \mu H \tag{10.71}$$

where $\mu = \mu_0\,(1 + \chi_m)$ is called the permeability of the medium. Therefore, if we assume that both, Eqs. (10.65) and (10.71), are valid, the last of Maxwell's equations [Eq. (10.45d)/Eq. (10.69)] becomes:

$$\nabla \times B = \mu\,(j + \varepsilon(\partial E/\partial t)) \tag{10.72}$$

The second [Eq. (10.45b)] and the third [Eq. (10.45c)] of Maxwell's equations are the same in a material medium as in free space. Therefore (in summary): when both, Eqs. (10.65) and (10.71), are valid, Maxwell's equations in the medium (of given ε and μ) are:

$$\nabla \cdot E = \rho/\varepsilon \tag{10.73a}$$

$$\nabla \cdot B = 0 \tag{10.73b}$$

$$\nabla \times E = -\partial B/\partial t \tag{10.73c}$$

$$\nabla \times B = \mu(j + \varepsilon(\partial E/\partial t)) \tag{10.73d}$$

which are identical with the equations in free space except that ε_0 is replaced by ε, and μ_0 by μ.

The first important conclusion to be drawn from the above is: Equations (10.73) have the same EM wave solutions in the medium as in free space, when $\rho = 0$ and $j = 0$; all formulae remain valid except that ε_0 and μ_0 must be replaced by ε and μ respectively. Accordingly the speed of propagation u of these waves in the medium is given by:

$$u^2 = (\varepsilon\,\mu)^{-1} = (\varepsilon_0(1+\chi)\,\mu_0(1+\chi_m))^{-1} = (\varepsilon_0\mu_0)^{-1}/((1+\chi)(1+\chi_m)$$

And therefore:

$$u = c/n \quad \text{where } n^2 = (1+\chi)(1+\chi_m) \tag{10.74}$$

And because χ and χ_m are positive numbers, we obtain $n > 1$, and therefore the speed of propagation of light in the medium is *smaller* than c (the speed of light in free space).

Let me conclude this section with a brief consideration of the reflection and refraction of a plane wave at a plane surface S between two media characterized by ε_1, μ_1 and ε_2, μ_2 respectively (as shown in Fig. 10.22). We remember that throughout this section we assume that the wavelength λ of the radiation is sufficiently large so that there is a large, practically constant, number of molecules in a volume of the order of λ^3 of the media involved (see Sects. 9.2.5 and 11.4).

The boundary conditions that the wave field must satisfy at the surface S follow from Maxwell's equations as follows: imagine a rectangle in the plane of incidence (as defined in the figure), one of its long sides in medium 1 and the other in medium 2. Let C be the closed line defined by this rectangle. Consider the integral

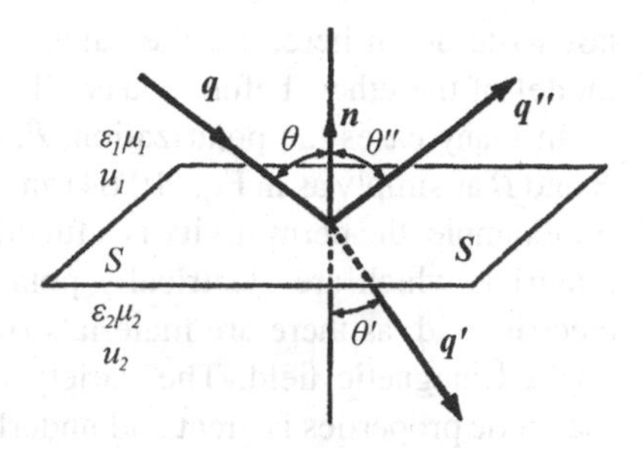

Fig. 10.22 The propagation vectors: $\boldsymbol{q}$ (incident wave), $\boldsymbol{q}'$ (refracted wave), and $\boldsymbol{q}''$ (reflected wave) are in the plane of incidence, which is determined by $\boldsymbol{q}$ and $\boldsymbol{n}$ (the normal to the reflecting surface)

$\oint_C \boldsymbol{E}\cdot d\boldsymbol{l}$. According to Stokes' theorem, this is equal to the surface integral $\oint_S (\nabla\times\boldsymbol{E})\cdot d\boldsymbol{S}$ over the area enclosed by the rectangle, which is equal (from Eq. 10.73c) to the integral $\oint_S(-\partial\boldsymbol{B}/\partial t)\cdot d\boldsymbol{S}$. Now if we let the width of the rectangle shrink to zero, the last integral goes to zero and therefore: $\oint_C \boldsymbol{E}\cdot d\boldsymbol{l} = 0$. But this integral is now reduced to two line integrals along the two sides of the rectangle (the one in medium 1 and the other in medium 2), the integration moving in opposite directions in the two media. If the two integrals are to cancel each other out whatever the length of the sides of the rectangle we must have: *the tangential (parallel to the surface S separating the two media) components of* **E** *in the two media must be equal at the surface S*. Applying the same method to the integral $\oint_C (\mathbf{B}/\mu)\cdot d\boldsymbol{l}$ and using Eq. (10.73d) with $\boldsymbol{j} = 0$, we find that: *the tangential components of* $\boldsymbol{H} = \boldsymbol{B}/\mu$ *in the two media must also be equal at the surface separating the media.*

Let $\boldsymbol{E}_0(\boldsymbol{q})cos(\boldsymbol{q}\cdot\boldsymbol{r}-\omega t)$ be the electric component of the incident wave (see Fig. 10.22); $\boldsymbol{E}_0'(\boldsymbol{q}')cos(\boldsymbol{q}'\cdot\boldsymbol{r}-\omega t)$ that of the refracted wave; and $\boldsymbol{E}_0''(\boldsymbol{q}'')cos(\boldsymbol{q}''\cdot\boldsymbol{r}-\omega t)$ that of the reflected wave. The continuity of the tangential fields at the surface S (we may assume this surface to be on the x–y plane at $z = 0$) is possible only if the cosines are the same on this surface ($z = 0$) for all three fields. This implies that the frequency ω does not change (we have assumed that much by writing the incident, reflected and refracted waves as above), but also that: $\boldsymbol{q}\cdot\boldsymbol{r} = \boldsymbol{q}'\cdot\boldsymbol{r} = \boldsymbol{q}''\cdot\boldsymbol{r}$ over the surface S; therefore the components of the wave vectors parallel to the surface must be equal:

$$\boldsymbol{q}_{//} = \boldsymbol{q}_{//}{}' = \boldsymbol{q}''_{//} \tag{10.75}$$

Also, the propagation vectors $\boldsymbol{q}$ and $\boldsymbol{q}''$ being in the same medium must have the same magnitude (we have: $\omega = u_1 q = u_1 q''$). This means (see Fig. 10.22) that $\theta = \theta''$. But with the speed of light in medium 2 being different, we have; $\omega = u_1 q = u_2 q'$; which, taken with Eq. (10.75), means that: $sin(\theta)/sin(\theta') = u_1/u_2$, which is of course Snell's law [see Eqs. (5.17) and (5.19)].

The continuity of the tangential components of the electric and magnetic fields at the surface between the two media allows us to determine also the amplitudes $\boldsymbol{E}_0'(\boldsymbol{q}')$ and $\boldsymbol{E}_0''(\boldsymbol{q}'')$ of the refracted and reflected waves from the known amplitude $\boldsymbol{E}_0(\boldsymbol{q})$ of the incident wave. The formulae obtained in this manner, which we need not write down here, are the same as those derived by Fresnel, using an elastic model of the ether, before Maxwell.

In many cases the polarization, $\boldsymbol{P}$, and the magnetization, $\boldsymbol{M}$, are not related to $\boldsymbol{E}$ and $\boldsymbol{B}$ as simply as in Eq. (10.64) and Eqs. (10.70 and 10.71). In the case of metals, for example, the permittivity is a function of the frequency: $\varepsilon = \varepsilon(\omega)$. And there are materials which are electrically polarized ($\boldsymbol{P} \neq 0$) in the absence of an applied electric field, as there are materials (magnets) where $\boldsymbol{M} \neq 0$ in the absence of an applied magnetic field. The variety of materials in respect to their electric and magnetic properties is great and underlies much of modern technology. The electric and magnetic properties of a material are determined by the microscopic structure of the material and, as a rule, a proper analysis of these properties requires a quantum–mechanical analysis (some examples will be given in Chap. 14).

Exercises

10.1. Four point charges q_1, q_2, q_3, q_4 are situated at the four corners $i = 1, 2, 3, 4$ respectively of a square of side a in the x–y plane with its centre at the origin of the coordinates. We have: $(x_1, y_1) = (-a/2, a/2)$, $(x_2, y_2) = (a/2, a/2)$, $(x_3, y_3) = (a/2, -a/2)$, $(x_4, y_4) = (-a/2, -a/2)$.

(a) Derive a formula for the electric field at the centre of the square.

Answer: $\boldsymbol{E} = A(q_1 - q_2 - q_3 + q_4)\boldsymbol{i} + A(-q_1 - q_2 + q_3 + q_4)\boldsymbol{j}$, where $A = (1/\sqrt{2})(2\pi\varepsilon_0 a^2)^{-1}$ and $\boldsymbol{i}$ and $\boldsymbol{j}$ are the unit vectors along the x and y directions.

(b) What is the force acting on a point charge $Q > 0$ at the centre of the square, (1) when $q_1 = q_2 = q > 0$ and $q_3 = q_4 = 0$, and (2) when $q_1 = q_2 = 0$ and $q_3 = q_4 = q < 0$.

Answer: (1): $\boldsymbol{F} = -\ 2QqA\boldsymbol{j}$; (2) $\boldsymbol{F} = -2QqA\boldsymbol{j}$.

10.2. Derive a formula for the electric field $\boldsymbol{E}(\boldsymbol{r})$ and for the (electrostatic) potential $V(\boldsymbol{r})$, for $0 < r < \infty$, due to a uniformly charged shell (total charge: Q) of radius R and infinitesimal thickness. The shell's centre is at the origin of the coordinates.

Hint: Use Gauss's law and the spherical symmetry of the charge distribution. And remember that $V(\boldsymbol{r})$ must be everywhere continuous.

Answer: $\boldsymbol{E}(r) = \frac{Q}{4\pi\varepsilon_0 r^2}\boldsymbol{e}_r$, for $r > R$; $= 0$, for $r < R$. $V(r) = \frac{Q}{4\pi\varepsilon_0 r}$ for $r > R$; $= \frac{Q}{4\pi\varepsilon_0 R}$ for $r < R$, where $\boldsymbol{e}_r$ is the unit vector in the radial direction.

10.3. (a) Derive a formula for the electric field $\boldsymbol{E}(\boldsymbol{r})$ and for the (electrostatic) potential $V(\boldsymbol{r})$, for $0 < r < \infty$, due to a charge Q uniformly distributed within a sphere of radius R. The sphere's centre is at the origin of the coordinates.

Answer: $E(r) = \frac{Q}{4\pi\varepsilon_0 r^2} e_r$, for $r > R$ $\quad V(r) = \frac{Q}{4\pi\varepsilon_0 r}$ for $r > R$

$= \frac{Q}{4\pi\varepsilon_0 r^2} \left(\frac{r}{R}\right)^3 e_r$, for $r<R$ $\quad V(r) = \frac{3Q}{8\pi\varepsilon_0 R} - \frac{Qr^2}{8\pi\varepsilon_0 R^3}$

(b) Calculate the potential energy of an electron at the centre of a sphere of radius R uniformly charged, when its total charge is equal to that of an electron.
Answer: $U = 3e^2/8\pi\varepsilon_0 R$

10.4. Derive a formula for the force per unit area ($f = F/A$) between the two plates of the charged condenser of Fig. 10.15, (1) directly from Eq. (10.3); (2) from the energy associated with the electrostatic field.

Answer: $f = \sigma^2/\varepsilon_0$, where $\sigma = Q/A$ is the charge density on the plates.

10.5. (a) An electron (charge $-e$) at a distance x from a metal surface induces a positive charge distribution on the surface of the metal (the total induced charge on the surface of the metal equals $+e$), which in turn attracts the electron towards the metal. Show that this force is $F = -e^2/(16\pi\varepsilon_0 x^2)$ and that the corresponding potential energy field (known as the image potential field) is $U(x) = -e^2/(16\pi\varepsilon_0 x)$.

Hint: Convince yourself that because of symmetry the electric field lines between the induced charge on the metal surface and the electron will be the same with those of two point charges: the electron at x and a positive charge ($+e$) at the position $-x$ (this being the mirror image of the electron with respect to metal surface).

(b) When we consider the force acting on an electron in the region between the plates of a condenser (see Fig. 10.15) we usually disregard the image force described in (a). Why and when is this justified?
Hint: See Fig. 11.10.

10.6. Using Eq. (10.9), show that the magnetic field due to current I in an infinitely long straight wire is that of Fig. 10.5 with $B = 2k_M I/L$.

Solution: Assume that the wire is along the z-axis extending from $-\infty$ to $+\infty$. Calculate $\boldsymbol{dB}$ at a point on the y-axis a distance L from the z-axis, due to Idz at point z of the wire: $\boldsymbol{dB} = -dB\boldsymbol{i}$, where $\boldsymbol{i}$ is the unit vector in the x-direction; $dB = k_M Ir\sin\theta dz/r^3$, where $\boldsymbol{r}$ is the vector from point z on the wire to point L on the y-axis, and θ is the angle between the z-axis and $\boldsymbol{r}$. We have: $L = r\sin\theta$ and $z = -r\cos\theta = L\cos\theta/\sin\theta$. We note that $\theta \to 0$ as $z \to -\infty$, and $\theta \to \pi$ as $z \to +\infty$. Therefore: $dB = k_M I\sin\theta/L$. Therefore, at the given point: $\boldsymbol{B} = -B\boldsymbol{i}$, and $B = \frac{k_M I}{L}\int_0^{\pi} \sin\theta d\theta = 2k_M I/L$. And because our choice of the y-direction is arbitrary, the above proves the result shown in Fig. 10.5.

10.7. A wire forms a square frame of side L. The frame is suspended vertically (it can rotate about the z-axis which defines the vertical direction) in the space between the poles of a permanent magnet where the magnetic field is given by: $\boldsymbol{B} = B\boldsymbol{j}$, where $\boldsymbol{j}$ is the unit vector in the y-direction. The *magnetic moment* $\boldsymbol{M}$ of the frame when a current I is flowing in it is defined by: $\boldsymbol{M} = I\boldsymbol{S}$, where $\boldsymbol{S}$ is a vector normal to the plane of the frame, with magnitude $S = L^2$, and pointing in the direction a right-turning screw advances when it rotates in the direction of the current flow in the frame.

(a) Show that the torque exerted on the frame is given by:

$$\boldsymbol{N} = N\boldsymbol{k}, \quad N = MB\sin\theta \tag{A}$$

Where $\boldsymbol{k}$ is the unit vector in the z-direction and θ is the angle between $\boldsymbol{M}$ and $\boldsymbol{B}$.

Note: The force on the different sides of the frame is obtained from Eq. (10.10b). Only the vertical sides contribute to $\boldsymbol{N}$.

(b) Show that

$$N = -\frac{dU}{d\theta}, \quad \text{where } U(\theta) = -\boldsymbol{M}\cdot\boldsymbol{B} = -MB\cos\theta \tag{B}$$

Accordingly, $U(\theta)$ can be taken as the potential energy associated with the orientation of the magnetic moment (dipole) $\boldsymbol{M}$ in the magnetic field $\boldsymbol{B}$. Convince yourself that $-dU = Nd\theta$ is the work done by the torque when the frame turns by an angle $d\theta$.

10.8. Derive a formula for the torque $\boldsymbol{N}$ exerted on a dipole, $\boldsymbol{p} = q\boldsymbol{l}$, in a uniform electrostatic field $\boldsymbol{E}$.

Answer: $\boldsymbol{N} = N\boldsymbol{k}$, where $\boldsymbol{k}$ is a unit vector normal to the plane defined by $\boldsymbol{p}$ and $\boldsymbol{E}$, and

$$N = -\frac{dU}{d\theta}, \quad \text{where } U(\theta) = -\boldsymbol{p}\cdot\boldsymbol{E} = -pE\cos\theta$$

where, θ is the angle between $\boldsymbol{p}$ and $\boldsymbol{E}$.

10.9. Calculate the energy dissipated as heat per unit time in a circuit of resistance R, when the current, I, is sustained by a voltage V.

Answer: $Q = VI = I^2R$.

10.10. Calculate the current $I(t)$ due to an *e.m.f.*, $\varepsilon = \varepsilon_0 \sin(\omega t)$, in a circuit consisting of an ohmic resistance R in series with a coil. This will be a circuit as in Fig. 10.17, but with the capacitor replaced by a coil. We assume that the *e.m.f.* induced by the changing magnetic flux through the coil is given according to Lentz law by: $-LdI/dt$, where L is an appropriate constant called the self-inductance L of the coil. The unit of self-inductance is the henry (H) introduced in Eq. (10.15).

Note: Ohm's law remains valid when the *e.m.f.* varies with the time, because the collision time (τ) in the formula for the conductivity [see Eq. (14.125)] is very much smaller than the period ($T = 2\pi/\omega$) of the *e.m.f.*. Therefore we can write:

$$\varepsilon_0 sin(\omega t) - LdI/dt = I(t)R$$

Which is the same as:

$$LdI/dt + I(t)R = \varepsilon_0 sin(\omega t) \qquad \text{(A)}$$

(a) Verify that: $I_c(t) = Asin(\omega t) + Bcos(\omega t)$, where $B = -(\omega L/R)A$ and $A = \varepsilon_0 R/(\omega^2 L^2 + R^2)$, satisfies Eq. (A).

(b) Consider the equation:

$$LdI/dt + I(t)R = 0 \qquad \text{(B)}$$

Verify that $I_h(t) = De^{-(R/L)t}$ satisfies Eq. (B) whatever the value of D.

(c) Hence show that: $I(t) = Asin(\omega t) + Bcos(\omega t) + De^{-(R/L)t}$ with A and B as above, and D any constant, satisfies Eq. (A).

(d) Determine D so that: $I(t = 0) = 0$. Answer: $D = -B$.

The term $De^{-(R/L)t} \approx 0$, for $t \gg L/R$, and for this reason it is usually neglected.

(e) Using the following formula of trigonometry,

$$A \sin x + B \cos x = \sqrt{A^2 + B^2} \sin(x + \theta), \quad \text{Where } \sin\theta/\cos\theta = B/A \qquad \text{(C)}$$

Show that: $I_c(t) = I_0 sin(\omega t - \theta)$, where $I_0 = \varepsilon_0/(R^2 + \omega^2 L^2)^{1/2}$ and $\sin\theta/\cos\theta = \omega L/R$.

We note the (negative) phase difference, θ, between the voltage and the current.

10.11. Calculate the current $I(t)$ due to an *e.m.f.*, $\varepsilon = \varepsilon_0 sin(\omega t)$, in a circuit consisting of an ohmic resistance R in series with a capacitor (as in Fig. 10.17).

Note: We proceed as in exercise (10.10), except that in place of $-LdI/dt$ we now have the potential difference ($-q/C$) generated by the charge q accumulating on the plates of the capacitor (its capacitance denoted by C). We write:

$$\varepsilon_0 sin(\omega t) - q/C = I(t)R$$

(a) By taking the time derivative of the above, show that it is equivalent to the following equation:

$$(R/\omega)dI/dt + (1/\omega C)I(t) = \varepsilon_0 cos(\omega t) \qquad \text{(A)}$$

(b) Verify that: $I_c(t) = A sin(\omega t) + B cos(\omega t)$, where $A = \varepsilon_0\omega^2C^2R/(1 + \omega^2C^2R^2)$ and $B = A/(\omega CR)$, satisfies Eq. (A).

(c) Consider the equation:

$$(R/\omega)dI/dt + (1/\omega C)I(t) = 0 \tag{B}$$

Verify that: $I_h(t) = De^{-t/CR}$ satisfies Eq. (B) whatever the value of D. Hence show that: $I(t) = A sin(\omega t) + B cos(\omega t) + De^{-t/CR}$ with A and B as above, and D any constant, satisfies Eq. (A).

(d) Determine D so that: $I(t = 0) = 0$. Answer: $D = -B$.

We note that the term $De^{-t/CR} \approx 0$, for $t \gg CR$, and for this reason it is usually neglected.

(e) Using formula (C) of the previous exercise, show that:

$$I_c(t) = I_0 sin(\omega t + \theta), \text{ where } I_0 = \varepsilon_0/(R^2 + 1/\omega^2C^2)^{1/2} \text{ and } \sin\theta/\cos\theta = 1/R\omega C$$

We note the (positive) phase difference between the current and the voltage (compare with the corresponding result of the previous exercise).

10.12. Proceeding in the manner of the previous two exercises, calculate the (steady) current $I_c(t)$ due to an *e.m.f.*, $\varepsilon = \varepsilon_0 \, sin(\omega t)$, in a circuit consisting of an ohmic resistance R in series with a capacitor (with capacitance C) and a coil (with self-inductance L).

Answer: $I_c(t) = I_0 sin(\omega t - \theta)$;

$$I_0 = \varepsilon_0/[R^2 + (\omega L - 1/\omega C)^2]^{1/2} \text{and } \sin\theta/\cos\theta = (\omega L - 1/\omega C)/R$$

10.13. Calculate I_0 and $sin\theta/cos\theta$ (and therefore θ) for the circuit of exercise 10.12 when: $\varepsilon_0 = 200$ V; $R = 300\ \Omega$; $C = 3.5 \times 10^{-6} F$; $L = 0.6$ H; $\omega = 380\ s^{-1}$.

Answer: $I_0 = 0.331$ A; $sin\theta/cos\theta = -1.746$, therefore $\theta = -60.2°$.

10.14. Calculate the current $I(t)$ in a circuit consisting of a capacitor (capacitance C) in series with a coil (of self-inductance L). This is the simplest form of a Hertz oscillator (see Sect. 10.4.3).

Note: In the absence of an external *e.m.f.*, the current is determined by the equation:

$$-LdI/dt - q/C = 0$$

Which is equivalent to: $\frac{d^2I}{dt} + \frac{1}{LC}I(t) = 0$ (A)

(a) Show that: $I(t) = Acos\omega t$ satisfies the above equation for $\omega = \sqrt{LC}$
(b) In reality the amplitude of the oscillating current diminishes with time. Why?

Answer: Energy is radiated in the form of EM waves, and energy is dissipated as heat because of the ohmic resistance of the circuit [which we disregarded in Eq. (A)].

(c) How is the above equation modified when the ohmic resistance of the circuit is taken into account?

Answer: $-LdI/dt - q/C = RI(t)$.

10.15. It is possible that a beam of light will be totally reflected at the interface between two media. When will this happen?

Hint: Consider the case of a beam incident on the interface between two media from within the medium with the greater refractive index.

Chapter 11
Cathode Rays and X-rays

11.1 Cathode Rays

The study of the passage of electricity through gases had lagged behind the corresponding study on metals and liquids. Faraday himself had engaged in research on electrical discharges in gases as early as 1838 but was hampered by the inefficiency of his vacuum pumps. In the years 1858–1862, the German physicist Plücher published a number of papers on his experiments on electrical discharges in low pressure vacuum tubes, the most important result of which was the discovery of *cathode rays*. He observed the faint blue glow which appeared next to the negative electrode, the cathode (see Fig. 11.1), and noticed that when it extended to the walls of the tube a greenish-yellow glow was produced on the glass. He established that the glow on the wall changed its position when a magnet was brought near the faint blue glow in the gas. In 1862 he spoke about the electric radiation which in rarefied gases diverged from particular points on the cathode and travelled to the walls of the tube. Some German scientists around Plücher believed that the *cathode rays* were waves of the luminiferous ether. However, in 1871 Varley suggested that the *cathode rays* were small negatively charged particles thrown off from the cathode. This view was confirmed in the 1870s by the English physicist William Crookes. The eldest of sixteen children of a canny tailor, who made a fortune through successful property investments, Crookes inherited enough money to be able to give up work and devote his life to science. Among other things he developed a method for producing vacuum tubes in which the air pressure was 75,000 times less than in previous vacuum tubes.[1] After a series of experiments using his own tubes and involving the deflection of cathode rays by a magnet, Crookes was convinced that *cathode rays* consisted of very small charged particles, which he called the *fourth state of matter* or an *ultra gas*. (It is worth remembering that though the facts of electrolysis were known, the concept of a positive or negative *atom* of electricity did not exist in the 1870s.) Further progress was made by the German physicist Philipp Edward Lenard, using

[1] We note that this discovery was instrumental in the later mass-production of incandescent bulbs by Thomas Edison.

A. Modinos, *From Aristotle to Schrödinger*,
Undergraduate Lecture Notes in Physics, DOI: 10.1007/978-3-319-00750-2_11,

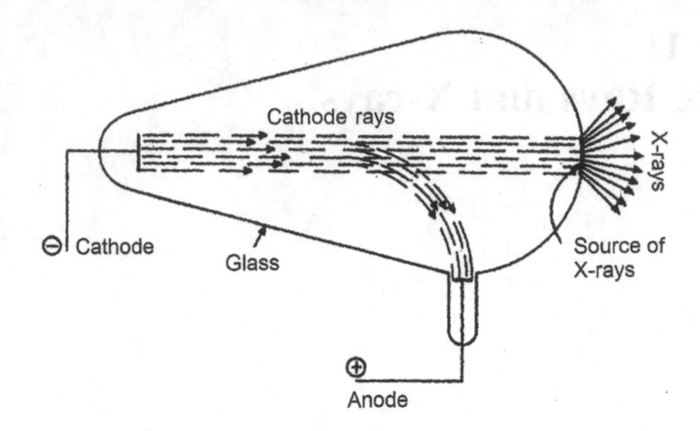

Fig. 11.1 A cathode-ray tube. The required high voltage between cathode and anode was produced by a Ruhmkorff induction coil: it consists of very finely-wound wire passing current from a battery, sudden and rapid interruption of which can give rise to a high voltage. We now know that electrons are knocked off the cathode by incident fast ions following a gas discharge

tubes, similar to those of Crookes, that he himself developed (as the one shown in Fig. 11.1). In particular, by having a thin aluminium window (transparent to cathode rays) opposite the cathode he was able to study the effects of the rays outside the tube. He observed that cathode rays caused phosphorescent bodies to light up in an arc around the aluminium window at distances up to 8 cm. A photographic plate was similarly blackened in these positions. And objects of any material with a thickness greater than 0.5 mm placed between the aluminium window and a phosphorescent screen cast shadows on the latter.[2] This is how things were when Roentgen entered the field in 1894.

11.2 The Discovery of X-rays

William Conrad Roentgen was born in 1845 in Lennep of Prussia. When he was 3 years old his family moved to Holland taking Dutch citizenship. After completing his secondary education, partly by studying at home (he was expelled from school for a minor reason of non cooperation), he studied at the University of Utrecht for about a year as a non-registered student before he obtained a place at the Zurich Polytechnic (in late 1865) to study Mechanical Engineering. One of his professors at Zurich was Rudolph Clausius. He got his degree in Mech. Eng. in 1868, and was awarded a Ph.D. in 1869 with a thesis entitled *Studies on gases*. On the strength of it, he became an assistant to August Kundt, who at the age of 29 succeeded Clausius in the chair of physics (when the latter moved to Würzburg).

[2] Lenard contributed significantly to the development of this field. For this, and his work relating to the photoelectric effect (see Sect. 13.1.2) Lenard was awarded the 1905 Nobel Prize for Physics.

In 1872 Roentgen married Bertha Ludwig, whom he had met and fell in love with when he was still a student, and with whom he lived happily for more than a half century until Bertha's death in 1919. They could not have children, but they adopted a child (a niece). They both loved nature and he particularly enjoyed hunting. In 1873 Roentgen got a chair in physics and mathematics at the Agricultural Academy at Hohenberg (about 70 miles from Strasburg), but he was not satisfied with the experimental facilities there, and in 1876 went to Strasburg as an associate professor, invited there by Kundt. It was during his years at Strasburg that Roentgen began to establish a reputation as a scientist. Eventually, at the age of 34, in 1879, he was made a professor at the University of Giessen in Hesse, where he stayed until 1888, when he accepted a chair at the University of Würzburg. It was at Würzburg that he would make his famous discovery of x-rays (in 1895). During his career, from 1868 to 1920, when he wrote his last paper (195 pages on the effect of x-rays and ordinary light on the electrical conductivity in crystals), Roentgen published 59 papers (he was the sole author of 46 of them) on a variety of topics (on the specific heat of gases; on electrical discharges in conducting and insulating crystals, on the viscosity, compressibility and surface tension of liquids, on oil films on water, etc.), but his fame rests on his three short papers announcing the discovery of X-rays. It happened as follows.

After acquiring his set of cathode-ray tubes (in 1894) Roentgen repeated many of Lenard's experiments, and by the summer of 1895, he was ready to try something new. He replaced Lenard's tube with the aluminium window with an *all-glass* tube through which cathode rays could not pass; and he also replaced the phosphorescent screen of Lenard (made of ketone) with one made from barium platinocyanide, which he knew to be only slightly fluorescent (directing ultraviolet light on it causes it to emit visible light). And then one evening in early November of 1895, the unexpected happened. When he switched on the Ruhmkorff coil connected to his cathode-ray tube, Roentgen observed a weak greenish light on a screen of barium platinocyanide several feet away from the all-glass tube. He was convinced that something was emanating from the tube, something other than cathode rays, which had the power to cause fluorescence at this large distance. For the next 8 weeks he very rarely left his laboratory, at times eating and sleeping there. Among the things he discovered was that the new rays had the ability to penetrate flesh but not the bone. He saw his finger bones shadowed on the screen as he was holding a disc of lead in front of a photographic plate which, he found, was sensitive to these rays. He persuaded his wife to place her hand on a cassette on top of a photographic plate and directed his new rays on it for 15 min. When developed, the photograph revealed the bone structure of his wife's hand, complete with shadows of the rings on her fingers. On the 25th December 1895 Roentgen delivered his first paper on the subject: *On a New Kind of Rays* (*Preliminary Communication*) to the secretary of the Würzburg Physico-Medical Society for publication in the Proceedings of the Society. And at the same time he sent copies of it and prints demonstrating the properties of the rays, which he called X-rays, to a number of distinguished scientists including Lord Kevlin (at Glasgow), Hendrik Lorentz (at Leiden in Holland), Poincare (in Paris), and Sir Arthur Schuster (in Manchester), who was soon taking photographs similar to those that Roentgen sent to him.

Roentgen wrote two more papers on X-rays, the second of which was published in March 1897 in the Proceedings of the Prussian Academy of Sciences. The properties of X-rays as summarized by Roentgen in his three papers are:

1. All substances are more or less transparent to X-rays; the transparency depending on density. Flesh is transparent to them, bones are not.
2. Many substances emit (fluorescent) light when acted upon by X-rays.
3. Photographic plates are sensitive to X-rays.
4. X-rays are generated by cathode rays in the glass of the discharge tube.
5. They are similar to light. They travel in straight lines. They can not be deflected by a magnet.
6. Interference effects of X-rays do not seem present.
7. Diffraction effects do not seem present (Fig. 11.2).

It took almost a quarter of a century before reflection, refraction and diffraction of X-rays were observed, contrary to what their discoverer had found (see Sect. 11.4), and their nature as electromagnetic waves of very short wavelength established. In the meantime some thought they were *longitudinal* light waves, some asserted that they were ultraviolet rays with an extremely short wavelength, while others believed

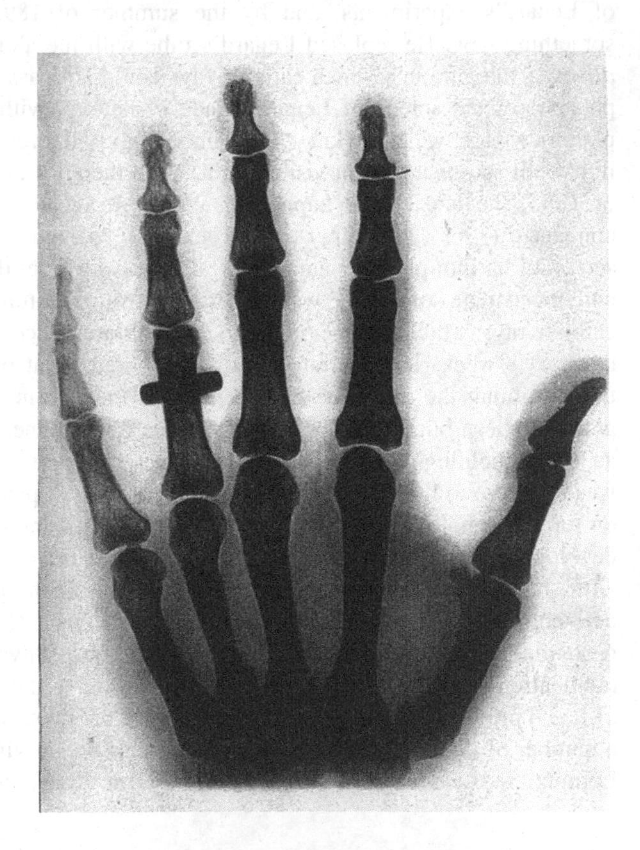

Fig. 11.2 X-ray photograph of a hand

that they had a very long wavelength like Hertzian or radio waves which were known to travel enormous distances and to pass through walls and bodies. And there were those that maintained that X-rays were not essentially different from cathode rays. Not knowing the exact nature of the X-rays did not prevent their application, particularly in medical diagnosis. Broken bones, tumours, and defects due to tuberculosis were indicated by them. It is not surprising that their discoverer became the most well known and admired scientist of his time. He received many honours (among them an honorary doctorate by his own University, and an honorary citizenship by his birthplace Lennep). And along with them came offers of commercial exploitation of his discovery. To the representative of the A.E.G. company he replied: *I am of the opinion that the discoveries and inventions of university professors belong to humanity and that they should not be hampered by patents, licenses or contracts, nor should they be controlled by any one group.* And when (in 1901) he was awarded the first Nobel prize in physics, he bequeathed the entire sum of the prize to the furtherance of scientific study at Würzburg University. Roentgen hated ceremony, he remained his usual self, staying away from publicity; working and enjoying nature in his spare time. In 1900 at the special request of the Bavarian Government he accepted the chair in physics and the directorship of the Physics Institute in Munich. As a director of the Institute, he helped re-establish a chair in theoretical physics there which was taken by Arnold Sommerfeld in 1906.[3] It was a member of Sommerfeld's group, Max von Laue and coworkers who, 6 years later, proved that X-rays diffracted, and established X-ray diffraction as a tool for determining the structure of crystals and, eventually of molecules. Roentgen was 75 years old when he finally retired. He died 3 years later in 1923.

11.3 The Discovery of the Electron

John Joseph Thomson was born in 1856, grew up in Manchester, and in due course enrolled at Owens College (the University of Manchester) to study engineering. When his father died, he switched, for financial reasons, to physics, chemistry and

[3] Arnold Sommerfeld was born in 1868 in Königsberg (East Prussia), the son of a physician. He studied mathematics at the university of his hometown graduating in 1891. After military service he worked at the mathematical institute of the University of Göttingen, where he delivered a paper on the mathematical theory of the diffraction of light. In 1897 he became a professor of mathematics at the Mining Academy of Clausthal, and in 1900 a professor of mechanics at the Technical Unversity of Aachen. In 1906 he was appointed to the chair of theoretical physics at the University of Munich, and occupied this post until he was awarded emeritus status in 1938. He was an excellent researcher and an esteemed and popular lecturer. He inspired a number of students including Wolfgang Pauli, Werner Heisenberg, Peter Debye, Hans Bethe and others. Today he is remembered mostly for his extension of Bohr's model of the atom (see Sect. 13.2.3) and his free-electron theory of metals (see Sect. 14.7.3). During his emeritus period he compiled his Lectures on Theoretical Physics, which were published in six volumes between 1943 and 1953. He died in 1951 following a traffic accident.

mathematics on the offer of a scholarship. And when 20-years old he won a second scholarship to study at Cambridge University. It is said that he was one of the few students there who could actually understand Maxwell's lectures. He eventually joined the Cavendish laboratory, becoming Cavendish Professor of experimental physics in 1884. He was apparently so clumsy with his hands that his colleagues, so the story goes, often tried to keep him out of the laboratory. But he was exceptionally gifted in devising experiments to test a theoretical model or hypothesis. He was convinced that cathode rays consisted of minute charged particles and he set out to prove that this is so experimentally. In 1897 he constructed a tube with a better vacuum than before, and he managed (the first to do so) to deflect the cathode rays by using an electric field rather than a magnetic field. Now, an electric field of constant value exerts a constant force on a moving charged particle and deflects it in the same way that the gravitational field deflects a projectile that would otherwise move horizontally (see Sect. 6.2.2). The deflection Y is proportional to the acceleration of the particle due to the electric field, which is proportional to the charge e of the particle and inversely proportional to the mass m of the particle: $Y = (e/m)\,A$, where A is a (known) quantity depending on the applied field. Thomson was able to measure Y with sufficient accuracy in different experiments with varying amounts of several residual gases in his vacuum tube. He consistently obtained the same value for the ratio e/m (very near its present-day value: 1.75890×10^{11} C/kg). The very large value of this ratio suggested a very small mass in agreement with Thomson's assumption that cathode rays consisted of minute particles. In a lecture he gave at the Royal Society in April 1897 he admitted that *the assumption of a state of matter more finely divided than the atom is a somewhat startling matter.* However, within 2 years the American Robert Millikan was able, by measuring the mass, velocity and charge of oil droplets falling in air between the horizontal plates of a capacitor, to determine the electric charge e of a single cathode-ray particle, to be known thereafter as the *electron.*[4] Knowing the values of e/m and e he could determine m. He found that the mass of the *negatively* charged electron is 1,800 times smaller than the mass of the lightest atom (hydrogen). The values obtained by Thomson and Millikan for e and m are not significantly different from the present-day values of these quantities. His discovery of the electron won Thomson the 1906 Nobel Prize for physics, and Millikan was awarded the same prize in 1923. After his great discovery Thomson devoted his time mostly to the administration of the Cavendish Laboratory.

[4] The forces acting on an oil drop, falling in air between the horizontal plates of a capacitor, are: its weight, a drag force due to the viscosity of the air, an upthrust force (which equals the weight of the air displaced by the oil drop), and the force qE, where q denotes the total charge on the drop and E is the electric field (in the vertical direction) between the plates of the capacitor. All the forces except the latter could be determined with reasonable accuracy by measuring the radius of the oil drop and the density of the oil. By adjusting the value (E) of the electric field so as to obtain a terminal (constant) velocity of the drop (implying a net zero force), one determines the charge q on the droplet. In every case, q was found to be an integral multiple of a certain unit, which was taken as the charge (e) of the electron.

Under his 35-year leadership the Laboratory became one of the most renowned centres of sub-atomic research. He died in 1940.

11.4 X-ray Diffraction

Between 1903 and 1905 Charles Barkla determined experimentally that X-rays scattered when passing through a body in proportion to its density and molecular weight, which led him to assert that X-rays were transverse vibrations as for light but of much higher frequency (much shorter wavelength), and that the increased scattering in the more massive substances was due to a larger number (per unit volume) of electrons in these substances. He also discovered in 1909 that X-ray tubes in which electrons impinge upon a metal target produce two types of X-radiation (as shown schematically in Fig. 11.3). The first is a continuous band of scattered X-radiation, and the second consists of one or more components at particular frequencies characteristic of the target material (for an explanation of this spectrum see Sect. 14.5 and footnote 9 of Chap. 14). The discussion concerning the nature of X-rays came to a definite conclusion with the work of Max von Laue.

Laue obtained his Ph.D. with Planck in 1903, and joined Sommerfeld's Institute for Theoretical Physics in 1903. He was already convinced that X-rays were electromagnetic waves. The clue that he required to prove it, was provided by Paul Ewald who was one of Sommerfeld's students. Ewald was concerned with the dispersion of light waves in crystals and came to Laue for a discussion. When the latter was informed about the interatomic distances in crystals he reasoned that scattering of X-rays by them would lead to some diffraction phenomena. They decided to experiment on that with the help of Walter Friedrich and Paul Knipping who had just finished his Ph.D. with Roentgen. In 1912 they obtained clear evidence of diffraction (of constructive interference) on a photographic plate behind a crystal when an X-ray beam passed through it from the other side. Laue subsequently developed a theory of diffraction, it won him the 1914 Nobel Prize for

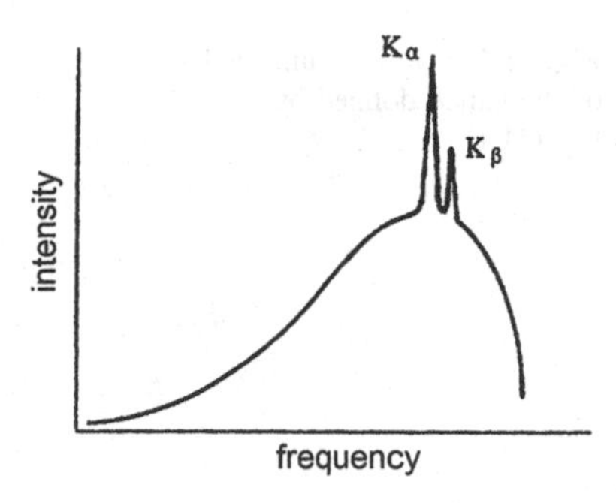

Fig. 11.3 A typical spectrum of the X-rays produced by electrons impinging on a metal target

Fig. 11.4 A crystal of touching *spheres* representing the atoms of the crystal. The lattice is a bcc lattice, with an atom on each lattice point

Physics, which made it possible to determine the structure of a crystal and in this way contributed significantly to the development of solid state physics.

One can summarize the essential results of his theory as follows. In a crystal the atoms are arranged periodically in space as shown, for example, in the picture of Fig. 11.4.

The lattice sites, $\boldsymbol{R}_n$, of a 3-dimensional lattice are defined by:

$$\boldsymbol{R}_n = n_1\boldsymbol{t}_1 + n_2\boldsymbol{t}_2 + n_3\boldsymbol{t}_3;\ n_1, n_2, n_3 = 0, \pm 1, \pm 2, \pm 3, \ldots \tag{11.1}$$

where, $\boldsymbol{t}_1$, $\boldsymbol{t}_2$ and $\boldsymbol{t}_3$ are three primitive (so-called) vectors which characterize the given lattice. Successive translations $\boldsymbol{t}_1$, $\boldsymbol{t}_2$ and $\boldsymbol{t}_3$ of the parallelepiped defined by $\boldsymbol{t}_1$, $\boldsymbol{t}_2$ and $\boldsymbol{t}_3$ (see Fig. 11.5) reproduce the entire lattice. In an actual crystal, there may be only one atom per lattice point (one atom in the primitive cell) in which case we can assume that it is centred on $\boldsymbol{R}_n$, or there may be more than one atom in the primive cell, in which case we have to specify the positions of the atoms in the primitive cell.

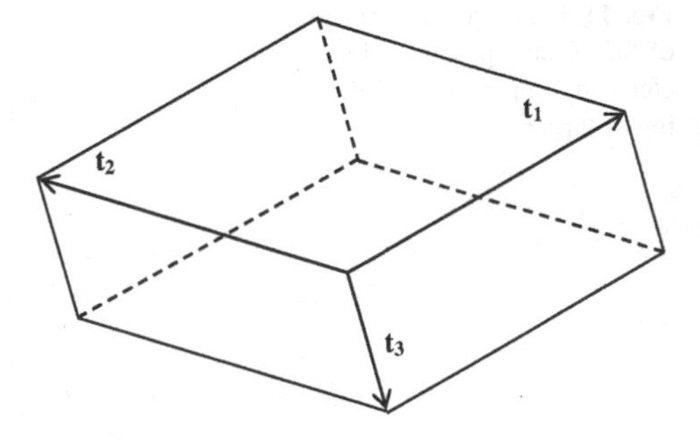

Fig. 11.5 Primitive unit cell of the lattice defined by Eq. (11.1)

The simplest lattice is the *simple cubic lattice* for which

$$t_1 = a\boldsymbol{i}, t_2 = a\boldsymbol{j},\ t_3 = a\boldsymbol{k}, \tag{11.2}$$

where $\boldsymbol{i}, \boldsymbol{j}, \boldsymbol{k}$ are the unit vectors along the x, y, and z directions of a Cartesian system of coordinates, and a is the lattice constant. One can visualize this lattice by imagining all space filled up with cubic boxes of the same volume a^3 and putting a lattice site at the centre of each cube. However, this lattice is not met in nature. In what follows I shall describe two lattices which are quite common among crystalline solids: these are the body-centred cubic lattice (denoted by bcc) and the face-centre cubic lattice (denoted by fcc).

The bcc lattice is defined by the following primitive vectors:

$$t_1 = (a/2)(-\boldsymbol{i}+\boldsymbol{j}+\boldsymbol{k}),\ t_2 = (a/2)(\boldsymbol{i}-\boldsymbol{j}+\boldsymbol{k}),\ t_3 = (a/2)(\boldsymbol{i}+\boldsymbol{j}-\boldsymbol{k}) \tag{11.3}$$

One can visualize this lattice by imagining all space filled up with cubic boxes of the same volume a^3 and putting a lattice site at the centre and on each corner of the cube (a corner is of course shared by eight cubes). This is demonstrated in Fig. 11.6. The volume of the parallelepiped defined by $\boldsymbol{t}_1$, $\boldsymbol{t}_2$ and $\boldsymbol{t}_3$ of Eq. (11.3) is: $a^3/2$, as expected: there are two lattice sites corresponding to the cube of volume a^3: one at the centre of the cube and a second one which we calculate as follows: there are 8 corners of the cube and each contributes 1/8 of a site (since the corner is shared by 8 cubes). When there is only one atom per lattice site we have a crystal like the one shown in Fig. 11.4.

The fcc lattice is defined by the following primitive vectors:

$$t_1 = (a/2)(\boldsymbol{j}+\boldsymbol{k}), t_2 = (a/2)(\boldsymbol{i}+\boldsymbol{k}), t_3 = (a/2)(\boldsymbol{i}+\boldsymbol{j}) \tag{11.4}$$

One can visualize the fcc lattice by imagining all space filled up with cubic boxes of the same volume a^3 and putting a lattice site at the centre of the cube and one on each of the midpoints of the twelve edges of the tube (an edge is of course shared by 4 cubes). This is demonstrated in Fig. 11.7.

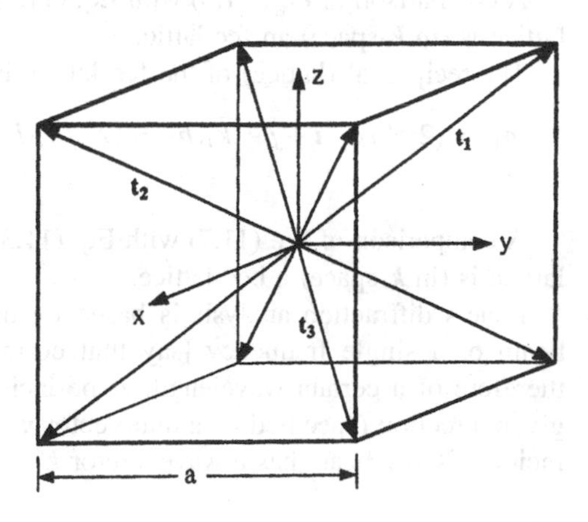

Fig. 11.6 Primitive vectors of the bcc lattice

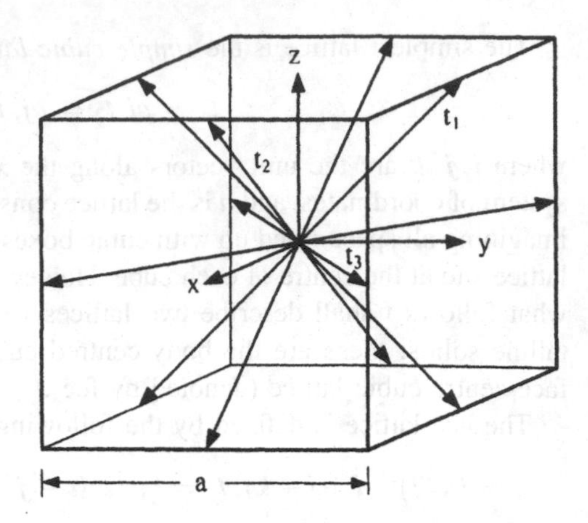

Fig. 11.7 Primitive vectors of the fcc lattice

The volume of the parallelepiped defined by t_1, t_2 and t_3 of Eq. (11.4) is: $a^3/4$, as expected: there are 4 lattice sites corresponding to the cube of volume a^3: one at the centre of the cube and three more which we calculate as follows: there are 12 edges of the cube and each contributes 1/4 of a site (since the edge is shared by 4 cubes).

There are a number of other crystal lattices which we need not describe here. The important thing that we need to know to describe Laue's essential result is that: to every lattice $\{R_n\}$ there corresponds a *reciprocal* lattice $\{K_h\}$ defined by:

$$K_h = h_1 b_1 + h_2 b_2 + h_3 b_3; h_1, h_2, h_3 = 0, \pm 1, \pm 2, \pm 3, \ldots \quad (11.5)$$

where b_1, b_2 and b_3 are vectors in k-space (wave vector space) uniquely determined by t_1, t_2, t_3 from the relations: $b_i \cdot t_j = 2\pi$ if $i = j$, and $b_i \cdot t_j = 0$ if $i \neq j$. We find that:

The reciprocal (lattice) of the bcc lattice is defined by the vectors:

$$b_1 = (2\pi/a)(j + k), b_2 = (2\pi/a)(i + k), b_3 = (2\pi/a)(i + j) \quad (11.6)$$

A comparison of Eq. (11.6) with Eq. (11.4) shows that the reciprocal of the bcc lattice is (in k-space) an fcc lattice.

The reciprocal (lattice) of the fcc lattice is defined by the vectors:

$$b_1 = (2\pi/a)(-i + j + k), b_2 = (2\pi/a)(i - j + k), b_3 = (2\pi/a)(i + j - k) \quad (11.7)$$

A comparison of Eq. (11.7) with Eq. (11.3) shows that the reciprocal of the fcc lattice is (in k-space) a bcc lattice.

Laue's diffraction analysis is based on the following theorem: Let an X-ray beam of a single frequency [say that corresponding to K_α of Fig. (11.3)] and therefore of a certain wavelength λ, be incident on a slab of the crystal along a given direction (specified by a unit vector n), which is the same as saying that the incident X-ray beam has a wave vector $k = (2\pi/\lambda)\, n$. *Then the scattered beam of wave vector* $\boldsymbol{k}'$ *will exhibit constructive interference only if* $\boldsymbol{k}' - \boldsymbol{k} = \boldsymbol{K}_h$, where $\boldsymbol{K}_h$ is a reciprocal vector of the crystal lattice under consideration.

We note that the magnitude of $\boldsymbol{k}'$ equals that of $\boldsymbol{k}$, and therefore: by observing the various directions of constructing interference we can determine the reciprocal lattice $\{\boldsymbol{K}_h\}$ and from it the space lattice $\{\boldsymbol{R}_n\}$ of the crystal under consideration.

The English physicist William Henry Bragg proposed a simpler way of looking at X-ray diffraction from crystals. This is demonstrated in Fig. (11.8). A crystal can be thought of as a sequence of planes of atoms normal to a given direction. The distance d between successive planes will of course be different depending on the direction; looking at the crystal of Fig. (11.4), we can see that d for the planes parallel to the side of the cube is different from d for the planes at 45° to this side. In every case, when an X-ray beam is directed at a particular face of the crystal the reflected (scattered backward) beams from the successive planes parallel to this face will interfere constructively when the difference in the distance travelled by the two rays shown in Fig. (11.8) is an integral multiple of the wavelength λ, i.e. when $n\lambda = 2d\sin\theta$, where n is an integer, assuming that λ is smaller than d (about 2×10^{-8} cm). Using this equation, sometimes referred to as Bragg's law, one can by measuring the angle θ at which constructive interference occurs, for a beam of known wavelength, determine the unknown inter-plane distance d.

In 1915 William Henry Bragg and his son William Laurence Bragg established through X-ray diffraction the crystalline structure of sodium chloride (common salt), showing that it did not consist of sodium chloride molecules but of periodically arranged ions of chlorine and sodium. This discovery won the Braggs (father and son) a shared Nobel Prize in Physics.

We should note that in the above analysis of X-ray diffraction we have assumed that the crystal extends over all space. Of course this is not true in reality, but in

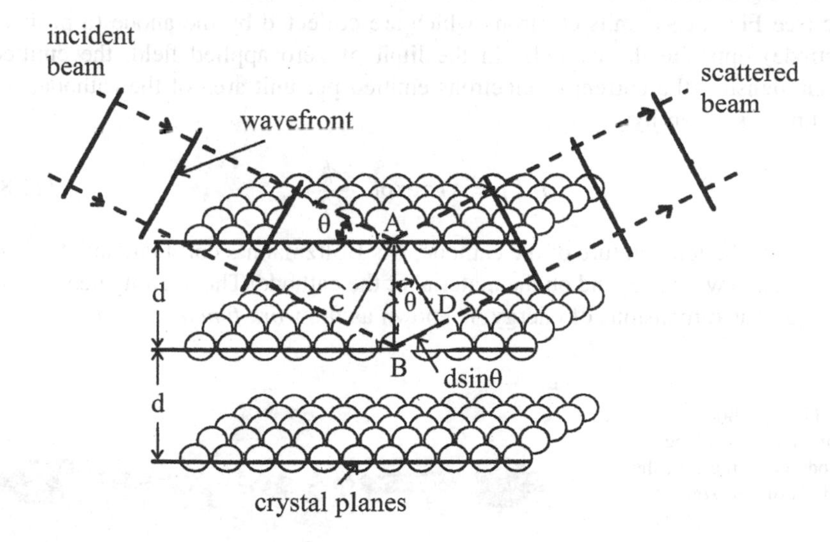

Fig. 11.8 Bragg's law of diffraction

relation to X-ray diffraction it is a good approximation when the spatial spread of the beam of X-rays is sufficiently smaller than the extension of the slab of the crystal it is incident upon.

In our discussion of X-ray diffraction we said nothing for the intensity of the diffracted/reflected beam. When data concerning the intensity of the diffracted/reflected beams are properly analysed, additional information is obtained about the contents of the unit cell of the crystal (the bit that extended periodically produces the macroscopic crystal), which is of particular importance if there are a number of different atoms in the unit cell (a number of different atoms per lattice site). Finally, we should mention that X-ray diffraction has been used extensively over the years for the determination of the structure of big molecules as well as for crystals. We need only mention that the *double-helix* structure of DNA was established (in 1953) on the basis of X-ray diffraction data.

11.5 Electron Emission and Electronic Valves

11.5.1 Thermionic Emission

In the early years of the 20th century it was discovered that a hot metallic wire emitted negatively charged particles which, following Thomson's discovery, were shown to be electrons. The first quantitative study of the phenomenon, known as thermionic emission, was made by O. N. Richardson. His paper, entitled *Negative radiation from hot platinum*, was published in 1902. Additional work by Richardson (in 1912) and by Max von Laue (in 1918) clarified the phenomenon which can be summarized as follows: A heated cathode (a metallic wire) in a vacuum tube (see Fig. 11.9) emits electrons which are collected by the anode (a positive electrode) opposite the cathode. In the limit of zero applied field, the emitted current density (the current of electrons emitted per unit area of the cathode, per unit time) is given by:

$$J(T) = AT^2 \exp(-\frac{\phi}{kT}) \tag{11.8}$$

where, T is the temperature of the cathode, k is Boltzmann's constant, and A and φ are constants which depend on the material of the cathode. The quantity denoted by φ, which has dimensions of energy, is known as the *work function* of the cathode.

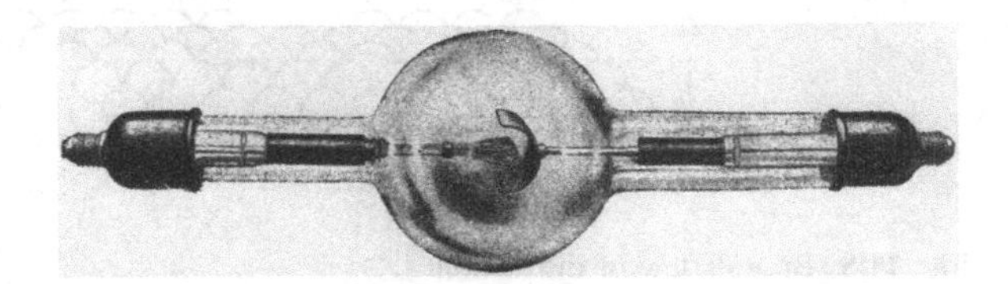

Fig. 11.9 A diode thermionic valve. The cathode is on the *left*; the anode is on the *right*

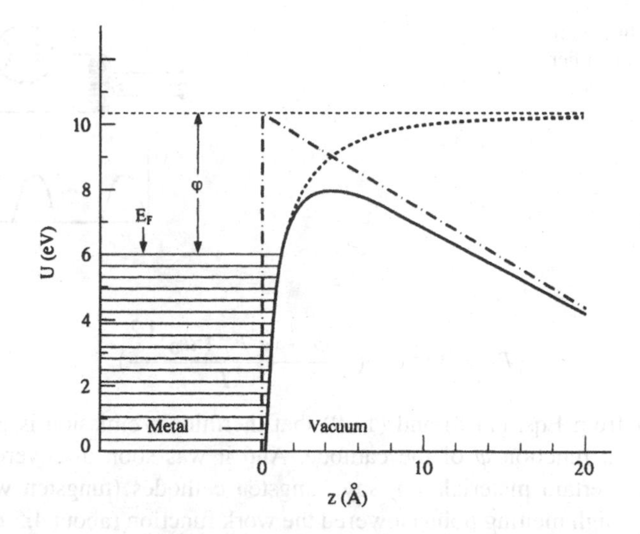

Fig. 11.10 The potential barrier the electron sees at the surface of the cathode: In the absence of an applied electric field, the potential energy rises (*dotted line*) from zero (inside the metal) to $E_F + \varphi$ (outside the metal), where E_F is a constant characteristic of the metal. The shape of the dotted line derives from the so-called image potential energy of the electron, $U_{im}(z) = -e^2/(16\pi\varepsilon_0 z)$, which arises as follows: the (negative) electron in front of the metallic surface induces a positive charge distribution (totaling +*e*) on the surface which pulls the electron towards it with a force $F = -\partial U_{im}/\partial z$. (The given formula for $U_{im}(z)$ is not valid very near the metal surface, but this need not worry us here.) The broken-solid line is the potential energy, $-eFz$, associated with the applied field. The solid line represents the total potential barrier seen by the electron: $U(z) = E_F + \varphi - e^2/(16\pi\varepsilon_0 z) - eFz$

It has a value between 2 and 5 eV depending on the material of the cathode. Eq. (11.8) is known as the Richardson-Laue-Dushman equation, although sometimes it is referred to simply as Richardson's equation.[5]

One can understand Eq. (11.8) by arguing that only electrons in the high energy tail of a Maxwellian-like distribution (similar to that of Fig. 9.10) at sufficiently high temperatures (in practice above $T = 1{,}000$ K) will be able to escape from the metallic cathode by going over the potential barrier at the metal-vacuum interface which keeps the vast majority of the electrons within the metal. And then, only a small fraction of high-energy electrons will escape from the metal: those which are incident normally or nearly normally to the surface. The essential correctness of this picture was verified by W. Schotky who showed (in a series of papers on electron emission between 1914 and 1923) that increasing the electric field *F* at the cathode lowers the barrier at the cathode surface (see Fig. 11.10) leading to increased emission, in accordance with the formula:

[5] S. Dushman gave the first quantum–mechanical derivation of Eq. (11.8), on the basis of Sommerfeld's free-electron theory of metals. According to this theory, the pre-exponential constant in Eq. (11.8) **is:** $A = emk^2/(2\pi^2\hbar^3) = 120$ A/(cm^{2}deg^2), where *e* and *m* are the charge and mass of the electron, *k* is Boltzmann's constant, and $\hbar$is Planck's constant.

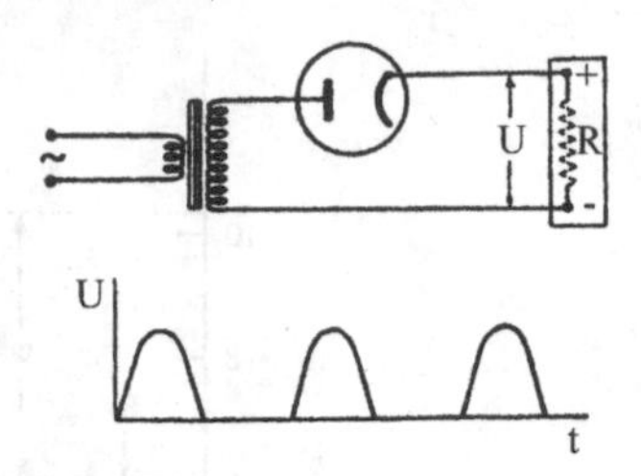

Fig. 11.11 A thermionic diode used as a rectifier

$$J(T) = AT^2 \exp(-\frac{\phi - (e^3F/4\pi\varepsilon_0)^{1/2}}{kT}) \qquad (11.9)$$

It is clear from Eqs. (11.8) and (11.9) that thermionic emission is greater, the lower the work function φ of the cathode. And it was soon discovered that the adsorption of certain materials on, say, tungsten cathodes (tungsten was chosen because of its high melting point) lowered the work function (about 4.5 eV) by one or more eV.

Thermionic emission is, of course, the process underlying the operation of the thermionic valves, the diode and the triode, used in the electronics industry, prior to the development of semiconductor technology. The diode, shown in Fig. 11.9, though replaced in most applications by the *p-n* junction (see Sect. 14.7.6), is still used as a rectifier of high-voltages (see Fig. 11.11). When the ac-voltage from a transformer is applied to the diode, current will flow only during the half cycle, when the cathode is negative and the anode positive, therefore the output voltage U, across R, will have the form shown in the figure.

The thermionic triode is essentially a thermionic diode with a metallic grid placed between the cathode and the anode. It was used as an amplifier of signals (ac currents or voltages) before the development of the transistor. A small ac-voltage applied between the grid (B) and the cathode (A), as shown in Fig. 11.12, results in a much larger variation in the current, i, going through the anode and, correspondingly a much larger variation in the output voltage (across the resistance R_a).

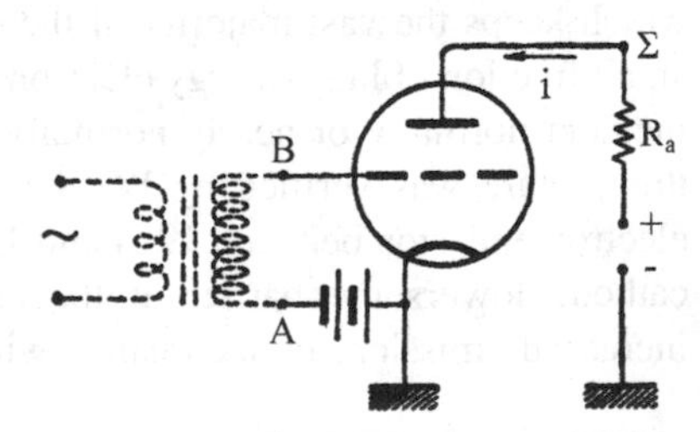

Fig. 11.12 A thermionic triode used as an amplifier

11.5.2 Field Emission

Evidence of electron emission from cold cathodes (at room temperature and lower) at high applied fields (resulting from a high voltage between the anode and a needle-shaped cathode) was obtained by the German physicist J. E. Lilienfeld, working in Leipzig, in 1910. The same published further on this subject in the early 1920s. It soon became clear that the phenomenon could not be explained by Schotky's formula [Eq. (11.9)]. Electron emission at low temperatures and high applied fields, known as field emission, was finally explained by R. H. Fowler and L. W. Nordheim in 1928, using the newly established quantum mechanics. Their theory is based on Sommerfeld's free electron theory of metals and the tunnelling phenomenon to be discussed in (Sect. 13.4.4). According to Sommerfeld's theory, at very low temperatures ($T \approx 0$) the more energetic electrons in the metal have energy about E_F (see Fig. 11.10). A fraction of these electrons will be incident normally, or nearly normally on the surface potential barrier, and according to quantum mechanics, a very small fraction will *tunnel through* the barrier giving rise to the observed field emission current. Fowler and Nordheim derived the following formula for the field-emitted current density, at $T = 0$, as a function of the applied field, F, at the emitting surface:

$$J(F) = AF^2 \exp(-B\phi^{3/2}/F) \tag{11.10}$$

where, $A \approx e^3/(16\pi^2\hbar\phi)$ and $B \approx (2m/\hbar^2)^{1/2}/e$. The above equation, now known as the Fowler-Nordhein equation, described well the experimental data, providing additional support for the new theory of quantum mechanics, and eventually won its authors a Nobel Prize for Physics.

Exercises

11.1. Using the relations (see text following Eq. (11.5):

$$\begin{aligned} b_i \cdot \mathbf{t}_j &= 2\pi, \quad \text{if } i = j \\ &= 0, \quad \text{if } i \neq j \end{aligned}$$

Show that the reciprocals of the bcc and the fcc lattices are given by Eqs. (11.6) and (11.7) respectively.

11.2. One can show that according to Sommerfeld's model of a metal (see Sect. 14.7.3) the number of electrons incident on the surface barrier of a metal, per unit area, per unit time, with normal energy (associated with the motion of the electron normal to the surface) between W and $W + dW$ is given by:

$$N(W,T)\,dW = \frac{mkT}{2\pi^2\hbar^3}\log[1+\exp(-\frac{W-E_F}{kT})]\,dW \qquad \text{(A)}$$

Using the above formula, remembering that the thermionically emitted electrons have energy $W > E_F + \varphi$, the top of the surface barrier (see Fig. 11.10), noting that $kT << \varphi \geq 2$ eV in practice, and that $log(1 + x) \approx x$ for small x, derive Eq. (11.8).

11.3. It turns out that the pre-exponential factor, A, in Eq. (11.8) is not the same for all metals, as suggested by Sommerfeld's model, but depends to some degree on the metal surface under consideration. Having this in mind, explain how we should plot the experimental data in order to determine the work function, φ, of the emitting surface on the basis of Eq. (11.8).

Hint: Plot $log(J/T^2)$ versus $1/T$.

11.4. (a) Show that for very low temperatures ($T \to 0$), Eq. (A) of exercise 11.2 reduces to:

$$N(W,T\to 0) = \frac{m(E_F - W)}{2\pi^2\hbar^3}, \quad \text{for } W < E_F$$
$$= 0, \ \text{for } E > E, \quad \text{for} W > E_F$$

(b) It can be shown (see exercise 13.13) that an electron with energy $W = E_F - x$, where $x > 0$, incident on the surface barrier of Fig. 11.10 has a transmission coefficient given by:

$$T(x) = \exp(-\frac{B}{F}(\phi + x)^{3/2}), \ \text{where } B = \frac{4\alpha}{3e}(\frac{2m}{\hbar^2})^{1/2}$$

where α is an *image potential* correction factor: $4\alpha/3 \approx 1$.

Using (a) and (b) derive Eq. (11.10) for the field-emission current density.

Solution: Because most of the emission comes from small x, we can put: $(\varphi + x)^{3/2} = \varphi^{3/2} + (3/2)\,\varphi^{1/2}x$ in which case:

$$J(T\to 0) = \frac{me}{2\pi^2\hbar^3}\exp(-\frac{B\phi^{3/2}}{F})\int_0^\infty xe^{-\lambda x}dx, \quad \text{where } \lambda = \frac{3B\phi^{1/2}}{F}$$

Note: $\int_0^\infty xe^{-\lambda x}dx = \frac{1}{\lambda^2}$. Convince yourself that integration to ∞ is a permissible approximation in the present case.

11.5. It turns out that the pre-exponential factor, in Eq. (11.10) is not the same for all metals, as suggested by Sommerfeld's model, but depends to some degree on the metal surface under consideration. Having this in mind, explain how we should plot the experimental data in order to determine the work function, φ, of the emitting surface on the basis of Eq. (11.10).

Hint: Plot J/F^2 versus $1/F$.

Chapter 12
Einstein's Theory of Relativity

12.1 Einstein's Early Life and Studies

Albert Einstein was born on the 14th of March 1879 in the city of Ulm in southern Germany the son of Herman and Pauline Einstein. In 1880 the family moved to Munich, where Herman joined the electrical and plumbing business of his younger brother Jacob. For a while the business went well and the family had a comfortable life. Albert had a younger sister Maja. In 1894 the Munich business collapsed and Einstein's family moved to northern Italy. Herman did set up an electrical factory in Milan but he was not very successful.

As a boy Albert Einstein liked music, in this he took after his mother, and started to play the violin at the age of five. His love of the violin and of music in general stayed with him throughout his life and was one of his main recreations, the other being sailing. At the age of seven he was sent to a public primary school in Munich. It was a Catholic school, and he was the only Jewish boy in his class. Einstein later recalled that the school was a liberal one and showed no discrimination, but that many of his classmates were anti-Semitic. His parents were far from devout and never discussed religion. He was given lessons on the Jewish religion by a relative at home, and for a while the boy Einstein got enthusiastic about it, and was irritated by his father's indifference to religion. He did well at school in contrast to what is often said about him. His mother, following his progress, would often say that one day Albert would become a great professor.

Einstein disliked his formal schooling at the Luitpold Gymnasium, which he entered when he was nine and a half years old, one of 1,300 pupils, but he was nevertheless a successful pupil. He was consistently excellent at mathematics. His uncle Jacob, who was a graduate of the Stuttgart Polytechnic Engineering School, introduced his nephew to geometry and algebra which the pupil Einstein thoroughly enjoyed. He also scored top marks in Latin and came close to doing so in Greek. His only weak spot was sports. He certainly enjoyed reading popular science books and, according to his own recollection, it was the reading of such books that ended his religiosity at the age of 12: *Out yonder there was this huge world, which exists independently of us human beings and which stands before us*

A. Modinos, *From Aristotle to Schrödinger*,
Undergraduate Lecture Notes in Physics, DOI: 10.1007/978-3-319-00750-2_12,

like a great, eternal riddle, at least partly accessible to our inspection and thinking. The contemplation of this world beckoned like a liberation, and I soon noticed that many a man whom I had learned to esteem and to admire had found inner freedom and security in devoted occupation with it.

When his family went to Italy in 1894, the 15-year old Einstein was left behind in Munich, in lodgings, to avoid interrupting his education. In the spring of the following year, without asking his parents, he suddenly abandoned school one and a half years before his final examination and went to Italy to join his family which he obviously missed. He effected his escape by presenting his teachers with a medical certificate, signed by a sympathetic doctor, stating that he was suffering from nervous disorders. At the same time he expressed his wish to give up German citizenship. This would free him from the obligation of military service that would be unavoidable after his 17th birthday. Einstein hated regimentation throughout his life. In any case in 1901 he presented himself for military service in Switzerland, only to be declared unfit due to varicose veins and flat feet.

Einstein wanted to pursue a more theoretical course, but giving into his father's wishes he agreed to study engineering at the Swiss Federal Polytechnic School at Zurich (now called the Eidgenössische Technische Hochschule or ETH). With the help of a family friend who knew the Polytechnic's director, Einstein was allowed to take the entrance examinations despite being two years younger than the normal age of eighteen and lacking a secondary school leaving certificate. Although he excelled in mathematics, he failed in his general knowledge questions. He had to go back to secondary school and try again the following year after obtaining his school certificate. He enrolled as a 3rd year pupil in the technical division of the cantonal school in Aarau, 20 miles west of Zurich. He stayed with the family of Jost Winteler, a professor of Greek, and was happy with them. His sister Maja eventually married the couple's son Paul, and his best friend Michele Besso married their daughter Anna, and apparently their daughter Marie was Einstein's first sweetheart.

Einstein was admitted to the Polytechnic at Zurich in 1896. One of his teachers at the Polytechnic was Hermann Minkowski who later made significant contributions to the theory of relativity. But it appears that Einstein educated himself mostly by independent study, by doing experiments in the laboratory, and by studying at home the works of Helmholtz, Kirchoff, Hertz and others. Among Einstein's student friends at the Polytechnic was Mileva Maritsch, a woman of Serbian and Greek Orthodox background who came from Hungary to study in Zurich. The two of them fell in love. They were both students in the section of the Polytechnic which primarily trained students to become science teachers. Einstein was hoping for an assistantship of some kind from one of the professors at the Polytechnic that would enable him to go on to advanced study, but his teachers did not recommend him for such a position after he graduated in 1900.

After short spells as a teacher in a technical high school and as tutor of young students in a boarding school, he was offered a job at a patent office in Berne in 1902, and he remained there for the next seven years. He now had enough financial security to marry Mileva, and so he did in 1903. They had two sons, Hans Albert born in 1904 and Eduard born in 1910. Ultimately their marriage was not successful, they separated in 1914 and finally divorced in 1919.

Einstein, isolated in the patent office in Berne, published three papers from 1902 to 1904 which consist mainly of his rediscovery of results that had been established already by Boltzmann and Gibbs. He later remarked that if he had known the works of Gibbs and Boltzmann he would never have published his own early papers. However, 1905 was a great year for him. During that year he published three great papers. One of them, with the title: *On the Movement of Small Particles Suspended in a Stationary Liquid Demanded by the Molecular Kinetic Theory of Heat*, established beyond any doubt the atomic theory of matter, as we have already pointed out in Sect. 9.2.5. A second paper entitled: *Concerning a Heuristic Point of View about the Creation and Transformation of Light* established the quantization of light by providing a convincing explanation of the photoelectric effect (as we shall see in Sect. 13.1.2). And in his third paper entitled: *Electrodynamics of Moving Bodies,* he introduced the Special Theory of Relativity. The achievement of Einstein is more remarkable for the fact that he wrote these three great papers working in almost complete isolation. The only person he was able to discuss his ideas with was his old friend Michele Besso, who was also working at the patent office, and whom he thanks for his loyal assistance and useful suggestions at the end of his relativity paper.

12.2 The Special Theory of Relativity

12.2.1 Preliminaries

We have seen, in Sect. 6.2.4, that Newton's law of motion has the same mathematical form in coordinate systems which travel with a constant velocity relative to each other. And accordingly one expected the mathematical expression of every fundamental law of physics to be the same in coordinate systems which travel with a constant velocity relative to each other, i.e. that it should be invariant under a Galilean transformation (Eq. (6.14)). But when this transformation was applied to Maxwell's equations (Eq. (10.45)), it was found that these equations are *not* invariant under a Galilean transformation. This difficulty was in fact noted in relation to electrostatics before Maxwell wrote down his famous equations. In the second volume of his book (*A Treatise on Electricity and Magnetism*) Maxwell quotes a work of Gauss in 1835 (which was not published during his lifetime) in which Gauss states the following in relation to Coulomb's law: *two elements of electricity in a state of relative motion attract or repel each other but not in the same way as if they are in a state of relative rest.* After Maxwell it was believed that, if the Galilean transformation was to remain operative, Maxwell's equations could only be valid in just one reference frame and this was taken to be that of the ether. And consequently the velocity of light, $c = (\varepsilon_0 \mu_0)^{-1/2}$, obtained from these equations [see text following Eq. (10.46c)] is the speed of light *relative to the ether*.

Let the coordinate system denoted by Σ in Fig. 12.1 be fixed in ether, and assume that the system denoted by Σ' moves relative to Σ with velocity V along the x direction. According to Galilean relativity the speed of light c' in the primed system, along the x' direction (same as the x direction), should be: $c' = c - V$. However, a number of experiments performed in the 19th century failed to detect any difference between c' and c. The most famous of these experiments were performed by Michelson in 1881 (*Am. J. Sci.* ***122****, 1220, 1881*) and by Michelson and Morley in 1887 (*Am. J. Sci.* ***134****, 333, 1887*). The second experiment being an improved version of the first one.

The basis of the experiment can be described by reference to Fig. 12.1b. Consider three positions (points) *A, B,* and *C* in the earth's coordinate system (Σ') and let the angle between *AB* and *AC* be a right angle (90°). There is a source of light at *A* and mirrors at *B* and *C*. The observer is of course located in the earth's system, which moves with velocity V through the ether along the x direction. Assuming that Galilean relativity holds, the time it takes a light signal from *A* to reach *B* and return (after reflection at the mirror at *B*) to *A* is:

$$t_{ABA} = l/(c - V) + l/(c + V) = (2l/c)/(1 - V^2/c^2) \qquad (12.1)$$

We wish to compare t_{ABA} with the time t_{ACA} it takes for a light signal from *A* to reach *C* and (after reflection there) to return to *A*. We obtain t_{ACA} as follows: Denote by t the time it takes the signal to go from *A* to *C*; during this time *C* moves a distance Vt to the right as shown in Fig. 12.1a; applying Pythagoras's theorem to the triangle of this figure we obtain:

$$c^2t^2 = l^2 + V^2t^2, \text{ therefore } t = (l/c)/\sqrt{(1 - V^2/c^2)}$$

The same reasoning applies to the return journey (from *C* to *A*); therefore:

$$t_{ACA} = 2t = (2l/c)/\sqrt{(1 - V^2/c^2)} \qquad (12.2)$$

The important thing to note is that t_{ACA} *is different* from t_{ABA}. The Michelson and Morley experiments were so designed as to measure directly the difference $(t_{ACA} - t_{ABA})$, if such existed, with sufficient accuracy. Their experiment was so

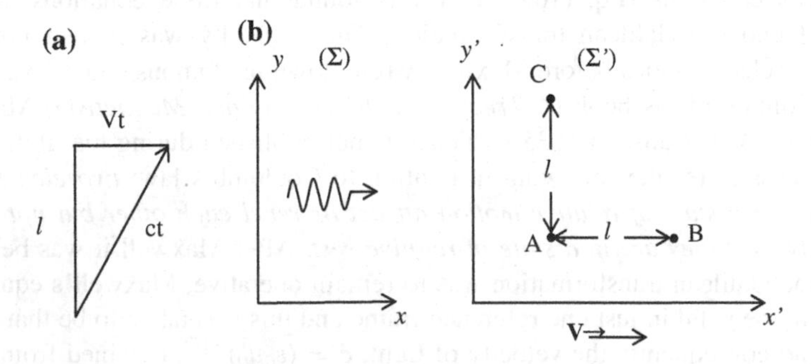

Fig. 12.1 The basis of the Michelson -Morley experiment

sensitive that a speed V of 10 km/s would be detectable. Yet in spite of the fact that the orbital speed of the earth around the sun by itself amounts to 30 km/s, no difference between t_{ACA} and t_{ABA} was observed. The speed of light appeared to be a constant, the same for all coordinate systems moving with a constant velocity relative to each other!

Hendrik Lorentz, perhaps the most prominent theoretical physicist of that time, who had studied the interaction of electrons assuming that this occurred via a static ether, tried to save the idea of the ether by suggesting that the diameter of the earth, and of all other lengths, are shortened in the direction of V by a factor $\sqrt{(1 - V^2/c^2)}$ but remain the same at right angles to V. In this way l in t_{ABA} is shortened by this factor and t_{ABA} becomes equal with t_{ACA}. Lorentz further suggested that not only the length of the apparatus (of the moving body) contracted, but that of the measuring rod as well, so that the contraction is not directly measurable. The physical reason for such a contraction was, according to Lorentz, the result of a modification of the electromagnetic force between bodies in relative motion to each other.

Hendrik Antoon Lorentz was born in Amhern (The Netherlands) in 1853. He entered the University of Leyden in 1870, obtained his B.Sc. degree in mathematics and physics two years later, and returned to Arnhem in 1872 to become a night-school teacher and to prepare his doctoral thesis on the reflection and refraction of light. He obtained his doctorate in 1875, and three years later was appointed to the Chair of Theorctical Physics at Leyden, which was created for him. Lorentz made important contributions to the theory of electromagnetism, and for these he was awarded the Nobel Prize for Physics for the year 1902. He died in 1928.

Although Lorentz believed in the existence of an immovable ether which penetrated all bodies, his theoretical work and results facilitated the development of the special theory of relativity. In a paper he wrote in 1904 (published in the *Proceedings of the Acacdemy of Science of Amsterdam*) Lorentz proved that Maxwell's equations (which are not invariant under the Galilean transformation) are invariant (they retain their form) in all coordinate systems in uniform motion relative to each other, if the Galilean transformation equations [Eqs. (6.14)] are replaced by the following equations (now known as the Lorentz transformation):

$$\begin{aligned} x' &= (x - Vt)/\sqrt{(1 - V^2/c^2)} \\ y' &= y \\ z' &= z \\ t' &= (t - Vx/c^2)/\sqrt{(1 - V^2/c^2)} \end{aligned} \tag{12.3}$$

where one assumes [as in Eq. (6.14) and Fig. 6.8] that the primed system of coordinates moves relative to the unprimed system with constant velocity V along the x-direction. The first thing to note is that in the Lorentz transformation the time changes (from t to t') beside the change in the spatial coordinate (from x to x'). In Newtonian mechanics time flows at the same rate however we choose the system of the spatial coordinates and whatever its velocity is relative to any other system of coordinates. *How are we to interprete the above equations*?

Let us first note that in the limit of small velocities ($V \ll c$) the above equations reduce to the Galilean transformation (Eq. (6.14)):

$$x' = x - Vt,\ y' = y, z' = z, t' = t \tag{12.4}$$

The above tells us that deviations, if any, from Newtonian physics will appear for large velocities of the physical bodies involved. But whatever the magnitude of such deviations, one still needs to have a convincing physical interpretation of the Lorentz transformation. Lorentz continued to believe in an absolute (Newtonian) time, and in the absolute reference frame of the ether, in spite of the fact, which he stressed, that its presence was undetectable. According to Lorentz, any change in time-intervals implied by Eq. (12.3) had its cause in electromagnetic interactions mediated by the ether, and would not be detetactable because the measuring clock would be similarly affected. And similarly any shortening of a length implied by Eq. (12.3) had its cause in the ether and would not be detectable, because the measuring rod would be similarly affected.

One scientist prior to Einstein, who expressed the need for a fundamentally new theory, was the French physicist and mathematician Henri Poincare (he was also a philosopher and politician), who in two papers he published in 1904 and 1905, he rejected the idea of the ether, and emphasised what is now known as the relativity principle: *the laws of physics must have the same mathematical form in all coordinate systems which move with constant velocity with respect to each other.* He also suggested that *no velocity exceeds that of light.*

We have already noted, in relation to the Michelson and Morley experiments, evidence that light propagates with the same speed in coordinate systems which move with constant velocity relative to each other. Additional evidence for the constancy of the speed of light came from astronomical observations. Astronomers showed, first, that the speed of light is the same for all colours: the minimum of emission was observed *simultaneously* for different colours during an eclipse of a star by a dark neighbour. And then observations of double stars showed that the speed of light does not depend on the motion of the source of the light. Therefore, the new theory demanded by Poincare had to satisfy both the relativity principle stated above, and to lead naturally to the observed constancy of the speed of light. Albert Einstein developed the required theory in his 1905 paper on the *Electrodynamics of Moving Bodies.*

The constancy of the speed of light meant that the Galilean transformation had to be replaced by some other, such as that of Lorentz, but in a way that made a physical interpretation and a test of the theory possible. Einstein based his theory on a proper analysis of the concept of *simultaneity*. It boils down to the following: we need a method by which we can decide whether or not two events occurred simultaneously. One may argue as follows:

If two flashes of lightning at points A and B, at some distance from each other along a straight line, reach an observer at point M halfway between A and B at the same moment of time the two events are obviously simultaneous. Time, according to Einstein, must be defined in the same manner as follows: clocks of identical construction are placed at various points of a system of coordinates and synchronized: the positions of their pointers are the same with each other simultaneously (in the above sense of simultaneity). In this way the 'time of an event' is the time indicated by the pointer of the clock at the place in space of the event.

Now, let two flashes of lightning at A and B viewed by an observer at M, halfway between them along a railway embankment (see Fig. 12.2), be simultaneous (they reach the observer at the same moment). And let M' be the mid-point of the distance A to B on a moving train. At the moment the flashes occur, M' coincides with M (as judged from the embankment). Now if an observer sitting at M' did not move with the train he would be permanently at M and the flashes from A and B would be simultaneous for him, but because he is moving along with the train towards B, he will see the flash emitted from B earlier than the one emitted from A. It follows that: events which are simultaneous with respect to the embankment are not simultaneous with respect to the train, and vice versa.

One may be tempted to assume that simultaneity should be decided by reference to the embankment. But this will not do, because the traveller on the train has no way of deciding whether it is the train or the embankment that moves. We must, therefore, conclude that *the interval of time between two events is not the same when measured from two different coordinate systems which are in relative motion with respect to each other.* We can say that each system has its own particular time.

If time intervals between events are not the same in two coordinate systems, a (primed) system moving with velocity V relative to an (unprimed) one along the x (x') direction, then the velocity dx'/dt of a body measured in the primed system will not be given by $(dx/dt - V)$, where dx/dt is its velocity in the unprimed system, as one would expect from Galilean relativity (see Sect. 6.2.4).

Similar conclusions are reached in relation to the measurement of length (of the distance between events). The distance between two points A' and B' on the train is measured in the usual way by an observer on the train. But an observer on the embankment who wishes to measure this distance is faced with more difficulty. For him both points move with velocity V along the embankment. He must therefore determine the points A and B of the embankment which coincide with A' and B' respectively at the same time t of the embankment clock. And this distance may not be the same as the one obtained by the observer on the train.

The question that now arises is: if we know the space coordinates x, y, z and the time t of an event in a given system of coordinates (the unprimed system in Fig. 12.3),

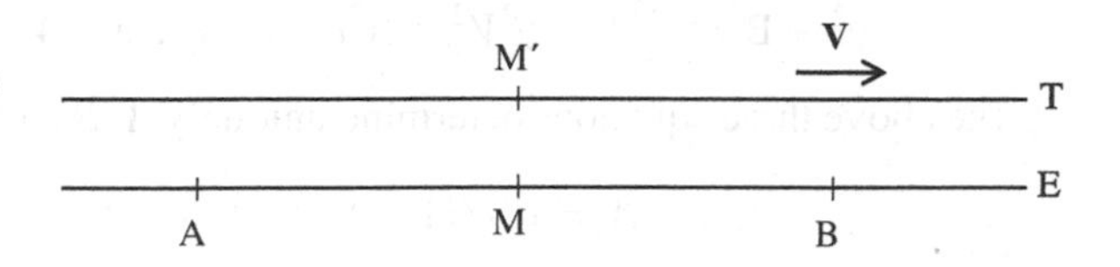

Fig. 12.2 A train (T) moves with velocity V relative to the embankment (E)

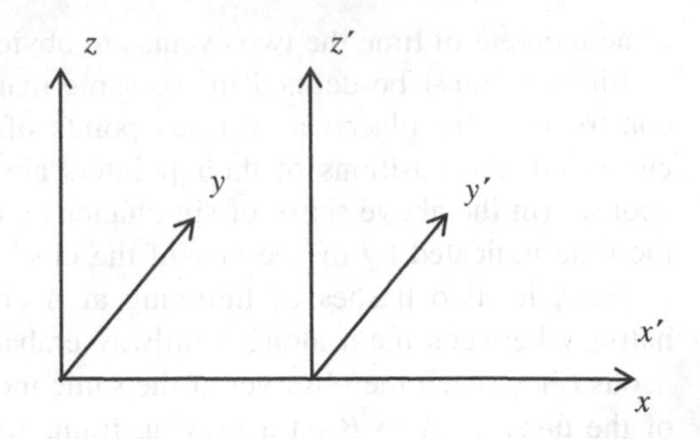

Fig. 12.3 The primed system moves relative to the unprimed one with constat velocity V along the x- axis (same as the x'- axis)

how do we obtain the coordinates x', y', z' and the time t' for the same event in a system of coordinates (the primed system of Fig. 12.3) which moves relative to the unprimed one with velocity V along the x-direction?

12.2.2 Derivation of the Lorentz Transformation

We may assume without any loss of generality that initially the two systems of Fig. 12.3 coincide: the point $x' = y' = z' = 0$ at the time $t' = t = 0$ coincides with the point $x = y = z = 0$. We then obtain the coordinates x', y', z' and t' corresponding to x, y, z and t as follows: Since the relative motion of the two systems is along the x -direction, we have $y' = y$, and $z' = z$. And because at low velocities ($V \ll c$) we expect the Galilean transformation Eq. (12.4) to be valid, we seek the rest of the transformation in the following form:

$$x' = \gamma(x - Vt) \text{ and } t' = At + Bx \tag{12.5}$$

where γ and A are close to unity and B close to zero when $V \ll c$. Since light from a source at the origin of coordinates propagates with the same speed c in both the primed and the unprimed systems, we must have (for propagation along the x- direction):

$$x'^2 - c^2t'^2 = 0 \text{ if } x^2 - c^2t^2 = 0, \text{ for any value of } t.$$

This implies that: $x'^2 - c^2t'^2 = x^2 - c^2t^2$. Substituting on the left of this equation the expressions for x' and t' of Eq. (12.5) we obtain:

$$(\gamma^2 - \mathrm{B}^2c^2)x^2 + (\gamma^2V^2 - A^2c^2)t^2 - 2(V\gamma^2 + \mathrm{AB}c^2) = x^2 - c^2t^2$$

The above equation will be satisfied only if

$$\gamma^2 - \mathrm{B}^2c^2 = 1, \quad \gamma^2V^2 - A^2c^2 = -c^2, \text{ and } V\gamma^2 + \mathrm{AB}c^2 = 0$$

The above three equations determine uniquely A, B and γ in terms of V. We find:

$$\gamma = A = 1/\sqrt{(1 - V^2/c^2)} \text{ and } B = -\gamma V/c^2 \tag{12.6}$$

Using the above values in Eq. (12.5) we finally obtain:

$$\begin{aligned} x' &= (x - Vt)/\sqrt{(1 - V^2/c^2)} \\ y' &= y \\ z' &= z \\ t' &= (t - Vx/c^2)/\sqrt{(1 - V^2/c^2)} \end{aligned} \tag{12.7}$$

This is of course the Lorentz transformation (Eq. (12.3)). It is worth noting that the derivation of the above transformation is based on the fact that the speed of light is the same, c, in any two coordinate systems which are in uniform relative motion with respect to each other. No other assumption has been made. One must therefore conclude (the value of c being a universal constant) that the Lorentz transformation is a property of space and time taken together and there is no more to be said about it. This is how nature works. We did not recognize this property of space–time in the past, because the values of V we have to deal with in practice are very much smaller than c. Even an astronaut orbiting the earth has a speed about 2×10^{-5} c. When $V \ll c$, the Lorentz transformation reduces to the Galilean transformation, and we have assumed (wrongly) that this holds more generally than it actually does. Now, after Einstein's ingenious analysis of the concept of simultaneity, we find it easier to disconnect from the idea of absolute time and accept the relativity of time demanded by nature.

Before discussing the consequences of the Lorentz transformation in measurements of space and time intervals, it is worth noting that, if we know the orbit $(x'(t'), y'(t'), z'(t'))$ of a particle in the primed system of coordinates of Fig. 12.3, then we can obtain, using Eq. (12.7), the orbit $(x(t), y(t), z(t))$ of the particle in the unprimed system. And we can of course calculate the velocity of the particle in the two systems given by $(dx'/dt', dy'/dt', dz'/dt')$ and $(dx/dt, dy/dt, dz/dt)$ respectively. We note that because $dt' \neq dt$ all three components of the velocity are different in the two systems; the relations between the two are easily established but we need not give them here. I may add that, when $V \ll c$, these relations are reduced to those we expect from Galilean relativity. I should also mention that: dx/dt is *less* than $V + dx'/dt'$ and is less than c if both V and dx'/dt' are less than c.

12.2.3 Length Contraction and Time Dilation

Let us obtain the length L of a rod in the unprimed system of Fig. 12.3 when its length in the primed system, which is moving with it (the rod is at rest in the primed system), is L'. We assume that the rod is oriented parallel to the relative motion of the two systems.

We denote by x_1' the beginning and by x_2' the end of the rod in the primed system. These values do not depend on t', therefore the length of the rod in the primed system is: $L' = x_2' - x_1'$. The length L of the rod in the unprimed system is

$L = (x_2 - x_1)$, where x_1 and x_2 denote the beginning and the end points of the rod in the unprimed system at time t. Using Eq. (12.7), we obtain:

$$x_1' = (x_1 - Vt)/\sqrt{(1 - V^2/c^2)} \text{ and } x_2' = (x_2 - Vt)/\sqrt{(1 - V^2/c^2)}.$$

Therefore: $x_2' - x_1' = (x_2 - x_1)/\sqrt{(1 - V^2/c^2)}$. Therefore:

$$L = L'\sqrt{(1 - V^2/c^2)} \tag{12.8}$$

This tells us that *the rod is shorter when in motion relative to the observer, when the rod lies in the direction of the motion*. But when the rod is normal to the direction of the motion (in the yz-plane) its length will be the same because $y' = y$ and $z' = z$ according to Eq. (12.7).

Let us next consider time measurements. Let there be a clock at the origin of the primed system of Fig. 12.3. Two successive ticks of this clock correspond to two events, both at the origin of coordinates ($x_2' = x_1' = 0$) separated by a time interval $\Delta t' = t_2' - t_1'$. And we can assume that $t_1' = 0$, so that $t_2' = \Delta t'$. In the unprimed system the two events occur, according to Eq. (12.7), the first at: $x_1 = 0$, $t_1 = 0$ and the second at $x_2 = V t_2$, $t_2 = t_2'/\sqrt{(1 - V^2/c^2)}$.[1] Therefore, the time interval between the two events in the unprimed system is: $\Delta t = t_2 - t_1 = t_2 = t_2'/\sqrt{(1 - V^2/c^2)} = \Delta t'/\sqrt{(1 - V^2/c^2)}$. In summary:

$$\Delta t = \Delta t' / \sqrt{(1 - V^2/c^2)} \tag{12.9}$$

This tells us that a clock in motion relative to the observer goes more slowly than when at rest.

Direct experimental confirmation of the above became possible with the advent of high-energy particle physics. The lifetimes of radioactively decaying mesons and pions have been measured in flight and at rest, and the results agreed with Eq. (12.9).[2]

12.2.4 The Law of Motion in Special Relativity

As we have already mentioned, Lorentz had proved that Maxwell's equations are invariant under the transformation that now bears his name. However, it is obvious that Newton's second law of motion is *not* invariant under this transformation. Einstein established that Newton's law Eq. (6.1) is the limiting form, valid when

---

1 We note that according to Eq. (12.7), x, y, z, t, are given by:
$x = (x' + Vt')/\sqrt{(1 - V^2/c^2)}$
$y = y'$
$z = z'$
$t = (t' + Vx'/c^2)/\sqrt{(1 - V^2/c^2)}$

2 Bailey J. and Picasso E.. *Progress in Nuclear Physics* **12**, *43 (1970)*.

the velocity υ of the accelerated particle is much smaller than the velocity of light, of the following law:

$$\frac{d}{dt}\boldsymbol{p} = \boldsymbol{F} \tag{12.10a}$$

$$\boldsymbol{p} = m\boldsymbol{\upsilon} \tag{12.10b}$$

$$m = m_0/\sqrt{(1 - \upsilon^2/c^2)}) \tag{12.10c}$$

where m_0 is the rest mass of the body. We note that $m = m_0$ in Newton's law, which implies that the velocity of a body under the action of a constant force $\boldsymbol{F}$ can increase indefinitely, according to Newtonian mechanics, eventually exceeding the velocity of light. In contrast, according to Eq. (12.10c) the effective inertial mass m of the body increases with υ, tending to infinity as υ approaches c, which implies that the velocity of the body can never exceed that of light.

Equation (12.10c) is justified as follows. This expression can be written as a series which, when $\upsilon^2/c^2 \ll 1$, reduces to:

$$m = m_0 + m_0\upsilon^2/(2c^2) \tag{12.11a}$$

if we keep only the first two terms in the series.[3] The above can be written as:

$$mc^2 = m_0c^2 + m_0\upsilon^2/2 \tag{12.11b}$$

We note that the second term on the right of the above equation is the ordinary kinetic energy of the body [defined by Eq. (7.30)]. Einstein assumed m_0c^2 to be part of the total energy of the body (the rest energy of the body) and he further assumed that mc^2 equals the total energy E of the body *whatever* the value of its velocity. Accordingly he writes:

$$E = mc^2 \tag{12.12}$$

And equates the rate of increase of the energy ($dE/dt = d(mc^2)/dt$), with the work done by the force per unit time $\left(\upsilon \cdot F = \upsilon \cdot \frac{d}{dt}(m\upsilon)\right)$:

$$d(mc^2)/dt) = \upsilon \cdot \frac{d}{dt}(m\upsilon)$$

Multiplying both sides of the above equation with 2 m one obtains:

$$c^2 2m(dm/dt) = 2m\upsilon \cdot \frac{d}{dt}(m\upsilon)$$

[3] We have: $m_0/\sqrt{(1 - \upsilon^2/c^2)}) = m_0(1 + \upsilon^2/(2c^2) + 3\upsilon^4/(8c^4) + \bullet\bullet\bullet)$.

Noting that $dm^2/dt = 2m(dm/dt)$ and that $d(m\upsilon)^2/dt = 2m\upsilon \cdot \frac{d}{dt}(m\upsilon)$, one writes the above as:

$$c^2 dm^2/dt = d(m\upsilon)^2/dt$$

Now, if the derivatives of two quantities are the same, the two quantities are the same apart from an additive constant C. Therefore:

$$m^2c^2 = m^2\upsilon^2 + C \tag{12.13}$$

The above holds, of course, for any value of υ. Putting $\upsilon = 0$, determines C in terms of the rest mass m_0 (the value of m for $\upsilon = 0$). One obtains: $C = m_0{}^2c^2$. When this is substituted in Eq. (12.13), the latter becomes: $m^2(c^2 - \upsilon^2) = m_0{}^2c^2$. This is the same as $m^2 = m_0{}^2c^2/(c^2 - \upsilon^2) = m_0{}^2/(1^2 - \upsilon^2/c^2)$. Finally, taking the square root of both sides of this equation one finds Eq. (12.10c).

Einstein elaborating on the physical significance of Eq. (12.12) he claimed that this equation is valid quite generally, which implies that the rest energy m_0c^2 of a body is partly due to the internal energy (motion) of the body. It is worth remembering, however, that under ordinary conditions the change in the mass m of the body as a result of an increase in its internal energy is negligibly small. But in the explosion of an atomic bomb, the energy of the explosion is enormous, and in such a case the mass after the explosion is certainly smaller than the initial mass. For example, in an atomic explosion equivalent to twenty kilotons of TNT, the mass is reduced by one gram.

12.3 On the Invariance of Physical Laws

12.3.1 The Minkowski 4-Dimensional Spacetime

Consider a vector $\boldsymbol{A}$ in ordinary 3-dimensional space. The components of this vector in the two Cartesian coordinates systems (primed and unprimed) of Fig. 12.4 will of course be different. We have:

$$A = A_x\boldsymbol{i} + A_y\boldsymbol{j} + A_z\boldsymbol{k} \text{ and } A = A'_x\boldsymbol{i}' + A'_y\boldsymbol{j}' + A'_z\boldsymbol{k}' \tag{12.14}$$

where $\mathbf{i}, \mathbf{j}, \mathbf{k}$ are unit vectors along the x, y, z directions and $\mathbf{i}', \mathbf{j}', \mathbf{k}'$ are unit vectors along the x', y', z' directions. The component $A_x{}'$ is of course the sum of the projections of $A_x\mathbf{i}$, $A_y\mathbf{j}$ and $A_z\mathbf{k}$ on the x' direction. And the other primed components are similarly obtained. We have:

$$\begin{aligned} A'_x &= cos(x', x)A_x + \cos(x', y)A_y + \cos(x', z)A_z \\ A'_y &= cos(y', x)A_x + \cos(y', y)A_y + \cos(y', z)A_z \\ A'_z &= cos(z', x)A_x + \cos(z', y)A_y + \cos(z', z)A_z \end{aligned} \tag{12.15}$$

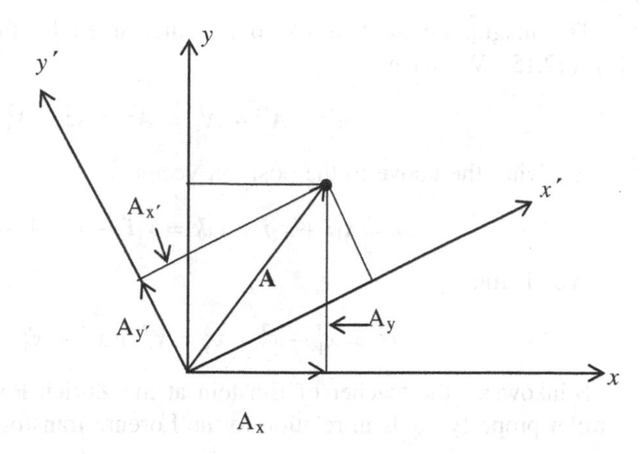

Fig. 12.4 The components of a vector A in two different Cartesian systems of coordinates. We have assumed for the sake of clarity that the z and z' directions coincide

where (x', x) denotes the angle between the directions of x' and x, and the same applies to (x', y) etc. In what follows we shall refer to the x, y, z directions as the x_1, x_2, x_3 directions, and we shall denote the corresponding components of $\boldsymbol{A}$ by A_1, A_2, and A_3. And we do the same for the primed system of coordinates. Accordingly we write Eq. (12.15) as follows:

$$\begin{aligned} A_1' &= c_{11}A_1 + c_{12}A_2 + c_{13}A_3 \\ A_2' &= c_{21}A_1 + c_{22}A_2 + c_{23}A_3 \\ A_3' &= c_{31}A_1 + c_{32}A_2 + c_{33}A_3 \end{aligned} \tag{12.15a}$$

where

$$c_{ij} = cos(x_i', x_j) \tag{12.15b}$$

In what follows we write Eq. (12.15a) more compactly as follows:

$$A_i' = \sum_{j=1}^{3} c_{ij}A_j, i = 1, 2, 3 \tag{12.15c}$$

In the same way we can obtain the components A_i of the vector in the unprimed system in terms of its components A_i' in the primed system:

$$A_i = \sum_{j=1}^{3} c_{ij}' A_j', i = 1, 2, 3 \tag{12.16}$$

where

$$c_{ij}' = cos(x_i, x_j') = cos(x_j', x_i) = c_{ji} \tag{12.17}$$

The magnitude of $\boldsymbol{A}$ is of course unchanged by the rotation described by Eq. (12.15). We have:

$$A_1'^2 + A_2'^2 + A_3'^2 = A_1^2 + A_2^2 + A_3^2 \tag{12.18}$$

Applying the above to the position vector

$$\boldsymbol{r} = x_1\boldsymbol{i} + x_2\boldsymbol{j} + x_3\boldsymbol{k} = x_1'\boldsymbol{i}' + x_2'\boldsymbol{j}' + +x_3'\boldsymbol{k}'$$

We obtain:

$$r^2 = x_1^2 + x_2^2 + x_3^2 = x_1'^2 + x_2'^2 + x_3'^2 = r'^2 \tag{12.19}$$

Minkowski, the teacher of Einstein at the Zurich Polytechnic, noticed that a similar property holds in relation to the Lorentz transformation Eq. (12.7):

$$x'^2 + y'^2 + z'^2 - (ct')^2 = x^2 + y^2 + z^2 - (ct)^2 \tag{12.20}$$

He then suggested that ict, where $i^2 = -1$ (see appendix A2 on complex numbers), should be formally treated as the *fourth* dimension of a 4—dimensional spacetime. The 'position' of an event in this spacetime is determined by a vector $\boldsymbol{s}$ with four components: x_i, $i = 1,2,3,4$:

$$\boldsymbol{s} = (\boldsymbol{r}, ict) = (x_1 = x, x_2 = y, x_3 = z, x_4 = ict) \tag{12.21}$$

The best way to think about spacetime is to imagine a clock at *every* point in ordinary space (or may be in every small region of space) as described by a Cartesian system of coordinates. The pointer of the clock determines the *fourth* coordinate, ict, at that point. If there are two coordinate systems which move relative to each other (as in Fig. 12.3) we imagine clocks at every point of both systems. The pointers of these clocks determine ict' and ict respectively at these points.

Let us consider how the (four) components of $\boldsymbol{s}$ transform under *rotations* of the 4-dimensional coordinate system. By a rotation we mean either a rotation of the space coordinates (as in Fig. 12.4) or a Lorentz transformation to a system moving with constant velocity relative to the initial one (as in Fig. 12.3). We are allowed to treat the latter as a rotation because, according to Eq. (12.20), the magnitude of $\boldsymbol{s}$ does not change under the transformation. We have:

$$\begin{aligned} s^2 &= r^2 - c^2t^2 = x_1^2 + x_2^2 + x_3^2 + x_4^2 \\ &= x_1'^2 + x_2'^2 + x_3'^2 + x_4'^2 = r'^2 - c^2t'^2 = s'^2 \end{aligned} \tag{12.22}$$

In general a transformation may consist of a rotation of the space coordinates followed by a Lorentz transformation or the other way round. But this need not concern us here. In what follows we assume that the transformation is either a rotation of the space coordinates or a Lorentz transformation. In either case the components of $\boldsymbol{s}$ (defined by Eq. 12.21) in the primed system are related to its components in the unprimed system as follows:

$$x'_\mu = \sum_{\nu=1}^{4} c_{\mu\nu} x_\nu, \ \mu = 1, 2, 3, 4 \tag{12.23}$$

We can think of $c_{\mu\nu}$ as the elements of a 4 × 4 matrix $\underline{C}$ (see appendix A4):

$$\underline{C} = \begin{pmatrix} c_{11} & c_{12} & c_{13} & c_{14} \\ c_{21} & c_{22} & c_{23} & c_{24} \\ c_{31} & c_{32} & c_{33} & c_{34} \\ c_{41} & c_{42} & c_{43} & c_{44} \end{pmatrix} \tag{12.24}$$

For a rotation in ordinary space we have:

$$\underline{C} = \begin{pmatrix} c_{11} & c_{12} & c_{13} & 0 \\ c_{21} & c_{22} & c_{23} & 0 \\ c_{31} & c_{32} & c_{33} & 0 \\ 0 & 0 & 0 & 1 \end{pmatrix} \tag{12.25}$$

With the matrix elements given by Eq. (12.15b). And for the Lorentz transformation (Eq. 12.7) we have:

$$\underline{C} = \begin{pmatrix} \gamma & 0 & 0 & i\beta\gamma \\ 0 & 1 & 0 & 0 \\ 0 & 0 & 1 & 0 \\ -i\beta\gamma & 0 & 0 & \gamma \end{pmatrix} \tag{12.26}$$

where $\gamma = 1/\sqrt{(1 - V^2/c^2)}$ and $\beta = V/c$.

We can similarly obtain the components of the vector in the unprimed system, x_μ (μ = 1, 2, 3, 4), in tems of its components in the primed system, x_μ' (μ = 1, 2, 3, 4). We obtain:

$$x_\mu = \sum_{\nu=1}^{4} c'_{\mu\nu} x'_\nu, s \tag{12.27}$$

and we note that:

$$c'_{\mu\nu} = c_{\nu\mu}, \nu, \mu = 1, 2, 3, 4 \tag{12.28}$$

which is the generalization of Eq. (12.17) to four dimensions.

12.3.2 Tensors and Relativistic Invariance

Relativistic invariance demands that the fundamental equations (laws) of physics must have the same form in two coordinate systems (the primed and unprimed systems of the previous section) which are obtained from each other by a rotation in 4-dimensional spacetime.

Tensors of rank zero: We know that some quantities (scalars) are numbers (with appropriate units attached to them) which are the same in all coordinate systems. Scalars are *tensors of rank zero.* The rest mass m_0 of a body is obviously a scalar (a tensor of rank zero). Another scalar quantity of great importance is the *proper time* interval between two events in spacetime. It is defined as follows:

Consider two events '1' and '2' at $s(1) = (x_1^{(1)}, x_2^{(1)}, x_3^{(1)}, x_4^{(1)})$ and $s(2) = (x_1^{(2)}, x_2^{(2)}, x_3^{(2)}, x_4^{(2)})$ in a given system of coordinates. The spacetime interval $\Delta s = s(2) - s(1)$ has a magnitude given [see Eq. (12.22)] by:

$$
\begin{aligned}
(\Delta s)^2 &= (\Delta r)^2 - (c\Delta t)^2 \\
(\Delta r)^2 &= \left(x_1^{(2)} - x_1^{(1)}\right)^2 + \left(x_2^{(2)} - x_2^{(1)}\right)^2 + + \left(x_3^{(2)} - x_3^{(1)}\right)^2 \\
(\Delta t)^2 &= \left(t^{(2)} - t^{(1)}\right)^2
\end{aligned}
\tag{12.29}
$$

Δr is the distance in ordinary space between the two events and Δt the time interval between the two in the given system of coordinates. And we remember [see Eq. (12.22)] that $(\Delta s)^2$ is a scalar quantity: the same in the primed and unprimed coordinate systems of Eq. (12.23).

The *proper time* interval $\Delta\tau$ between the two events is defined by:

$$(\Delta\tau)^2 = -(\Delta s)^2/c^2 = (\Delta t)^2 - (\Delta r)^2/c^2 \tag{12.30}$$

And because c is a constant, $\Delta\tau$ is a scalar quantity like Δs. We note that when $(\Delta t)^2 - (\Delta r)^2/c^2 > 0$, $(\Delta\tau)^2$ is a positive quantity and therefore $\Delta\tau$ is a real quantity (described by a real number); we then say that $\Delta\tau$ is a *time-like* interval. When $(\Delta t)^2 - (\Delta r)^2/c^2 < 0$, its square root is an imaginary quantity: $i\sqrt{((\Delta r)^2/c^2 - (\Delta t)^2)}$, where i is the imaginary unit ($i^2 = -1$); we then say that $\Delta\tau$ is a *space-like* interval. Of course, two events separated by a spacelike interval can *not* be connected: the one can not be the cause of the other, because no signal propagates faster than light.

Let $s(1)$ and $s(2)$ be the positions in spacetime of a body (a particle). If we then divide by $(\Delta t)^2$ both sides of Eq. (12.30) we obtain:

$$(\Delta\tau/\Delta t)^2 = 1 - (\Delta r/\Delta t)^2/c^2$$

which in the limit of small (infinitesimal) changes becomes:

$$(d\tau/dt)^2 = 1 - \upsilon^2/c^2$$

where $\upsilon = dr/dt$ is the velocity of the particle. Therefore:

$$d\tau/dt = \sqrt{(1 - \upsilon^2/c^2)} \text{ and } d\tau = dt\sqrt{(1 - \upsilon^2/c^2)} \tag{12.31}$$

Tensors of rank one: A quantity of *4* components is a 4-dimensional vector *if* under rotations in 4-dimensional space it transforms like the position vector s, i.e. according to Eq. (12.23). Vectors are *tensors of rank one*.

The infinitesimal displacement of a particle in spacetime

$$ds = (dx_1 = dx, dx_2 = dy, dx_3 = dz, dx_4 = icdt) \tag{12.32}$$

is a vector (a tensor of rank one), because it obviously transforms according to Eq. (12.23). Multiplying the above with the rest mass m_0 of the particle and dividing by $d\tau = dt\sqrt{(1-\upsilon^2/c^2)}$, both of them scalar quantities, we obtain a 4-dimensional momentum-energy vector:

$$\boldsymbol{P} = (\boldsymbol{p}, p_4) = (\boldsymbol{p}, im_0c/\sqrt{(1-\upsilon^2/c^2)}) = (\boldsymbol{p}, iE/c) \tag{12.33}$$

where $\boldsymbol{p}$ is the momentum of the particle defined by Eq. (12.10b), and E is its energy defined by Eqs. (12.10c) and (12.12).

We note that, like $(ds)^2$, $P^2 = p^2 - E^2/c^2$ is a scalar quantity: it does not change under a rotation in space–time. Looking at $\boldsymbol{P}$ as a function of the proper time τ, we write:

$$d\boldsymbol{P}/d\tau = \boldsymbol{F}^{(4)} \tag{12.34}$$

where $\boldsymbol{F}^{(4)}$ is a 4-dimensional vector, known as the Minkowski force. It is given by

$$\boldsymbol{F}^{(4)} = (\boldsymbol{F}, \boldsymbol{\upsilon}\cdot\boldsymbol{F}/c)/\sqrt{(1-\upsilon^2/c^2)} \tag{12.34a}$$

In the above, $\boldsymbol{F}$ is the ordinary force [e.g. for a charged particle in an E.M. field it will be given by the Lorentz force of Eq. (10.58)], and the fourth term, $\boldsymbol{\upsilon}\cdot\boldsymbol{F}/c$, relates to the work done by this force per unit time [see text following Eq. (12.12)]. According to Eqs. (12.33) and (12.34) we have: $d\boldsymbol{p}/d\tau = (d\boldsymbol{p}/dt)/\sqrt{(1-\upsilon^2/c^2)} = \boldsymbol{F}/\sqrt{(1-\upsilon^2/c^2)}$, which means that: $d\boldsymbol{p}/dt = \boldsymbol{F}$, which is, of course, the same as Eq. (12.10a), as we should expect. The important difference between Eqs. (12.34) and (12.10b) lies in the fact that Eq. (12.34), by the way it is written (in tensor form), makes the relativistic (Lorentz) invariance of the equation obvious. That an equation written in vector (tensor) form is Lorentz invariant (it has the same form in the primed and unprimed system of Fig. 12.3) is easily demonstrated as follows. Eq. (12.34) means that:

$$dP_\mu/d\tau - F_\mu^{(4)} = 0, \mu = 1,2,3,4 \tag{12.35}$$

The components of the same vector in the primed system (we remember that $d\tau$ is the same in both systems) is given [see Eq. (12.23)], by:

$$dP'_\mu/d\tau - F_\mu^{(4)\prime} = \sum_{\nu=1}^{4} c_{\mu\nu}(dP_\nu/d\tau - F_\nu^{(4)}) = 0$$

The second of the above equations follows from Eq. (12.35). Therefore, the equation of motion in the primed system is:

$$d\boldsymbol{P}'/d\tau = \boldsymbol{F}^{(4)\prime}$$

which has the same form as Eq. (12.34). Having established the Lorentz invariance of the equation of motion of a body in the above manner, we can safely use Eq. (12.10c) in actual calculations.

Tensors of rank two: There are physical quantities with $4 \times 4 = 16$ components: $A_{\mu\nu}$, $\mu, \nu = 1,2,3,4$ (we remember that these numbers correspond to the four 'directions' of spacetime), which we call tensors of rank two *if* they transform as follows under rotations in space time:

$$A'_{\mu\nu} = \sum_{k=1}^{4}\sum_{q=1}^{4} c_{\mu k} c_{\nu q} A_{kq} \tag{12.36}$$

where the c_{ij} are given by Eqs.(12.24) to (12.26); and

$$A_{\mu\nu} = \sum_{k=1}^{4}\sum_{q=1}^{4} c'_{\mu k} c'_{\nu q} A'_{kq} \tag{12.37}$$

$$c_{ij}\prime = c_{ji} (\text{as in Eq.}(12.28)$$

A tensor of rank two is often written as a matrix of $4 \times 4 = 16$ elements (components of the tensor). It often happens that some of these elements are the same (as when $A_{\mu\nu} = A_{\nu\mu}$) or zero.

Tensors of rank three: There are physical quantities with $4 \times 4 \times 4 = 64$ components: $A_{\lambda\mu\nu}$, $\lambda.\mu, \nu = 1, 2, 3, 4$, which we call tensors of rank three *if* they transform as follows under rotations in space time:

$$A'_{\lambda\mu\nu} = \sum_{k=1}^{4}\sum_{q=1}^{4}\sum_{r=1}^{4} c_{\lambda k} c_{\mu q} c_{\nu r} A_{kqr} \tag{12.38}$$

where the c_{ij} are given by Eqs.(12.24)–(12.26); and

$$A_{\lambda\mu\nu} = \sum_{k=1}^{4}\sum_{q=1}^{4}\sum_{r=1}^{4} c'_{\lambda k} c'_{\mu q} c'_{\nu r} A'_{kqr} \tag{12.39}$$

where $c_{ij}{}' = c_{ji}$ as in Eq.(12.28).

Tensors of higher rank are defined in similar fashion but these need not concern us here.

Finally, tensors of any rank can be functions of the position $\boldsymbol{s}$ (as defined by Eq. 12.21) in spacetime; we often refer to them as tensor fields. They transform in accordance with the equations given above. For example: a tensor field of rank two transforms as follows:

$$A'_{\mu\nu}(s') = \sum_{k=1}^{4}\sum_{q=1}^{4} c_{\mu k} c_{\nu q} A_{kq}(s) \tag{12.40}$$

where $s' = (x_1', x_2', x_3', x_4')$ is obtained from $s = (x_1, x_2, x_3, x_4)$ according to Eq. (12.23).

Derivatives of tensors with respect to x_i, $i = 1,2,3,4$ transform in similar fashion. We note that the derivatives: $\partial f/\partial x_i$, $i = 1, 2, 3, 4$, of a function $f(s)$, a tensor field of rank zero, constitute a 4-dimensional vector field: they transform like a vector (a tensor of rank one). To show this, we first observe that, if we look at the primed coordinates x_i' as functions of the unprimed coordinates x_i, we can write [see Eq. (12.23)]:

$$dx_i' = \sum_{j=1}^{4} (\partial x_i'/\partial x_j)dx_j = \sum_{j=1}^{4} c_{ij}dx_j \tag{12.41a}$$

and similarly [see Eq. (12.27)]:

$$dx_i = \sum_{j=1}^{4} (\partial x_i/\partial x_j')dx_j' = \sum_{j=1}^{4} c_{ij}'dx_j' \tag{12.41b}$$

and we remember that $c_{ij}' = c_{ji}$ [according to Eq.(12.28)].

Using the above equations and the chain rule of differentiation Eq. (5.38) we obtain:

$$\partial f/\partial x_i' = \sum_{j=1}^{4} (\partial x_j/\partial x_i')\partial f/\partial x_j = \sum_{j=1}^{4} c_{ji}'\partial f/\partial x_j = \sum_{j=1}^{4} c_{ij}\partial f/\partial x_j \tag{12.42}$$

which shows that $f_i = \partial f/\partial x_i$, $i = 1, 2, 3, 4$, is a vector (a tensor of rank one); it transforms in the same way as the position vector Eq.(12.23).

Similarly one can show that if f_i, $i = 1,2,3,4$ is a tensor field of rank one, the quantity f_{ij} defined by

$$f_{ij} = \partial f_i/\partial x_j, i, j = 1, 2, 3, 4 \tag{12.43}$$

is a tensor field of rank two.

And if f_{ij}, $i.j = 1,2,3,4$ is a tensor field of rank two, the quantity f_{ijk} defined by

$$f_{ijk} = \partial f_{ij}/\partial x_k, i, j = 1, 2, 3, 4 \tag{12.44}$$

is a tensor field of rank three.

We can similarly define tensor fields of higher ranks, but we need not do so here.

In what follows [see Eq. (12.48a)] we shall need the following formula, relating to Eq. (12.44). *By convention* a double index implies summation:

$$\partial f_{ij}/\partial x_j = \sum_{j=1}^{4} \partial f_{ij}/\partial x_j = g_i, \tag{12.45}$$

and the quantity, g_i, $i = 1,2,3,4$ obtained in this manner is a tensor of rank one. The operation is analogous to taking the divergence of a vector [see Eq. (10.28)].

The great advantage of using tensors lies in the following: A set of equations written in tensor form have the same form in all coordinate systems which are obtained from each other by a rotation in spacetime [in accordance with Eq. (12.23)]. To see that this is so consider the following example: assume that the following equation holds in the unprimed system of Fig. 12.3:

$$A_{kqr}(\boldsymbol{s}) - B_{kqr}(\boldsymbol{s}) = 0, k, q, r = 1, 2, 3, 4 \tag{12.46}$$

where A and B are tensors of rank three and functions of $\boldsymbol{s} = (x_1, x_2, x_3, x_4)$. In the primed system $\boldsymbol{s}$ becomes $\boldsymbol{s}' = (x_1', x_2', x_3', x_4')$ according to Eq. (12.23), and A and B become

$$A'_{\lambda\mu\nu}(\boldsymbol{s}') = \sum_{k=1}^{4}\sum_{q=1}^{4}\sum_{r=1}^{4} c_{\lambda k}c_{\mu q}c_{\nu r}A_{kqr}(\boldsymbol{s}), \lambda, \mu, \nu = 1, 2, 3, 4$$

$$B'_{\lambda\mu\nu}(\boldsymbol{s}') = \sum_{k=1}^{4}\sum_{q=1}^{4}\sum_{r=1}^{4} c_{\lambda k}c_{\mu q}c_{\nu r}B_{kqr}(\boldsymbol{s}), \lambda, \mu, \nu = 1, 2, 3, 4$$

Therefore:

$$A'_{\lambda\mu\nu}(\boldsymbol{s}') - B'_{\lambda\mu\nu}(\boldsymbol{s}') = \sum_{k=1}^{4}\sum_{q=1}^{4}\sum_{r=1}^{4} c_{\lambda k}c_{\mu q}c_{\nu r}(A_{kqr}(\boldsymbol{s}) - B_{kqr}(\boldsymbol{s}))$$

We note that every term in the above sum vanishes because of Eq. (12.46). Therefore, in the primed system we have:

$$A'_{\lambda\mu\nu}(\boldsymbol{s}') - B'_{\lambda\mu\nu}(\boldsymbol{s}') = 0, \lambda, \mu, \nu = 1, 2, 3, 4 \tag{12.46a}$$

which has the same form as Eq. (12.46).

Let me conclude this section by writing Maxwell's equations in tensor form, which proves that they are indeed invariant under rotations in 4-dimensional spacetime.

It turns out that the electric field $\boldsymbol{E}$ and the magnetic field $\boldsymbol{B}$ are the components of a tensor field of rank two: $f_{\mu\nu}(\boldsymbol{s})$, $\mu, \nu = 1, 2, 3, 4$, defined as follows:

$$f_{\mu\nu}(\boldsymbol{s}) = \begin{pmatrix} 0 & B_z & -B_y & -iE_x/c \\ -B_z & 0 & B_x & -iE_y/c \\ B_y & -B_x & 0 & -iE_z/c \\ iE_x/c & iE_y/c & iE_z/c & 0 \end{pmatrix} \tag{12.47}$$

where μ enumerates the rows and ν the columns of the matrix, and we remember that $\boldsymbol{s}$ is defined by Eq. (12.21). We note the *antisymmetry* of the above tensor: $f_{\nu\mu} = -f_{\mu\nu}$ and its consequent property: $f_{\nu\mu} = 0$ *if* $\mu = \nu$. This means that the tensor is completely determined by $\boldsymbol{E}$ and $\boldsymbol{B}$. And it is equally important to note that this property holds under rotations in spacetime. It is true for the primed and the unprimed coordinate system of Fig. 12.3.

Maxwell's equations Eq. (10.45) are written in tensor form as follows [in reading Eq. (12.48a) we must remember Eq. (12.45)]:

$$\frac{\partial f_{\mu\nu}}{\partial x_{\nu}} = \mu_0 J_{\mu} \tag{12.48a}$$

$$\frac{\partial f_{\nu\sigma}}{\partial x_{\alpha}} + \frac{\partial f_{\sigma\alpha}}{\partial x_{\nu}} + \frac{\partial f_{\alpha\nu}}{\partial x_{\sigma}} = 0 \tag{12.48b}$$

where μ, ν, σ, α take the values 1, 2, 3, 4 corresponding to the four coordinates of spacetime. $J_{\mu}(s)$, $\mu = 1, 2, 3, 4$ is defined by

$$(J_1, J_2, J_3, J_4) = (\boldsymbol{j}, ic\rho) \tag{12.49}$$

where $\boldsymbol{j}$ is the ordinary current density and ρ the ordinary charge density of Sect. 10.4.3 One can show (the proof of this is straightforward but need not concern us here) that J_{μ} is indeed a 4-dimensional vector (a tensor field of rank one).

In conclusion let me note that, if we know $\boldsymbol{E}$ and $\boldsymbol{B}$ in one system of coordinates, say the unprimed one in Fig. 12.3, we can obtain the same in the primed system of coordinates by constructing the tensor $f_{\mu\nu}(s)$ as above, transform this tensor into $f_{\mu\nu}{}'(s')$, using the rules of tensor transformation, and read from it the components of the electric and the magnetic field in the primed system. For example, we can obtain the electric and magnetic fields due to a charge (q) moving with constant velocity (v) with respect to a given coordinate system as follows: We begin with the Coulomb field Eq. (10.2) of q in the coordinate system where q is at rest, and calculate the corresponding to it electric and magnetic fields in the given system of coordinates (with respect to which q is moving) by the transformation we described.

An alternative (Lorentz invariant) formulation of Maxwell's equations is summarised in Appendix A.

12.4 The General Theory of Relativity

12.4.1 The Principle of General Relativity

The demand that the laws of physics be invariant under rotations in spacetime, and in particular under a Lorentz transformation of the coordinates, plays an important role not only in classical physics but also in quantum mechanics, and has led to important discoveries. Einstein went further in this direction by inquiring as to whether the laws of nature must have the same form in any reference system whatever its motion. The question arises more naturally in relation to gravity. We have seen in Sect. 6.2.3 that the inertial mass, m, of a body, which appears in

Newton's law of motion, appears also in the law of gravity Eq. (6.12). It is because of this equality of the inertial mass with the gravitational mass that bodies fall with the same acceleration, g, towards the centre of the earth. Einstein was not prepared to accept this equality as a mere coincidence. Consider, he argued, a box, something like an elevator, in a region of space far away from any star, so that there is no gravitational force acting on it, and imagine that the box is pulled by a 'being' with constant force, so that it acquires an acceleration, γ, relative to a frame (coordinate system K) outside the box. Now assuming that there is a man in the box, he will observe that an object that he lets go from his hand falls to the floor of the box with a constant acceleration, γ, as if it were in a field of gravity. On the other hand, an observer outside the box will interprete the same phenomenon as follows: while the object was held by the man in the box a force was acting on it by his hand and that gave it the same acceleration as that of the box, when he let it go this force vanished and the velocity of the object increased no further and was therefore approached by the still accelerating floor of the box. Both interpretations are perfectly reasonable, and if that is the case then the laws of physics should be stated in such a way as to be applicable in both systems, that of the man in the box as well as the inertial one of the outside observer.

And there is, of course, the awkward fact, which was noted by Newton and worried some of his contemporaries that the action of gravity between bodies at a distance from each other occurs instantaneously, when according to special relativity no signal and therefore no interaction should propagate faster than light.

For the above reasons Einstein came to believe that the principle of general relativity, i.e., that the laws of physics must have the same form in all reference systems whatever their relative motion is, would lead to an improved theory of gravity, that would not require instantaneous action at a distance and would also explain why the inertial mass of a body equals its gravitational mass. Well before he was able to formulate such a theory (in 1916), he published a paper (in 1911) in which he argued that the apparent equivalence of gravitation to an accelerated frame of reference implied that light would bend in a strong gravitational field.[4] For example: light emitted by a star would bend slightly as it passes nearby the sun, and this bending though very small would be discernible during an eclipse of the sun. We can see how this bending of light would arise by reference to his accelerated (elevator-like) box in gravity-free space. Some one in the rest frame outside the elevator generates a beam of light in such a way that the light enters the elevator (through a small window) parallel to the floor of the elevator. To the observer in the accelerating upward elevator the light will appear on a curved path bending toward the floor. He may justifiably conclude that there is a gravity field in his space which bends the light downward. The fact that according to Newton's theory gravitational attraction occurred only between bodies that had mass, did not contradict with the above interpretation because light carries energy, and energy is

[4] A. Einstein, On the Effect of Gravitation on the Propagation of Light, Annalen der Physik, **35**, 898 (1911).

equivalent to mass according to Eq. (12.12) of special relativity. Using the said relation and Newton's formula, Einstein estimated the angle by which light passing tangentially to the surface of the sun is bent by it, and that came to be one half of the right answer, which he later obtained from his own theory of gravity (the general theory of relativity published in 1916), and which was confirmed experimentally in 1919. The bending of light in a gravitational field is obviously a most important phenomenon because it implies that the velocity of light varies with position, and this is contrary to the special theory of relativity, which assumes that the velocity of light in vacuum is constant. One is now forced to assume that the special theory of relativity applies as long as we are able to disregard the effect of gravitational fields on the phenomena under consideration. In the words of Einstein: *No fairer destiny could be allotted to any physical theory, than that it should of itself point the way to the introduction of a more comprehensive theory, in which it lives on as a limiting case.*

The mathematical formulation of a theory of gravity which accords with the demands of general relativity is not an easy task. Let me indicate the difficulties one is faced with, through an example that Einstein himself uses in his exposition of relativity.[5] Assume that the world is a disc rotating uniformly in its own plane about its centre. An observer at rest anywhere on the disc can measure distances in its vicinity, and time intervals, using a measuring rod and a clock as usual. Now his measurements must correspond (via a certain well defined transformation) to the measurements of the same distances and time intervals by an observer at the centre of the disc, if the principle of general relativity holds. We note that a clock at a distance from the centre, moves relative to the centre, and according to special relativity it goes at a rate permanently slower than the clock at the centre when viewed by the observer at the centre [see Eq. (12.9)]. Moreover, the rate at which this clock goes (viewed by the observer at the centre of the disc) depends on its position (because its velocity relative to the centre depends on its position). Similarly a measuring rod applied in the direction of motion of a place A on the disc (i.e. normal to its radial distance from the centre) will be shorter when viewed from the centre of the disc, in accordance with Eq. (12.8). But if the rod is applied in the radial direction (normal to the motion of A) no shortening of the rod will be observed by the observer at the centre of the disc. It follows from the above that, when measured by the observer at the centre of the disc, the circumference of a circle is *not* equal to πD, where D is the diameter of the circle, which implies that the familiar to us Euclidian geometry does not apply in this case. And this means, if we accept the principle of equivalence between accelerating frames and gravitational fields, that Euclidian geometry does not apply in a gravitational field. It appears that we are in an impossible situation, for unless we can define the space coordinates and the time consistently in the different frames of reference we can not formulate the laws of physics in accordance with the principle of general relativity.

[5] A. Einstein, The Meaning of Relativity, 5th edition, Princeton University Press, N.J. (1953); Relativity (A popular exposition), 15th edition, Methuen & Co Ltd, Great Britain (1954).

And yet there is a way out of our difficulties if we recognize that the Cartesian coordinates of the Minkowski space we are familiar with, and the Euclidian geometry that underlies them, is not the only way spacetime can be described. The geometry that meets the demands of general relativity was conceived by Gauss and perfected by Riemann and was available, though not much known to physicists, at the time Einstein formulated his theory.

Gauss considered the 2-dimensional space of a curved 'surface', drawing curves on it as in Fig. 12.5: two x^1—curves never cross each other, and form a continuum in the sense that any point on the surface lies on a x^1—curve or other, but are otherwise arbitrary. This implies, for example, that between the $x^1 = 1$ and the $x^1 = 2$ curves there are infinitely many x^1—curves. And the same holds in relation to the x^2—curves. Therefore, a point on the surface has definite values of x^1 and x^2, which can therefore serve as its coordinates (now known as Gaussian coordinates).

It is evident that two points close to each other will have coordinates (x^1, x^2) and $(x^1 + dx^1, x^2 + dx^2)$ respectively, where dx^1 and dx^2 are infinitesimal (very small). Then, according to Gauss, the distance ds between them, as measured by a given measuring rod, is given by:

$$(ds)^2 = g_{11}(dx^1)^2 + 2g_{12}dx^1dx^2 + g_{22}(dx^2)^2 \qquad (12.50)$$

where, the quantities g_{11}, g_{12} and g_{22} are definite functions of the coordinates (x^1, x^2). It is these quantities that will determine the length of the given rod at the point (x^1, x^2). If the space under consideration is Euclidian, and only then, we can choose the coordinates in such a way that: $(ds)^2 = (dx^1)^2 + (dx^2)^2$. In this case the Gaussian coordinates become our familiar Cartesian coordinates.

The generalization of the Gaussian method to n-dimensions ($n > 2$) is straightforward and obvious. In the case of the 4-dimensional spacetime that concerns us here, we have four Gaussian coordinates: (x^1, x^2, x^3, x^4), the first three of which cover ordinary space: x^1 and x^2 will be as in Fig. 12.5 and x^3 will be a

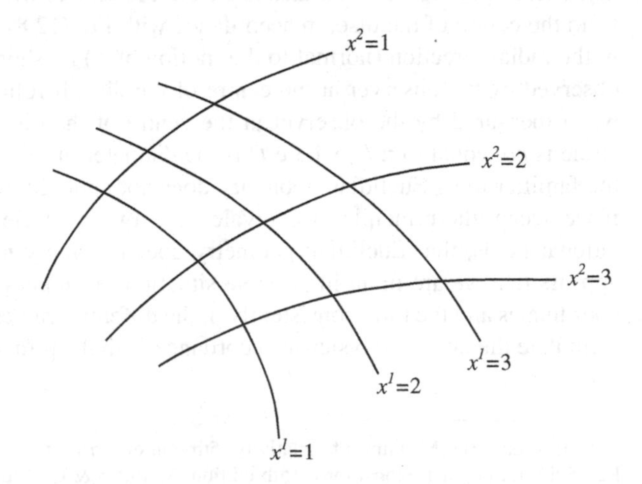

Fig. 12.5 Gaussian coordinates in two dimensions

third independent coordinate, so that each point in ordinary space corresponds to a point (x^1, x^2, x^3) and vice versa, but otherwise, x^3, like the other two coordinates, can be chosen arbitrarily.

The fourth coordinate ($x^4 = ict$) is determined by the reading of an (imagined) clock at (x^1, x^2, x^3); the rate by which it goes can be arbitrary as long as the times of adjacent clocks vary continuously. Because c is a constant, we can think of the fourth coordinate as the time coordinate [as in Eq. (12.21)]. The spacetime interval ds between two adjacent events, a scalar quantity, is now given by:

$$(ds)^2 = \sum_{\mu=1}^{4}\sum_{\nu=1}^{4} g_{\mu\nu} dx^\mu dx^\nu \tag{12.51}$$

where, $g_{\mu\nu}$ are definite functions of the coordinates $(x^1, x^2\ x^3, x^4)$. They can always be chosen to be symmetric: $g_{\mu\nu} = g_{\nu\mu}$. A corresponding *proper time* interval is defined by analogy to Eq. (12.30) as follows:

$$(d\tau)^2 = -(ds)^2/c^2 \tag{12.52}$$

It turns out that a Gaussian description of spacetime implies that spacetime is *locally* Euclidian. This means that, there is about every point $s = (x^1, x^2, x^3, x^4)$ of spacetime a small region over which special relativity is valid. We can say this using metaphorical language as follows: though as a whole spacetime is curved, it is locally flat: the way a small bit on the surface of a large sphere is locally flat. A measuring rod will not in general sit flat (every bit of it touching the surface) of a sphere, but if the rod is sufficiently small relative to the radius of the sphere it will do so. Having convinced ourselves that in our locality (where we are in the universe) this is indeed so, we can assume that the universe is locally flat everywhere.

Finally the geometry developed by Gauss and Riemann allows us to determine (calculate) the *metric* tensor $g_{\mu\nu}$ in any (arbitrarily chosen) system of Gaussian coordinates if we know it in one such system. This is made possible through the use of curvilinear tensors. And of course, as we have already noted, $(ds)^2$, given by Eq. (12.51), being a scalar quantity is the same in all coordinate systems.

12.4.2 Curvilinear Tensors

The coordinates $s' = (x^{1\prime}, x^{2\prime}, x^{3\prime}, x^{4\prime})$ of a point in spacetime in a Gaussian system of coordinates, which we refer to as the primed system, are related to its coordinates $s = (x^1, x^2, x^3, x^4)$ in another such system, which we refer to as the unprimed system of coordinates, through a set of functions as follows:

$$x^{\mu\prime} = x^{\mu\prime}(x^1, x^2, x^3, x^4),\ \mu' = 1, 2, 3, 4 \tag{12.53}$$

For each μ' the expression on the right is a function of x^1, x^2, x^3, x^4. Therefore, the components of an infinitesimal displacement in the primed system will relate to its components in the unprimed system as follows:

$$dx^{\mu\prime} = \sum_{\nu=1}^{4} (\partial x^{\mu\prime}/\partial x^{\nu}) dx^{\nu} = \sum_{\nu=1}^{4} c^{\mu\nu}(s) dx^{\nu}, \quad \mu = 1,2,3,4 \tag{12.54}$$

which is similar to Eq. (12.41a) for Cartesian coordinate systems, except that now the c's are not constants but functions of the position. The inverse transformation [corresponding to Eq. (12.41b)] is obtained in similar fashion, but now the coefficients of the inverse transformation do not relate to those in Eq. (12.54) in the easy manner of Eq. (12.28).

The derivatives $\partial f/\partial x^{\mu}$, $\mu = 1, 2, 3, 4$, of a (scalar) function $f(s)$ transform as follows:

$$\partial f/\partial x^{\mu\prime} = \sum_{\nu=1}^{4} (\partial x^{\nu}/\partial x^{\mu\prime}) \partial f/\partial x^{\nu} = \sum_{\nu=1}^{4} c_{\mu\nu}(s) \partial f/\partial x^{\nu} \tag{12.55}$$

where $\partial f/\partial x^{\mu}$ are evaluated at s and $\partial f/\partial x^{\mu'}$ at s' in accordance with Eq. (12.53). We note that, contrary to what is the case for Cartesian coordinates, the c's are functions of the position and the $c_{\mu\nu}$ are *different* from the $c^{\mu\nu}$.

A 4-component quantity $A^{\mu}(s)$, $\mu = 1, 2, 3, 4$, which transforms like dx^{μ}, $\mu = 1, 2, 3, 4$, [of Eq.(12.54)] is a (curvilinear) *contravariant tensor (field) of rank one* (or a contravariant vector). We have:

We have: $A^{\mu\prime}(s') = \sum_{\nu=1}^{4} c^{\mu\nu}(s) A^{\nu}(s), \mu = 1,2,3,4$

A 4-component quantity $A_{\mu}(s)$, $\mu = 1, 2, 3, 4$, which transforms like $\partial f/\partial x^{\mu}$, $\mu = 1,2,3,4$, of Eq. (12.55) is a (curvilinear) *covariant tensor of rank one* (or a covariant vector). We have:

$$A'_{\nu}(s') = \sum_{\nu=1}^{4} c_{\mu\nu}(s) A_{\nu}(s), \mu = 1,2,3,4$$

The reader should note that contravariant tensors are denoted by *upper* indexes, and covariant ones by *lower* indexes. Contravariant and covariant tensors of higher rank are defined in similar fashion. And we note that a curvilinear tensor may be contravariant with respect to some indices and covariant with respect to others. For example:

$$F_{\lambda}^{\mu\nu\prime}(s') = \sum_{k=1}^{4} \sum_{q=1}^{4} \sum_{r=1}^{4} c_{\lambda k} c^{\mu q} c^{\nu r} F_{k}^{qr}(s)$$

The need to introduce contravariant and covariant tensors complicates matters, but it is not the main difficulty one is faced with, when dealing with curvilinear coordinates (Gaussian coordinates). One finds that unlike the case of Cartesian coordinates (see Eq. 12.43), a straightforward differentiation of a vector, or a tensor of rank greater than one, does *not* produce a tensor of higher rank. However, the mathematicians established a modified version of ordinary differentiation, called *covariant* differentiation, which produces derivatives of tensors (they are called *covariant* derivatives) which *are* tensors. And it turns out (not surprisingly may be) that these covariant derivatives are the quantities of physical interest. The derivation of the mathematical formula which transforms ordinary derivatives ($\partial/\partial x$) to covariant derivatives (D/Dx) is too complicated to concern us here. Let me just give an example which shows the complexity arising thereof. The covariant derivative of a tensor $A_{\mu\nu}$ *derives* from its ordinary derivative as follows:

$$\frac{DA_{\mu\nu}}{Dx^{\rho}} = \frac{\partial A_{\mu\nu}}{\partial x^{\rho}} - \sum_{\tau=1}^{4} \Gamma^{\tau}_{\mu\rho} A_{\tau\nu} - \sum_{\tau=1}^{4} \Gamma^{\tau}_{\nu\rho} A_{\mu\tau} \tag{12.56}$$

$\Gamma^{\tau}_{\nu\rho}$ is known as the *metric connection* or the *Christoffer symbol*; it is a function of the $g_{\mu\nu}$:

$$2\sum_{\tau=1}^{4} g_{\nu\tau}\Gamma^{\tau}_{\mu\rho} = \partial g_{\mu\nu}/\partial x^{\rho} - \partial g_{\rho\mu}/\partial x^{\nu} + \partial g_{\nu\rho}/\partial x^{\mu}$$

$\Gamma^{\tau}_{\nu\rho}$ and the ordinary derivative $\partial A_{\mu\nu}/\partial x^{\rho}$, are *not* tensors, but $DA_{\mu\nu}/Dx^{\rho}$ *is* a tensor.

The reader probably has guessed already the reason we have introduced curvilinear tensors: Like in the case of Cartesian tensors, a set of differential equations written in tensor form has the same form in all Gaussian systems of coordinates and therefore satisfies the requirements of general relativity.

12.4.3 Einstein's Equation

Let us first see how the motion of a particle is described in a Gaussian system of coordinates (x^1, x^2, x^3, x^4). If the particle existed for just one moment, its position in spacetime would be given by four numbers for x^1, x^2, x^3 and x^4 respectively. Its continued existence is accordingly described by a sequence of such tetrads of numbers which constitute a continuous line in the 4-dimensional spacetime (described by the above four coordinates). And if there are N particles, there will be N such lines: one for each particle. And the same will be true if the particle is replaced by a light signal. When it comes to the motion of a particle in a gravitational field, we must remember that, according to Einstein's theory, gravity is a property of space–time taken into account by the metric tensor $g_{\mu\nu}$, and therefore if no other force is acting on the particle, it must be considered a *free* particle, whose motion (the unidimensional line in the continuum of the selected system of Gaussian coordinates mentioned above) is fully determined by $g_{\mu\nu}$ (which are functions of x^1, x^2, x^3, x^4) and the initial state of the particle. It turns out that any two points on the line of its motion determine the entire motion of the particle. This is what we should expect. In the flat (Euclidian) space of special relativity a free particle moves from point A to point B along the path *which minimizes the length $c\Delta\tau$ between A and B*, and this is a straight line. The same is true in the curved spacetime of general relativity: a free particle moves from point A to point B along the path *which minimizes the length $c\Delta\tau$ between A and B*, which for a curved space is given, according to Eq. (12.51), by

$$s = \int_A^B ds = \int_A^B \sqrt{\left(\sum_{\mu=1}^{4}\sum_{\nu=1}^{4} g_{\mu\nu} dx^{\mu} dx^{\nu}\right)} \tag{12.57}$$

The line so defined is known as the *geodesic* line (in the given *4*-dimensional space) passing through points A and B.

[The meaning of a geodesic line in curved space is made clearer by reference to the geodesic lines of the two dimensional space of a spherical surface, say the surface of the earth. The shortest distance between two points A and B on the surface of the earth is obtained along the *great* circle on the surface of the earth that passes through these points. To obtain this great circle consider the plane defined by these two points and the centre of the earth: the circle cut by this plane on the surface of the earth is the required great circle.]

The simplest way to describe the geodesic which desribes the motion of a free particle in spacetime is to describe the points that make up this line as a function of a continuous parameter (say τ) as follows: $s(\tau) = (x^1(\tau), x^2(\tau), x^3(\tau), x^4(\tau))$. And if required, we can easily obtain the coordinates x^1, x^2, x^3 as functions of the time coordinate x^4 from the known $s(\tau)$. It turns out that the $x^1(\tau), x^2(\tau), x^3(\tau), x^4(\tau)$ which minimize s of Eq. (12.57) satisfy a system of coupled partial differential equations (we need not write them down here) which are in scope similar to the equations which determine $x(t)$, $y(t)$, $z(t)$ of a particle moving in a given potential field $U(x,y,z)$ in ordinary space according to the laws of Newtonian mechanics. In the present case $g_{\mu\nu}$, which are given functions of x^1, x^2, x^3, x^4, play the role of the potential field. Therefore, in order to calculate the motion of a particle in space–time, we must first know (calculate) the metric tensor $g_{\mu\nu}(x^1, x^2, x^3, x^4)$ of spacetime in the chosen system of coordinates. For this purpose we need an equation that will allow us to determine $g_{\mu\nu}$ in a manner consistent with the principle of general relativity.

We have already noted that the geometry of curved space differs from that of flat (Euclidian) space. But how is one to measure the 'curvature' of a given curved space? In the case of a two-dimensional spherical surface we can do this as follows: Imagine a spherical triangle (on the surface of a sphere) whose sides are bits of great circles of the sphere. The sum of the internal angles α, β and γ of this triangle is greater than π and not equal to π (as it happens for triangles on a flat surface). In this case, a measure of the (local) curvature is given by the (local) radius of curvature, R, defined by:

$$1/R^2 = (\alpha + \beta + \gamma - \pi)/\mathrm{A} \tag{12.58}$$

where A denotes the area of the triangle. We see that when $(\alpha + \beta + \gamma)$ is close to π, the corresponding R becomes very large, and when $(\alpha + \beta + \gamma) = \pi$, R goes to infinity, which corresponds to a flat surface. We note that the above measure of curvature can be applied to any (two-dimensional) surface; but in an arbitrary two-dimensional surface, R will depend on the position on the surface. The real difficulty arises when one attempts to define curvature in spaces of three or more dimensions. Riemann had established that a full description of the curvature of a 4-dimensional space (x^μ, $\mu = 1, 2, 3, 4$) is provided by a tensor of rank four, the components of which are functions of the $g_{\mu\nu}$ and their first and second derivatives with respect to x^μ, $\mu = 1, 2, 3, 4$. This tensor is now called the *Riemann curvature tensor*. We shall not need the explicit expression of this tensor in our present discussion.

Einstein argued that the curvature of spacetime as measured by Riemann's curvature tensor (or some other tensor that derives from it) should be determined by the distribution of mass (and energy) in spacetime. But since, according to special relativity, mass-energy and momentum are components of the same quantity [see Eq. (12.33)], a tensor that somehow measures the curvature of spacetime, must relate to a tensor which is obtained directly from the distribution of momentum-energy in spacetime. Einstein postulated a form for this tensor, which is known as the stress-energy tensor and denoted by $T_{\mu\nu}$; it is a symmetric tensor: $T_{\mu\nu} = T_{\nu\mu}$; and it is a divergenceless tensor meaning that

$$\sum_{\nu=1}^{4} \frac{\partial T_{\mu\nu}}{\partial x_\nu} = 0, \mu = 1, 2, 3, 4 \qquad (12.59)$$

Equation (12.59) makes sure that the distribution of momentum-energy in space–time satisfies the requirement of momentum and energy conservation. And we note that the symmetry of $T_{\mu\nu}$ and the above equations tell us that $T_{\mu\nu}$ has only 6 independent components (we remember that an arbitrary tensor of rank two has 16 components).

Let me write down, as an example, $T_{\mu\nu}$ for a cloud of dust, in the Cartesian system of coordinates of special relativity. Assuming that in the rest frame (static cloud) there are n_O particles per unit volume, each particle having a rest mass m_O, we can write:

$$T_{\mu\nu} = n_O m_O \upsilon_\mu \upsilon_\nu \qquad (12.60)$$

where $m_O \upsilon_\mu$, $\mu = 1, 2, 3, 4$ are the four components of ($\boldsymbol{p}$, iE/c) of a particle in accordance with Eq. (12.33), and υ_ν, $\nu = 1, 2, 3, 4$ are the four components of its velocity (in traditional notation: $dx/d\tau$, $dy/d\tau$, $dz/d\tau$, $icdt/d\tau$). We can now interprete $T_{\mu\nu}$ as follows: $T_{\mu\nu}$ is the flow of the μth component of the 4-component momentum along the νth direction. For example, $T_{41}(x,y,z,t)$ represents the flow of energy (say heat) along the x-direction at the point (x,y,z,t) of space–time; $T_{44}(x,y,z,t)$ gives the change in the energy density with time at the point (x,y,z,t), etc. Because of Eq. (12.59) no energy-mass or momentum is lost or generated out of nothing.

Finally, we note that the above procedure can be applied to a Gaussian system of coordinates appropriate to general relativity. One need only replace the ordinary derivatives appearing in Eq. (12.59) by covariant derivatives [see text preceding Eq. (12.56)].

Einstein established a relation between the stress-energy tensor and Riemann's curvature tensor as follows. We have noted that the latter is a rank four tensor. However, by contraction [a well defined mathematical operation in tensor analysis, similar to that of Eq. (12.45)] it reduces to a tensor of rank two which we denote by $G_{\mu\nu}$ which is, like $T_{\mu\nu}$, symmetric and divergenceless. This led Einstein to postulate a relation between the two tensors, which we now know as Einstein's equation:

$$G_{\mu\nu} = -KT_{\mu\nu} \tag{12.61a}$$

where K is a universal constant. The equation is often written in the following form:

$$R_{\mu\nu} - 1/2Rg_{\mu\nu} = -KT_{\mu\nu} \tag{12.61b}$$

where $R_{\mu\nu}$ is known as the Ricci tensor and R as the Ricci scalar. Einstein showed that the above equation reduces to Newton's law of gravity for weak gravitational fields and when all masses move relative to the coordinate system with velocities small compared to that of light, if

$$K = 8\pi G/c^4$$

where G is the ordinary constant of gravitation given in Sect. 6.2.3. This is how he determined the value of this constant.

Einstein's equation (12.61) written, as it is, in tensor form holds for an arbitrary Gaussian system of coordinates, and therefore accords with the principle of general relativity. But it is indeed very difficult to solve, however we choose the system of coordinates. It is worth remembering that Eq. (12.61), stands for a system of nonlinear coupled partial differential equations which the $g_{\mu\nu}$ must satisfy, and that these equations involve $g_{\mu\nu}$ and their first and second derivatives with respect to the four coordinates of spacetime. An exact solution of these equations was obtained by Schwarzchild (in 1916) for the empty spacetime around a spherically mass distribution. The $g_{\mu\nu}$ obtained by Schwarzchild (known as the Schwarzchild metric) is adequate for the calculation of general relativistic effects in the solar system, and for the simplest type of black holes (see Sect. 15.3.2). However, sufficiently accurate approximate solutions can be obtained in a number of cases, and moreover some important results can be obtained by considering in depth the structure of the equations themselves.

Let me now describe briefly some results of general relativity, and in particular those obtained by Einstein himself. These are:

In relation to the motion of the planets about the sun the theory gives practically identical results with those of Newtonian mechanics except for Mercury (the planet that lies nearest to the sun). It has been known for some time before Einstein that the elliptical orbit of Mercury about the sun, after it has been corrected for the influence of the other planets, was still not in complete agreement with the prediction of Newtonian theory. The ellipse corresponding to the orbit of this planet rotates very slowly in the plane of the orbit and in the sense of the orbital motion, by 43 s of a degree per century (with an accuracy of a few seconds of a degree). According to Einstein's theory of gravity the ellipse of every planet rotates in the manner described above, but for all planets except mercury this rotation is too small to be measured. In the case of Mercury, Einstein's theory gives a rotation in perfect agreement with the observed value noted above.

In relation to the bending of light by gravity, Einstein's theory, as we mentioned earlier, gave a result for the bending of light by the sun which was twice his original rough estimate. The measured value of this deflection (1.7 s of a degree), obtained by the English astronomer Arthur Eddington during the solar eclipse of 29 May 1919 was in perfect agreement with the prediction of the theory.

The existence of **gravitational waves** is another important prediction of Einstein's theory. One assumes that a weak gravitational field can be expressed by the following metric tensor: $g_{\mu\nu} = g_{\mu\nu}^{(0)} + h_{\mu\nu}$, where $g_{\mu\nu}^{(0)}$ are the components of a flat space [according to Eq. (12.29): $g_{\mu\nu}^{(0)} = 1$ for $\mu = \nu$ and $g_{\mu\nu}^{(0)} = 0$ for $\mu \neq \nu$], and the magnitudes of the $h_{\mu\nu}$ are all much smaller than unity. In this case Einstein's equation is reduced to the d'Alenbert equation:

$$\frac{\partial^2 h_{\mu\nu}}{\partial x^2} + \frac{\partial^2 h_{\mu\nu}}{\partial y^2} + \frac{\partial^2 h_{\mu\nu}}{\partial z^2} = \frac{1}{c^2}\frac{\partial^2 h_{\mu\nu}}{\partial t^2}$$

which has solutions of the type: $h_{\mu\nu} = h_{\mu\nu}(t - x/c)$ which represent gravity waves propagating with the speed of light (c) along a certain direction (the x-direction in this example). When such a wave crosses a region of space, it is spacetime itself that vibrates. As a result the separation between two points in space will change by a small amount. The most energetic gravitational waves likely to reach the earth will emanate from stellar collapse in our galaxy, and are expected to produce stresses in spacetime of amplitude about 10^{-18}, which represents a displacement of about one nuclear diameter in a length of one meter. Researchers are confident that present day technology is approaching a state of development that will make it possible to detect such events in the near future.

We note in passing that since the 1960s it has been possible to compare the rate of atomic clocks, one on earth and one in a rocket 10,000 km above and the difference between them is in agreement with Einstein's theory.

One feature of Einstein's equation, which is currently of great interest, relates to the so-called cosmological constant. When Einstein's equation was used in calculations of cosmology (concerning the universe on the large scale), it became apparent that it favoured an expanding universe. This conflicted with the prevailing opinion at the time, that the universe was static. Einstein found that he could extend his equation in only one way: by adding a term $\Lambda g_{\mu\nu}$, where Λ is a constant, known as the cosmological constant, to be determined empirically. So he modified his equation as follows:

$$G_{\mu\nu} = -K\mathrm{T}_{\mu\nu} + \Lambda g_{\mu\nu} \tag{12.62}$$

One can readily see that the presence of the new term would lead to a curvature of spacetime even in the absence of matter or energy ($T_{\mu\nu} = 0$). It is therefore possible to obtain a static universe by choosing Λ appropriately. However, in 1931, Hubble and Humason established that the universe is currently expanding, and Einstein happily rejected the cosmological constant. But may be he should not have done so. Present day cosmologists brought it back into the equation in an effort to accommodate new data on the expansion of the universe (see Sect. 15.3.4).

12.5 The Famous Scientist

In 1908 Einstein was 29 and still working at the patent office in Berne. His work had gone mostly unnoticed up to this time. However, in 1909 he was appointed 'professor extraordinary' at the Zurich Polytechnic, which was something like an assistant professor, with a salary not greater than the one he had at the patent office. In 1911 he moved to Prague, but in the autumn of 1912 was back in Zurich, this time as a (full) professor at the Polytechnic, which signifies that by this time he had gained the respect of outstanding scientists of that time. In 1913 he was elected a member of the Prussian Academy of Science, and in 1914, a little before the First World War began, he was appointed director of the newly formed Kaiser Wilhelm Research Institute for Physics with the title of professor at the University of Berlin. He had no official academic duties and could teach or research in any way he wanted. At that time and until the mid 1920s (when the centre of gravity shifted to younger men in Göttingen and Copenhagen) Berlin was at the frontier of research in theoretical physics (Planck, Einstein and, later, Schrödinger were there). The Berlin physics colloquium was attended regularly by outstanding scientists. Einstein was happy to be there. Unfortunately his marital life had been in difficulty for some time and soon after their arrival in Berlin, he and Mileva separated, and she returned to Zurich with their two sons. His biographer, Philipp Frank, who saw him often during that period, tells us that Einstein had no interest in much of the social life organized around the University. According to Frank, he resembled no one so much as some of those Bohemian violinists who frequented the coffee houses where both Einstein and Frank spent a good deal of their spare time. It happened also that in Berlin, Einstein rediscovered parts of his family. In earlier times they looked down upon him as an irresponsible Bohemian, but now that he was a professor and an academician, they were happy to receive him. Einstein ate well at his uncle's house and enjoyed the company of his cousin Elsa, recently widowed with two daughters. She had a friendly maternal character, interested in creating a pleasant home, and in time she became Einstein's second wife.

Eddington's confirmation (in 1919) of the bending of light by the sun proved the validity of Einstein's theory of general relativity and established the reputation of its author as a great scientist. By now, of course, his other work on special relativity and statistical physics was also well known and appreciated. In 1921 he was awarded the Nobel Prize for Physics, notably not for his theory of relativity but for his interpretation of the photoelectric effect (see Sect. 13.1.2). Apparently the Swedish Academy, while recognizing Einstein's genius was trying to avoid the 'speculative' character of the theory of relativity (that is how it was perceived by many at the time), when in reality his interpretation of the photoelectric effect was if anything more speculative. The award was announced in November 1922, and he collected the prize in April 1923 (he gave all the money to Mileva for looking after their two sons).

Eddington's confirmation of the bending of light coincided with the first of the well-organised political attacks against the theory of relativity, apparently for no reason other than the fact that Einstein, now a public figure, was Jewish. The madness of these attacks was of course the least of what happened in the years that followed. The horrors of Germany under the Nazis are well known.

In the spring of 1933 Einstein took refuge in the Belgian sea resort of Le-Coq-sur-Mer. His summer home near Berlin had been sacked by the Gestapo and his property confiscated with the excuse that it would be used in support of a Communist revolt. He left Belgium for the United States in October 1933. With him were his wife Elsa, and his faithful secretary Helen Dukas. He becamc the first professor at the newly founded Institute for Advanced Studies at Princeton. The Institute was conceived as a centre where young scholars could have informal contact with a small number of first rate permanent members. There would be no formal classroom lectures and the Institute would offer no degrees. In 1935 Einstein bought a modest house, not far from the Institute, where he lived until his death. By 1939 the fate of Europe and his fellow Jews weighed so heavily on him that though a life-long pacifist he came to the conclusion that war was the only way to deal with Hitler. After the war (in 1946) when Arnold Sommerfeld, who had been an anti-Nazi but who had, like Planck, stayed in Germany, wrote to Einstein to suggest that he might be interested to renew his membership of the Bavarian Academy from which he had been expelled in 1933, he replied: *The Germans slaughtered my Jewish brethren; I will have nothing to do with them, not even with a relatively harmless academy. I feel different about the few people who, in so far as it was possible, remained steadfast against Nazism. I am happy to learn you were among them.* Einstein was of course aware that not a few scientists joined Hitler's party. The Nobel laureate Philipp Edward Lenard was in fact a prominent Nazi.

After 1917, Einstein continued to work on aspects of the general theory of relativity but, while still in Germany, he made some important contributions to statistical physics (we refer to the Bose–Einstein distribution in Sect. 14.4.2), and to the theoretical treatment of the interaction of light with matter (see Sect. 13.1.3). But his main preoccupation from the 1920s to the day he died, had to do with the unification of gravity and electromagnetism, and with the physical interpretation of quantum mechanics. Einstein tried and tried again to produce a unified theory of gravitation and electromagnetism but without success. Although one can easily write Maxwell's equations in a form valid in every Gaussian system of coordinates (one need only replace the ordinary derivatives in Eqs. (12.48) and (12.49) by covariant derivatives), these equations are not obtainable from a unified source (giving rise to both gravity and electromagnetism). His reaction to quantum mechanics (which he and Planck initiated) derived from his deeply felt views about the nature of science. We shall have occasion to refer to his doubts about this theory in the next chapter.

His wife, Elsa, died in 1936. His sister Maya moved to Princeton and stayed with him until her death in 1951. And Helen Dukas stayed with him and looked after him until his own death in 1955. In the last years of his life he was the most well known living scientist. His face recognized all over the world by scientists and millions of non scientists as well. Most scientists now agree that he was one of the greatest scientists of all time.[6]

Exercises

12.1. Two events occur at different points in space, (x_1, y_1, z_1) and (x_2, y_2, z_2) respectively, but at the same time, t_o, in the unprimed system of Fig. 12.3. Show that in the primed system of Fig. 12.3 the two events are separated by:

$$\Delta t' = t_2' - t_1' = -\frac{V(x_2 - x_1)}{c^2\sqrt{1 - V^2/c^2}}$$

12.2. An electron moves with a velocity of 0.75 c. Show that its relativistic momentum is 50 % greater than its classical momentum.

12.3. A photon has a momentum-energy vector: $(\hbar \boldsymbol{q}', iE'/c)$, where $E' = \hbar\omega' = \hbar c q'$, in the primed system of coordinates of Fig. 12.3. We have: $q_x' = q' cos\theta$, where θ is the angle between $\boldsymbol{q}'$ and the x' direction (which is the same with the x direction). And we remember that $cos\theta < 0$ when $q_x' < 0$. Use the transformation law for the momentum-energy vector to obtain $(\hbar \boldsymbol{q}, iE/c)$, where $E = \hbar\omega = \hbar c q$, in the unprimed system of coordinates of Fig. 12.3, and hence show that:

$$\omega' = \frac{\omega}{(1 + \beta\cos\theta)\gamma}$$

where $\beta = V/c$ and $\gamma = 1/\sqrt{(1 - V^2/c^2)}$. The above shift in the frequency of the photon, known as the Doppler shift, is of great importance in astronomy. We note

6 The interested reader will find a good selection of Einstein's ideas and opinions on a variety of topics including freedom, education, war and politics, as well as short but very informative pieces on theoretical physics and its relation to mathematics, and on the basic assumptions of relativity in: Albert Einstein: *Ideas and Opinions*, Souvenir Press Ltd, London (1973).

that a shift in the frequency of the photon implies a corresponding shift in its wavelength, λ, since $\lambda = (2\pi/\omega)c$.

12.4. How fast does a body travel if red light ($\lambda = 7{,}000$ a.u.) appears blue ($\lambda = 4{,}000$ a.u.) to an observer in its approach? (1 a.u $= 0.529 \times 10^{-10}$ m)

Answer: 0.51 c.

12.5. Find the velocity $\boldsymbol{u} = (u_x = dx/dt,\ u_y = dy/dt,\ u_z = dz/dt)$ of a particle in the unprimed system of coordinates of Fig. 12.3, in terms of its velocity $\boldsymbol{u}' = (u_x' = dx'/dt',\ u_y' = dy'/dt',\ u_z' = dz'/dt')$ in the primed system of coordinates.

Answer: $u_x = \frac{u_x'+V}{1+u_x'V/c^2},\ u_y = \frac{u_y'}{\gamma(1+u_x'V/c^2)},\ u_z = \frac{u_z'}{\gamma(1+u_x'V/c^2)}$ where $\gamma = 1/\sqrt{(1 - V^2/c^2)}$.

12.6. Two particles travel in opposite directions with a speed of *0.99c*. What is the speed of one particle relative to the other?

Answer: 0.99995 c.

12.7. The mean lifetime of a muon in its own rest system of coordinates is 2.0×10^{-6} s. What is the average distance the particle travels in vacuum before it decays when measured in coordinate systems in which its velocity is 0, 0.6c and 0.99c, where c is the speed of light?

Answer: 0, 450 m, 4,200 m.

12.8. Evaluate the amount of energy, ΔE, released in a nuclear reaction when the total mass after the reaction is 10^{-3} g less than the initial mass.

Answer: $\Delta E = 8.9875 \times 10^{10}$ J.

12.9. In Chap. 8 we stated that: matter can neither be created nor destroyed. How should this statement be corrected in view of the special theory of relativity?

12.10. Show that the symmetry character of a given Cartesian tensor is the same in all coordinate systems.

12.11. In a collision between two particles the total energy (the sum of the energies of the two particles) and the total momentum (the sum of the momenta of the two particles) are conserved (they do not change) in accordance with Newton's third law. A particle of rest mass M with speed $\beta_O c$ after colliding with a particle at rest (with a rest mass m) sticks to it. What is the speed of the composite particle after the collision?

Answer: $[\gamma_O M/(\gamma_O M + m)]\ \beta_O c$, where $\gamma_0 = \left(1 - \beta_0^2\right)^{-1/2}$.

12.12. Convince yourself that Eqs. (12.48) are equivalent to Maxwell's equations [Eqs. (10.45)] by writing explicitly a sufficient number of the components of these equations.

12.13. By applying the transformation law of a second rank tensor to $f_{\mu\nu}$ of Eq. (12.47), show that the Lorentz transformation equations for the electric and magnetic field are:

$$E_x' = \mathrm{E}_x, E_y' = \gamma(\mathrm{E}_y - \beta c B_z), E_z' = \gamma(\mathrm{E}_z + \beta c B_y)$$
$$B_x' = B_x, B_y' = \gamma(B_y + \beta E_z/c), B_z' = \gamma(B_z - \beta E_y/c)$$

Chapter 13
Quantum Mechanics

13.1 The Quantization of Light

13.1.1 Black Body Radiation and Planck's Law

By the end of the 19th century it had been established that the electromagnetic (EM) radiation within a cavity at thermal equilibrium (viewed from a hole on one of its walls) is the same for all cavities. Its overall intensity and its spectral distribution [the spread of the radiation over the observed range of wavelengths (frequencies)] at a given temperature does not depend on the shape of the cavity or the material its walls are made of. This radiation is called isothermal cavity radiation and because it manifests itself as the radiation of a black body it is also known as the black-body radiation. Its spectral distribution is shown in Fig. 13.1 for three different temperatures. $I(\lambda, T)d\lambda$ is the electromagnetic energy radiated from a black body at absolute temperature T per unit area per unit time in wavelengths between λ and $\lambda + d\lambda$. The total amount of energy radiated per unit time per unit surface of the black body (the integral over all frequencies of $I(\lambda, T)$) is given by the Stefan law:

$$I(T) = \sigma T^4, \sigma = 5.7 \times 10^{-8} \mathrm{J\ s^{-1} m^{-2} K^{-4}} \tag{13.1}$$

which was discovered by Joseph Stefan, who lived in the years 1853–1893. Botzmann was able to derive the above law, its form but *not* the actual value of σ, by treating the radiation inside the cavity as a thermodynamic fluid whose properties he obtained from the laws of electromagnetism; for this reason, Eq. (13.1) is also known as the Stefan-Boltzmann law. However, when statistical mechanics was applied to the problem, notably by Rayleigh and Jeans, it failed to reproduce the observed spectral distribution (of Fig. 13.1).

The first step in any theory of the black body radiation must be the determination of the independent states of the EM field in a given cavity. The most general state of the EM field in the cavity is a linear superposition (sum) of these independent states (often called the *normal modes* or *eigenmodes* or simply the *modes*

A. Modinos, *From Aristotle to Schrödinger*,
Undergraduate Lecture Notes in Physics, DOI: 10.1007/978-3-319-00750-2_13,

Fig. 13.1 $I(\lambda, T)$ is the electromagnetic energy radiated from a black body at absolute temperature T (measured in degrees Kelvin, K) per unit area per unit time (in units of joule per second per square centimeter: J s^{-1} cm^{-2}) per unit wavelength. The wavelength is measured in microns (1 micron (μ) = 10^{-6}m). We note that the visible region of the electromagnetic radiation extends from 0.4 (*violet*) to 0.8 μ (*red*)

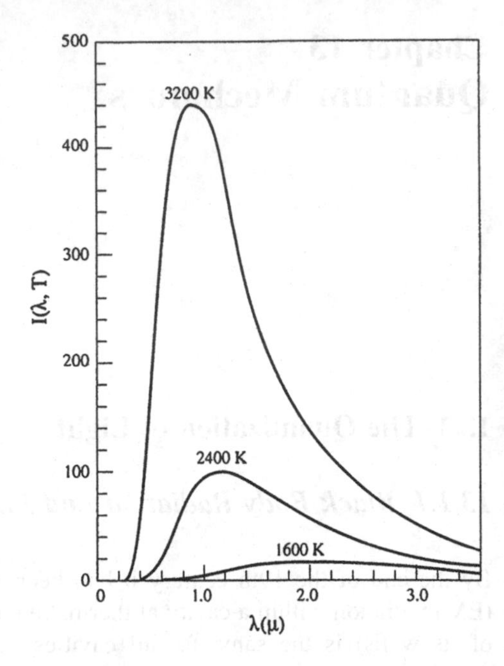

of the cavity), in the same way that the most general state of a vibrating string is a linear superposition of the normal modes of the string (see Sect. 7.2).

We have seen that in empty unbounded space *any* linear superposion of plane waves (we can think of them as the modes of the EM field in unbounded free space) is a possible state of the EM field (see Eq. 10.49). This is not the case in a cavity. The modes of the EM field within the cavity are solutions of Maxwell's equations in the empty space of the cavity which satisfy the appropriate boundary conditions at the walls of the cavity, i.e. the components of the electric and the magnetic field tangential to the walls of the cavity must vanish there. Fortunately, we are here concerned not with the exact form of the eigenmodes but with how many of them are there. It turns out that we can get the number of eigenmodes of the EM field in the cavity right, if we count the plane waves of Eq. (10.48), which are *periodic* in all three directions (x,y,z) of space, with a period L, where L is the edge length of a cubic cavity with the same volume $L^3 = V$ as that of the cavity under consideration. [We remember that the measured properties of black body radiation do not depend on the shape of the cavity]. In other words we demand that:

$$E(x, y, z, t) = E(x + L, y, z, t) = E(x, y + L, z, t) = E(x, y, z + L, t) \qquad (13.2)$$

We argue as follows: a periodic field in unbounded space satisfying the above equation will not change significantly if we insert walls separating space into (infinitely many) boxes, all identical with the cubic cavity under consideration. If the boxes are (as we assume they are) sufficiently large, the EM field will be distorted relative to that of free space only within a very small distance from each wall (negligibly small compared to L), but otherwise it will not change. The above implies that for our purposes we can think of *any* arbitrary field in the cavity as part of a periodic field in all space. And for the description of such a field we need to include in the sum of Eq. (10.49) only plane waves which satisfy Eq. (13.2). Looking at the form of the cosine function (Fig. 5.6) we can see that these plane waves have wavectors $\boldsymbol{q}$ given by:

$$q = (q_x, q_y, q_z) = (2\pi n_x/L, 2\pi n_y/L, 2\pi n_z/L) \quad (13.3)$$

where n_x, n_x, n_z are any three integers (positive or negative; only the point $q_x = q_y = q_z = 0$ is not included). It follows from Eq. (13.3) that, there is an eigenmode of the EM field in the cavity for every bit of q-volume equal to $(2\pi/L)^3$. In fact there are two eigenmodes to this bit of q-volume. We remember (see Eq. 10.48) that the electric field of a plane wave is normal to $\boldsymbol{q}$, and because there are two independent directions normal to $\boldsymbol{q}$, there are two independent eigenmodes of the EM field for every $\boldsymbol{q}$ satisfying Eq. (13.3). We need to know the number of eigenmodes with frequency (as defined by Eq. 10.47) between ω and $\omega + d\omega$. Because $\omega = cq$, the q-volume corresponding to frequencies between ω and $\omega + d\omega$ is the volume between two spherical surfaces of radii q and $q + dq$, which is given by $4\pi q^2 dq$, where $q = \omega/c$. According to what we said above, to obtain the number of eigenmodes of the EM field in the cavity we must divide this volume by $(2\pi/L)^3$ and multiply the result by two:

$$2(4\pi q^2 dq)/(2\pi/L)^3 = V\rho(\omega)d\omega$$

where V is the volume of the cavity, and

$$\rho(\omega)d\omega = (\omega^2/\pi^2 c^3)d\omega \quad (13.4)$$

is the quantity we are interested in: the number of eigenmodes of the EM field in the cavity, per unit volume of the cavity, with frequency between ω and $\omega + d\omega$. Using the relation $\omega = 2\pi c/\lambda$ (see text after Eq. 10.47), we can translate $\rho(\omega)d\omega$ into the following equation:

$$N(\lambda)d\lambda = (8\pi/\lambda^4)d\lambda \quad (13.4a)$$

which is the number of eigenmodes of the EM field in the cavity, per unit volume of the cavity with wavelength between λ and $\lambda + d\lambda$.

In the thermodynamic treatment of the EM field by Rayleigh and Jeans it is assumed that the eigenmodes of the EM field behave like harmonic oscillators. One thinks of $E_0(\boldsymbol{q})$ in Eq. (10.55) as the amplitude of an oscillator. According to the equipartition theorem (see Sect. 9.2.4), in a gas of oscillators at thermal equilibrium each oscillator has energy (on average) equal to kT. Rayleigh and Jeans assumed that the same will apply to the EM field in a cavity: that at thermal equilibrium each eigenmode of the EM field in the cavity will carry energy equal to kT. Accordingly, the energy in the cavity (per unit volume) carried by wavelengths between λ and $\lambda + d\lambda$ should be: $kTN(\lambda)d\lambda = kT(8\pi/\lambda^4)d\lambda$. Accordingly, the electromagnetic energy radiated from a black body at absolute temperature T per unit area per unit time with wavelengths between λ and $\lambda + d\lambda$ should be given by:

$$I(\lambda, T)\, d\lambda = (c/4)\, kTN(\lambda)d\lambda = (2\pi ckT/\lambda^4)d\lambda \tag{13.5}$$

The factor ¼ multiplying the speed of light accounts for the fact that half of the eigenmodes are waves propagating away from the surface, and those which propagate toward the surface will have on average a normal to the surface component equal to $c/2$.

It turns out that the above, the Rayleigh-Jeans formula, describes well the observed distribution (of Fig. 13.1) at very long wavelengths ($\lambda \to \infty$) but it fails badly at all other frequencies. We note that it goes to infinity as λ approaches zero in apparent contradiction to the observed distribution.

By the time Planck became interested in this problem it had been established that the distribution of Fig. 13.1 was described by the following empirical formula:

$$I(\lambda, T) = \frac{c_1}{4} \frac{1}{\lambda^5 (e^{c_2/\lambda T} - 1)} \tag{13.6}$$

where c_1 and c_2 are appropriate constants. In 1900 Lummer and Pringsheim measured the spectral distribution and the total amount of the emitted radiation with sufficient accuracy for a reliable determination of these constants.

Max Planck was born in Kiel in 1858, and grew up at the time of German unification under Otto von Bismarck, in a family which believed in orderliness and hard work. The son of a successful law professor he chose to study physics. At seventeen he went to Berlin where he studied under Hermann Helmholtz and Gustav Kirchoff, both of them renowned for their work in thermodynamics and electromagnetism. He excelled in his studies and in 1892 he was a professor at Berlin University. He became interested in black-body radiation in the mid-1890s and in 1900 he published his famous work on this topic, introducing 'quantization' and the constant, h, that now bears his name. This work earned him the 1918 Nobel Prize for Physics, and it is for that work that he is now remembered. He despised the Nazis, but he remained in Germany. His son was involved in an assassination attempt against Hitler and was executed by the regime a few days before the end of the war. After the war Planck helped to reconnect German scientists with their colleagues in Europe and the rest of the world. He died in 1947.

We can summarise Planck's theory of black-body radiation as follows.[1] He argued that since this radiation is the same whatever the material of the cavity, the observed distribution would be obtained from a set of electrically charged

[1] A detailed exposition of Planck's theory is given in his book with M. Masius, *The Theory of Heat Radiation*, McGraw-Hill Book Company, Inc., N.Y., 1914.

one-dimensional harmonic oscillators in thermal equilibrium with the EM field. He further assumed that:

1. Each oscillator absorbs energy from the radiation field continuously in accordance with the laws of Newton and Maxwell, but radiates *only* when its (total) energy, E, is given by:

$$E = nhf, \quad n = 1, 2, 3, \ldots \tag{13.7}$$

where f denotes the frequency of the oscillation and h is a constant to be determined.[2]

2. When an oscillator radiates, under condition (1), it puts off *all* its energy.
3. Whether an oscillator radiates or not, when its energy passes through a critical value, given by Eq. (13.7), is determined by statistical chance: the ratio of the probability of non emission to the probability of emission is proportional to the intensity of the radiation that excites the oscillator.

Using the above hypotheses and known formulae of statistical thermodynamics Planck derived the following formula for the spectral distribution of the black-body radiation:

$$I(\lambda, T) = \frac{2\pi c^2 h}{\lambda^5 (e^{ch/\lambda kT} - 1)} \tag{13.8}$$

where c is the speed of light, k is Boltzmann's constant and h is a new constant, now known as Planck's constant. We note that the above expression is identical in form with the empirical formula of Eq. (13.6) which reproduces accurately the experimental data. One can therefore, by comparison, determine the value of Planck's constant:

$$h = 6.626176 \times 10^{-34} \mathrm{J\ s} (\mathrm{or\ m^2\ kg\ s^{-1}}) \tag{13.9a}$$

One often uses, instead of h, the following quantity:

$$\hbar = h/(2\pi) = 1.05459 \times 10^{-34} \mathrm{J\ s} \tag{13.9b}$$

As we shall see, Planck's constant is of great importance in quantum mechanics.

Planck, commenting on the hypotheses he made in order to derive Eq. (13.8), wrote: (see Footnote 1).

---

[2] One can show that, when the linear harmonic oscillator described by Eq. (6.6) is acted upon by a periodic force $F_0 cos(\omega t)$, it will execute a periodic motion $X(t) = X_0 cos(\omega t)$ with the frequency ω of the external force. To show this one must add the above periodic force (divided by m) to the right of Eq. (6.6), and then go on to show that $X(t)$, with an appropriate value for X_0, satisfies the modified equation. In the present case the thing to note is that, the electrically charged one-dimensional harmonic oscillators postulated by Planck, in thermal equilibrium with the EM field in the cavity, will oscillate at the frequencies of the EM eigenmodes of this cavity.

> These hypotheses do not necessarily represent the only possible or even the most adequate expression of the elementary dynamical law of the vibration of oscillators. On the contrary I think it very probable that the theory may be greatly improved as regards form and contents. There is, however, no method of testing its admissibility except by the investigation of its consequences, and as long as no contradiction in itself or with experiment is discovered in it, and as long as no better theory can be advanced to replace it, it may justly claim a certain importance.

The next step toward a better understanding of the nature of light was made by Einstein in his 1905 paper: *Concerning a Heuristic Point of View about the Creation and Transformation of Light*. At the beginning of this paper he writes:

> The wave theory, operating with continuous spatial functions, has proved to be correct in representing purely optical phenomena and will probably not be replaced by any other theory. But one must however, keep in mind that the optical observations are concerned with temporal mean values and not with instantaneous values, and it is possible that the theory of light that operates with continuous functions may lead to contradictions with observation if it is applied to the phenomena of generation and transformation of light.
>
> It appears to me in fact, that the observations on black-body radiation, the generation of cathode rays with ultraviolet light and phenomena related to the generation and transformation of light can be understood better on the assumption that the energy in light is distributed discontinuously in space. According to the presently proposed assumption the energy in a beam of light emanating from a point source is not distributed continuously over larger and larger volumes of space but consists of a finite number of energy quanta localized at points of space, which move without subdividing and which are absorbed and emitted only as units.

Most people are familiar with Einstein's use of the above ideas in relation to the photoelectric effect (see Sect. 13.1.2), but they are equally significant in relation to the black-body radiation. Einstein, inspired, no doubt, by the work of Planck proposed that the energy associated with an eigenmode of the EM field in the black-body cavity is quantized: it can have one of a set of discrete values and none other. (Note the difference with Planck's model where only the energy at which emission is possible is quantized). Let us then, following Einstein, *assume* that the energy U_{qe} of the eigenmode corresponding to a given $\boldsymbol{q}$ (of Eq. 13.3) and a given polarization, e, of the electric field ($e = 1$ *or* 2, corresponding to the two independent directions of the electric field normal to $\boldsymbol{q}$) can have one of the following values (and none other)[3]:

$$U_{qe}(n) = (n + 1/2)\hbar\omega_q, \quad n = 0, 1, 2, 3, \ldots \qquad (13.10)$$

where $\omega_q = cq$ (see Eq. 10.47). We note that $\hbar\omega_q$ has the correct dimensions (of energy), and we note also that $U_{qe}(n)$ does not depend on the polarization. We can

[3] I have included a ground-state energy, $\hbar\omega_q/2$, corresponding to $n = 0$, to facilitate comparison with results to be presented in later sections. It is of no importance in the present discussion.

obtain the average value, $U_{qe}(T)$, of the $\boldsymbol{q}$,e eigenmode of the EM field at temperature T, by the methods of statistical physics. Using Eqs. (9.33) and (9.34) we write:

$$U_{qe}(T) = \sum_n U_{qe}(n) P_{qe}(n, T)$$

$$P_{qe}(n, T) = (1/Z) \exp(-U_{qe}(n)/kT), Z = \sum_n \exp(-U_{qe}(n)/kT)$$

The sums (over all n) are easily evaluated (we omit the details of the calculation here) and we find that

$$U_{qe}(T) = \hbar\omega_q/2 + \frac{\hbar\omega_q}{e^{\hbar\omega_q/kT} - 1} \tag{13.11}$$

The sentence 'The $\boldsymbol{q}$,e eigenmode of the EM field is in its *nth* excited state ($n = 1,2,3, \ldots$), meaning that its energy is given by Eq. (13.10) with $n > 0$, can be stated in a different way; we can, following Einstein, say: We have n *quanta of light* (we now call them photons),[4] of definite polarization e, definite wavevector $\boldsymbol{q}$ and definite energy $\hbar\omega_q = \hbar cq$. We can, accordingly, think of

$$n_{qe}(T) = \frac{1}{e^{\hbar\omega_q/kT} - 1} \tag{13.12}$$

which appears in Eq. (13.11), as the (average) number of $\boldsymbol{q}$,e photons in the black-body cavity at temperature T. [The distribution of Eq. (13.12) is a special case of the so-called *Bose–Einstein* distribution, which we shall introduce in Sect. 14.4.2 (it is obtained from Eq. (14.77a) by putting $\mu = 0$)].

Noting that $n_{qe}(T)$ depends only on the frequency ω of the eigenmode, we conclude that the number of photons per unit volume of the cavity, at temperature T, with frequency between ω and $\omega + d\omega$ is given by

$$W^T(\omega)d\omega = \frac{\rho(\omega)d\omega}{e^{\hbar\omega/kT} - 1} = \frac{\omega^2 d\omega}{\pi^2 c^3 (e^{\hbar\omega/kT} - 1)} \tag{13.13}$$

where for $\rho(\omega)$ we used the quantity given by Eq. (13.4). To obtain the energy radiated from a unit surface of the black body per unit time, with frequency between ω and $\omega + d\omega$, we multiply the above quantity with $c/4$ to obtain the number of the emitted photons (see text after Eq. 13.5), and multiply the result with the energy $\hbar\omega$ of each photon. The final result, when we translate ω into λ, is the $I(\lambda,T)$ of Eq. (13.8).

---

[4] The reality of the photon as a particle with definite momentum $\hbar\boldsymbol{q}$ and definite energy $\hbar cq$, was demonstrated by the American physicist A. Compton, who in 1923 observed the conservation of momentum in the scattering of photons by electrons (see Sect. 15.2.2).

Let me conclude this section with the following observation concerning the average energy of an eigenmode of given frequency $\omega_q = cq = 2\pi/\lambda$. When $kT \gg \hbar\omega_q$, which is true for sufficiently large wavelengths, we can write: $exp(\hbar\omega_q/kT) = 1 + \hbar\omega_q/kT$, and therefore Eq. (13.11) gives: $U_{qe}(T) - \hbar\omega_q/2 = kT$, in accordance with the classical theory of Rayleigh and Jeans (the constant ground energy of the eigenmode is unimportant and can be dropped in this case), which is the reason why the latter theory agrees with the observed spectral distribution at large wavelengths. Not much energy (*kT*) is required to excite an eigenmode of very large wavelength (very small $\hbar\omega_q$), but this is not the case generally: an eigenmode of frequency ω will be excited (photons of frequency ω will be created) only if kT is at least as large as $\hbar\omega$.

13.1.2 The Photoelectric Effect

This effect was discovered by Hertz in 1887. He noticed that a spark would occur in a small gap more easily if the negative terminal of the gap was illuminated with ultraviolet light. Subsequent experiments by Hallwachs, Elster and Geitel in the years 1889–1891 established that the phenomenon involved negatively charged particles forcibly ejected from the negative terminal by the illuminating light, and that this photocurrent was proportional to the intensity of the incident light. After Thomson's discovery of the electron, it was established by him, and by Philipp Lenard, in 1899, that the ejected particles were electrons. Lenard further established (in 1902) that:

1. The kinetic energies of photo-emitted electrons do *not* depend on the intensity of the incident light; only the number of the emitted electrons does (it is proportional to the intensity of the incident light).
2. The emitted electrons have a maximum kinetic energy which is greater, the higher the frequency, f, of the incident light. And no electrons are emitted at all, if f falls below a threshold value, f_0.

The above two observations can not be understood within the framework of classical physics. Classically we expect the effect on the electron by the incident light, to increase with the intensity (the magnitude of the electric field) associated with the incident wave and not with the frequency of the wave. Moreover, it should take some time to build up the energy of an electron to the value required to break its bond to the emitting surface, especially so when the incident energy is spread over an area containing a great number of electrons. But in the experiments it often happened that an electron would be emitted almost immediately, 3×10^{-9} s after illumination began, when delays of a few microseconds (for high intensity illumination) to several minutes or even days (for weak illumination) would be expected from classical considerations. Only for very weak illumination some considerable delay was observed.

In his 1905 paper, Einstein proposed a physical explanation of the above observations based on the idea that the EM radiation whether in thermal equilibrium (as in the black-body radiation) or not (as in light incident on a surface from a source) consists of individual quanta (photons) as described in the previous section. A beam of light, described by a plane wave (Eq. 10.48) of given wave-vector ($\boldsymbol{q}$) and given polarization (e) of the electric field, consists (at a microscopic level) of a flow of individual quanta (photons) of definite polarization e, definite momentum $\hbar\boldsymbol{q}$, and definite energy $\hbar\omega_q = \hbar cq$. The intensity of the beam (classically determined by the magnitude of the electric field) is now defined by *the number* of photons in the beam (the number of photons crossing a unit area normal to the direction of propagation (that of $\boldsymbol{q}$). Einstein made two further assumptions:

1. A photon (the quantum of energy) is taken up *as a whole* by an electron in the emitting surface, the absorption occurring with some probability determined by the interaction of the EM field with the electrons in the emitting surface.
2. There is no possibility of an electron absorbing more than one photon at the same time.

Assuming that a minimum of energy, φ, is required for an electron to escape the emitting surface, we must conclude (if Einstein's model is correct) that for an electron to escape, it must absorb a photon of energy hf greater than, or at least equal to φ. Therefore *no emission* will occur if the frequency of the incident light is: $f < f_0 = \varphi/h$, where h is Planck's constant. Of course an electron may absorb that much energy and still not escape because it moves away from the surface toward the interior of the emitting material (usually a metal), or because it looses much of its energy in a collision with another electron before it reaches the surface. And by the same token the kinetic energy ($mv^2/2$) of the emitted electron will vary from just above zero to a *maximum* value given by the difference $(hf - \varphi)$ between the energy hf of the absorbed photon and the minimum energy φ required for the electron to escape. And this is of course what has been observed by the experimentalists. Einstein's model explains also why the delay between illumination and the onset of emission is much smaller than what we expect classically. When a photon is absorbed, its (whole) energy hf is absorbed instantly by the electron, whose energy is therefore increased instantly to the energy required for its escape from the surface. Whatever delay there is in the *quantum* mechanical picture arises from the process which determines the *probability* of the absorption happening in a unit of time. And this is, of course, proportional to the intensity of the incident light (the number of photons incident on a unit of surface per unit time).

We may conclude that Einstein's model explains all the main features of the photoelectric effect. Experiments performed after the publication of his paper confirmed the validity of his model beyond any doubt. Moreover, it turned out that the assumptions made by Einstein in his phenonomenological theory of the photoelectric effect follow from the basic principles of the quantum theory developed in the 1920s.

13.1.3 The Einstein Coefficients

Einstein returned to black-body radiation in a paper he published in 1917. It is a good example of Einstein's ability to extract important results from seemingly simple observations. It goes as follows. Einstein considers a gas of identical atoms (N per unit volume) in a radiation cavity. He assumes, following Bohr (see Sect. 13.2.3), that the energy levels of the atoms (this means the values of energy an atom can have) are discrete, and makes a further simplifying assumption, namely that there are only two such energy levels: E_1 and E_2, and that there are g_1 different states of the atom with energy E_1 and g_2 different states of the atom with energy E_2; the difference between the two energy levels defines a frequency by the relation (see Fig. 13.2): $E_2 - E_1 = hf = \hbar\omega$, where h is Planck's constant.

Let an atom be in a state 1. If there is no radiation (no photons) of frequency ω, there is no possibility of the atom transiting to a state 2, because in any transition energy must be conserved. But if there is a density of photons $W(\omega)$, meaning that there are $W(\omega)d\omega$ photons with frequency between ω and $\omega + d\omega$ per unit volume of the radiation cavity, the transition $1 \rightarrow 2$ can proceed with the absorption of a photon ($E_1 + \hbar\omega = E_2$). We denote the corresponding transition rate (the probability of the transition happening in a unit of time) by: $B_{12}W(\omega)$, where B_{12} is a constant. Next we consider an atom in a state 2. It is reasonable to assume that the atom may *spontaneously* fall to a state 1 by emitting a photon $\hbar\omega$.[5] We denote the corresponding transition rate by: A_{21}. The above two processes are shown by respective arrows in Fig. 13.2. Einstein introduced a third process, whereby the radiation density $W(\omega)$ *induces* a transition from state 2 to a state 1; the corresponding transition rate is denoted by: $B_{21}W(\omega)$.

Now assume that at time t there are N_1 atoms (per unit volume) in states 1, and N_2 atoms (per unit volume) in states 2. These numbers may change with time (we do not assume equilibrium) because of the above mentioned transitions, but their sum, $N = N_1 + N_2$, is constant. Having this in mind and what we have said above about possible transitions, we write:

$$dN_1/dt = -dN_2/dt = N_2A_{21} - N_1B_{12}W(\omega) + N_2B_{21}W(\omega) \qquad (13.14)$$

The first term on the right of the above equation gives us the increase of N_1 per unit time due to atoms in states 2 falling to states 1 by emitting spontaneously a photon $\hbar\omega$ and is of course proportional to N_2. The second term due to transitions *from* states 1 to states 2 with photon absorption is proportional to both N_1 (which is diminished by these transitions) and to the radiation density. Finally the last term representing induced emission from states 2 to states 1 is propotional to both N_2

---

[5] One can argue as follows: A classical atom, consisting of negatively charged electrons moving about a positive nucleus, is similar to an oscillating electric dipole, and as such it emits continuously radiation (see text about Eq. 10.57). In the present case the process changes to one where the atom emits a quantum of energy $\hbar\omega$ in an abrupt transition which occurs with some probability per unit time.

Fig. 13.2 Transitions in an atom of two energy levels: E_1 and E_2

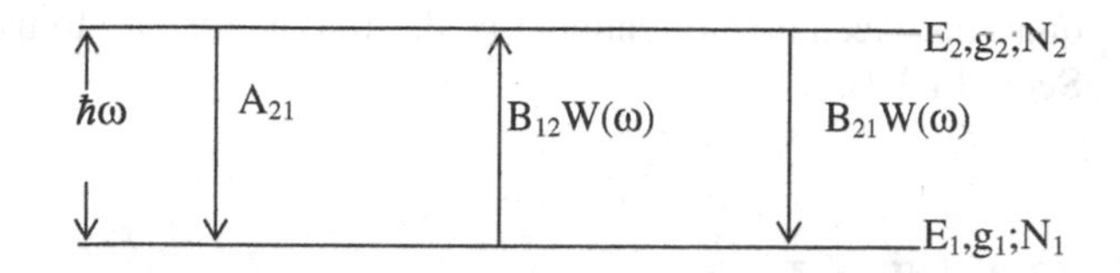

and the radiation density. We remember that A_{21}, B_{12} and B_{21}, now known as the *Einstein coefficients*, do not depend on the radiation density but they do depend on the properties of the atoms involved. We should also remember that Eq. (13.14) is valid whether the system (atoms and radiation) is in thermal equilibrium or not. In the absence of equilibrium, the photon density will of course depend on the time (t). However, the case of thermal equilibrium is of particular importance, because it allows us to establish a definite relation between the A_{21}, B_{12} and B_{21} coefficients. At thermal equilibrium we have $dN_1/dt = 0$, therefore:

$$N_2A_{21} - N_1B_{12}W^T(\omega) + N_2B_{21}W^T(\omega) = 0 \qquad (13.15)$$

where $W^T(\omega)$ is the equilibrium photon density distribution at temperature T, given by Eq. (13.13). It follows from Eq. (13.15) that:

$$W^T(\omega) = \frac{A_{21}}{(N_1/N_2)B_{12} - B_{21}} \qquad (13.16)$$

According to the rules of statistical physics, at thermal equilibrium the number of atoms in any of the g_1 states with energy E_1 will be proportional to $exp(-E_1/kT)$, and the number of atoms in any of the g_2 states with energy E_2 will be proportional to $exp(-E_2/kT)$, with the same proportionality constant, therefore: $N_1/N_2 = (g_1/g_2)exp((E_2 - E_1)/kT) = (g_1/g_2)exp(\hbar\omega/kT)$. Therefore Eq. (13.16) becomes:

$$W^T(\omega) = \frac{A_{21}}{(g_1/g_2)B_{12}e^{\hbar\omega/kT} - B_{21}} \qquad (13.17)$$

Now the above expression must be identical with the expression of Eq. (13.13). This means that:

$$(g_1/g_2)B_{12} = B_{21} \quad \text{and} \quad \left(\omega^2/\pi^2c^3\right)B_{21} = A_{21} \qquad (13.18)$$

Which tells us that the three Einstein coefficients are related to each other and therefore they can be expressed in terms of a single coefficient. It is also worth emphasizing that the consistency of the above theory with Planck's law could not be obtained without the introduction of the induced emission term in Eq. (13.14). The practical importance of induced emission (also called stimulated emission) will be discussed briefly in the next section. At this point I should simply say that the relations of Eq. (13.18) are obeyed by the exact formulae derived by a full quantum mechanical treatment of the interaction of atoms with the EM field (see Sect. 14.3.3).

13.1.4 The Laser

A laser is a device that produces highly directional monochromatic coherent light. Here coherent means that practically all photons have the same polarization and are in phase so as to reinforce each other. The word LASER is in fact an acronym for *light amplification by stimulated emission of radiation*, and as such it makes it clear that its operation depends on stimulated emission. Although the concept of stimulated emission was introduced by Einstein in 1917, it took about 35 years before someone realized the possibilities inherent in this process.[6] The basic idea of the laser can be summarized as follows. We note to begin with, that at thermal equilibrium, the ratio of the stimulated (induced) emission to the spontaneous emission is very small. For example, $N_2 B_{21} W^T(\omega) \ll N_2 A_{21}$ in the two-level system of Eq. (13.14). This is because $W^T(\omega)$ is very small. A laser operates away from equilibrium and an essential part of it is an *optical resonant cavity* in which the photon density about a certain ω and a certain polarization e can build up by multiple internal reflections. However, to have more stimulated emission than absorption we must have (in the example of Eq. 13.14):

$$\frac{B_{21} N_2 W(\omega)}{B_{12} N_1 W(\omega)} = \frac{B_{21} N_2}{B_{12} N_1} > 1 \tag{13.19}$$

Therefore if stimulated emission is to dominate we must have $N_2 > N_1$, i.e., more atoms in the higher energy level E_2 than in level E_1, which is contrary to what happens at equilibrium. And this why, the above requirement is known as *population inversion*. In Fig. 13.3 we show how this is achieved in a ruby laser. The diagram shows the relevant energy levels of Cr atoms in the crystal. The level E_2 is that of a metastable state (the atom stays in this state for a relatively long time, about 5 ms, before it falls to the ground state). If by pumping energy into the system, atoms are excited from the ground level E_1 to an excited level E_3 at a rate faster than they fall from E_2 to E_1, then the population of E_2 becomes larger than that of E_1, and the required population inversion is achieved. We have assumed that the atoms fall from E_3 to E_2 in a very short, negligibly small, time interval. We need not go into the details of laser operation here. There is in fact a great variety of them and their applications are many.

---

[6] The first working lasers were built by American and, independently, by Russian scientists in the early 1960s.

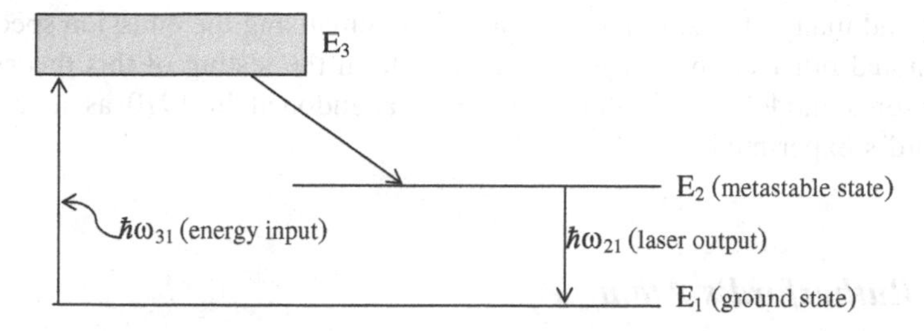

Fig. 13.3 A ruby laser (schematic). *Input* $\hbar\omega_{31} = E_3 - E_1$; *Output* $\hbar\omega_{21} = E_2 - E_1$

13.2 Atomic Structure

13.2.1 Thomson's Static Model

By 1905 practically everyone accepted the atomic theory of matter, but nobody knew what an atom looked like. After his discovery of the electron, and because electrons appeared in a variety of phenomena, Thomson proposed a model of the atom along the following lines: It consists of a massive jellylike sphere with positive charge uniformly distributed through it. This, the atomic nucleus, accounts for practically all the mass of the atom. And there are a number of electrons which are fixed at certain positions on the surface or within the nuclear mass; so many of them as is required to make the atom electrically neutral. X-ray scattering experiments indicated that the number of electrons in an atom was of the same order as its atomic weight. Hydrogen appeared to be the simplest of atoms with just one electron. The model had certain merits: the electrons could vibrate about their (fixed) equilibrium positions in the manner of a harmonic oscillator (see Sect. 6.2.2); once excited by an incident radiation they would themselves emit radiation, in the manner of an oscillating electric dipole (see Sect. 10.4.3), at certain frequencies determined by the elastic properties of the atom. That could explain, at least in principle, the observed spectra of atoms: the different frequencies at which atoms emit light. This idea was elaborated upon by Lorentz and Zeeman, who went on to show how such radiation would be modified if the atom was placed in a magnetic field; they showed that each spectral frequency would split into three separate frequencies: $\omega_1 = \omega_0 + eB/m$, ω_0, and $\omega_2 - eB/m$, where ω_0 is the observed frequency for the undisturbed atom, B is the magnitude of the magnetic field, and e and m denote the charge and the mass of the electron respectively, in qualitative agreement with Zeeman's observation (in 1896) of this phenomenon in relation to certain emission-lines of sodium. The phenomenon, now known as the Zeeman effect, would play a crucial role in the analysis of emission spectra using quantum mechanics.

It is worth noting at this point the important contribution that the spectroscopists in general made to the development of quantum mechanics. The painstaking work during the last two decades of the 19th century of Balmer, Lyman, Paschen,

Rydberg and many others, in measuring and systematising the emission spectra of hydrogen and other atoms, played a crucial role in the testing of this theory.

Thomson's model of the atom had to be abandoned in 1910 as a result of Rutherford's experiments.

13.2.2 Rutherford's Atom

Ernest Rutherford was born in 1871 in New Zealand, the fourth child in a family of twelve children. In 1889 he won a scholarship to study at the University of New Zealand in Wellington. Four years later he graduated with a double first in Mathematics and Physical Science. In 1895 he won another scholarship, and joined J. J. Thomson's Cavendish laboratory at Cambridge as a research student. He spent three years with Thomson, with whom he had a cordial relationship, first working on X-ray experiments and then on the uranium and radium rays that had been recently discovered by Beckerel and the Curies (see Sect. 15.1.1).

After three years at Cambridge Rutherford was looking for a job that would pay him sufficiently to enable him to marry Mary Newton, his fiancé waiting in New Zealand. When he was offered a professorship at McGill University in Montreal (Canada), he moved there, married Mary Newton and carried on with his work on α-particles, helped by the packages of uranium and radium sent to him by Madame Curie.

In 1899, he made his first important discovery. Assisted by Frederick Soddy, a British chemist he teamed up with in Montreal, he discovered that the radiation of a sample of uranium could be stabilized by wrapping it in three layers of aluminium foil. After putting more and more layers of aluminium foil about it, the radiation would at some point begin decreasing again. This suggested to him that there were at least two types of radiation involved: one which is easily stopped, which he called alpha (α) radiation, and a second more penetrating one which he called beta (β) radiation. Soon afterwards it was established by him, and by Beckerel and the Curies, that α-particles were relatively heavy and positively charged, and that the β-particles were electrons. A year later Rutherford observed a third type of radiation, more penetrating, which involved photons of very high frequency which he called gamma (γ) rays.

Having established that for any given radioactive sample half of its atoms decay in a certain time, with a corresponding reduction in the intensity of the radiation, Rutherford realized that he had the means of knowing how long a radioactive substance had been inside a given rock. He calculated that pitchblende must be about 500 million years old, which was much older than people thought at the time.

Rutherford returned to England in 1907, when he was appointed director of the physics laboratory at the University of Manchester. In 1908 he and his assistant, a visiting German physicist named Hans Geiger, established the identity of α-particles: They placed a sample of radium in a vacuum tube with thin walls so that α-particles could pass through them. Then they put the vacuum tube in another vacuum tube with thick walls that could block the α-particles. After running the experiment for a few days, they passed an electric current (electrons) through the space that the α-particles were trapped. When they looked at the subsequent faint glow through the spectroscope, they determined that it was characteristic of light emitted by helium atoms. They discovered that α-particles are the nuclei of helium atoms.

Rutherford's discoveries earned him the 1908 Nobel Prize for Chemistry. He was very happy to receive the Prize, but was not impressed by the classification. He remarked that in his work he saw many rapid transformations of radioelements but none so rapid as his transformation from a physicist into a chemist.

In 1911 Rutherford published a paper (*Phil. Mag.*, **21**, 669 (1911)) in which he proposed a revolutionary model for the structure of the atom. He based his model on a thorough analysis of measurements made in his laboratory in the preceding 2 years by Hans Geiger, and a young undergraduate researcher Ernest Marsden. In the experiments, α-particles, from the decay of radioactive atoms were made to pass through a thin metal foil. The deflection of the α-particles from their original direction of motion was measured by the scintillations produced by their impacts upon a zinc sulfide screen (see Fig. 13.4).

If Thomson's model was correct, the α-particles would deflect very little: a very few degrees from their initial direction. In practice a large number of α-particles were deflected by large angles, even greater than 90°. This meant that a much stronger force was acting on the α-particle at some stage of its passage through the foil. The only way Rutherford could account for the observed large deflections, given the size of the atoms in the metallic foil (estimated from crystal lattice

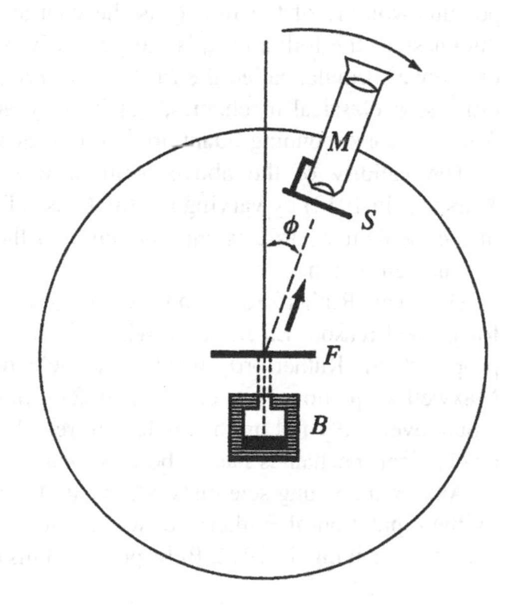

Fig. 13.4 The Rutherford apparatus. The angle φ by which an α-particle is deflected on passing through the foil F is measured by its impact on the screen S, which is attached to the microscope M, which can turn by as much as 150°. [From G. Holton and S. G. Brush, *loc. cit.*]

spacings to be about 4×10^{-10} m) was to assume that the whole mass of the positively charged nucleus was concentrated within so small a volume that could be reasonably approximated by a point at the centre of a spherical shell of electrons of radius about 4×10^{-10} m. The analogy between Rutherford's model of the atom and of the solar system is striking. There is a lot of empty space between the massive sun and the planets that move about it and, on a different scale, the same is true in Rutherford's atom; there is a lot of empty space between the nucleus and the electrons about it. And there is a further analogy: A body passing the solar system at its periphery experiences a force from the sun and the nearby planets, and as a result it is deflected from its direction of motion by a relatively small angle; but if the body passes close to the sun it experiences a much greater force from the sun (the force from the planets is neglible by comparison in this case) and as a result is deflected much more from its original direction of motion. The same thing happens when the α-particle is scattered by the Rutherford atom. The (Coulomb) force exerted by the point-nucleus on the α-particle, when the latter passes the nucleus at a short distance from it, is responsible for the observed large deflections of the α-particles in Rutherford's experiments. Assuming that this is the case, i.e. disregarding the effect of the electronic cloud on the passing α-particle, Rutherford derived the following formula for the number, $dN/d\Omega$, of α-particles scattered into a unit solid angle at an angle φ with respect to the incident direction:

$$\frac{dN}{d\Omega} = \frac{N_0 n t Q^2 e^2}{16\pi^2 \varepsilon_0^2 m^2 \upsilon_0^4 \sin^4(\phi/2)} \tag{13.20}$$

where N_0 is the number of incident α-particles, m is the mass of the α-particle, $2e$ is the charge of it, and υ_0 is its initial velocity; n is the number of scattering centres per unit volume of the foil, Q is the charge of a scattering centre, and t is the thickness of the foil; and ε_0 is the permittivity of vacuum. It is worth noting that the above formula, called the *Rutherford law of single scattering*, was derived by him using classical mechanics, but it is by good luck not significantly different from the corresponding quantum mechanical law.

The validity of the above formula was thoroughly tested by Geiger and Marsden (in 1913) by varying the thickness of the foil, its composition, the energy of the incident α-particles, and the angle of the zinc sulfide screen with respect to the incident beam.

However, Rutherford's model of the atom was not immediately accepted and for a good reason. Electrons moving about the nucleus in circular-like orbits, as proposed by Rutherford, would emit electromagnetic radiation according to Maxwell's equations (see, e.g., Fig. 10.21), and by loosing energy in this way, they would eventually fall on the nucleus to rest there. So either Rutherford was wrong or classical mechanics had to be modified.

Among the young scientists who visited Manchester at the time, attracted there by the reputation of Rutherford, was a young Danish fellow called Niels Bohr. A year after arriving, in 1912, Bohr proposed his quantum hypothesis, which kept the

Rutherford model of the atom at the expense of classical mechanics (see next Section).

In his later research Rutherford, who returned to Cambridge as Cavendish Professor in 1919, tried to split the atomic nucleus by artificial means. The spontaneous emission of α-particles from radioactive materials was obviously a nuclear phenomenon, and as such it suggested that the nucleus was not one compact indivisible entity but was made of perhaps a number of smaller particles bound together by a force or other. In 1919 Rutherford was able to show that this is the case by bombarding a gas of nitrogen with α-particles: an α-particle would knock a *proton* (the nucleus of hydrogen) off a nitrogen nucleus and attach itself to the nitrogen to create an oxygen nucleus. However his nuclear reaction was not an efficient one. Only one out of 300,000 α-particles would collide successfully with a nitrogen nucleus. Rutherford continued to direct research in nuclear physics until the end of his life (see Sect. 15.1). He died in 1937.

13.2.3 Bohr's Model

Niels Bohr was born in 1885. He had just obtained his Ph.D. from the University of Copenhagen when he went to Cambridge to work as a visiting scientist with J. J. Thomson. After a few months there, he moved to Manchester. Rutherford liked him. He found him to be *quite the most intelligent man I ever met.*

In April 1913 Bohr published his famous paper *On the constitution of atoms and molecules*, produced with Rutherford's support.[7] In it he postulated that:

1. The hydrogen atom consists of an electron in circular motion about a positive nucleus (a proton) under the action of the Coulomb force between the two.
2. The atom exists for extended periods only in certain states (we call them stationary states) for which the angular momentum, L, of the atom is an integral multiple of $\hbar$:

$$L = n\hbar, \cdots n = 1, 2, 3, \ldots \tag{13.21}$$

3. When the atom jumps from a state n (with energy E_n) to a state n′ (with lower energy $E_{n'}$) it emits radiation of frequency f determined by:

$$hf = E_n - E_{n'} \tag{13.22}$$

where h is Planck's constant.

[7] The paper appeared in volume 26 of the *Philosophical Magazine*, 1913.

Bohr calculated the allowed (by Eq. (13.22) energy levels of the hydrogen atom as follows[8]: The angular momentum L of an electron (mass m) in circular motion of radius r with velocity υ is: $L = m\upsilon r = m\omega r^2$, where $\omega = \upsilon/r$ is the angular velocity of the electron. According to Eq. (13.21), we must have: $m\omega r^2 = n\hbar$, where n is an integer. Therefore:

$$m\omega^2 r^4 = n^2\hbar^2/m \tag{13.23}$$

On the other hand, Newton's law tells us that the acceleration (in the radial direction) of circular motion ($\upsilon^2/r = \omega^2 r$) times the mass m of the electron must equal the force on the electron, which in this case is given by the Coulomb force: $e^2/(4\pi\varepsilon_0 r^2)$. Therefore: $m\omega^2 r = e^2/(4\pi\varepsilon_0 r^2)$ which means that:

$$m\omega^2 r^4 = [e^2/(4\pi\varepsilon_0)]r \tag{13.24}$$

Comparing Eqs. (13.23) and (13.24) we obtain:

$$r = r_n = (4\pi\varepsilon_0\hbar^2/me^2)n^2, \quad n = 1, 2, 3, \ldots \tag{13.25}$$

Therefore the stationary states of the hydrogen atom are defined by the circular orbits corresponding to the radii given by the above formula.

The total energy E of the atom is the sum of the kinetic energy of the electron, $K = m\upsilon^2/2$, and its potential energy, $U(r) = -[e^2/(4\pi\varepsilon_0)]/r$, in the field of the nucleus (see Sect. 7.3). We have $K = m\upsilon^2/2 = m\omega^2 r^2/2 = [e^2/(8\pi\varepsilon_0)]/r$, after using Eq. (13.24). Therefore:

$$E = K + U(r) = [e^2/(8\pi\varepsilon_0)]/r - [e^2/(4\pi\varepsilon_0)]/r = -[e^2/(8\pi\varepsilon_0)]/r$$

Therefore the allowed energies of the hydrogen atom (called the *energy eigenvalues* of the atom) are determined by the allowed values of r given by Eq. (13.25). These are:

$$\begin{aligned} E_n &= -\frac{e^2}{8\pi\varepsilon_0 r_n} = -\frac{m_r e^4}{32\pi^2\varepsilon_0^2\hbar^2 n^2}, n = 1, 2, 3, \ldots \\ r_n &= a_0' n^2, \; a_0' = \frac{4\pi\varepsilon_0\hbar^2}{m_r e^2} = 5.29172 \times 10^{-11} m \end{aligned} \tag{13.26}$$

---

[8] Here, for the sake of simplicity, I derive Bohr's formula assuming that the centre of mass of the atom concides with the centre of the nucleus which is reasonable (the nucleus is about 2,000 times heavier than the electron). The exact formula for the energy levels is obtained by replacing the electronic mass m with the reduced mass $m_r = Mm/(M + m)$, where M is the mass of the proton, in the final formula (see Sect. 14.1.2). It is also worth remembering that the energy we are interested in here, is the internal energy of the atom: its energy in the centre-of-mass system of coordinates (the origin of coordinates is at the centre of mass of the atom). In any other system of coordinates the (total) energy of a free atom (one not acted upon by external forces) is the sum of its internal energy and its translational (kinetic) energy, $MV^2/2$, where M is the total mass of the atom and V the velocity of its centre of mass.

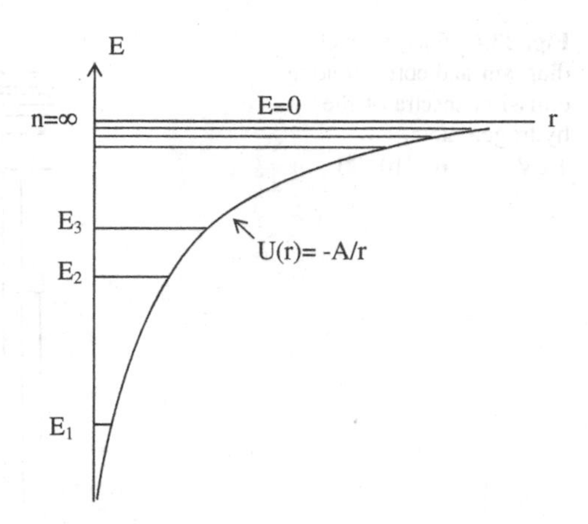

Fig. 13.5 Schematic representation of the energy levels of a hydrogen atom

where, a_0' is the radius of the circular orbit of the electron when the atom is in its ground state (the state of minimum energy). And we note that in the final formulae m has been replaced by m_r to take into account the contribution of the nucleus to the angular momentum and the energy of the atom (see Footnote 8).

Observing (see Fig. 13.5) that the potential energy $U(r)$ is negative and tends to zero as r increases, we can deduce that if an atom in its *nth* state is to be ionized (the electron removed away from the nucleus to infinity), it needs to absorb an amount of energy greater or at least equal to $(-E_n)$. If the atom is in its ground state the ionization energy is: $-E_1 = 13.598$ eV, in excellent agreement with the measured value.

Bohr's model of the hydrogen atom was accepted above all because it reproduced successfully the emission spectra of this atom, which were accurately known many years before he conceived his theory. The *wave number* $\bar{f}$ of emitted radiation is defined by $\bar{f} = f/c = 1/\lambda$, where λ denotes as usual the wavelength of the radiation. Using Eq. (13.26) in Eq. (13.22), we obtain the following formula for the wavenumber $\bar{f}$ of the radiation generated by a transition of the atom from a state n to a state n' of lower energy:

$$\bar{f} = R_H(1/n'^2 - 1/n^2)$$

where $R_H = m_r e^4/(8\varepsilon_0^2 h^3 c) = 10,967,757.6\, m^{-1}$ is known as the Rydberg number. When the above formula is applied to transitions to state $n' = 2$ from the states $n = 3,4,5$ etc. it gives the following formula for the wavelength of the emitted radiation (we remember that $\lambda = 1/\bar{f}$):

$$\lambda = bn^2/(n^2 - 4)$$

where $b = 1/R_H$. The above result of Bohr's theory was of critical importance. A formula of exactly the same form had been established empirically from spectroscopic measurements by Balmer in 1885 (the year Bohr was born). The Balmer series is shown schematically in Fig. 13.6. Bohr's theory gave not only the correct

Fig. 13.6 Energy-level diagram and corresponding emission spectra of the hydrogen atom.
1 eV = 1.6×10^{-19}J

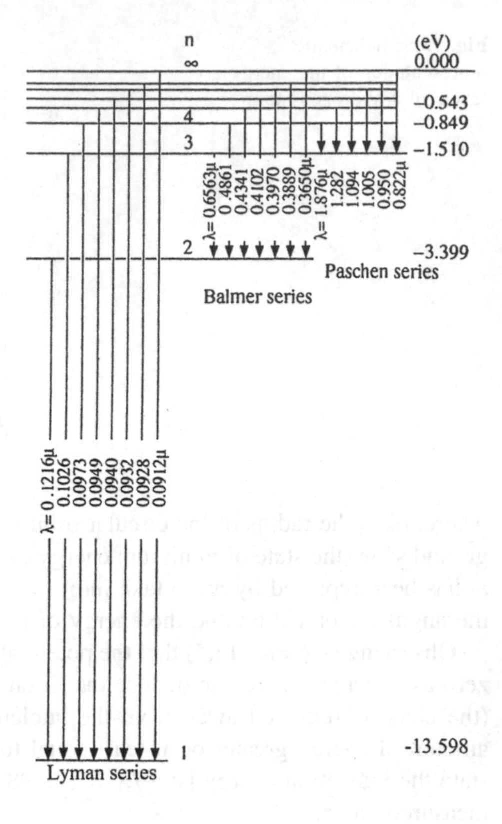

form for λ as a function of n, but also an accurate value for b. It also gave correct values for the other spectral series shown in Fig. 13.6.

Bohr returned to Copenhagen a rising star. In March of 1914 he wrote to the Danish Ministry of Education: *The undersigned takes the liberty of petitioning the Ministry to bring about the founding of a professorship in theoretical physics at the university (of Copenhagen) and in addition to possibly entrust me with the position.* It did not happen, and Bohr returned to Manchester where he stayed throughout the war (1914–1918). Denmark remained neutral in that terrible conflict and it was therefore a good place to start an international centre of scientific research after the madness of the war. Bohr was a likeable person and that made it easier for him to obtain financial assistance from commercial organizations as well as from state ones. In 1921 he inaugurated the Institute of Theoretical Physics, a three-storey purpose-built building in Copenhagen, funded by the Danish Government and the Carlsberg brewing company. Bohr, his wife Margarethe and their two sons lived on the second floor, and residential guests stayed on the third. The ground floor had the library, the lecture rooms and offices, and in the basement were laboratories for experimental work. The Institute became a centre of attraction to scientists from around the world, a happy and productive place where people, including the great Heisenberg, would come to present and discuss their ideas about various subjects and especially about the new quantum mechanics. An idea of the Institute's influence in shaping the accepted physical interpretation of the new theory is the fact that it is now known as the Copenhagen interpretation of quantum mechanics. Another measure of its international reputation is the fact that, when at some stage it run short of money, it received 40,000 U.S. dollars (equivalent to about half a million in today's money) from the Rockfeller Foundation (in New York) to expand the building of the Institute. As for Bohr himself, it is said that the way he spoke was so convoluted that it was often difficult for his listeners to follow his arguments, but those who had the patience to hear him were invariably impressed by his intelligence and the depth of his thinking.

In Bohr's original model of the atom, described above, the orbits of the electron are circular. In 1916 Sommerfeld extended the model to include elliptical orbits. Similar attempts were made by other scientists, who aimed at a systematic description of the spectra of other atoms beside hydrogen, and of the helium atom in particular. Sommerfeld found that a number of orbits, corresponding to different angular momenta, may have the same energy, but the allowed energy levels were the same with those of the Bohr model. But when slight corrections were made to account for the variation of mass with velocity (as required by special relativity) a small separation occurred between energies of different states of the hydrogen atom that were previously identical, and these small separations were in very good agreement with the observed fine structure of these spectra. However, the proposed theories could *only* account for the observed spectra of one-electron atoms (hydrogen and ionized helium).

Bohr approved Sommerfeld's extension of his model and used it in an attempt he made (in 1926) to interprete the *Periodic Table* of the elements (atoms). This is an arrangement of the known atoms in rows and columns resulting in: atoms in the same column having similar chemical behaviour. A periodic table of the elements of this kind was originally proposed in 1869 (on the basis of empirical evidence alone) by the Russian Dimitri Mendeleev, using relative atomic masses for the arrangement of the atoms, and was later modified by the English chemist Henry Moseley, who arranged the atoms according to increasing atomic number (number of protons in the nucleus of the atom). It was Bohr's attempt to explain the periodic table that led Wolfgang Pauli (a friend of Bohr) to formulate (in 1925) his famous *Exclusion Principle*, according to which there can not be two electrons in the same orbit. Bohr's basic idea in relation to the Periodic Table and the exclusion principle survive (and are more easily understood) in the framework of quantum mechanics (see Sect. 14.5.1); I mention them here to make the point that both were advanced prior to the development of formal quantum mechanics.

Bohr, who was awarded the Nobel Prize for Physics in 1922, continued to contribute to physics as a researcher (especially in theoretical nuclear physics) and as a director of research until the end of his life in 1962.

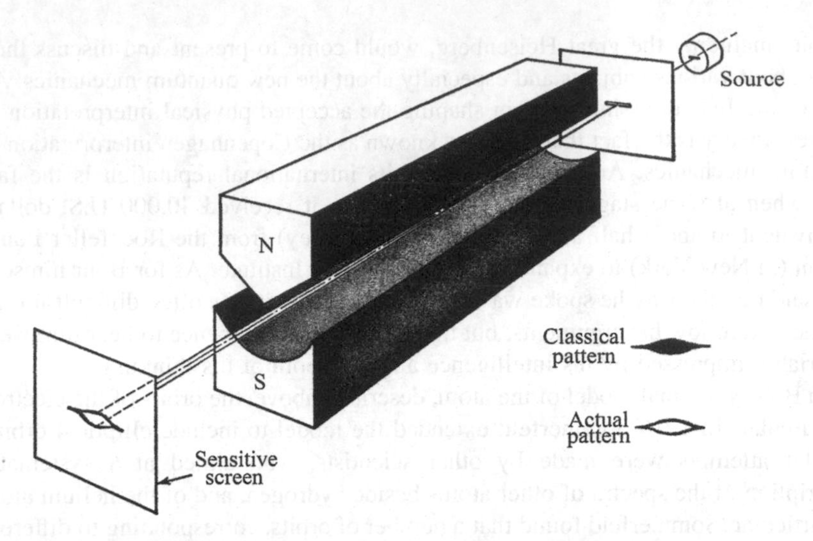

Fig. 13.7 The Stern-Gerlach experiment (schematic). [Reproduced from R.B. Leighton, *Principles of Modern Physics*, McGraw-Hill, N. Y. 1959.] The inhomogeneous magnetic field between the magnet's poles (*N* and *S*) exerts a transverse force upon the atoms and causes a splitting of the atomic beam

13.2.4 The Discovery of Spin

The first direct experimental evidence of quantized angular momentum was obtained by Stern and Gerlach in 1922.[9] In their experiment (see Fig. 13.7) a beam of neutral silver atoms defined by two consecutive narrow slits passes through an inhomogeneous magnetic field, which acting on the magnetic-dipole moment $\boldsymbol{\mu}$ of the atoms produces a transverse force on each atom: $F_z = \mu_z \partial B_z / \partial z$, where z is the direction of maximum inhomogeneity of the magnetic field.[10] Classically, the random orientation of the dipole moments would be expected to result in a continuous spreading of the beam. Stern and Gerlach observed a splitting of the beam into *two discreet* components. [According to quantum mechanics (see Sect. 14.1.1) orbital angular momentum is quantized, but that should lead to an odd number $(2\,l + 1)$ beams, not the two beams seen in the experiment].

The idea that the electron has an *intrinsic* angular momentum (spin) with only two possible orientations was put forward by G. Uhlenbeck and S. Goudsmit in 1925,[11] and was introduced formally in quantum mechanics by Pauli in 1927.[12]

---

[9] *Z. Physik*, **8**, 110 (1922); **9**,349 (1922).

[10] An electron orbiting around the nucleus corresponds to a circular current and as such gives rise to a magnetic moment.

[11] *Natuwiss*, **13**, 953 (1925); *Nature*, **117**, 264 (1926).

[12] *Z. Physik*, **43**, 601 (1927).

In the same year, Phips and Taylor repeated the Stern and Gerlach experiment using a beam of hydrogen atoms. Hydrogen atoms in their ground state have *zero* orbital angular momentum according to quantum mechanics (see Sect. 14.1.2), and therefore have no orbital magnetic moment. But the beam still split in two. One had to conclude that the electron has indeed an intrinsic angular momentum, a spin, and that it is quantized: it can have one or the other of two possible orientations.

13.3 The Wave-Particle Duality

Louis de Broglie was born in 1892. A French aristocrat so wealthy, that he could afford to build his own laboratory not far from the Champs Elysée. He lived a long life (he died in 1987), he won a Nobel Prize in 1929, and taught at the Sorbonne in Paris for 34 years. But he is best known for a paper he wrote in 1924 (*Philosophical Magazine*, **47**, 446) based on his doctoral thesis. In it he argued that, *since photons have wave and particle characteristics, perhaps all forms of matter have wave as well as particle properties*. He proposed that, since a photon of energy $\hbar\omega$ is assumed to have a momentum $\boldsymbol{p} = \hbar\boldsymbol{q}$, where $q = 2\pi/\lambda = \omega/c$ (see Sect. 13.1.2), it was reasonable to assume that, *a particle of momentum* $\boldsymbol{p} = mv$ *should in some way behave like a wave with wavevector* $\boldsymbol{q} = \boldsymbol{p}/\hbar$, with a corresponding wavelength $\lambda = 2\pi/q = h/p$.

Assuming that his hypothesis was true, de Broglie was able to derive Bohr's rule for the quantization of angular momentum (Eq. 13.21), by arguing that the circumference of an allowed circular orbit, $2\pi r$, should be an integral multiple of the wavelength, $\lambda = h/p$, of the wave associated with the orbiting electron, as one would expect for any circular stationary wave. Therefore it must be that: $2\pi r = n\lambda = nh/p$, where n is an integer. Which can be written as: $mvr = nh/2\pi$, which is the same as Eq. (13.21).

Although de Broglie's idea was a revolutionary one, with no experimental evidence to support it, it received immediate attention by his fellow scientists and played an important role in the development of quantum mechanics. If de Broglie was right and particles had wave characteristics, then they would under the right conditions exhibit interference phenomena. Three years after de Broglie's thesis, in 1927, the Americans C. J. Davisson and L. H. Germer succeeded in confirming this prediction. It happened at first by accident. During an experiment involving the scattering of low energy electrons (about 54 eV) from a nickel target in vacuum, they noticed that after the surface was cleaned by heating to remove an oxide coating, the scattered electrons exhibited intensity maxima and minima at specific angles. They realized that the nickel formed large crystals after heating and that the regularly spaced atom planes of the crystal served as a diffraction grating for electron matter waves. Subsequent experiments by Davisson and Germer, and by G. P. Thomson in Scotland, who observed electron diffraction by passing electrons

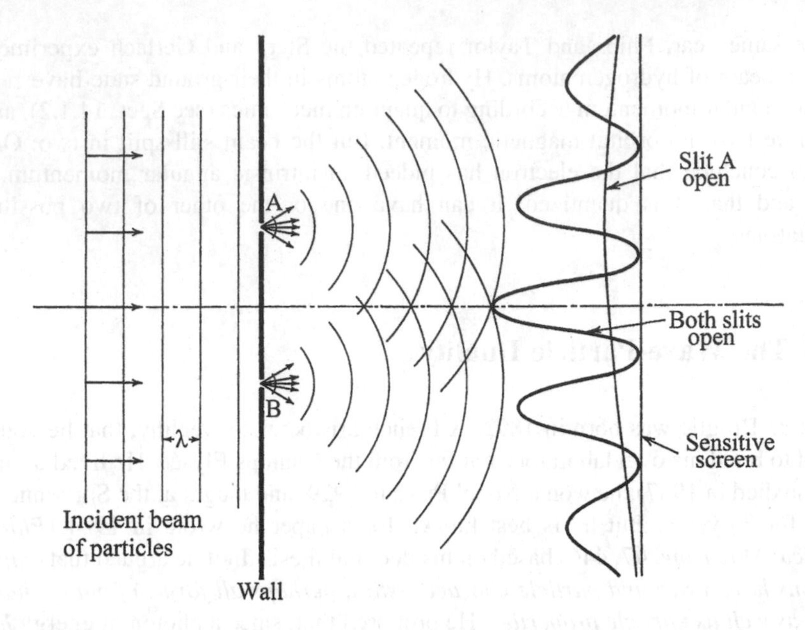

Fig. 13.8 Diffractive interference of particle-waves. [From R. B. Leighton, *loc. cit*]

through a very thin gold film, and by other scientists, confirmed de Broglie's hypothesis.[13]

One way, often used, of demonstrating the wave character of material particles is the imagined experiment described in Fig. 13.8. Assume that a beam of monoenergetic particles (e.g. electrons) is incident upon a wall containing two parallel slits (A and B) near each other. The particles that go through the slits are detected (counted) by their impacts on a sensitive screen at an appropriate distance from the wall. When only slit A is open, the number of impacts, per unit area of the screen per unit time, is described by a curve (shown in the figure) which accords with what we expect classically (the particles hitting the screen gradually diminish with the distance from the point directly opposite slit A), but when both slits are open, what we observe is far from the classical picture, it accords with what we should expect if the particles were waves (compare with Fig. 10.13).

It is certainly difficult to comprehend what is going on. One thing is certain: the moment the particle hits the screen it is just a particle and nothing else. But if it were all the time a particle it would arrive at the screen having gone through slit A *or* through slit B, but *not through both*; and there would be no interference. But interference there is! The reader who is not already familiar with quantum mechanics is, at this point, confused, a reader who has sufficient familiarity with quantum mechanics is resigned to the fact that he will never 'understand' the

[13] Thomson and Davisson were awarded the 1937 Nobel Prize 'for their experimental discovery of the diffraction of electrons by crystals'.

phenomenon in any sense that would satisfy our classical way of thinking. It is a 'puzzle' that one learns to live with. One must first learn how to calculate the wave associated with a particle in any given situation, and must then learn how to interprete it physically, in order to get answers that can be tested experimentally.

13.4 Formal Quantum Mechanics

13.4.1 Schrödinger's Equation

Erwin Schrödinger was born in Vienna in 1887. His father ran a small linoleum factory which he had inherited from his own father. His mother was half English, and the child Erwin learned English as well as German. He took lessons at home from a private tutor until the age of eleven, when he entered the traditional *Akademishes Gymnasium*. He was an excellent student. In 1906 he entered the University of Vienna, and in 1910 he was awarded his doctorate for a thesis on the effect of damp air on electrical conductors and insulators. After his voluntary military service he was appointed to an assistantship at the University and did mostly experimental work. In 1914 he joined the war, experienced combat and was awarded a citation for his command of a gun battery. After the war he worked at a number of places. He was a professor at Stuttgart, then Breslau, and then at Zurich, where he remained for 6 years and produced his best known work, on the foundation of quantum mechanics. In 1927 he succeeded Max Planck as head of physics at the Kaiser Wilhelm Institute in Berlin, and in 1929 he shared the Nobel Prize for Physics with de Broglie. In the same year, Schrödinger, who was a Catholic Austrian, left Germany, because the Nazis and their pathological anti-Semitism disgusted him. He was offered a position at Princeton, in the USA, but the authorities there were not happy about a man who planned to arrive with both his wife Annmarie and his mistrss Hilde, who was at the time pregnant with his first child. He gladly accepted a fellowship at Magdalen College, Oxford (in England). In 1936 he went back to Austria, although by now the Nazis considered him a traitor. Then he went for a while to Rome, then to Dublin (in Ireland), where the Prime Minister Eamon de Valera tempted him by founding an Institute for Advanced Studies. Then he went back to Oxford, then to Belgium, and finally back to Dublin again in 1939, where at last he settled for the next two decades, and had two more children by two different Irish women. Schrödinger was undoubtedly a womanizer. *I put beauty before science*, he wrote to Max Born, *we are always longing for our neighbour's wife and for the perfection we are least likely to achieve.* He died in 1961.

Schrödinger presented his famous theory in a series of papers in 1926 (*Annal. Physik*, **79**, 361, 489,784; **80**, 437; **81**, 109). *My theory*, he said, *was stimulated by de Broglie's thesis and by short but infinitely far seeing remarks by Einstein.* In his papers Schrödinger introduced the equation that bears his name, used it to obtain the energy levels and corresponding states of the hydrogen atom, and also showed how to deal with transitions between states using perturbation theory. His theory can be summarized as follows.

If de Broglie's wave, which from now on we shall call a *wavefunction*, is to have a central role in physical theory, it must satisfy a partial differential equation and appropriate boundary conditions. Schrödinger postulated that the wavefunction $\psi(\boldsymbol{r},t)$ of a particle, of mass m, moving in a potential field $U(\boldsymbol{r})$ (as defined in Sect. 7.3) satisfies the following equation:

$$-\frac{\hbar}{i}\frac{\partial\psi}{\partial t} = -\frac{\hbar^2}{2m}\nabla^2\psi(r,t) + U(\boldsymbol{r})\psi(\boldsymbol{r},t) \qquad (13.27)$$

where $i = \sqrt{(-1)}$ and ψ is a *complex* function of the position $\boldsymbol{r}$ and the time t. [A complex function has the form: $\psi(\boldsymbol{r},t) = \psi_R(\boldsymbol{r},t) + i\psi_I(\boldsymbol{r},t)$, where ψ_R and ψ_I are ordinary (real) functions of $\boldsymbol{r}$ and t. The reader will find an introduction to complex numbers, sufficient for the purposes of this book, in Appendix A]. The Laplace operator ∇^2 is defined by Eq. (7.46).

Schrödinger's equation (this is how Eq. (13.27) is presently known) is often written as follows:

$$-\frac{\hbar}{i}\frac{\partial\psi}{\partial t} = \hat{H}\psi \qquad (13.28)$$

The mathematical operator

$$\hat{H} = -\frac{\hbar^2}{2m}\nabla^2 + U(\boldsymbol{r}) \qquad (13.28a)$$

is known as the *Hamiltonian* of the particle; we can think of it as the mathematical operator for the (total) energy of the particle. The potential energy term $U(\boldsymbol{r})$ has a clear physical meaning and we need not say anything more about it here.[14] The operator

$$-\frac{\hbar^2}{2m}\nabla^2 = -\frac{\hbar^2}{2m}\left(\frac{\partial^2}{\partial x^2} + \frac{\partial^2}{\partial y^2} + \frac{\partial^2}{\partial z^2}\right) \qquad (13.28b)$$

is known as the *kinetic energy* operator (for the particle).

Let us see how the above relates to de Broglie's idea of a wave associated with a particle moving with velocity υ along the x-direction. This wave has a wave-vector (along the x-direction) $q = p/\hbar = m\upsilon/\hbar$, and a frequency $\omega = \mathrm{E}/\hbar$, where E is the energy of the particle (all this by analogy to the photon case). And we

[14] We have assumed that the force acting on the particle can be written as the gradient of a potential energy field $U(\boldsymbol{r})$ as described in Sect. 7.3. We note that a Hamiltonian operator can also be defined when the force acting on the particle is the Lorentz force (Eq. 10.58), which involves the magnetic field and can not be written as the gradient of a potential energy field. Schrödinger's equation remains valid, but the Hamiltonian operator takes a slightly different form.

remember that $E = mv^2/2 = \hbar^2q^2/(2m)$. It is therefore not unreasonable to identify this de Broglie wave with the following expression (according to Eq. (A2.14), $exp(i\varphi) = cos(\varphi) + isin(\varphi)$):

$$\psi_q(x,t) = Ae^{iqx - iE_qt/\hbar} \tag{13.29}$$

where $E_q = \hbar^2q^2/(2m)$ is the (kinetic) energy of the particle, and A is a normalization constant. We can now easily verify that $\psi_q(x,t)$ satisfies Schrödinger's Eq. (13.28): $(-\hbar/i)\partial\psi_q/\partial t = (-\hbar/i)(-iE_q/\hbar)\psi_q(x,t) = E_q\psi_q(x,t)$. On the other hand, $U(\mathbf{r}) = 0$ in the present case, and the motion is along the x-direction only, therefore: $\hat{H}\psi_q(x,t) = (-\hbar^2/2m)\partial^2\psi_q/\partial x^2 = (-\hbar^2/2m)(iq)^2\psi_q(x,t) = (\hbar^2q^2/2m)\psi_q(x,t) = E_q\psi_q(x,t)$. Therefore $(-\hbar/i)\partial\psi_q/\partial t = \hat{H}\psi_q(x,t)$. De Broglie's wavefunction for a free particle in the form of Eq. (13.29) satisfies Schrödinger's equation. Finally, it is worth noting that there are two states of the particle with the same energy, $\hbar^2q^2/(2\mathrm{m})$, corresponding to $q > 0$ (the particle moves to the 'right') and $q < 0$ (the particle moves to the 'left').

Before we consider the wavefunction for a bound particle, let me introduce in a parenthesis some useful concepts that arise naturally from the above discussion. We note that:

$$\frac{\hbar}{i}\frac{\partial}{\partial x}Ae^{iqx} = \hbar qAe^{iqx} \tag{13.30}$$

This prompts us to think of

$$\widehat{p}_x = \frac{\hbar}{i}\frac{\partial}{\partial x} \tag{13.31}$$

as the operator for the momentum along the x-direction. We say that $Aexp(iqx)$ is the *eigenfunction* (or *eigenstate*) of $\widehat{p}_x$corresponding to the eigenvalue $\hbar q$. We note that q can take any value: $-\infty < q < +\infty$. But for practical reasons we may assume that

$$q = q_n = (2\pi/L)n, n = 0, \pm 1, \pm 2, \pm 3, \ldots \tag{13.32}$$

where L is a very large length, so that successive values of q are infinitesimally close to each other but discrete.

The set of functions

$$u_q(x) = \frac{1}{\sqrt{L}}e^{iqx} \tag{13.33}$$

q given by Eq. (13.32), have the following important properties [known to mathematicians since Fourier's time (see Sect. 7.2)]. They are *orthogonal* to each other:

$$\int_{-L/2}^{L/2} u_q^*(x) u_{q'}(x) dx = 1 \text{ if } q = q' \tag{13.34}$$

$$= 0 \text{ if } q \neq q'$$

where $u_q^*(x)$ is the complex conjugate of $u_q(x)$. We note that: $u_q^*(x) = u_{-q}(x)$.

And they are a *complete* set of functions: Any well behaved (i.e. that has physical meaning) $f(x)$, $-L/2 < x < \cdot L/2$, can be written as a sum of the functions of Eq. (13.33):

$$f(x) = \sum_q c_q u_q(x); \quad c_q = \int_{-L/2}^{L/2} u_q^*(x) f(x) dx \tag{13.35}$$

We have assumed that the particle moves along the x-direction, but similar arguments apply to any direction. The extension of Eqs. (10.30) and (10.31) to 3-dimensions is obvious:

$$(\hbar/i)\nabla A e^{i\boldsymbol{q}\cdot\boldsymbol{r}} = \hbar \boldsymbol{q} A e^{i\boldsymbol{q}\cdot\boldsymbol{r}} \tag{13.36}$$

$$\boldsymbol{p} = \frac{\hbar}{i} \cdot \nabla = \frac{\hbar}{i}\left(\frac{\partial}{\partial x}, \frac{\partial}{\partial y}, \frac{\partial}{\partial z}\right) \tag{13.37}$$

We write the normalized eigenfunctions of the momentum operator, $\boldsymbol{p} = (\hbar/i)\nabla$, by analogy to Eq. (13.33), as follows

$$u_q(\boldsymbol{r}) = \frac{1}{\sqrt{L^3}} e^{i\boldsymbol{q}\cdot\boldsymbol{r}}, \tag{13.38}$$

$$\boldsymbol{q} = (2\pi/L)(n_x, n_y, n_z), n_x, n_y, n_z = 0, \pm 1, \pm 2, \pm 3, \ldots$$

The $u_{\boldsymbol{q}}(\boldsymbol{r})$ are an orthogonal, and complete set: any function $f(\boldsymbol{r})$, where $-L/2 < x, y, z < \cdot L/2$, can be written as a sum of these functions (as in Eq. 13.35). The integrals in Eqs. (10.34 and 10.35) are replaced by volume integrals over $-L/2 < x, y, z < \cdot L/2$.

Finally, the normalized wavefunction associated with a particle of momentum $\boldsymbol{p} = \hbar\boldsymbol{q}$ is:

$$\psi_{\boldsymbol{q}}(\boldsymbol{r}, t) = \frac{1}{\sqrt{L^3}} exp(i\boldsymbol{q}\cdot\boldsymbol{r} - iE_q t/\hbar) \tag{13.39}$$

which satisfies Eq. (13.28) when $U(\boldsymbol{r}) = 0$. And we note that the energy, $E_q = \hbar^2 q^2/(2m)$ is the same for any wavevector of magnitude q whatever its direction].

Let me rewrite Eq. (13.29) as follows:

$$\psi_q(x,t) = u_q(x)e^{-iE_q t/\hbar} \tag{13.40}$$

$$\hat{H}u_q(x) = E_q u_q(x) \tag{13.41}$$

Where $\hat{H} = (-\hbar^2/2m)\frac{\partial^2}{\partial x^2}, E_q = \hbar^2 q^2/(2m)$ and $u_q(x)$ is given by Eq. (13.33). A solution of Schrödinger's equation in the form of Eq. (13.40), corresponding to a definite value of its energy, is called an (energy) eigenstate of the particle. It is also, some times, called a stationary state for reasons that will become apparent later in our discussion. At this stage let me emphasize the *completeness* property of these states. We note that Schrödinger's equation (Eq. 13.28) is a linear equation, which implies that, a linear sum of the above stationary waves:

$$\psi(x,t) = \sum_q c_q u_q(x)e^{-iE_q t/\hbar} \tag{13.42}$$

where c_q are any constants and the sum is over all q of Eq. (13.32), is also a solution of Schrödinger's equation (for the free particle). And it is the most general one. Suppose that we know the wavefunction at $t = 0$: $\psi(x,0) = f(x)$. Then Eq. (13.35) allows us to determine uniquely the coefficients c_q, and, therefore, $\psi(x,t)$ at any time $t > 0$.

It turns out that what we established for a free particle ($U(\boldsymbol{r}) = 0$) is valid generally. Given a field $U(\boldsymbol{r})$, the energy eigenvalues, E_v, and corresponding eigenstates, ψ_v, of a particle moving in it are given by:

$$\psi_v(\boldsymbol{r},t) = u_v(\boldsymbol{r})e^{-iE_v t/h} \tag{13.43}$$

$$\hat{H}u_v(\boldsymbol{r}) = E_v u_v(\boldsymbol{r}) \tag{13.44}$$

Where v stands for one or more indexes which specify the eigenstate; $u_v(\boldsymbol{r})$ are smoothly varying functions, with continuous derivatives, which satisfy appropriate boundary conditions. We shall see that eigenstates of a particle bounded in space are possible *only* for certain discreet values, E_v, of the energy.

One easily verfies that $\psi_v(\boldsymbol{r},t)$ satisfies Schrödinger's equation. We have: $(-\hbar/i)\partial\psi_v/\partial t = (-\hbar/i)(-iE_v/\hbar)\psi_v(\boldsymbol{r},t) = E_v\psi_v(\boldsymbol{r},t)$, and $\hat{H}\psi_v(\boldsymbol{r}) = (\hat{H}u_v(\boldsymbol{r}))$ $exp(-iE_v t/\hbar) = E_v\psi_v(\boldsymbol{r},t)$. Therefore: $(-\hbar/i)\partial\psi_v/\partial t = \hat{H}\psi_v(\boldsymbol{r},t)$.

Schrödinger's theory assumes that, the set of functions, $u_v(\boldsymbol{r})$, is an orthogonal and complete set of functions in the same way that the functions $u_q(x)$ are (see Eqs. (13.34) and (13.35); the integrals over x are replaced by volume integrals, and dx by $dV = dxdydz$). Accordingly any function $f(\boldsymbol{r})$ of physical significance can be written as a linear sum of these functions:

$$f(\boldsymbol{r}) = \sum_v c_v u_v(\boldsymbol{r}) \tag{13.45}$$

Where the sum includes all allowed states (solutions of Eq. 13.44). By analogy to Eq. (13.42) the most general wavefuntion for the particle in the field $U(\boldsymbol{r})$ is given by:

$$\psi(\boldsymbol{r},t) = \sum_{v} c_v u_v(\boldsymbol{r}) exp(-iE_v t/\hbar) \tag{13.46}$$

The coefficients c_v are uniquely dermined from $\psi(\boldsymbol{r},0)$, the wavefunction at $t = 0$, in the same way as in Eq. (13.42).
Equation (13.44) is known as *Schrödinger's time-independent equation.*

13.4.2 The Stationary States of a Bound Particle

Particle in a one-dimensional box: Imagine a free particle moving along the x-axis between two perfectly reflecting walls at $x = 0$ and $x = a$. Between the walls no force is acting on it, therefore we can put its potential energy, U, equal to zero there, and since the particle remains between the walls, whatever its energy, we can assume that U rises to ∞ at the walls:

$$\begin{aligned} U(x) &= 0 \quad \text{for} \quad 0<x<a \\ &= \infty \quad \text{for} \quad x<0 \,\text{and for}\, x>a \end{aligned} \tag{13.47}$$

Schrödinger's time-independent equation (Eq. 13.44) for a particle of mass m in the above potential well is: $-\frac{\hbar^2}{2m}\frac{d^2u}{dx^2} = Eu(x)$, which can be written as:

$$\frac{d^2u}{dx^2} = -q^2 u(x) \tag{13.48}$$

where $0 < x < a$; and $E = \hbar^2 q^2/2m$. The general solution of the above equation is:

$$u(x) = Ae^{iqx} + Be^{-iqx} \text{for} 0<x<a \tag{13.49}$$

The above is what we expect from our discussion of the previous section for a wave associated with a free particle. The first term represents a wave with energy E propagating to the right, and the second term represents a wave of the same energy propagating to the left. The constants A and B are to be determined so that $u(x)$ is *continuous* everywhere *including* the endpoints at $x = 0$ and $x = a$. Now we must assume that $u(x) = 0$ for $x < 0$ and for $x > a$; i.e. that the wave does not penetrate into the region of infinite U. That this is so is confirmed when one looks at the infinitely deep potential well of Eq. (13.47) as the limit of a very deep well (see next section). Therefore the continuity of $u(x)$ demands that:

$$u(0) = 0 \,\text{and}\, u(a) = 0 \tag{13.50}$$

We proceed as follows: using thc identity $exp(i\varphi) = cos(\varphi) + isin(\varphi)$, we write $u(x)$ of Eq. (13.49) in the form:

$$u(x) = Dcos(qx) + Csin(qx) \tag{13.51}$$

Looking at the *cos* and *sin* functions of Fig. 5.6 we see that: for $u(0)$ to be zero we must have: $D = 0$; i.e. $u(x) = Csin(qx)$. And in order to have solutions which satisfy also the requirement $u(a) = 0$, with $C \neq 0$, we must restrict the values of q so that: $sin(qa) = 0$. This means (see Fig. 5.6) that $qa = n\pi, n = 1, 2, 3$, etc.; which, in turn, means that the energy of the particle, $E = \hbar^2 q^2/2m$, in the potential well of Eq. (13.47) can take one or the other of the following eigenvalues (but none other):

$$E_n = \frac{\hbar^2\pi^2 n^2}{2ma^2}, n = 1, 2, 3, 4, \ldots \tag{13.52}$$

contrary to what we expect from classical physics. We remember that, classically, the particle can have any (kinetic) energy: it has a continuous energy spectrum, unlike the *discrete* energy spectrum of Eq. (13.52).

The stationary wave (state) corresponding to the *nth* energy level is, accordingly, given by:

$$\begin{aligned} \psi_n(x,t) &= u_n(x)exp(-iE_n t/\hbar) \\ u_n(x) &= Csin(n\pi x/a) \end{aligned} \tag{13.53}$$

The constant C is dermined by normalization: we demand that

$$\int_0^a u_n^*(x)u_n(x)dx = C^2 \int_0^a \sin^2(\frac{n\pi x}{a})dx = C^2 a/2 = 1$$

Therefore: $C = \sqrt{(2/a)}$. And $\int_0^a u_n^*(x)u_m(x)dx = 0$ if $n \neq m$ as expected (see text preceding Eq. 13.45).

The reader should note that *zero* energy is not an allowed energy level. The classical picture of a motionless particle sitting somewhere between $x = 0$ and $x = a$ does not exist in the real (quantum-mechanical) world. Certainly not, if we are dealing with atomic systems. The least energy (the ground-state energy) of the particle in the considered potential well is:

$$E_1 = \frac{\hbar^2\pi^2}{2ma^2} \tag{13.54}$$

For example: putting m = mass of the electron, and $a = 2 \times 10^{-8}$ cm (about the diameter of an atom) we obtain $E_1 \approx 10$ eV. On the other hand if a equals a few centimeters, E_1 is practically zero. We note also that the difference between two successive energy levels,

$$E_n - E_{n-1} = \frac{\pi^2 \hbar^2}{2ma^2}(2n - 1) \tag{13.55}$$

tends to zero when a becomes sufficiently large.

At this point it is worth noting the similarity between the stationary states of the particle in the potential well of Eq. (13.47) and the normal modes of the vibrating string discussed in Sect. 7.2. And the similarity between the two goes further in the following sense: In the same way that any vibration of the string can be written as a linear sum of the normal modes of the string, the most general state $\psi(x,t)$ of the particle in the potential well can be written (according to Eq. 13.46) as a linear sum of its stationary states (defined by Eq. 13.53) :

$$\psi(x,t) = \sum_n c_n u_n(x) \exp(-iE_n t/\hbar)$$

where the coefficients c_n are determined in the same way: by the state (wave-function $\psi(x,0)$) of the particle at $t = 0$.

Particle in a rectangular potential well: Consider a particle of mass m in a potential field described by:

$$\begin{aligned} U(x) &= -U_0 \quad \text{for} \quad -a<x<a \\ &= 0 \quad \text{for} \quad x< -a \,\text{and for}\, x > a \end{aligned} \tag{13.56}$$

as shown schematically in Fig. 13.9. In the present case the spectrum of the energy levels (eigenvalues) consists of a continuum part extending from $E = 0$ to $E = +\infty$, corresponding to free states of the particle (we shall not be concerned with it here), and a discrete part consisting of a number of negative energy levels corresponding to bound (localized in and about the well) states of the particle, which we can obtain as follows:

Schrödinger's time-independent equation for a particle in the potential well of Eq. (13.56) is:

$$\begin{aligned} -\frac{\hbar^2}{2m}\frac{d^2u}{dx^2} - U_0 u(x) &= Eu(x) \quad \text{for} -a<x<a \\ -\frac{\hbar^2}{2m}\frac{d^2u}{dx^2} &= Eu(x) \quad \text{for}\, x< -a \text{ and for} x > a \end{aligned} \tag{13.57}$$

where $-U_0<E<0$ (for bound states).

One can easily verify by direct substitution that each of the following expressions satisfies the above equation in the region of its definition:

$$\begin{aligned} u(x) &= Ae^{\beta x} + Be^{-\beta x} \text{for}\, x< -a \\ &= Ce^{i\alpha x} + De^{-i\alpha x} \text{for} -a<x<a \\ &= Fe^{\beta x} + Ge^{-\beta x} \text{for}\, x > a \end{aligned} \tag{13.58}$$

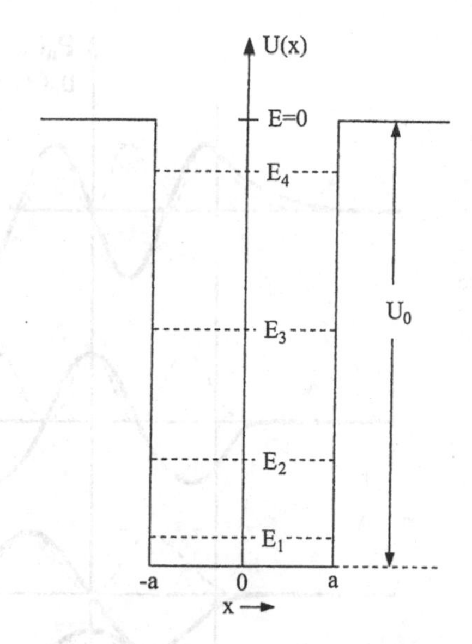

Fig. 13.9 Rectangular potential well with $U_0 = 49h^2/(128ma^2)$, where h is Planck's constant. The negative energy levels are: $E_1 = -0.94U_0, E_2 = -0.77U_0, E_3 = -0.50U_0, E_1 = -0.16U_0$

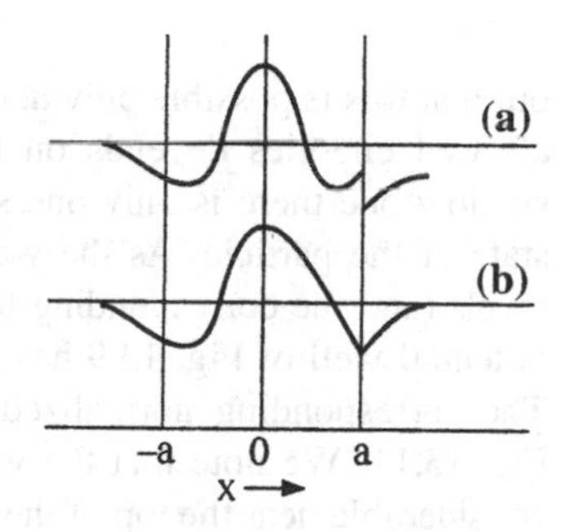

Fig. 13.10 The wave function and its derivative must be continuous. A function (**a**) would not be acceptable because of its discontinuity at $x = a$. A function (**b**) would not be acceptable because of the discontinuity of its derivative at $x = a$

where *A*,*B*,*C*,*D*,*F*,*G* are constants to be determined so that $u(x)$ and du/dx are every everywhere, including at the points $-a$ and $+a$, continuous (see Fig. 13.10).

The wavefunction must also be normalizable: we must have:

$$\int_{-\infty}^{+\infty} u^*(x)u(x)dx = Q \tag{13.59}$$

where Q is a finite number. [The above integral is to be understood as a limit, as in footnote 11 of Chap. 9]. The normalization requirement demands that B and F in Eq. (13.58) are zero, since the corresponding terms increase exponentially away from the potential well, rendering $u(x)$ *unnormalizable*. The remaining constants must be such that $u(x)$ and du/dx are continuous at the points $-a$ and $+a$. It turns

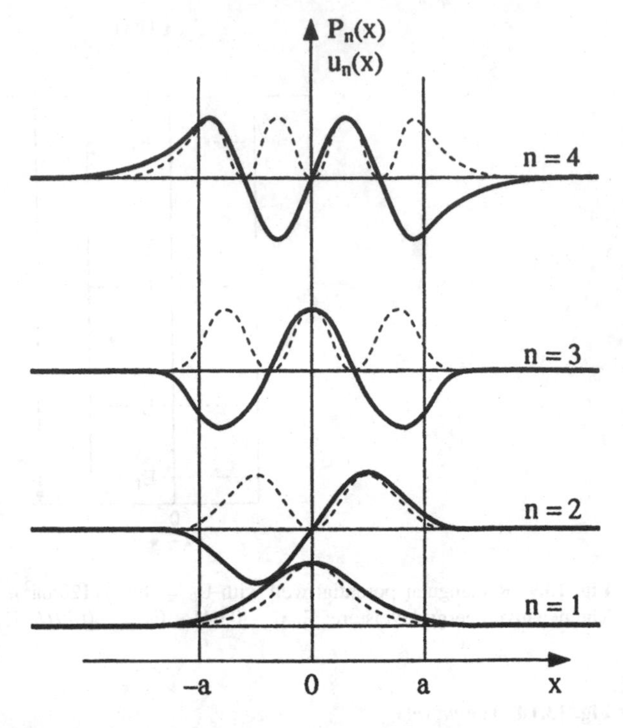

Fig. 13.11 The *solid lines* describe the bound states $u_n(x)$, $n = 1, 2, 3, 4$ of the particle in the rectangular potential well of Fig. 13.9. The *broken lines* give the corresponding probability distribution: $P_n(x) = |u_n(x)|^2$. The wave functions are normalized: $\int_{-\infty}^{+\infty} P_n(x)dx = 1$

out that this is possible only at certain energies (see exercise 13.4). The number of allowed energies depends on the depth (U_0) of the potential well. For a very shallow one there is only one such energy and therefore there is only one bound state of the particle. As the well becomes deeper the number of allowed energy levels (and the corresponding to them bound states of the particle) increases. The potential well of Fig. 13.9 has four bound states at the shown four energy levels. The corresponding normalized wave functions are shown by the solid lines in Fig. 13.11. We note that the wavefunctons spread out of the well; this spread is considerable near the top of the well and is very small at the bottom of the well. At the bottom of a very deep well this spread practically vanishes (the situation is in this case similar to that of a particle in a box. We shall have more to say about the wavefunctions of Fig. 13.11 in the following section.

The previous examples were meant to convince the reader that Schrödinger's theory leads, via a well-defined mathematical formalism, to the quantization of the energy spectrum of a bound (in space) particle (and as we shall see the same applies equally well to a system of particles). It is important to note at this point that Schrödinger applied his theory to the hydrogen atom and that the results he obtained were in excellent agreement with the experimental data (apart from minute relativistic corrections) and with Bohr's theory (where the latter agreed with the experiments). We shall summarize these results in Sect. 14.1.2. The theory could also describe how an atom may transit from one state to another under the influence of an external purturbation (say due to the atom being exposed to

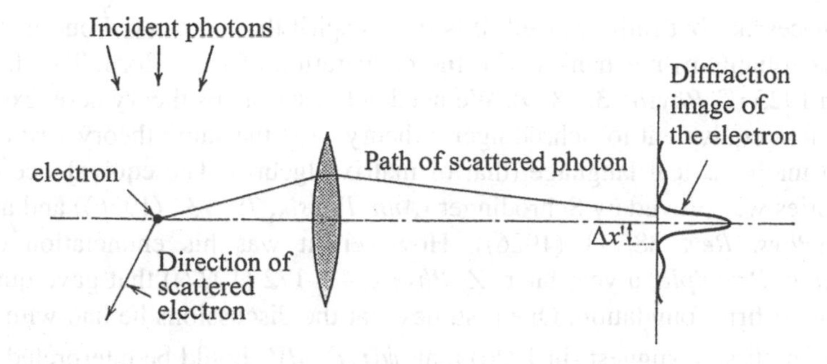

Fig. 13.12 An illustration of the Uncertainty Principle. $\Delta x' = \Delta x/2$

electromagnetic radiation). But there was something missing, and that something was very important. Schrödinger could not say what the *physical meaning* of the wavefunction, $\psi(\boldsymbol{r},t)$, was. He believed that there was some hidden physics behind it, that would allow one to keep the clarity of classical physics (as in a particle moving in space–time along a well defined path), but though he tried he could not find a satisfactory answer. The resolution of this problem we owe, above all, to Heisenberg.

13.4.3 Heisenberg's Uncertainty Principle

Werner Heisenberg was born in Wurzburg (Germany) in 1901, with a twin brother Erwin. Apparently the two did not get along well, and had very little to do with each other during their lives. In 1910 the family moved to Munich, where his father had been appointed to the prestigious Chair of Greek Philology at the University of Munich. He introduced the boy Werner to the Greek philosopher-scientists and encouraged his talented playing of the piano. After the first world war the 18-year old Heisenberg became the leader of a youth movement called the 'German New Boy Scouts'. They were romantics who spent time camping in the hills and yearned for a simpler and moral way of life. Such movements would later be exploited by the Nazis, but at the time they were quite innocent, and certainly not anti-Semitic. In 1920 Heisenberg entered the University of Munich intending to study mathematics but very quickly he transferred to physics. While still a student he attended a lecture given by Bohr at Göttingen and was fascinated by Bohr's theory of the atom. He got his first academic job at the University of Göttingen, where he became an assistant to Max Born. At Göttingen he acquired a 'positivist' view of science, which roughly speaking meant that, what matters in scientific theory is its ability to predict the outcome of an experiment, and that any attempt to understand the *unseen* workings of nature between observations is not only unnecessary but futile as well. It is in this spirit that he worked out his theory of matrix (quantum) mechanics with the cooperation of Max Born. This he published in 1925 (Z. *Physik*, **33**, 879). We need not describe his theory here, except to say that it is equivalent to Schrödinger's theory. It is the same theory written in a different mathematical language (that of matrix algebra). The equivalence of the two theories was proved by Schrödinger (*Ann. Physik*, **79**, 734 (1926)) and also by Eckart (*Phys. Rev.* **28**,711 (1926)). However, it was his enunciation of the *Uncertainty Principle*, a year later (Z. *Physik*, **43**, 172 (1927)) that gave quantum mechanics a firm foundation. One assumes that the discussions he had with Born, who was the first to suggest (in 1926) that $|\psi(\boldsymbol{r},t)|^2 dV$ should be interpreted as the *probability* of finding the particle within the infinitesimal volume dV about $\boldsymbol{r}$ at time t, and the discussions he had with Bohr helped him along the way, but the Uncertainty Principle is certainly Heisenberg's idea. And it was a revolutionary idea. Heisenberg went on to make significant contributions in a variety of fields, including nuclear physics, fluid dynamics, statistical and solid state physics, but his fame rests above all on his contribution to the foundation of quantum mechanics, and it is for this contribution that he shared with Dirac the Nobel Prize for Physics in 1933.

In 1939 Heisenberg, who was by this time married with Elisabeth Schumaher, travelled to the United States, where he lectured at Ann Arbor and in Chicago, and met many old friends. Enrico Fermi tried to persuade him to stay in America, but he chose to return to Berlin, where the Kaiser Wilhelm Institute for Physics was now the German centre for nuclear research. Its director at the time, Peter Debye, travelled to Switzerland for a lecture tour and did not return, and Heisenberg emerged as the Institute's leader. Nobody knows how close Heisenberg and his team came to building a nuclear bomb, or how hard they tried. The prevailing view among historians appears to be that Heisenberg always believed that he could build a uranium reactor, but not in the short time that the Nazis would have demanded of him if he promised to deliver one. After the war he claimed that he did not ask for large sums of money because he wanted to delay the development of the bomb. Many believe that if he had the time and the money he would not shy away from making the bomb, or from seeing it used. In a philosophical paper he wrote in 1942, he expressed a fatalist's view of history: *Reality is transformed without our influence*, he wrote, … *the individual can contribute nothing to this, other than to prepare internally for the changes that will occur anyway.*

Heisenberg remained a keen philosopher and among his other works he later wrote a book entitled: *Physics and Philosophy* (Penguin Books, New York, 1958). After the war he organized and became director of the Max Planck Institute for Physics and Astrophysics at Göttingen, later (in 1958) moving with the Institute to Munich. In 1970 he became director emeritus of the Institute. He died in 1976.

Heisenberg's *uncertainty principle* tells us that: However hard we try, whatever the method and the quality of our instruments, it is impossible to know exactly the position *and* the velocity of a particle at any given moment of time. If Δx is the uncertainty in our knowledge of the position coordinate x (in the sense that the best we can say is that the particle is to be found between $x - \Delta x/2$ and $x + \Delta x/2$), and Δp_x is the uncertainty in our knowledge of the momentum of the particle along the same direction, then:

$$\Delta x \Delta p_x \cdot \geq \hbar/2 \tag{13.60}$$

where $\hbar$ is Planck's constant. As an illustration of the principle consider the measurement of the position of a small particle (say an electron) using a microscope, as shown schematically in Fig. 13.12.

At least a single photon must strike the electron if we are to 'see' it. The photon is scattered by the electron and after going through the microscope hits a point on the diffraction pattern (on a screen). The resolution of the microscope is at best equal to the wavelength λ of the incident photon: $\Delta x \geq \lambda$, so one must use photons of the smallest possible wavelength to minimize Δx. However, a photon of wavelength λ has a momentum, $\hbar q = \hbar(2\pi/\lambda)$, and in colliding with the electron it will transfer to it an amount of momentum of this order, therefore the momentum of the electron immediately after measurement can only be known with an accuracy $\Delta p_x \approx \hbar/\lambda$. We have: $\Delta x \Delta p_x \approx \hbar$, in agreement with Eq. (13.60). If we wish to know x with high accuracy, we must accept a greater uncertainty in our knowledge of its momentum. We remember that the above is only an illustration of the principle. The exact form of the uncertainty principle (Eq. 13.60) is obtained from the theory after certain rules for calculating measurable quantities from $\psi(r, t)$ have been established. As we shall see, the rules by which we obtain, from the known (calculated) $\psi(r, t)$, the values of quantities that can be measured are straightforward. But we must first understand what constitutes a meaningful (answerable) question in quantum mechanics.

The uncertainty principle forces us to abandon for good the classical concept of the orbit of a particle. Knowing the orbit of a particle means: we know the position $r(t)$ of the particle at *any* moment of time, and this implies the simultaneous knowledge of the velocity $v(t)$ of the particle, and this is impossible according to the uncertainty principle. It is not surprising that many scientists, including Schrödinger and Einstein, were reluctant to accept it. The irony is that Heisenberg attributed his original inspiration of his *matrix* mechanics to Mach's view, which, he believed, underlied the special theory of relativity, namely that, all quantities that entered a physical theory must have *operational definitions* in terms of measuring instruments (clocks, measuring rods, etc.). So when he met Einstein in 1926, he was surprised when Einstein asked:[15] *But you don't seriously believe that none but observable magnitudes must go into a physical theory?* To which Heisenberg answered, *Isn't that precisely what you have done with relativity? After all you did stress the fact that it is impermissible to speak of absolute time, simply because absolute time can not be observed; that only clock readings, be it in the moving reference system or the system at rest are relevant to the determination of time.*

---

[15] The story appears in Jeremy Bernstein's book, Einstein, published by Fontana (U.K.), 1973; p. 155.

As Heisenberg recalled, Einstein replied: *Possibly I did use this kind of reasoning but it is nonsense all the same. Perhaps I could put it more diplomatically by saying that it may be heuristically useful to keep in mind what one actually observed. But on principle it is quite wrong to try founding a theory on observable magnitudes alone. In reality the very opposite happens. It is the theory which decides what we can observe.* This phrase '*It is the theory which decides what we can observe*' remained in Heisenberg's mind and eventually led to his invention of the uncertainty principle.

The question now arises: If we can not know the orbit of a particle what is there to know, encoded in the wavefunction $\psi(\boldsymbol{r},t)$, that can be meaningfully compared with observations? The answer to this question was hammered out by the so-called Copenhagen School with Heisenberg and Bohr as its protagonists.[16] It goes like this:

Following Born we interprete $P(\boldsymbol{r},t)dV = |\psi(\boldsymbol{r},t)|^2 dV$ as the *probability* of finding the particle under consideration within a volume dV about $\boldsymbol{r}$ at time t. We imagine a large number of identical systems, by which we mean that they are all described by the given $\psi(\boldsymbol{r},t)$. At a given moment we observe (measure) the position of the particle in every one of these systems. The particle is found in a volume dV about $\boldsymbol{r}$ in a fraction, equal to $P(\boldsymbol{r},t)dV$, of the measurements. [We have assumed that the integral of $|\psi(\boldsymbol{r},t)|^2 dV$ over all available space has been normalized to unity.] In the case of a stationary state, the wavefuction has the form: $\psi_E(\boldsymbol{r},t) = u_E(\boldsymbol{r})exp(iEt/\hbar)$ and therefore $|\psi_E(\boldsymbol{r},t)|^2 dV = |u_E(\boldsymbol{r})|^2 dV$ does not depend on the time. An example of the above is provided in Fig. 13.11. $P_n(x)dx$ is the probability of finding the particle between x and $x + dx$ (at any time) when the particle is in the *nth* stationary state. We note that the particle can be with some probability in a region of space (outside the potential well) where its potential energy U is *higher* than its total energy E. In classical mechanics that would be impossible, because $E - U = mv^2/2$ (the kinetic energy of the particle) is of course always positive. However, it is important to remember that the $P_n(x)$ shown in Fig. 13.11 do *not* imply that the particle can be outside the well *with* negative kinetic energy. The phrase 'when the particle is at position x its velocity and therefore its kinetic energy is so much', is *never* true in quantum mechanics, because we can not know the velocity (and therefore the kinetic energy) of the particle *and* its position. Perhaps a better way of putting it would be: when the particle is at x, *it* does not know what its velocity is. The reality of quantum mechanics can be summarized as follows: the physical state of the particle is encoded in the wavefunction $\psi(x,t)$. Knowing the wavefunction, we can, using

[16] A more systematic (axiomatic) presentation of quantum mechanics (according to the Copenhagen School) will be found in: R. B. Leighton, Principles of Modern Physics, McGraw-Hill Book Company, Inc., N.Y., 1959. Other (tentative) interpretations of quantum mechanics, which accept its mathematical predictions but attempt to propose a physical reality behind the mathematics, nearer to our direct experience of the physical world, have been proposed, but none has been accepted by the scientific community. A most interesting book on this subject has been written by David Bohm: Wholeness and the Implicate Order, Routlege Classics, 2002. N.Y.

well defined rules, answer only certain questions, such as: What is the probability of finding the particle between x and $x + dx$? What is the probability of the particle having a kinetic energy (always positive) between K and $K + dK$? etc.

The mean position $< x(t) >$ of the particle, when its wavefuction is $\psi(x, t)$ is defined by[17]:

$$<x(t)> = \int_{-\infty}^{+\infty} \psi^*(x,t) x \psi(x,t) dx = \int_{-\infty}^{+\infty} x P(x,t) dx \tag{13.61}$$

and the uncertainty Δx (a measure of the spread about the mean value) is defined by[18]

$$\Delta x = \left[\int_{-\infty}^{+\infty} \psi^*(x,t) (x - <x>)^2 \psi(x,t) dx \right]^{\frac{1}{2}} \tag{13.62}$$

and we note that both $< x >$ and Δx, which are in general functions of time, do *not* depend on the time when $\psi(x, t)$ is a stationary state.

Let us next consider what one can say about the momentum of a particle when its wavefunction, $\psi(x, t)$, is known.[17] We remember that $u_q(x)$, defined by Eq. (13.33), is associated with a definite momentum, $\hbar q$, and we remember also that any function can be written as a linear superposition of the $u_q(x), q$ given by Eq. (13.32), in accordance with Eq. (13.35). Therefore we can write:

$$\psi(x,t) = \sum_q c_q(t) u_q(x) = \sum_q c_q(t) \frac{1}{\sqrt{L}} e^{iqx} \tag{13.63}$$

$$c_q(t) = \int_{-L/2}^{+L/2} u_q^*(x) \psi(x,t) dx = \int_{-L/2}^{+L/2} \frac{1}{\sqrt{L}} e^{-iqx} \psi(x,t) dx$$

where L is very large so that ψ vanishes outside the region $-L/2 < x < L/2$. We assume that $\psi(x,t)$ is normalized, therefore:

$$\int_{-L/2}^{+L/2} \psi^*(x,t) \psi(x,t) dx = 1 \tag{13.64}$$

Substituting in the above the expression of Eq. (13.63) and using the orthogonality relation of Eq. (13.34), we obtain:

[17] For the sake of simplicity we give the formulae for motion along the x-direction. The extension to three dimensions is straightforward.

[18] Formulae (13.61) and (13.62) are analogous to the formulae one uses to obtain the mean of a series of measurements (e.g. of the temperature of a body) and the spread of these measurements about the mean.

$$\sum_q c_q^*(t)c_q(t) = \sum_q |c_q(t)|^2 = 1 \qquad (13.65)$$

Equation (13.63) tells us that, $\psi(x,t)$ is made of waves, $u_q(x)$, each of which corresponds to a definite value, $\hbar q$, of the momentum. It is then reasonable to expect, in view of Eq. (13.65), that $|c_q(t)|^2$ is the probability that a measurement of the momentum of the particle at time t will give the value $\hbar q$. The way to check this is the same as with the measurement of position. We construct a large number of identical systems, all of them described by the given $\psi(x,t)$. At a given moment we observe (measure) the momentum of the particle in every one of these systems. In a fraction of the measurements, equal to $|c_q(t)|^2$, the particle will have momentum between $\hbar q$ and $\hbar(q+\Delta q)$, where $\Delta q = 2\pi/L$ is a very small number.

It follows from the above discussion that, the mean momentum $<p_x(t)>$ of the particle described by $\psi(x,t)$ is:

$$<p_x(t)> = \sum_q \hbar q |c_q(t)|^2 \qquad (13.66)$$

We note that the same is obtained from the following formula:

$$<p_x(t)> = \int_{-L/2}^{+L/2} \psi^*(x,t)\widehat{p}_x\psi(x,t)dx = \sum_q \hbar q|c_q(t)|^2 \qquad (13.67)$$

where $\widehat{p}_x$is the momentum operator defined by Eq. (13.31). To obtain the above, we replace $\psi(x,t)$ by the sum of Eq. (13.63), use Eq. (13.30), and then the orthogonality relation of Eq. (13.34).

The uncertainty Δp_x is obtained (by analogy to Eq. 13.62) by:

$$\Delta p_x = \left[\int_{-L/2}^{L/2} \psi^*(x,t)(\widehat{p}_x - <p_x>)^2\psi(x,t)dx\right]^{\frac{1}{2}} \qquad (13.68)$$

And needless to say that Δx and Δp_x, defined by Eqs. (13.62) and (13.68), satisfy the uncertainty principle (Eq. 13.60).

Finally, we note that for a stationary state, $\psi_E(x,t) = u_E(x)exp(iEt/\hbar)$, the coefficients $c_q(t) = c_q(0)exp(iEt/\hbar)$, and therefore $|c_q(t)|^2 = |c_q(0)|$ are constants independent of time; and the same holds for $<p_x>$ and Δp_x.

It turns out that, for each physical quantity $\Omega(\boldsymbol{r},\boldsymbol{p})$ we can write down an operator $\widehat{\Omega} = \widehat{\Omega}(\boldsymbol{r},\widehat{p})$ with a set of eigenvalues ω_n and corresponding eigenstates $f_n(\boldsymbol{r})$ such that: $\widehat{\Omega} f_n(\boldsymbol{r}) = \omega_n f_n(\boldsymbol{r})$ for $n = 1, 2, 3$, etc.; and that through them we can determine what the result of a measurement of Ω would be (one or the other of the eigenvalues ω_n) and with what probability, as long as we know the wave function $\psi(\boldsymbol{r},t)$ of the particle. The procedure is analogous to the one we followed for the momentum of the particle. [We assume that the functions $f_n(\boldsymbol{r}), n = 1, 2, 3, \ldots$ constitute an orthogonal and complete set of functions, i.e. that it is possible to write any wavefunction as follows: $\psi(\boldsymbol{r}, t) = \sum_n c_n(t) f_n(\boldsymbol{r})$]. One important quantity, the angular momentum of a particle, will be dealt with in the next chapter.

13.4.4 Scattering States and the Tunnelling Phenomenon

A free particle (of mass m) with definite momentum $\hbar q$ is described by the wavefunction of Eq. (13.29). The probability of finding this particle between x and $x + dx$ at time t is $P(x)dx = \left|A exp(iqx - iE_q t/\hbar)\right|^2 = |A|^2$, i.e. the particle can be *anywhere* with equal probability. This we should expect. When a particle is described by Eq. (13.29) its momentum is definite ($\hbar q$), and the corresponding uncertainty $\Delta p_x = 0$. Therefore, according to the uncertainty principle (Eq. 13.60), $\Delta x = \infty$. In reality a free particle is described by a superposition of momentum states, by a wavepacket (so called), as in Eq. (13.42). An appropriate form of this wavepacket is:

$$\psi_k(x, t) = \sum_q c_q(k) \frac{1}{\sqrt{L}} e^{iqx - iE_q t/\hbar} \tag{13.69}$$

Where $\hbar k$ is a given momentum (it can have any value between $-\infty$ and $+\infty$), the sum is over all q (of Eq. 13.32), $E_q = \hbar^2 q^2/2m$, and the coefficients $c_q(k)$ are defined by:

$$c_q(k) = \frac{(2\alpha)^{1/2} \pi^{1/4}}{\sqrt{L}} e^{-i(q-k)x_0} e^{-\alpha^2 (q-k)^2/2} \tag{13.69a}$$

We note that $\left|c_q(k)\right|^2$, which is the probability of the particle having momentum $\hbar q$, has its maximum value for $q = k$, and practically vanishes outside the region $(k - 1/\alpha) < q < (k + 1/\alpha)$, as shown schematically in Fig. 13.13. We may conclude that the mean value of the momentum (along the x-direction) of the particle described by the wavefunction of Eq. (13.69) is $<p> = \hbar k$, with a corresponding uncertainty $\Delta p \approx \hbar/\alpha$.

In Fig. 13.14 we show the real part of $\psi_k(x, t)$ of Eq. (13.69), and $|\psi_k(x, t)|$, at three moments of time $t = 0, t = t_0$ and $t = 3t_0$; where $t_0 = m\alpha^2/\hbar$ and $\alpha = \pi/k$. The corresponding probability function $P_k(x, t) = |\psi_k(x, t)|^2$ is given by:

$$P_k(x, t) = \frac{1}{\sqrt{\pi\left(\alpha^2 + \frac{\hbar^2 t^2}{m^2 \alpha^2}\right)}} \exp\left\{ -\frac{(x - x_0 - pt/m)^2}{(\alpha^2 + \hbar^2 t^2/m^2 \alpha^2)} \right\} \tag{13.70}$$

where $p = \hbar k$. And we note that $\int_{-\infty}^{+\infty} P_k(x, t) dx = 1$ at any time.

At time t, the particle is found with certain probability about a mean value

$$<x(t)> \; = x_0 + pt/m \tag{13.71}$$

which is the same with the position at which a classical particle would be (without any uncertainty), if it was moving along the x-axis with velocity $\upsilon = p/m$ and if it were initially (at $t = 0$) at x_0. In the real world (of quantum mechanics) the particle is found with (not always negligible) probability within a region about $<x(t)>$. The uncertainty $(\Delta x)_t$ in the position of the particle is determined by the extension of this region and it is, according to Eq. (13.70), given by:

$$(\Delta x)_t \approx (\alpha^2 + \hbar^2 t^2/m^2\alpha^2)^{1/2} \tag{13.72}$$

We see that the uncertainty in the position of the particle, at $t = 0$, is $\Delta x \approx \alpha$. We established that $\Delta p \approx \hbar/\alpha$. Therefore at $t = 0$, Δx is determined by the

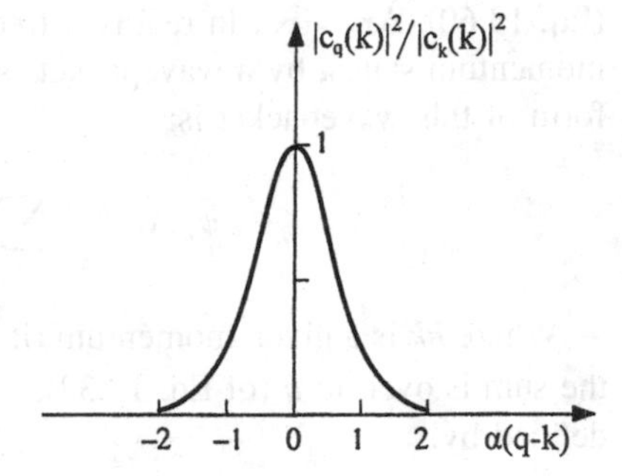

Fig. 13.13 Contribution of different momentum waves to the wavepacket of Eq. (13.69). It does not depend on the time

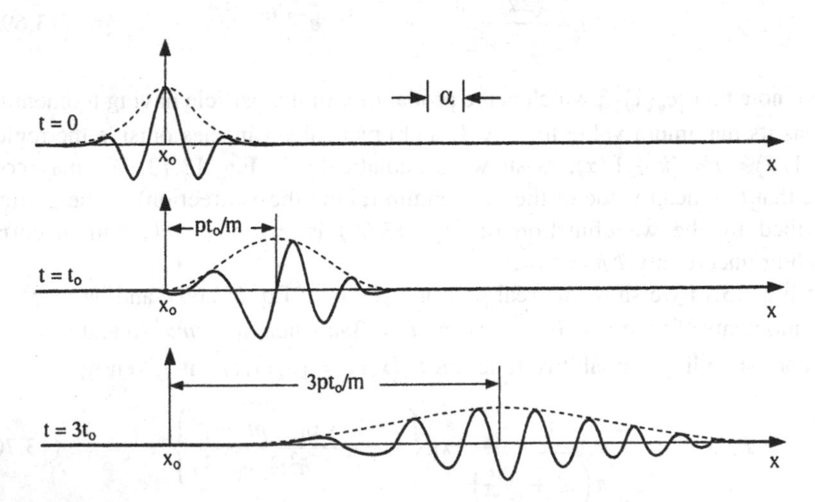

Fig. 13.14 Schematic description of the wavepacket $\psi_k(x,t)$ of Eq. (13.69).The *solid lines* represent the real part of $\psi_k(x,t)$, and the *broken lines* its magnitude $|\psi_k(x,t)|$, at three different times. $t_0 = m\alpha^2/\hbar$ and $\alpha = \pi/k$. [From R. B. Leighton, *loc. cit.*]

Fig. 13.15 Scattering by a rectangular potential barrier: $U(x) = U_0$ for $-a<x<a$, and $U(x) = 0$ for $x < -a$ and for $x > a$. (α) incident wave: $|\psi_i(x,t)|^2$, (β) reflected wave: $|\psi_r(x,t)|^2$, (γ) transmitted wave: $|\psi_t(x,t)|^2$

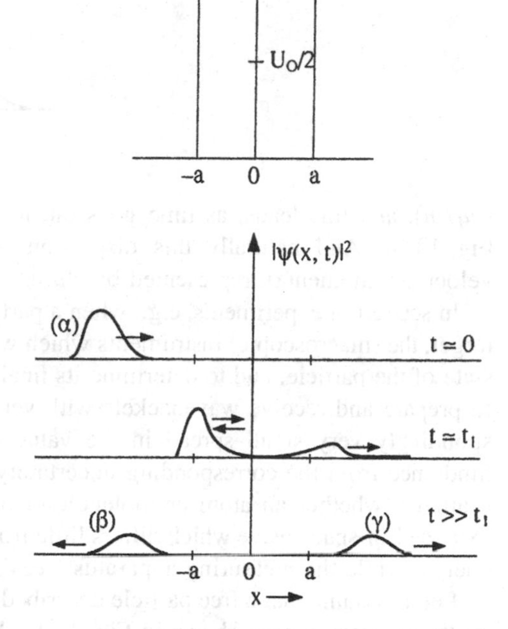

Fig. 13.16 Transmission coefficient for the rectangular barrier of Fig. 13.15 with $U_0 = \hbar^2\pi^2/2ma^2$

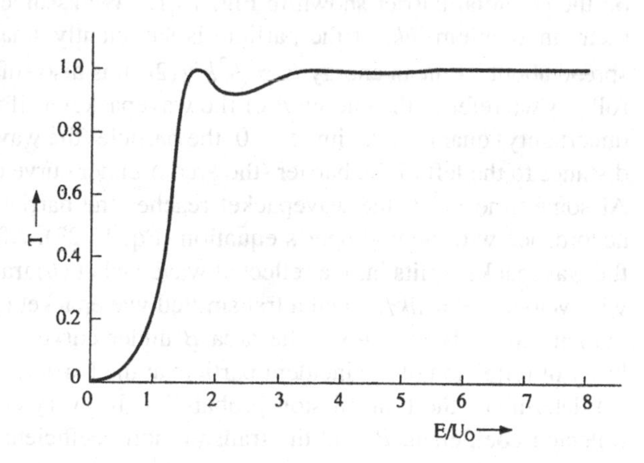

uncertainty principle: $\Delta x \approx \alpha \approx \hbar/\Delta p$. However, as time goes on the second term in the parenthesis of the above equation dominates and we obtain: $(\Delta x)_t \approx \hbar t/m\alpha \approx (\Delta p/m)t$, which tells us that the uncertainty in the position of the particle increases with time and that the spread about its mean value is greater the larger the value of Δp. We can understand this as follows: the various momentum waves that contribute to the wavepacket propagate with different velocities

Fig. 13.17 A potential barrier which varies slowly with the position

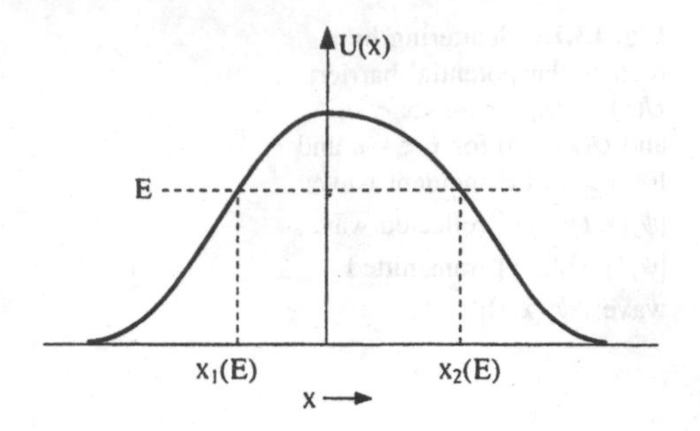

$(\hbar q/m)$, and this leads, as time goes on, to the dispersion of $|\psi_k(x,t)|$, seen in Fig. 13.14. And naturally this dispersion is greater when the spread of the velocities (momenta), represented by Δp, is larger.

In scattering experiments, e.g., when a particle is incident on and scattered by a target, the (macroscopic) instruments which we use to prepare the initial (incident) state of the particle, and to determine its final (after the scattering) state, allow us to prepare and receive wavepackets with very small values of Δp (with a correspondingly very small spread in the value of the kinetic energy) without any hindrance from the corresponding uncertainty in the position of the particle. The scatterer (whether an atom or molecule or an area of a solid surface) 'sees' an extended-in-space wave which differs little from a wave of definite momentum and energy, while the measuring apparatus 'sees' in effect a classical particle.

Let us assume that a free particle described by $\psi_k(x,t)$ of Eq. (13.69) is incident on the potential barrier shown in Fig. 13.15. We assume that the spread about the mean momentum, $\hbar k$, of the particle is sufficiently small and that, therefore, the spread about its mean energy $E = \hbar^2k^2/(2m)$, is also sufficiently small, so in what follows we refer to the energy E of the wavepacket as if it were a definite (without uncertainty) quantity. At time $t = 0$, the particle (the wavepacket) is found at some distance to the left of the barrier (the area A under curve (α) is normalized: $A = 1$). At some time $t \approx t_1$ the wavepacket reaches the barrier and is scattered by it, in accordance with Schrödinger's equation (Eq. 13.27). After the scattering ($t \gg t_1$) the wavepacket splits into: a reflected wavepacket (β) travelling towards $x = -\infty$, with velocity $\upsilon = \hbar k/m$, and a transmitted wavepacket (γ) traveling with the same velocity towards $x = +\infty$. The area B under curve (β) detetermines the probability of reflection of the incident particle at the barrier, and the area C under curve (γ) determines the transmission probability. In every case: $B + C = A = 1$. The reflecton coefficient, R, and the transmission coefficient, T, are defined by:

$$R = \frac{B}{A} \quad \text{and} \quad T = \frac{C}{A} \tag{13.73}$$

And, of course, $R + T = 1$; a particle incident on a potential barrier will either be reflected by it or transmitted through it, it can not be lost.

In Fig 13.16 we show the calculated transmission coefficient through the potential barrier of Fig. 13.15 for a specific value of U_0, as a function of the energy

E of the particle. [A theorem of quantum mechanics allows one to obtain T and R by solving the time-independent Schrödinger equation for a given energy E of the particle by imposing appropriate boundary conditions on the wavefunction for $x \ll -a$ and $x \gg a$; we need not go into the technical details here (see exercise 13.11)].

We note that the transmission coefficient does not vanish when the energy of the incident particle is smaller than the height of the barrier (as would be the case in classical mechanics), and it does not equal unity, except when $E \gg U_0$, when its energy lies above the barrier height (as would be the case in classical mechanics). The transmission of a particle through a potential barrier when its energy lies below the top of the barrier, known as the tunnelling phenomenon, is not restricted to a rectangular barrier; it applies to any shape of barrier, and plays an important role in many phenomena that could not be otherwise explained (we have already cited tunnelling as the cause of field electron emission, in Sect. 11.5.2).

It can be shown that the transmission coefficient, T, of a particle tunnelling through a slowly varying potential barrier, such as the one shown in Fig. 13.17, is given to a good approximation by:

$$T = \exp(-2(\frac{2m}{\hbar^2})^{1/2} \int\limits_{x_1(E)}^{x_2(E)} (U(x) - E)^{1/2} dx) \tag{13.74}$$

where m is the mass of the particle, E is its energy, and $x_1(E)$ and $x_2(E)$ are the classical turning points shown in the figure.

It is worth repeating how one interprets the result shown in Fig. 13.16 in relation to an (real or imagined) experiment: one shoots a particle (of mass m) with energy E on the given barrier; at some distance to the left of the barrier there is a receiver (a sensitive screen) to collect the reflected particle, and at some distance to the right of the barrier there is another receiver (another sensitive screen) to collect the transmitted particle. We repeat the experiment a hundred times (for the same energy of the particle) and we find that in, say, 95 (out of a 100) times the particle was reflected (we know it because on hitting the screen to the left of the barrier the particle left a spot on it), and in 5 (out of the 100) times the particle was transmitted (we know it because on hitting the screen to the right of the barrier the particle left a spot on it). The experimentally determined reflection and transmission coefficients (for the given energy) are: $R = 0.95$ and $T = 0.05$. We repeat the experiment for different energies of the particle. The experimentally determined $T(E)$ agrees with the calculated one of Fig. 13.16. The thing to note is that our measuring instruments record the arrival of a whole particle, the splitting of the incident wave (α) into (β) and (γ), shown in Fig. 13.15 is a theoretical construction that allows us to predict correctly (in the statistical manner described) the result of the measurements. This is how quantum mechanics works.

Consider, for fun, the following possibility: assume that the receiver on the right of the barrier lies much further away from the barrier than the receiver on its left. Then a reflected particle will arrive at the left receiver much sooner than the time it would take a transmitted electron to arrive at the right receiver, in which case the observer at the left receiver, having observed the arrival of the particle, can tell the observer at the right receiver not to wait in vain. The measurement (observation) of (β) of Fig. 13.15 destroys the entire wavefunction including the far away (γ). A person who is not familiar with quantum mechanics will find the above confusing, to say the least. It appears that an abstract entity, the wavefunction, carries all that there is to know about the particle, but the moment we extract concrete information from it (concrete in classical terms) the wavefuction is destroyed, to be replaced by a new wavefunction (after the event of measurement).

13.4.5 Transitions

Consider a particle in a given potential field; it could be an electron moving about a nucleus as in the hydrogen atom or a particle moving in the potential well of Fig. 13.9. We denote the (unperturbed) Hamiltonian of the particle by H^0, and we assume that we already know the stationary states of the particle for this Hamiltonian:

$$H^0 u_n(x) = E_n u_n(x), n = 1, 2, 3, 4, \text{ etc} \tag{13.75}$$

And let us further assume that the particle is initially in the *ith* ($n = i$) stationary state (a bound state):

$$\psi_i = u_i(x) e^{-iE_i t/\hbar} \tag{13.76}$$

The particle will remain in this state for ever if not acted upon by some external force. If the particle is charged (like an electron) that could arise when light is thrown on it. For example when an atom is placed in the field of an EM wave (say visible or ultra-violet light) it will experience a force, due to the electric field of the wave, which varies like $sin(\omega t)$ with the time.[19] To take account of this force we must add to the Hamiltonian a perturbing term $H'(x, t)$. In order to find the effect on the particle of this perturbation, we must obtain the wavefunction, $\psi(x, t)$, of the particle by solving Schrödinger's Eq. (13.28), with $\hat{H} = H^0 + H'$, subject to the condition that $\psi(x, t = 0) = u_i(x)$, in accordance with Eq. (13.76).

Because the functions $u_n(x)$ of Eq. (13.75) are a complete set (any function can be written as a linear sum of the $u_n(x)$), we seek $\psi(x, t)$ in the following form:

$$\psi(x, t) = \sum_n c_n(t) u_n(x) e^{-iE_n t/\hbar} \tag{13.77}$$

---

[19] An alternative way of treating the interaction of an atom with light is presented in Sect. 14.3.3. In both treatments the interaction of the atom with light is treated by first order perturbation theory.

Substituting the above in Schrödinger's equation we obtain (after some algebra that need not concern us here) the following equations for the $c_n(t)$ coefficients:

$$\frac{dc_k}{dt} = -\frac{i}{\hbar}\sum_n c_n(t)H'_{kn}(t)e^{i\omega_{kn}t/\hbar}, k = 1, 2, 3, etc \tag{13.78}$$

where $\omega_{kn} = (E_k - E_n)/\hbar$, and $H'_{kn}(t) = \int_{-\infty}^{+\infty} u_k^*(x)H'(x,t)u_n(x)dx$.

Now, if the perturbation is weak, as it usually is, we can put, to begin with, on the right-hand-side of the above equation: $c_n(t) = 0$ for all n, except for $n = i$ (the *ith* state being the initial state of the system and the one it would stay in if the weak perturbation was not there); we put $c_i(t) = 1$. We can then integrate Eq. (13.78) to obtain the following formula for $c_k(t)$ for any $k \neq i$:

$$c_k(t) = -\frac{i}{\hbar}\int_0^t H'_{ki}(t')e^{i\omega_{ki}t'/\hbar}dt' \text{for } k \neq i \tag{13.79}$$

It follows from Eq. (13.77) that, $|c_k(t)exp(-iE_kt/\hbar)|^2 = |c_k(t)|^2$ is the probability of finding the particle, at time *t*, in a state *k* other than its initial state *i*. One can go beyond the above (first order) approximation, by substituting on the right of Eq. (13.78) the result of the first order approximation for $c_k(t)$, and integrate again to obtain the second order approximation to $c_k(t)$. And one can repeat the process as many times as one wishes to. One expects that the result will soon converge to definite values for the $c_k(t)$. It turns out that in most cases of practical interest we do not need to go beyond the first order approximation of Eq. (13.79).

13.4.6 The Relation Between Classical and Quantum Mechanics

One can show that $<x(t)>$ and $<p_x(t)>$, as defined by Eqs. (13.61) and (13.67), satisfy the equation:

$$\frac{d<x(t)>}{dt} = \frac{<p_x(t)>}{m} \tag{13.80}$$

So that when the uncertainties Δx and Δp_x, defined by Eqs. (13.62) and (13.68), can be ignored, we can replace $<x(t)>$ by $x(t)$, and $<p_x(t)>$ by $p_x(t)$ of classical mechanics (denoting the position and the momentum of the particle, respectively, at time *t*). Moreover, one can show that:

$$\frac{d<p_x(t)>}{dt} = <-\frac{\partial U}{\partial x}> \tag{13.81}$$

where $<-\frac{\partial U}{\partial x}> = \int(-\frac{\partial U}{\partial x})|\psi(x,t)|^2dx$ is the negative of the *mean* value of the gradient of the potential energy $U(x)$ of the particle. Now if we can disregard Δx (which determines the spread of $|\psi(x,t)|^2$ about its mean value $<x(t)>$) we can replace $< -\partial U/\partial x >$ by $-\partial U/\partial x$ at $<x(t)>$). Therefore, when Δx and Δp_x can be ignored, Eq. (13.81) reduces to:

$$\frac{dp_x(t)}{dt} = -\frac{\partial U}{\partial x} \tag{13.82}$$

Which is, of course, Newton's law of motion. We remember that, $-\partial U/\partial x$ equals the force acting on the particle when at $x(t)$. Therefore: when Δx and Δp_x can be ignored, classical mechanics is equivalent to quantum mechanics. This is known as the *correspondence principle*, introduced by Bohr. A good example of the validity of this principle in relation to microscopic particles is provided by Thomson's analysis of the motion of an electron in a vacuum tube under the action of an electric or magnetic field (see Sect. 11.3).[20] And, obviously, when we consider the motion of macroscopic bodies we can ignore the deviations from the classical (mean) values demanded by the uncertainty principle. There is, in principle, a probability that a macroscopic body will be away from where Newton's law tells us it should be, but this is so very small that the actual probability of this happening before the universe comes to an end is zero.

The above is what we expect. When a theory (Newton's theory in our case) is superseded by a new theory, the old theory is contained in the new one as a limiting case. However, classical physics can not be regarded entirely in this way. It is more than that, it must be there, because in interpreting the results of any quantum mechanical calculation, we have to use classical concepts in order to connect with the results of experiments. We do not have an entirely quantum mechanical language with which to describe the physical world.

Exercises

13.1. (a) Prove the Stefan law given by Eq. (13.1): $I(T) = \sigma T^4$; $\sigma = 2\pi^5 k^4/(15c^2h^3)$.

Note:

$$\int_0^\infty \frac{x^3 dx}{e^x - 1} = \frac{\pi^4}{15}$$

(b) Estimate the temperature of the surface of the sun, if the radiated per unit time energy from the sun is 3.74×10^{26}W. The radius of the sun is $6.96 \times 10^8 m$.

13.2. Using the uncertainty principle obtain a rough estimate of the ground-state energy (E_1) of a particle of mass m in the one-dimensional box of Eq. (13.47).

[20] Other examples of seemingly classical trajectories are provided by the tracks of subnuclear particles observed in a Wilson chamber (see, e.g., Fig. 15.2). *Exact* measurements of the positions of the particles would result in points very near but randomly distributed about the shown trajectories.

Hint: Assume that $p_x \geq \Delta p_x$ and $\Delta x \approx a$.
Answer: $E_1 \approx \hbar^2/8ma^2$.

13.3. The state of a particle in the one-dimensional box of Eq. (13.47) at $t = 0$ is given by: $u(x) = \frac{1}{\sqrt{2a}} \sin\frac{\pi x}{a} + \sqrt{\frac{3}{2a}}[\sin\frac{2\pi x}{a}\cos\frac{\pi x}{a} + \cos\frac{2\pi x}{a}\sin\frac{\pi x}{a}]$

Find the wavefunction, *u(x,t)*, which describes the state of the particle for $t > 0$.
Hint: $sin(a+b) = sina\,cosb + cosa\,sinb$
Answer: $u(x,t) = \frac{1}{\sqrt{2a}} \sin\frac{\pi x}{a} \exp(-\frac{E_1 t}{\hbar}) + \sqrt{\frac{3}{2a}} \sin\frac{3\pi x}{a} \exp(-\frac{E_3 t}{\hbar})$;

$$E_n = \frac{\pi^2\hbar^2}{2m}\frac{n^2}{a^2}, n = 1, 2, 3, \ldots$$

13.4. Determine the energy eigenvalues and the corresponding eigenfunctions of the bound states of a particle of mass *m* in a rectangular potential well defined by Eq. (13.56).

Solution: We note (see, e.g., Fig. 13.11)) that the eigenfunctions are either *even* (have *even parity*): $u_n(-x) = u_n(x)$, or *odd (have odd parity)*: $u_n(-x) = -u_n(x)$. The interested reader can show that this property derives from the symmetry of the potential field: $U(-x) = U(x)$, and the fact that the energy eigenvalues of one-dimensional bound states are not *degenerate*: to a given energy eigenvalue corresponds only one eigenstate. Using this property show that the even eigenstates have the form:

$$\begin{aligned} u(x) &= Ae^{\beta x}, \quad x < -a \\ &= C\cos\alpha x, -a < x < a \\ &= Ae^{-\beta x}, \quad x > a \end{aligned} \tag{1}$$

And the odd eigenstates have the form:

$$\begin{aligned} u(x) &= Ae^{\beta x}, \quad x < -a \\ &= C\sin\alpha x, -a < x < a \\ &= -Ae^{-}, \quad x > a \end{aligned} \tag{2}$$

where:

$$\alpha = [2m(E + U_0)]^{1/2}/\hbar; \beta = [2m|E|]^{1/2}/\hbar; -U_0 < E < 0 \tag{3}$$

We remember that bound states of the particle may exist only within the above energy region. The reader will verify that the *u(x)*, given by (1) and (2) above, satisfy the time-independent Schrödinger equation for the given *U(x)* in the regions: $x < -a, -a < x < a, x > a$; and that they are normalizable. However, *u(x)* and *du/dx* must also be continuous everywhere including at $x = \pm a$. The reader will show that this is possible only if:

$$\text{For even states: } \cos\alpha a = (\alpha/\beta)\sin\alpha a \tag{4}$$

$$\text{For odd states: } \sin \alpha a = -(\alpha/\beta)\cos \alpha a \tag{5}$$

The above can be written in a more convenient form as follows:

$$\text{For even states: } \frac{\cos k}{\sin k} = \frac{k}{(b^2 - k^2)^{1/2}} \tag{4'}$$

$$\text{For odd states: } -\frac{\sin k}{\cos k} = \frac{k}{(b^2 - k^2)^{1/2}} \tag{5'}$$

where:

$$k = \alpha a = a[2m(E + U_0)]^{1/2}/\hbar, -U_0 < E < 0 \tag{6}$$

$$b = a[2mU_0]^{1/2}/\hbar; \tag{7}$$

We note that k varies from $k = 0$ (corresponding to $E = -U_0$) to $k = b$ (corresponding to $E = 0$). Accordingly, $k/(b^2 - k^2)^{1/2}$ varies (increases) continuously from *zero* (obtained at $k = 0$) to $+\infty$ (obtained asymptotically as k approaches b). The energy eigenvalues of the even states correspond to those values of k at which a plot of $\cos k/\sin k$ versus k crosses the $k/(b^2 - k^2)^{1/2}$ curve. One can easily verify that$\cos k/\sin k$ decreases continuously from $+\infty$ (at $k = 0$) to *zero* (at $k = \pi/2$), and therefore the two curves will cross at some point, k_1, which determines, through Eq. (6) the ground state energy E_1. We note that $\cos k/\sin k$ is negative for $\pi/2 < k < \pi$, so no crossing with the $k/(b^2 - k^2)^{1/2}$ curve occurs over this region. However, over this region, $-\sin k/\cos k$ decreases continuously from $+\infty$ (at $k = \pi/2$) to *zero* (at $k = \pi$), therefore a crossing between this curve and $k/(b^2 - k^2)^{1/2}$ will occur at some point, k_2, in this region, which determines through Eq. (6) the energy eigenvalue E_2. And then again $\cos k/\sin k$decreases continuously from $+\infty$ (at $k = \pi$) to *zero* (at $k = 3\pi/2$), and its crossing with the $k/(b^2 - k^2)^{1/2}$ curve determines the eigenvalue E_3, and so on until k reaches its maximum possible value: b. The reader should verify that the number (N) of energy levels (eigenvalues) of the particle in a given rectangular potential well (which are obtained in the manner described) is given by:

$$N = \textit{least integer} \geq \frac{2a}{\pi\hbar}(2mU_0)^{1/2} \tag{8}$$

Once an energy level, E_n, has been determined, the values of α and β, to be denoted by α_n and β_n in the corresponding eigenfunction $u_n(x)$ (Eq. (1) or Eq. (2)), are known. The coefficient C is then determined in terms of the coefficient A, and finally A is determined by the normalization of $u_n(x)$.

The reader may wish to apply the above to the rectangular potential well of Fig. 13.9 to verify the results noted in the text.

13.5. Let $u_v(x)$ be a normalized bound (in space) energy-eigenstate of a particle of mass m in a potential field $U(x)$. We have:

$$\widehat{H}u_v(x) = E_v u_v(x); \tag{1}$$

$$\int_{-\infty}^{+\infty} u_v^*(x)u_v(x)dx = 1 \tag{2}$$

$\widehat{H} = \widehat{T} + U(x)$, where $\widehat{T}$ is the kinetic energy operator: $\widehat{T} = -\frac{\hbar^2}{2m}\frac{d^2}{dx^2}$

(a) Show that the mean value of the kinetic energy of the particle in the v_{th} state is obtained by either of the following two formulae:

$$<T>_v = \int_{-\infty}^{+\infty} u_v^*(x)\widehat{T}u_v(x)dx \tag{3}$$

$$<T>_v = \int_{-\infty}^{+\infty} u_v^*(x)(E_v - U(x))u_v(x)dx \tag{4}$$

(b) Write the corresponding formulae for a particle moving in 3-dimensions.

Answer: Eqs. (3) and(4) with: $x \to r$; $\widehat{T} = -\frac{\hbar^2}{2m}\left(\frac{d^2}{dx^2} + \frac{d^2}{dy^2} + \frac{d^2}{dz^2}\right)$, $dx \to dV$ and the integrals extend over all space.

(c) Show (this follows from Eqs. (1) and (2) and their 3-dimensional analogues) that:

$$E_v = \int u_v^*(\boldsymbol{r})\widehat{H}u_v(\boldsymbol{r})dV \tag{5}$$

The above formula suggests that if somehow we can write (guess) to a good approximation the wavefunction $u_v(\boldsymbol{r})$, we can obtain the corresponding energy-eigenvalue through Eq. (5), i.e., we avoid solving the time-independent Schrödinger equation for this purpose.

13.6. Verify the correctness of Eqs. (3) and (4) of the previous exercise, by applying the formulae to the ground state of the particle in the rectangular well of Fig. 13.9.

13.7. Find the energy eigenvalues, and the corresponding eigenstates (normalized eigenfunctions) of a particle of mass m moving in a potential field $U(\boldsymbol{r})$ given by:

$$\begin{aligned} U(x,y,z) &= 0, \text{ when } 0<x<a,\ 0<y<b,\ 0<z<c \\ &= \infty, \end{aligned} \tag{1}$$

when $\boldsymbol{r}$ is anywhere else in space.

We may say that the particle moves in a 3-dimensional box with dimensions a, b, c along the x, y, z directions respectively. We assume that a, b and c are different from each other.

Answer:

$$E_{npl} = \frac{\pi^2\hbar^2}{2m}(\frac{n^2}{a^2}+\frac{p^2}{b^2}+\frac{l^2}{c^2}), n = 1,2,3,\ldots; p = 1,2,3,\ldots; l = 1,2,3,\ldots; \quad (2)$$

$$u_{npl}(x,y,z) = \sqrt{\frac{8}{abc}}\sin\frac{n\pi x}{a}\sin\frac{p\pi y}{b}\sin\frac{l\pi z}{c}, 0<x<a; 0<y<b; 0<z<c \quad (3)$$

The reader will verify that u_{npl} satisfies the time-independent Schrödinger equation:

$$-\frac{\hbar^2}{2m}(\frac{d^2}{dx^2}+\frac{d^2}{dy^2}+\frac{d^2}{dz^2})u_{npl}(x,y,z) = E_{npl}u_{npl}(x,y,z) \quad (4)$$

and the boundary conditions: it vanishes at the 'walls' of the box at: $x = 0$, a; $y = 0$, b; and $z = 0$, c. It is also normalized:

$$\int_0^a (\int_0^b (\int_0^c |u_{npl}(x,y,z)|^2 dz)dy)dx = 1 \quad (5)$$

We note the following: the first (n), second (p), and third (l) term of $E_{n,p,l}$ give the kinetic energy of the particle along the x, y, and z direction respectively. Finally, we remember that the stationary states of the particle in the box are given by:

$$u_{npl}(x,y,z,t) = \sqrt{\frac{8}{abc}}\sin\frac{n\pi x}{a}\sin\frac{p\pi y}{b}\sin\frac{l\pi z}{c}\exp(-iE_{npl}t/\hbar) \quad (6)$$
$$n = 1,2,3,\ldots; p = 1,2,3,\ldots; l = 1,2,3,\ldots$$

13.8. Comment on the energy eigenvalues and corresponding eigenstates of a particle in a box (as defined in the previous exercise): (1) when $b = a$, and (2) when $c = b = a$.

Answer: (1): We note that the two *different* eigenstates (we assume that $n \neq p$):

$$u_{npl}(x,y,z) = \sqrt{\frac{8}{a^2c}}\sin\frac{n\pi x}{a}\sin\frac{p\pi y}{a}\sin\frac{l\pi z}{c} \quad (1)$$

and

$$u_{pnl}(x,y,z) = \sqrt{\frac{8}{a^2c}}\sin\frac{p\pi x}{a}\sin\frac{n\pi y}{a}\sin\frac{l\pi z}{c} \quad (2)$$

have the same energy:

$$E_{npl} = E_{pnl} = \frac{\pi^2\hbar^2}{2m}(\frac{n^2+p^2}{a^2} + \frac{l^2}{c^2}) \tag{3}$$

We say that the given energy eigenvalue $(E_{npl} = E_{pnl})$ is doubly *degenerate* (meaning that there are two different eigenstates corresponding to this energy). The physical picture behind the mathematics is obvious: Assume that $n > p$: in the $u_{npl}(x, y, z)$ state the particle moves faster parallel to the x-direction than parallel to the y-direction, while the opposite is true in the $u_{pnl}(x, y, z)$. And we note that not all energy levels are degenerate. The energy levels E_{nnl}, where n and l are any two positive integers, are obviously not degenerate (expressions (1) and (2) above are identical in this case).

(2) In a cubic box ($a = b = c$): the energy levels E_{lll} are not degenerate; on the other hand, the eigenstates $u_{nll}, u_{lnl}, u_{lln}$, which are given by Eq. (3) of exercise 13.7, with $a = b = c$ and are of course different from each other when $l \neq n$, correspond to the same energy: $E_{nll} = E_{\mathrm{ln}\,l} = E_{l\,\mathrm{ln}} = \frac{\pi^2\hbar^2}{2m}(\frac{n^2+l^2+l^2}{a^2})$ which is, therefore, triply degenerate. Finally we have energy levels with a six-fold degeneracy: the eigenstates: $u_{npl}, u_{nlp}, u_{pnl}, u_{pln}, u_{lnp}, u_{lpn}$, which are different from each other when $n \neq p \neq l \neq n$, correspond to the same energy level:

$$E_{npl} = E_{nlp} = E_{pnl} = E_{pln} = E_{lnp} = E_{lpn} = \frac{\pi^2\hbar^2}{2m}(\frac{n^2+p^2+l^2}{a^2}),$$

It is known that the degeneracy of an energy level may be greater the greater the symmetry of the potential field is in space. We have just seen that in the more symmetrical box ($a = b = c$) the maximum degeneracy (six-fold) is higher than the maximum degeneracy (two-fold) in the less symmetric box ($a = b \neq$ c), which is in turn higher than that of the asymmetric box ($a \neq b \neq$ c) where none of the energy levels is degenerate. It follows from the above that in a spherically symmetric field, we should expect, the maximum degeneracy to be much greater.

13.9. Calculate the first five (in order of increasing energy) energy-eigenvalues of an electron in a 3-dimensional box, when the dimensions of the box are given by (1) and (2) below (in angstroms). Give the results in eV. Note the degeneracy of each energy level and write down the corresponding eigenfuctions.

(1): $a = 1$, $b = 1.5$, $c = 2$

(2): $a = b = c = 1.5$

To begin with, verify that $\hbar^2/2m = 3.82$ eV *angstrom*^{2}

Answer: (1) $E_{111} = 63.86; E_{112} = 92.14; E_{121} = 114.08; E_{113} = 139.26;$ $E_{211} = 176.96$; all are *non-degenerate*. The corresponding eigenfunctions are given by Eq. (3) of exercise 13.7.

(2) $E_{111} = 50.22$ (*non-degenerate*); $E_{112}(= E_{121} = E_{211}) = 100.44$ (*triply-degenerate*); $E_{122}(= E_{212} = E_{221}) = 150.68$ (*triply-degenerate*); $E_{113}(= E_{131} = E_{311}) = 184.14$ (*triply-generate*); $E_{222} = 200.88$ (*non-degenerate*). The corresponding eigenfunctions are given by Eq. (3) of exercise 13.7 with $c = b = a$.

13.10. Quite often it happens that the Hamiltonian, $\widehat{H}$, of a particle can be written as:

$$\widehat{H} = \widehat{H}_0 + V(\boldsymbol{r}) \tag{1}$$

where: $\widehat{H}_0 = -\frac{\hbar^2}{2m}\nabla^2 + U(\boldsymbol{r})$ is referred to as the unperturbed Hamiltonian, and $V(\mathbf{r})$ is a *small* (relative to $U(\mathbf{r})$) addition to the potential field, known as the perturbation term. We assume that the energy eigenvalues, $E_v^0$and the corresponding (normalized) eigenfunctions, $u_v^0(\boldsymbol{r})$, where v stands for an appropriate set of quantum numbers (such as n, p, l for the particle in a box discussed in exercises 13.7 and 13.8), are known:

$$\widehat{H}_0 u_v^0(\boldsymbol{r}) = E_v^0 u_v^0(\boldsymbol{r}) \tag{2}$$

The addition of V to the potential field often makes an exact determination of the eigenvalues and eigenfuncions of the particle impossible. However, the smallness of V allows one to obtain the correction ΔE_v of the energy level E_v^0due to V as follows. (a) LetE_v^0be *non-degenerate*. We can estimate the *first order* correction, ΔE_v, to this energy-eigenvalue due to $V(\mathbf{r})$ from the formula:

$$\Delta E_v = \int u_v^0(\boldsymbol{r})^* V(\boldsymbol{r}) u_v^0(\boldsymbol{r}) dxdydz \tag{3}$$

Therefore, to a first order approximation the energy level in the presence of V is given by:

$$E_v = E_v^0 + \Delta \mathrm{E}_v \tag{4}$$

Equations (3) and (4) follow directly from Eq. (5) of exercise 13.5, if we assume that the unperturbed wavefunction does not change significantly because of V.

Problem: An electron moves in the rectangular potential well of Fig. 13.9. The well is placed in the space of a *weak* constant electric field F in the x-direction, resulting in a perturbation $V = eFx$. Show that to a first order approximation the energy levels of the electron in the given well do *not* change.

(b) LetE_v^0be *doubly-degenerate*. We have:

$$\widehat{H}_0 u_{vi}^0(\boldsymbol{r}) = E_v^0 u_{vi}^0(\boldsymbol{r}), i = 1, 2. \tag{5}$$

However, any linear combination of the two eigenstates satisfies the same equation: $\widehat{H}_0(c_1 u_{v1}^0(\boldsymbol{r}) + c_2 u_{v2}^0(\boldsymbol{r})) = E_v^0(c_1 u_{v1}^0(\boldsymbol{r}) + c_2 u_{v2}^0(\boldsymbol{r}))$; if the eigenfunction is to be normalized we must have: $|c_1|^2 + |c_2|^2 = 1$, but otherwise c_1 and c_2 can be chosen arbitrarily. Therefore, Eq. (3) is applicable *only* if we know what the right combination is. Sometimes symmetry considerations tell us what this combination is, but in general this is not possible. The way to proceed is as follows: we seek an approximate solution of the time-independent Schrödinger equation for the perturbed system in the following manner:

$$(\widehat{H}_0 + V(\boldsymbol{r})) \cdot (c_1 u_{v1}^0(\boldsymbol{r}) + c_2 u_{v2}^0(\boldsymbol{r})) = E_v(c_1 u_{v1}^0(\boldsymbol{r}) + c_2 u_{v2}^0(\boldsymbol{r})) \tag{6}$$

The reader will verify that if we use Eq. (5) we can replace $\widehat{H}_0$ by E_v^0. He must then verify that if we multiply the above equation from the left with $u_{v1}^0(\boldsymbol{r})^*$ and integrate over all space we obtain Eq. (7a), and if we multiply the above equation from the left with $u_{v2}^0(\boldsymbol{r})^*$ and integrate over all space we obtain Eq. (7b):

$$(E_v^0 + V_{11} - E_v)c_1 + V_{12}c_2 = 0 \tag{7a}$$

$$V_{21}c_1 + (E_v^0 + V_{22} - E_v)c_2 = 0 \tag{7b}$$

where: $V_{ij} = \int u_{vi}^0(\boldsymbol{r})^* V(\boldsymbol{r}) u_{vj}^0(\boldsymbol{r}) dV$. We know (see Eqs. A4.8 to A4.10) that a solution of the above equations where at least c_1 or c_2 is different from zero is possible only if the determinant of the coefficients vanishes:

$$\begin{vmatrix} E_v^0 + V_{11} - E_v & V_{12} \\ V_{21} & E_v^0 + V_{22} - E_v \end{vmatrix} = 0 \tag{8}$$

The above constitutes an algebraic equation the solution of which provides us with the required eigenvalues of the energy: $E_v = E_{v1}, E_{v2}$. Putting $E_v = E_{v1}$ in Eqs. (7) we determine c_1 in terms of c_2 in the corresponding eigenfunction, and finally we determine c_2 by normalizing the eigenfunction. And we do the same for $E_v = E_{v2}$.

Example: Assuming that $V_{11} = V_{22} = 0$ and $V_{12} = V_{21} = \upsilon$, show that:

$$E_{v1} = E_v^0 + \upsilon \text{and} E_{v2} = E_v^0 - \upsilon$$

and that the corresponding eigenfunctions are given by:

$$u_{v1}(\boldsymbol{r}) = \frac{1}{\sqrt{2}}(u_{v1}^0(\boldsymbol{r}) + u_{v2}^0(\boldsymbol{r})) \quad \text{and} \quad u_{v2}(\boldsymbol{r}) = \frac{1}{\sqrt{2}}(u_{v1}^0(\boldsymbol{r}) - u_{v2}^0(\boldsymbol{r}))$$

Problem: An electron moves in a 3-dimensional box with dimensions a, $b = a$, c along the x, y, and z directions. A weak perturbation $V(x)$ is added to the potential field of the electron in the box. Show that as a result each of the doubly-degenerate energy levels of the electron splits into two separate levels.

(c) The perturbation treatment we described in relation to doubly-degenerate energy-eigenvalues, can be extended to multiply-degenerate energy-eigenvalues as well. The treatment is a bit more laborious but straightforward. Do it.

13.11. A theorem of quantum mechanics tells us that the scattering of a wavepacket, such as that of Eq. (13.69), with a small spread Δp_x about its mean momentum, $\hbar k$, can be described by solving the following time-independent Schrödinger equation:

$$-\frac{\hbar^2}{2m}\frac{d^2u}{dx^2}+U(x)u(x)=Eu(x) \tag{1}$$

where $E=\hbar^2k^2/2m$ is the energy of the particle (of mass m) incident on the scatterer represented by $U(x)$. We assume that:

$$\begin{aligned} U(x) &= 0, \quad \text{for } x < -a \\ &= \textit{any shape}, \ \text{for } -a < x < a \\ &= 0, \quad \text{for } x > a \end{aligned}$$

If the particle is incident on the scatterer from the left, the wave function must satisfy Eq. (1) and the following boundary conditions:

$$\begin{aligned} u(x) &= e^{ikx}+re^{-ikx}, \quad \text{for } x < -a \\ &= t\,e^{ikx}, \quad \text{for } x > a \end{aligned} \tag{2}$$

If the particle is incident on the scatterer from the right, the wave function must satisfy Eq. (1) and the following boundary conditions:

$$\begin{aligned} u(x) &= e^{-ikx}+re^{ikx}, \quad \text{for } x > a \\ &= t\,e^{-ikx}, \quad \text{for } x < -a \end{aligned} \tag{2'}$$

The first term on the top line of Eq. (2) (and the same applies to (2)′) represents the incident wave (we put its amplitude equal to unity); the second term represents the reflected wave. The term on the second line of the above equations represents the transmitted wave. We note that the above expressions do satisfy Eq. (1) for $x < -a$ and $x > a$. The form of $u(x)$ for $-a < x < a$ will of course depend on the given $U(x)$; and the requirement that both $u(x)$ and du/dx be continuous everywhere including at $x = \pm a$, determines t and r. The transmission coefficient (T) and the reflection coefficient (R), defined by Eq. (13.73), are given by:

$$T=|t|^2 \text{ and } R=|r|^2 \tag{3}$$

(a) Consider a particle, of mass m and energy $E > U_0$, incident on the potential barrier of Fig. 13.15 from the left. (1): write $u(x)$ for $-a < x < a$; (2): obtain the equations expressing the continuity of $u(x)$ and du/dx at the points $x = -a$ and $x = a$. (3): obtain a formula for $T(E)$.

Answer: (1): $u(x) = De^{iqx}+Fe^{-iqx}$, where $q = (2m(E-U_0)/\hbar^2)^{1/2}$.

(2):

$$\begin{aligned} &e^{iqa}D+e^{-iqa}F-te^{ika}=0 \\ &qe^{iqa}D-qe^{-iqa}F-tke^{ika}=0 \\ &-e^{ika}r+e^{-iqa}D+e^{iqa}F=e^{-ika} \\ &ke^{ika}r+qe^{-iqa}D-qe^{iqa}F=ke^{-ika} \end{aligned}$$

The solution of the above system of equations determines the values of D, F, r and t

(3): $T(E) = (1 + \frac{U_0^2 \sin^2([8m(E-U_0)]^{1/2} a/\hbar)}{4E(E-U_0)})^{-1}$

(b) Consider a particle, of mass m and energy $E < U_0$, incident on the potential barrier of Fig. 13.15 from the left. (1): write $u(x)$ for $-a < x < a$; (2): obtain the equations expressing the continuity of $u(x)$ and du/dx at the points $x = -a$ and $x = a$, and hence obtain a formula for $T(E)$.

Answer: (1): $u(x) = De^{-qx} + Fe^{qx}$, where $q = (2m|E - U_0|/\hbar^2)^{1/2}$.

(2): $T(E)$is given by the same formula as for $E > U_0$: shown in (3) above. We note that: according to Eq. (A2.14), when $\theta > 0$, we have:

$$\sin(\sqrt{-\theta}) = \sin(i\sqrt{\theta}) = (e^{-\sqrt{\theta}} - e^{\sqrt{\theta}})/2i$$

(c) Verify the graph shown in Fig. 13.16.

(d) Finally, show that the same transmission coefficient, $T(E)$, is obtained when the particle is incident on the given potential barrier from the right.

13.12. A particle of mass m and energy E is incident from the left on the step-barrier defined by:

$$\begin{aligned} U(x) &= 0, \quad \text{for } x < 0 \\ &= U_0, \quad \text{for } x > 0 \end{aligned}$$

(a) Assume that $E > U_0$. Proceeding as in the previous exercise show that the scattering of the particle by the step barrier is described by the wavefunction:

$$\begin{aligned} u(x) &= e^{ikx} + re^{-ikx}, \text{for} x < 0 \\ &= te^{iqx}, \qquad \text{for } x > 0 \end{aligned}$$

where $k = (2mE/\hbar^2)^{1/2}$ and $q = (2m[E - U_0]/\hbar^2)^{1/2}$.

(b) Show that r and t, determined by the continuity of $u(x)$ and du/dx at $x = 0$, are given by: $t = \frac{2k}{k+q}$ and $r = \frac{k-q}{k+q}$

(c) Evaluate the reflection coefficient $R(E) = |r(E)|^2$ and the transmission coefficient $T(E) = |t(E)|^2(q/k)$, where the factor (q/k) takes into account the fact that the velocity $(hq/m)$of the transmitted particle is different from the velocity $(\hbar k/m)$of the incident one. Verify that: $R(E) + T(E) = 1$.

(d) Assume that $E < U_0$. Show that, in this case the scattering of the particle by the step barrier is described by the wavefunction:

$$\begin{aligned} u(x) &= e^{ikx} + re^{-ikx}, \quad \text{for } x < 0 \\ &= te^{-q'x}, \qquad \text{for } x > 0 \end{aligned}$$

where $k = (2mE/\hbar^2)^{1/2}$ and $q' = (2m[U_0 - E]/\hbar^2)^{1/2}$; and that r and t, determined by the continuity of $u(x)$ and du/dx at $x = 0$, are given by: $t = \frac{2k}{k+iq'}$ and $r = \frac{k-iq'}{k+iq'}$.

In the present case there is obviously no transmitted wave since $u(x) \to 0$ as $x \to \infty$. Therefore we expect the reflection coefficient $R(E) = |r(E)|^2$ to be unity. Verify that this is indeed so.

13.13. Evaluate the transmission coefficient, $T(E_F)$, of an electron incident from the left, with energy E_F, on the potential barrier:

$$U(x) = 0, \text{for} x < 0$$
$$= E_F + \varphi - eFx, \text{for } x > 0$$

Note: This is the potential barrier shown in Fig. 11.10, when the image potential correction is neglected.

Hint: According to Eq. (13.74) we have:

$$T(E_F) = \exp(-2\left(\frac{2m}{\hbar^2}\right)^{1/2} \int_0^{x_2} (\varphi - eFx)^{1/2} dx$$

where $x_2 = \varphi/eF$. To evaluate the integral, put: $w = \varphi - eFx; dw = -eFdx$.

Answer: $T(E_F) = \exp\left(-\frac{4}{3}\left(\frac{2m}{\hbar^2}\right)^{1/2} \frac{\varphi^{3/2}}{eF}\right)$.

Note: The image correction modifies the above formula by multiplying the exponent with $\alpha \approx 3/3$.

Chapter 14
Atoms, Molecules and Solids

14.1 The One-Electron Atom

When dealing with systems of spherical symmetry, as in the case of a particle moving in a potential energy field, $U(r)$, which depends only on the distance, r, from a centre (which we assume to be at the centre of coordinates), it is convenient to use, instead of the familiar to us Cartesian coordinates (x, y, z), a set of spherical coordinates (r, θ, φ) to describe the position of the particle. These are introduced in Fig. 14.1.

14.1.1 Orbital Angular Momentum

Classically, when a particle moves in a central (spherically symmetric) field $U(r)$, its angular momentum, defined by Eq. (7.1), is a constant of the motion; the force corresponding to $U(r)$ is in the direction of $\boldsymbol{r}$ and therefore the time derivative of the angular momentum [Eq. (7.3)] vanishes. As we shall see the same is true in quantum mechanics.

According to what we have said in the previous chapter (see Sect. 13.4.3), we need, to begin with, the quantum mechanical operators $\widehat{l}_x$, $\widehat{l}_y$, $\widehat{l}_z$ for the three components of the orbital angular momentum, $\boldsymbol{l} = \boldsymbol{r} \times \boldsymbol{p}$, of the particle. The classical expressions for them are [see Eq. (7.2)]:

$$l_x = yp_z - zp_y, \quad l_y = zp_x - xp_z, \quad l_z = xp_y - yp_x \tag{14.1}$$

We obtain the corresponding quantum mechanical operators by replacing (p_x, p_y, p_z) by their respective operators:

$$\widehat{l}_x = y\widehat{p}_z - z\widehat{p}_y, \quad \widehat{l}_y = z\widehat{p}_x - x\widehat{p}_z, \quad \widehat{l}_z = x\widehat{p}_y - y\widehat{p}_x \tag{14.2}$$

$$(\widehat{p}_x = \frac{\hbar}{i}\frac{\partial}{\partial x}, \widehat{p}_y = \frac{\hbar}{i}\frac{\partial}{\partial y}, \widehat{p}_z = \frac{\hbar}{i}\frac{\partial}{\partial z})$$

A. Modinos, *From Aristotle to Schrödinger*,
Undergraduate Lecture Notes in Physics, DOI: 10.1007/978-3-319-00750-2_14,

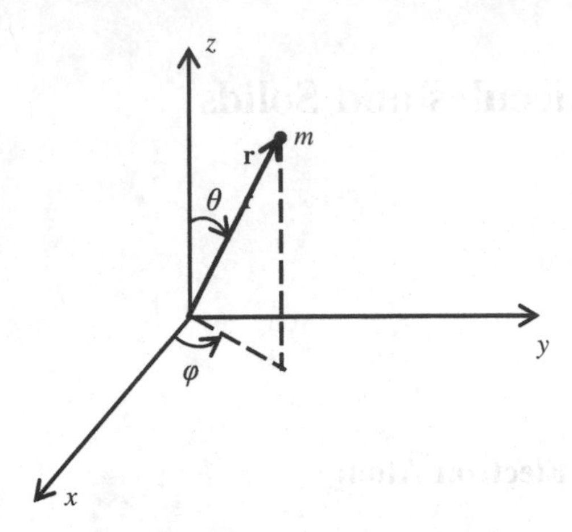

Fig. 14.1 Spherical coordinates: The position vector $\boldsymbol{r}$ of the particle (m) is determined by: its distance r from the origin of the coordinates: $0 \le r < \infty$, the polar angle θ between the z-axis and $\boldsymbol{r}$: $0 \le \theta \le \pi$; and the azimuthal angle φ between the x-axis and the projection of $\boldsymbol{r}$ on the xy-plane: $0 \le \varphi < 2\pi$. The spherical coordinates are related to the Cartesian coordinates as follows: $x = r\,\sin\theta\,\cos\varphi$, $y = r\,\sin\theta\,\sin\varphi$, $z = r\,\cos\theta$

It follows from our discussion of Sect. 13.4.3 that a state of a physical system [e.g., a particle in $U(r)$] will A stands for: correspond to a definite value of a physical observable (it could be the energy, or the momentum, or a component of the orbital angular momentum of the particle, or some other observable) *if* the state of the particle is an *eigenstate* of the corresponding operator corresponding to an eigenvalue of the given observable. And it follows that two different observables will have definite values only if the state of the system is an eigenstate of both observables, which is possible only if the two observables have a *common* set of eigenstates. We have seen, for example, that the energy of a free particle and its momentum have a common set of eigenstates [the plane waves of Eq. (13.39)], and therefore a free particle described by one of these plane waves has definite values for both the energy and the momentum: on measuring one or the other of these quantities, we shall definitely (without any uncertainty) obtain the corresponding eigenvalue.

It turns out that the stationary states of a particle in a spherically symmetric field, $U(r)$, can be chosen so that they are also eigenstates of the square of the orbital angular momentum, $\widehat{l}^{2} = l_x^2 + l_y^2 + l_z^2$ and of one of its components (but only one of its components). And because of the way the spherical coordinates have been defined, it is convenient to choose the z-component. Another way of saying the above is: we can know without any uncertainty the magnitude of the angular momentum of a particle rotating about an axis, and the angle that this axis makes with the z-axis, but the projection of the axis on the xy-plane can take any value according to some probability distribution.

[An important theorem of quantum mechanics states that two operators, $\hat{A}$ and $\hat{B}$, have a common set of eigenstates only if they *commute*, i.e. if $\left[\hat{A}, \hat{B}\right] \equiv \hat{A}\hat{B} - \hat{B}\hat{A} = 0$. This means that for any function $f(r)$, we obtain: $\hat{A}\left(\hat{B} f(r)\right) = \hat{B}\left(\hat{A} f((\mathbf{r}))\right)$, which is not always true. One can also show that the reverse theorem is also true: if $\hat{A}$ and $\hat{B}$ have a common set of eigenstates, they *commute*.

Now, $\hat{l}_x, \hat{l}_y, \hat{l}_z$ and $\hat{l}^2 = \hat{l}_x^2 + \hat{l}_y^2 + \hat{l}_z^2$ satisfy the following commutation relations:

$$\begin{gathered} \left[\hat{l}^2, \hat{l}_x\right] = 0, \quad \left[\hat{l}^2, \hat{l}_y\right] = 0, \quad \left[\hat{l}^2, \hat{l}_z\right] = 0 \\ \left[\hat{l}_x, \hat{l}_y\right] = i\hbar \hat{l}_z, \quad \left[\hat{l}_y, \hat{l}_z\right] = i\hbar \hat{l}_x, \quad \left[\hat{l}_z, \hat{l}_x\right] = i\hbar \hat{l}_y \end{gathered} \tag{14.3}$$

Therefore $\hat{l}^2$ and just one component $\left(\hat{l}_z\right)$ do have a common set of eigenfunctions (eigenstates).

There is more information to be had from the commutation relations (14.3). A most remarkable theorem of quantum mechanics states that: given any set of (angular momentum) operators: $\hat{J}_x, \hat{J}_y, \hat{J}_z$ and $\hat{J}^2 = \hat{J}_x^2 + \hat{J}_y^2 + \hat{J}_z^2$, which satisfy (14.3) with l replaced by J, we can be certain that the eigenvalues of $\hat{J}^2$ will be given by:

$$j(j+1)\,\hbar^2, \tag{14.4a}$$

where j is an integer ($j = 0, 1, 2, \ldots$) or half-integer ($j = 1/2, 3/2, 5/2, \ldots$).

And that for given j, the z-component of the angular momentum takes the values (eigenvalues of $\hat{J}_z$):

$$m_j\hbar, \text{ where } m_j = -j, \ -j+1, \ldots, j-1, \ j. \tag{14.4b}$$

We note that for given j, the quantum number m_j, takes (in all) $2j + 1$ values].

In what follows we shall need to know $\hat{l}^2$ and $\hat{l}_z$ in spherical coordinates. These are:

$$\hat{l}^2 = -\hbar^2\left[\frac{\partial^2}{\partial\theta^2} + \frac{\cos\theta}{\sin\theta}\frac{\partial}{\partial\theta} + \frac{1}{\sin^2\theta}\frac{\partial^2}{\partial\phi^2}\right], \quad \hat{l}_z = \frac{\hbar}{i}\frac{\partial}{\partial\phi}, \tag{14.5}$$

The eigenstates of $\widehat{l}^2$ and $\widehat{l}_z$ will have the form: $R(r)\,f(\theta, \varphi)$, where $R(r)$ is any function of r, since these operators do not involve r. Moreover, the simple form of $\widehat{l}_z$ allows us to write: $f(\theta, \varphi) = X(\theta)\,\Phi(\varphi)$. In which case the eigenvalue equation for $\widehat{l}_z$ reduces to:

$$\frac{\hbar}{i}\frac{\partial \Phi_m}{\partial \phi} = m\hbar\Phi_m(\phi) \tag{14.6}$$

The eigenvalue, $m\hbar$, and $\Phi_m(\varphi)$, are obtained by noting that $\Phi_m(\varphi) = A exp(im\varphi)$, with A and m constants, satisfies the above equation. However, if $\Phi_m(\varphi)$ is to be a single-valued function of the position, we must have: $\Phi_m(\varphi + 2\pi) = \Phi_m(\varphi)$, which means that m must be an integer or zero. Therefore the (normalized) eigenstates of $\widehat{l}_z$ are:

$\Phi_m(\phi) = \frac{e^{im\varphi}}{\sqrt{2\pi}}$ corresponding to the eigenvalue: $m\hbar$

$$m = 0,\ \pm 1,\ \pm 2,\ \pm 3, \tag{14.7}$$

We note that: $\int_0^{2\pi} |\Phi_m(\phi)|^2 d\phi = 1$.

We must next determine $X(\theta)$ so that $f(\theta.\varphi) = X(\theta)\Phi_m(\varphi)$ is an eigenstate of $\widehat{l}^2$ as well. We must have:

$$\widehat{l}^2\, X(\theta)\Phi_m(\varphi) = \hbar^2 l(l+1)\, X(\theta)\Phi_m(\varphi) \tag{14.8}$$

where, without any loss of generality, we have written the eigenvalues of $\widehat{l}^2$ in the form: $\hbar^2 l(l+1)$. Writing $\widehat{l}^2$ explicitly [Eq. (14.5)], and using Eq. (14.7), one obtains the following eigenvalue equation for $X(\theta)$:

$$\frac{d^2X}{d\theta^2} + \frac{\cos\theta}{\sin\theta}\frac{dX}{d\theta} - \frac{m^2}{\sin^2\theta}X + l(l+1)X = 0 \tag{14.9}$$

Fortunately, the above equation and its solutions were studied by the French mathematician Adrien-Marie Legendre in the 1780s. He established that the above equation has finite, continuous and single-valued solutions only for *positive integer values of* $l \geq |m|$. These are denoted by:

$$X_{lm}(\theta) = N_{lm}P_l^m(\cos\theta) \tag{14.10}$$

where N_{lm} are normalization constants, and P_l^m are the so-called Legendre polynomials defined by:

$$P_l^m(\cos\theta) = \frac{1}{2^l l!}(\sin\theta)^m \frac{d^{l+m}}{(d\cos\theta)^{l+m}}(\cos^2\theta) - 1)^l \tag{14.11}$$

where $\cos^2\theta$ stands for $(\cos\theta)^2$; and we remember [see Eq. (5.42)] that $d^n f/dx^n$ is the nth derivative of $f(x)$. Because Eq. (14.9) depends on m^2 (not on m): $P_l^{-m}(\cos\theta) = P_l^m(\cos\theta)$.

It is customary to denote the eigenstates of $\widehat{l}^2$ and $\widehat{l}_z$ as follows:

$$Y_{lm}(\theta,\varphi) = X_{lm}(\theta)\Phi_m(\varphi) \tag{14.12}$$

$$l = 0, 1, 2, 3, \ldots; m = -l, -l+1, \ldots, l-1, l$$

$Y_{lm}(\theta,\varphi)$ are known as the spherical harmonics. We have:

$$\widehat{l}^2 Y_{lm}(\theta,\varphi) = \hbar^2 l(l+1) Y_{lm}(\theta,\varphi)$$

$$\widehat{l}_z Y_{lm}(\theta,\varphi) = m\hbar Y_{lm}(\theta,\varphi), \tag{14.13}$$

It is worth remembering that for given l (a positive integer or zero), the quantum number m takes in all $2l + 1$ values (as shown above).

The following are examples of spherical harmonics:

$$Y_{00} = (1/4\pi)^{1/2}$$

$$Y_{1-1} = (3/8\pi)^{1/2} sin\theta e^{-i\varphi}, Y_{10} = (6/8\pi)^{1/2} cos\theta, Y_{11} = -(3/8\pi)^{1/2} sin\theta e^{i\varphi}$$

$$Y_{2-2} = (15/32\pi)^{1/2} sin^2\theta e^{-2i\varphi}, Y_{2-1} = (15/8\pi)^{1/2} sin\theta cos\theta e^{-i\varphi},$$

$$Y_{20} = (10/32\pi)^{1/2}(3cos^2\theta - 1), Y_{21} = -(15/8\pi)^{1/2} sin\theta cos\theta e^{i\varphi}$$

$$Y_{22} = (15/32\pi)^{1/2} sin^2\theta e^{2i\varphi} \tag{14.14}$$

It follows from Eqs. (14.12) and (14.7) that $|Y_{lm}(\theta,\varphi)|^2 = (N_{lm}P_l^m(\cos\theta))^2/2\pi$ does not depend on the angle φ. In Fig. 14.2 we show schematically the variation

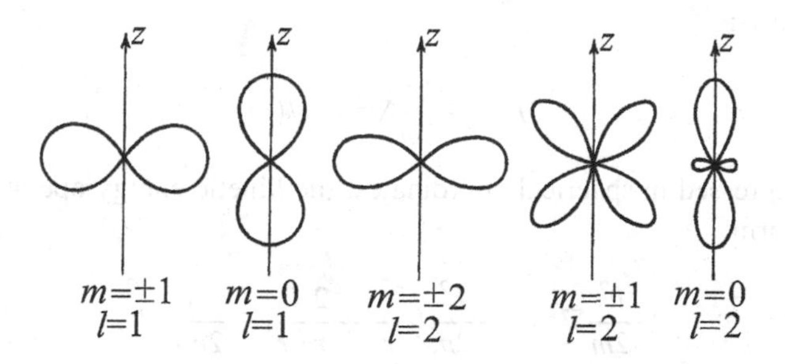

Fig. 14.2 Variation of $(P_l^m(\cos\theta))^2$ with the angle θ in polar form: imagine a vector from the origin to a point on the diagram, the angle this vector makes with the z-axis defines the angle θ; the length of the vector measures the magnitude of $(P_l^m(\cos\theta))^2$

of $(P_l^m(\cos\theta))^2$ with the angle θ. In what follows we shall see that such diagrams help us visualize the distribution of the electronic cloud, so called, about the nucleus of an atom.

The spherical harmonics have the following important properties. They are an orthogonal set of functions:

$$\int_0^{2\pi}\int_0^{\pi} Y^*_{l'm'}(\theta,\phi)Y_{lm}(\theta,\phi)\sin\theta\, d\theta\, d\phi = 1, \text{ if } l = l' \text{ and } m = m' \\ = 0, \text{ otherwise} \tag{14.15}$$

where $\sin\theta\, d\theta\, d\varphi$ is the element of solid angle about the direction (θ, φ) (see Fig. 14.3). They are also a complete set of functions, i.e. any function $f(\theta, \varphi)$, where $0 \le \theta \le \pi$ and $0 \le \varphi \le 2\pi$, can be written as a sum of spherical harmonics:

$$f(\theta,\varphi) = \sum_{l=0}^{\infty}\sum_{m=-l}^{l} c_{lm}Y_{lm}(\theta,\phi); \quad c_{lm} = \int_0^{2\pi}\int_0^{\pi} Y^*_{lm}(\theta,\phi)f(\theta,\phi)\sin\theta d\theta d\phi \tag{14.16}$$

Finally, they satisfy the following relation:

$$\sum_{m=-l}^{l} |Y_{lm}(\theta,\varphi)|^2 = (2l+1)/4\pi \tag{14.17}$$

14.1.2 Stationary States of a Particle in a Central Field

The stationary states (energy eigenstates) of a particle of mass m moving in a central field, $U(r)$, are determined by Eqs. (13.43) and (13.44). The corresponding Hamiltonian is:

$$H = -\frac{\hbar^2}{2m}\nabla^2 + U(r) \tag{14.18}$$

When expressed in spherical coordinates, the kinetic energy operator takes the following form:

$$-\frac{\hbar^2}{2m}\nabla^2 = -\frac{\hbar^2}{2m}\left(\frac{\partial^2}{\partial r^2} + \frac{2}{r}\frac{\partial}{\partial r}\right) + \frac{1}{2mr^2}\hat{l}^2 \tag{14.19}$$

where $\hat{l}^2$ is the angular momentum operator of Eq. (14.5). It is obvious that $\hat{l}^2$ and $\hat{l}_z$ commute with H and, therefore [see text preceding Eq. (14.3)] the stationary

Fig. 14.3 Surface element dS (on the *spherical* surface of radius r) and volume element dV at the point P (with spherical coordinates r,θ,φ); $d\Omega \equiv \sin\theta\, d\theta\, d\varphi$ is the element of the solid angle subtended to dS

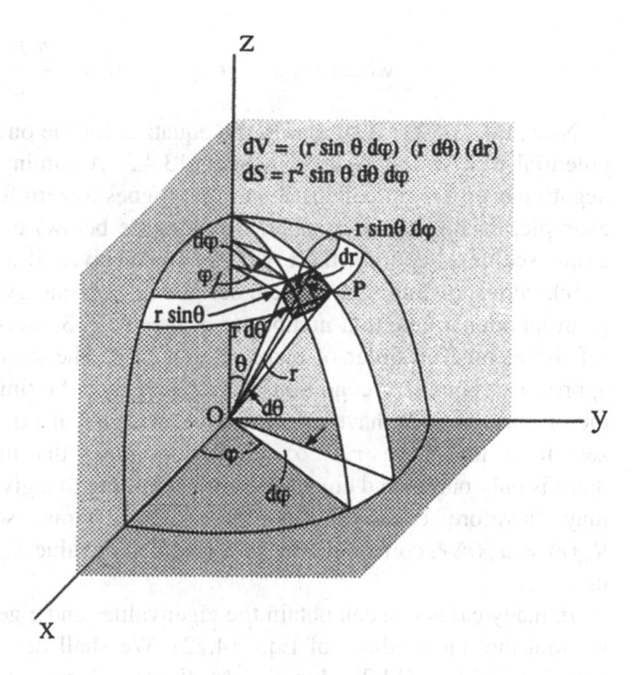

states of the particle are eigenstates of $\hat{l}^2$ and $\hat{l}_z$ as well as of $\hat{H}$. We seek the corresponding eigenfunctions in the form: $R(r)Y_{lm}(\theta,\varphi)$, in which case Eq. (13.44) becomes:

$$\left\{-\frac{\hbar^2}{2m}\left(\frac{\partial^2}{\partial r^2}+\frac{2}{r}\frac{\partial}{\partial r}\right)+\frac{1}{2mr^2}\hat{l}^2+U(r)\right\}R(r)Y_{lm}(\theta,\varphi)=ER(r)Y_{lm}(\theta,\varphi) \quad (14.20)$$

which, on using Eq. (14.13), reduces to an eigenvalue-equation for $R(r)$:

$$-\frac{\hbar^2}{2m}\left(\frac{d^2R}{dr^2}+\frac{2}{r}\frac{dR}{dr}\right)+\left(\frac{\hbar^2}{2m}\frac{l(l+1)}{r^2}+U(r)\right)R(r)=ER(r) \quad (14.21)$$

for given l. We demand, of course, that $R(r)$ be continuous with a continuous derivative and, in the case of bound states, it should be squarely integrable [see Eq. (14.24a)].

[We can transform Eq. (14.21) into a more familiar form by putting:

$$R(r)=u(r)/r \quad (14.22)$$

Substituting the above in Eq. (14.21) we obtain the following equation for $u(r)$:

$$-\frac{\hbar^2}{2m}\frac{d^2u}{dr^2}+U_l(r)u(r)=Eu(r) \quad (14.23)$$

$$\text{where}: \qquad U_l(r) = U(r) + \frac{\hbar^2 l(l+1)}{2mr^2} \qquad (14.23\text{a})$$

Now, Eq. (14.23) is similar to the equation for the one-dimensional rectangular potential well we considered in Sect. 13.4.2. Assuming that $U_l(r)$ is sufficiently negative over a region of space and that it goes to zero for large values of r as, for example, in the case of a spherical well (see below) or the Coulomb field of an atom (see next section), there will be (for a given l) a number of bound states, $u_{nl}(r)$, corresponding to a discrete set of negative energy eigenvalues E_{nl}, where n is an additional quantum number (e.g. $n = 1, 2, 3$, etc.) denoting the eigenstates (of the given l) in order of increasing energy. The $u_{nl}(r)$ satisfy Eq. (14.23) and appropriate boundary conditions: since $R(r)$ must be finite everywhere (including the origin), we must have $u(0) = 0$; and because of Eq. (14.24a), $u(r)$ must go to zero faster than $1/r$ as $r \to \infty$. Finally, we know that in one-dimensional motion there is only one bound eigenstate corresponding to a given energy eigenvalue; we may therefore conclude that there is only one solution of Eq. (14.21), $R_{nl}(r) = u_{nl}(r)/r$, corresponding to a given eigenvalue E_{nl}].for a one-electron atom in

In many cases one can obtain the eigenvalues and eigenfunctions of Eq. (14.20) without the intermediary of Eq. (14.22). We shall denote the normalized eigenfunctions of Eq. (14.20) by: $\psi_{nlm}(r, \theta, \varphi)$, where n is the additional quantum number, required to specify the radial part of the eigenfunction, $R_{nl}(r)$, and the corresponding eigenvalue, E_{nl}.

In summary: The stationary states of the particle described by the Hamiltonian of Eq. (14.18) are given by:

$$\begin{aligned} \psi_{nlm}(\mathbf{r}, t) &= \psi_{nlm}(r, \theta, \varphi) exp(-iE_{nl}t/\hbar) \\ \psi_{nlm}(r, \theta, \varphi) &= R_{nl}(r) Y_{lm}(\theta, \varphi) \\ \hat{H}\psi_{nlm}(r, \theta, \varphi) &= E_{nl}\psi_{nlm}(r, \theta, \varphi) \\ \int |\psi_{nlm}(r, \theta, \phi)|^2 dV &= 1 \end{aligned} \qquad (14.24)$$

The integration in the last equation is over all space [dV is defined in Fig. (14.3)]. When use is made of Eq. (14.15), the last equation reduces to:

$$\int_0^\infty (R_{nl}(r))^2 r^2 dr = 1 \qquad (14.24\text{a})$$

We note that the radial part of the eigenfunction and the energy eigenvalue do not depend on the quantum number m. We should expect that much: the energy eigenvalues of the particle in the given field can *not* depend on the choice (arbitrary) of the z-direction. Now since for a given l (which may be zero or an integer) m takes $2l + 1$ values [see Eq. (14.12)], there are $2l + 1$ eigenstates corresponding to a given E_{nl}. We say that E_{nl} has a degeneracy of $2l + 1$.

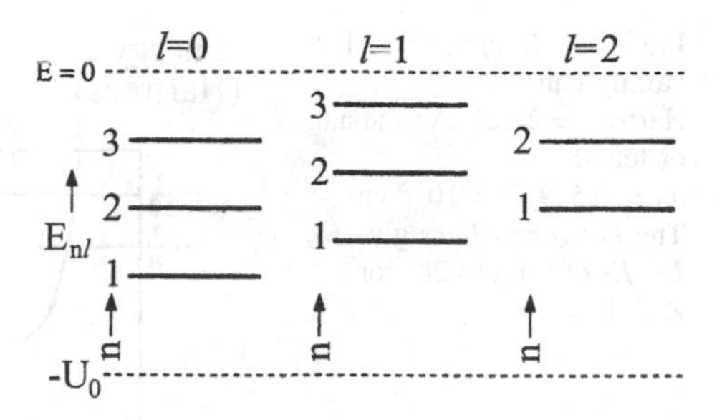

Fig. 14.4 Bound energy levels of a particle in a *spherical* well: $U(r) = -U_0$ for $0 < r < a$ and $U(r) = 0$ for $r > a$

In Fig. 14.4 we show schematically the bound energy levels of a particle in a finite spherical well. As the depth, U_0, of the well increases, we obtain more bound states for the shown l, and in addition we obtain bound states corresponding to higher values of the angular momentum ($l > 2$). Similarly, when the depth of the well decreases we obtain fewer bound states. The situation is similar to what we found in relation to the one-dimensional potential well. It is also worth noting that the energy levels of given l are quite different from those of another l. This is generally so, except for the case of the Coulomb field of a one-electron atom to be discussed in the next section.

14.1.3 Stationary States of the Hydrogen Atom

The Hamiltonian $\widehat{H}$ for a one-electron atom in the centre-of-mass system of coordinates is[1]:

$$\widehat{H} = -\frac{\hbar^2}{2m_r}\nabla^2 + U(r) \tag{14.25}$$

$$U(r) = -\frac{Ze^2}{4\pi\varepsilon_0 r} \tag{14.26}$$

We note that m_r in the kinetic energy term of the Hamiltonian, is the reduced mass of the atom defined by:

$$m_r = \frac{mM_N}{m + M_N} \tag{14.27}$$

[1] Consider a system of two particles of masses m_1 and m_2 at $\boldsymbol{r}_1(t)$ and $\boldsymbol{r}_2(t)$ respectively. Their centre of mass (*CM*) is at $\boldsymbol{R}(t) = (m_1/M)\boldsymbol{r}_1 + (m_2/M)\boldsymbol{r}_2$, where $M = m_1 + m_2$. The separation between the two particles is defined by $\boldsymbol{r} = \boldsymbol{r}_2 - \boldsymbol{r}_1$. In the *CM* system of coordinates $\boldsymbol{R}(t) = 0$, and therefore: $\boldsymbol{r}_1(t) = -(m_2/M)\boldsymbol{r}(t)$ and $\boldsymbol{r}_2(t) = (m_1/M)\boldsymbol{r}(t)$. Therefore the kinetic energy of the two particles in the *CM* system is: $m_1(d\ \boldsymbol{r}_1/dt)^2/2 + m_2(d\boldsymbol{r}_2/dt)^2/2 = m_r\upsilon^2/2$, where $\upsilon = d\boldsymbol{r}/dt$ and $m_r = m_1m_2/(m_1 + m_2)$.

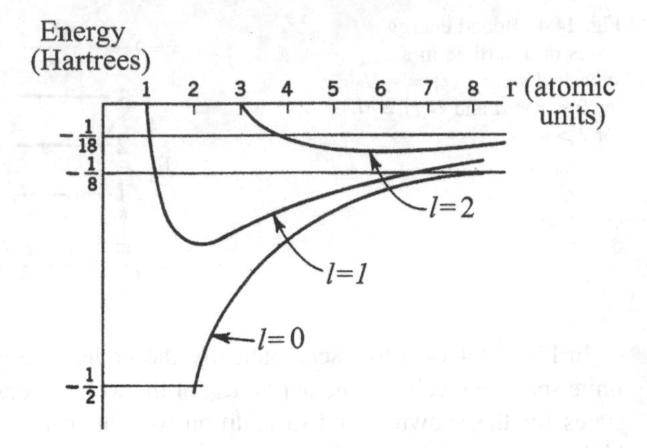

Fig. 14.5 $U_l(r)$ for $Z = 1$ in atomic units: 1 Hartree = 27.21 eV; the a.u. of length is $a_0 = 0.52917 \times 10^{-8}$ cm. The *horizontal lines* give E_1, E_2, E_3 of Eq. (14.28) for $Z = 1$

where m is the mass of the electron and M_N is the mass of the nucleus. We note that $m_r = m(1 + m/M_N) \approx m$, because $m<$. In relation to $U(r)$: Ze is the (positive) charge of the nucleus, and $-e$ is as usual the charge of an electron.

Figure 14.5 shows the effective potential energy: $U_l(r) = -e^2/(4\pi\varepsilon_0 r) + \hbar^2 l(l+1)/(2m_r r^2)$ for the hydrogen atom ($Z = 1$). It turns out that the stationary states of given l, defined by Eq. (14.24), of an electron in the field of Eq. (14.26) have energy levels, E_{nl}, given by:

$$E_n = -\frac{m_r Z^2 e^4}{32\pi^2 \varepsilon_0^2 \hbar^2 n^2}, \quad n = l+1,\ l+2,\ l+3, \ldots \tag{14.28}$$

The energy levels of the hydrogen atom are obtained for $Z = 1$. We note that, when $Z = 1$, the above is identical with Bohr's formula [Eq. (13.26)], which as we have stated already describes well the experimental data. And the same is true for ionized helium ($Z = 2$).

The $R_{nl}(r)$ of the corresponding eigenfunctions [see Eq. (14.24)] are described in terms of the so-called *associated Laguerre polynomials*. In Fig. 14.6 we show schematically some $R_{nl}(r)$ and in the figure caption we give their analytic expressions. We need no more for our purposes.

It is customary to call the eigenstates of the atom with $l = 0$, *s-states*; those with $l = 1$ we call *p-states*; those with $l = 2$ are called *d-states*; those with $l = 3$ are called *f-states*, etc. An eigenstate of given quantum numbers n and l is accordingly denoted as follows:

$$\begin{array}{ll} 1s,\ 2s,\ 3s,\ \ldots & \text{if } l = 0 \text{ and } n = 1,\ 2,\ 3, \ldots \\ 2p,\ 3p,\ 4p,\ \ldots & \text{if } l = 1 \text{ and } n = 2,\ 3,\ 4,\ \ldots \\ 3d,\ 4d,\ 5d,\ \ldots & \text{if } l = 2 \text{ and } n = 3,\ 4,\ 5,\ \ldots \\ 4f,\ 5f,\ 6f,\ \ldots & \text{if } l = 3 \text{ and } n = 4,\ 5,\ 6,\ \ldots \end{array} \tag{14.29}$$

The top diagram of Fig. 14.6 shows the radial part of the $1s$ state, the ground state of the atom corresponding to the lowest energy level (E_1), the middle diagram

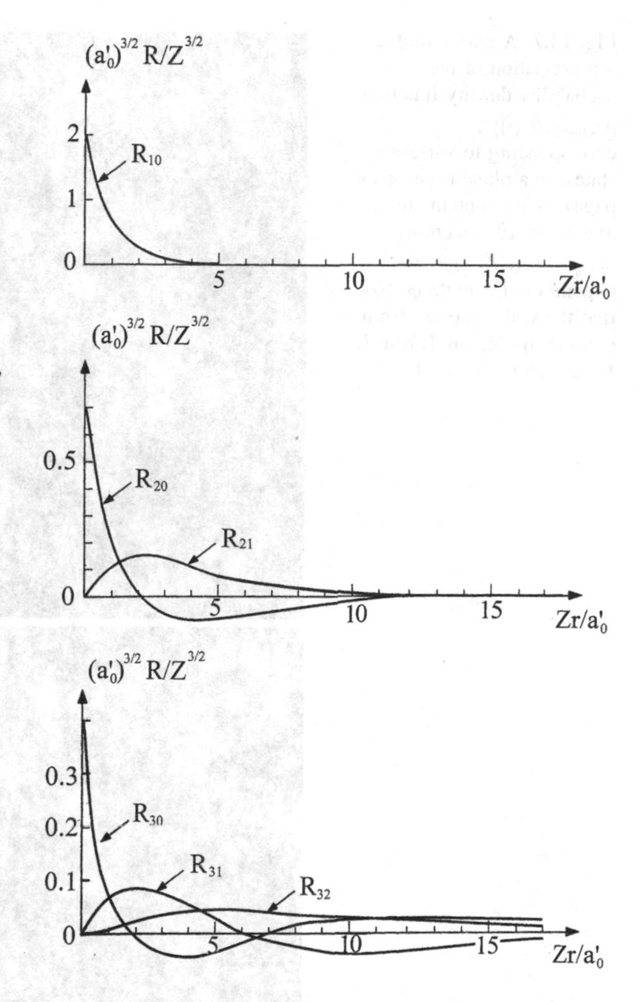

Fig. 14.6 $R_{nl}(r)$ of a one-electron atom. [From R.B. Leighton, *loc. cit.*], $R_{10} = 2(\frac{Z}{a_0'})^{3/2}\exp(-\frac{Zr}{a_0'})$; $R_{20} = \frac{1}{\sqrt{8}}(\frac{Z}{a_0'})^{3/2}(2-\frac{Zr}{a_0'})$ $\exp(-\frac{Zr}{2a_0'})$, $R_{30} = \frac{2}{81\sqrt{3}}(\frac{Z}{a_0'})^{3/2}$ $(27 - \frac{18Zr}{a_0'} + 2(\frac{Zr}{a_0'})^2)$ $\exp(-\frac{Zr}{3a_0'})$;, $R_{21} = \frac{1}{\sqrt{24}}(\frac{Z}{a_0'})^{3/2}$ $(\frac{Zr}{a_0'})\exp(-\frac{Zr}{2a_0'})$; $R_{31} = \frac{4}{81\sqrt{6}}$ $(\frac{Z}{a_0'})^{3/2}(6-\frac{Zr}{a_0'})(\frac{Zr}{a_0'})\exp(-\frac{Zr}{3a_0'})$, $R_{32} = \frac{\sqrt{8}}{81\sqrt{15}}(\frac{Z}{a_0'})^{3/2}(\frac{Zr}{a_0'})^2$ $\exp(-\frac{Zr}{3a_0'})$, $a_0' = (1+m/M)$ a_0; $a_0 = 4\pi\varepsilon_0\hbar^2/me^2$

shows the radial parts of the $2s$ and $2p$ states, and the bottom diagram the radial parts of the $3s$, $3p$, $3d$ states. We note that the radial part of the eigenfunction corresponding to the lowest energy (smallest n) for given l has no zeros: $R_{10}(r)$, $R_{21}(r)$ and $R_{32}(r)$ have no zeros. As n (and the energy E_n) increases the number of zeros increases with it: $R_{20}(r)$ and $R_{31}(r)$ have one zero; and $R_{30}(r)$ has two zeros, and so on. When n goes one step up the ladder, $R_{nl}(r)$ acquires one more zero. And this is what we expect from the similarity of Eq. (14.23) to motion in a one-dimensional well (see Fig. 13.11). We note also that as n (and therefore the energy) increases, $R_{nl}(r)$ extends to larger values of r (the electron can be found with appreciable probability further away from the nucleus). This is so because as the energy increases the space in which the electron is allowed to move increases (the region where it could move in classical terms increases, and the region where it can penetrate further because of the uncertainty principle also increases).

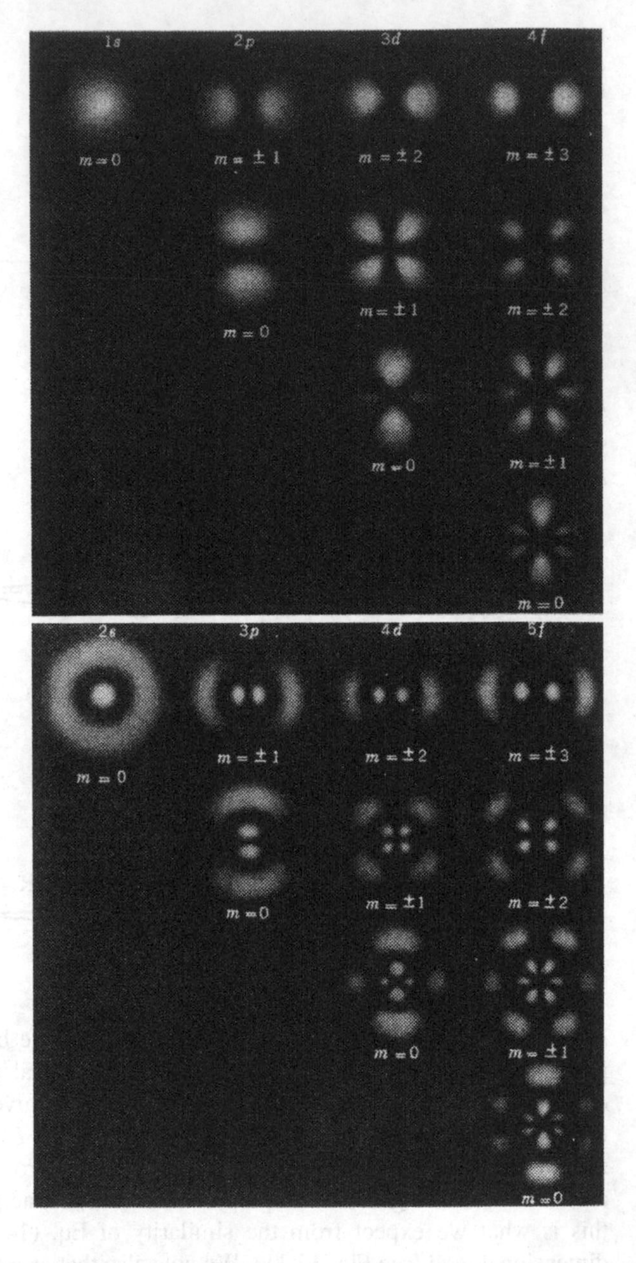

Fig. 14.7 A photographic representation of the probability density function $|\psi_{nlm}(r, \theta, \varphi)|^2$, corresponding to various states, in a plane (that of the page) which contains the *z*-axis (*vertical direction*). $|\psi_{nlm}(r, \theta, \varphi)|^2$ does *not* depend on the angle φ. Note that the scale changes from column to column. [From R. B. Leighton, *loc. cit.*]

Finally we must remember that there are a number of eigenstates: $\psi_{nlm}(r, \theta, \varphi) = R_{nl}(r)Y_{lm}(\theta, \varphi)$ corresponding to a given energy level (eigenvalue) E_n. We remember that for given l, we have the $2l + 1$ states corresponding to the different values of m for the given l. And in the present case, the degeneracy of E_n ($n = 1, 2, 3, \ldots$) given by Eq. (14.28), increases because the states with $l = 0, 1, 2, \ldots. n - 1$, have all the same energy. Some people call this additional degeneracy, a peculiarity of the Hamiltonian of Eq. (14.26), an accidental degeneracy. We shall see in Sect. 14.2 that the degeneracy is further increased (doubled) because of the spin of the electron.

Let me close this section with a note on the probability density defined by $|\psi_{nlm}(r,\theta,\varphi)|^2$. Noting that the mass of the electron is much smaller than the mass of the nucleus, we can assume that the latter lies, to a very good approximating, at the origin of coordinates, in which case we can say that $|\psi_{nlm}(r,\theta,\varphi)|^2\,dV$ is the probability of finding the electron within volume dV about the point (r,θ,φ) (see Fig. 14.3). We remember that $|\psi_{nlm}(r,\theta,\varphi)|^2=(R_{nl}(r))^2|Y_{lm}(\theta,\varphi)|^2$ and that $|\boldsymbol{Y_{lm}(\theta,\varphi)}|^2=(\boldsymbol{N_{lm}P_l^m}(\cos\boldsymbol{\theta}))^2/2\boldsymbol{\pi}$ does not depend on the angle φ. Therefore, the probability density $|\psi_{nlm}(r,\theta,\varphi)|^2$ does not depend on φ. In Fig. 14.7 we present a photographic representation of this probability density (often referred to as the electronic cloud) for a number of states: $1s$, $2p$, etc.

14.2 Electron Spin

14.2.1 Fine Structure of the Energy Levels of Hydrogen

The energy levels given by Eq. (14.28) describe very well the spectroscopic data for hydrogen as summarized in Fig. 13.6. However, the spectroscopists established, well before the development of Schrödinger's theory, that the energy levels of hydrogen have a *fine structure* as shown in Fig. 14.8. We note that the displacement of the actual energy levels (solid lines) relative to the levels of Eq. (14.28) (broken lines) has been overemphasized in the figure. The scale for the displacement is ten thousand times larger than that which defines the difference between the broken lines. Beside the displacement of the energy levels one observes a doubling of all energy levels which correspond to states with angular momentum $l>0$. Both effects are less pronounced the larger the values of the quantum numbers n and l.

The fine structure of the energy levels shown in the figure derives from relativistic corrections to Schrödinger's non relativistic theory. Sommerfeld had in fact shown, in relation to the old Bohr-Sommerfeld theory of the hydrogen atom that, a lowering of the energy levels is to be expected from the fact that the actual kinetic energy of the electron (in accordance with the special theory of relativity) is lower than the one obtained from Newtonian mechanics. When $p\ll mc$, where m is the *rest* mass and p the momentum of the particle, one obtains the following formula for the kinetic energy, K, of the particle:[2]

[2] The formula follows from $p^2-E^2/c^2=-m^2c^2$, which is obtained from Eq. (12.33) using the invariance of the magnitude of a 4-vector under a Lorentz transformation: $-m^2c^2$ is the value of p^2-E^2/c^2 in the rest frame (where $p=0$ and $E=mc^2$). We note that here m denotes the *rest*

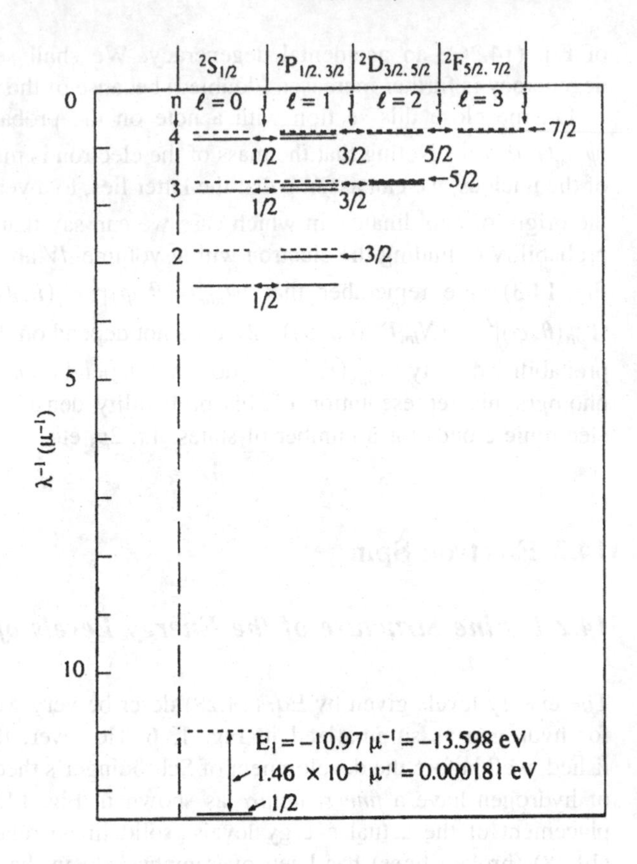

Fig. 14.8 Fine structure of the energy levels of the hydrogen atom

$$K = \frac{p^2}{2m} - \frac{p^4}{8m^3c^2} \tag{14.30}$$

In Schrödinger's theory one obtains the correction ΔE_r to the energy level E_n, due to the second term in the above equation, from the following perturbation formula (see exercise 13.10):

$$\Delta E_r = \int \psi^*(r,\theta,\phi)\widehat{V}_p\psi(r,\theta,\phi)dV \tag{14.31}$$

$$\widehat{V}_p = -\frac{1}{8m^3c^2}\widehat{p}^4$$

where $\widehat{p} = -(\hbar/i)\nabla$ is the momentum operator (Eq. (13.37)), and ψ the appropriate unperturbed wavefunction.

(Footnote 2 continued)
mass of the particle. We have: $K = E - mc^2 = (m^2c^4 + c^2p^2)^{1/2} - mc^2 \approx p^2/2m - p^4/8m^3c^2$ when $p \ll mc$.

In relation to the doubling of the energy levels, an explanation which turned out to be essentially correct was suggested by Uhlenbeck and Goudsmit in 1925 (see also Sect. 13.2.4). They argued that the electron possesses a *spin* (an intrinsic angular momentum) $\boldsymbol{\sigma}$ and, associated with it, a magnetic moment $\boldsymbol{\mu}$. Assuming (as seemed to be the case) that $\boldsymbol{\sigma}$, and therefore $\boldsymbol{\mu}$, is allowed only *two* orientations, parallel and antiparallel to a given direction, the interaction of $\boldsymbol{\mu}$ with the magnetic moment associated with the orbital motion of the electron would lead to a splitting (doubling) of the energy levels of states with $l > 0$.

14.2.2 Formal Description of the Electron Spin

The spin angular momentum (usually called, simply, spin) was formally introduced in quantum mechanics by Pauli (in 1927). We can summarize Pauli's theory as follows. The fact that the electron spin is allowed only two orientations, say, parallel and antiparallel to the z-axis, implies (see the parenthesis in Sect. 14.1.1) that the square of its magnitude is given by Eq. (14.4a) with $j = 1/2$:

$$\sigma^2 = \frac{1}{2}(\frac{1}{2}+1)\hbar^2 = \frac{3}{4}\hbar^2 \tag{14.32}$$

so that: $\sigma_z = -\hbar/2$ or $\hbar/2$

It follows that for a complete description of the state of an electron, whether it is free or moving in a potential energy field, the corresponding wavefunction must change from $\psi(\boldsymbol{r};t)$ to $\psi(\boldsymbol{r},s;t)$ where s takes the values: $s = 1$ (corresponding to spin *up*, meaning $\sigma_z = \hbar/2$) and $s = 2$ (corresponding to spin *down*, meaning $\sigma_z = -\hbar/2$). Accordingly, $|\psi(\boldsymbol{r},1;t)|^2\, dV$ is the probability of finding the electron within volume dV about $\boldsymbol{r}$ with spin *up*, and $|\psi(\boldsymbol{r},2;t)|^2\, dV$ is the probability of finding the electron within volume dV about $\boldsymbol{r}$ with spin *down*. And we have assumed that:

$$\int \left(|\psi(\boldsymbol{r},1;t)|^2 + (|\psi(\boldsymbol{r},2;t)|^2\right) dV = 1 \tag{14.33}$$

A wave function $\psi(\boldsymbol{r},s;t)$ of the above form (see also Eq. (14.34) is called a spinor).

[In many cases it is convenient (and instructive) to write $\psi(\boldsymbol{r},s;t)$ in the form of a column matrix with two rows, as follows (see Appendix A):

$$\begin{pmatrix} \psi(\vec{r},1,t) \\ \psi(\vec{r},2,t) \end{pmatrix} = \psi(\boldsymbol{r},1,t)\begin{bmatrix} 1 \\ 0 \end{bmatrix} + \psi(\boldsymbol{r},2,t)\begin{bmatrix} 0 \\ 1 \end{bmatrix} \tag{14.34}$$

where $\psi(\boldsymbol{r}, 1, t)$ and $\psi(\boldsymbol{r}, 2, t)$ may be the same or different functions of $\boldsymbol{r}$: $\psi_1(\boldsymbol{r}, t)$ and $\psi_2(\boldsymbol{r}, t)$ respectively. The above suggests that we seek the operators $\widehat{\sigma}_x$, $\widehat{\sigma}_y$, $\widehat{\sigma}_z$ for the three components of the spin, and $\widehat{\sigma}^2 = \widehat{\sigma}_x^2 + \widehat{\sigma}_y^2 + \widehat{\sigma}_z^2$ for the square of its magnitude, in the form of 2 × 2 matrices. The matrices introduced by Pauli, now known as the Pauli matrices, are:

$$\hat{\sigma}_x = \frac{\hbar}{2}\begin{bmatrix} 0 & 1 \\ 1 & 0 \end{bmatrix}, \hat{\sigma}_y = \frac{\hbar}{2}\begin{bmatrix} 0 & -i \\ i & 0 \end{bmatrix}, \hat{\sigma}_z = \frac{\hbar}{2}\begin{bmatrix} 1 & 0 \\ 0 & -1 \end{bmatrix}, \tag{14.35}$$

and therefore: $\hat{\sigma}^2 = \frac{3\hbar^2}{4}\begin{bmatrix} 1 & 0 \\ 0 & 1 \end{bmatrix}$

Using the rules of matrix multiplication one can easily verify that the above operators satisfy the required commutation relations [Eq. (14.3)]. For example:

$$\begin{aligned}\hat{\sigma}_x\hat{\sigma}_y - \hat{\sigma}_y\hat{\sigma}_x &= \begin{bmatrix} 0 & w \\ w & 0 \end{bmatrix}\begin{bmatrix} 0 & -iw \\ iw & 0 \end{bmatrix} - \begin{bmatrix} 0 & -iw \\ iw & 0 \end{bmatrix}\begin{bmatrix} 0 & w \\ w & 0 \end{bmatrix} \\ &= \begin{bmatrix} iw^2 & 0 \\ 0 & -iw^2 \end{bmatrix} - \begin{bmatrix} -iw^2 & 0 \\ 0 & iw^2 \end{bmatrix} = \begin{bmatrix} i\hbar w & 0 \\ 0 & -i\hbar w \end{bmatrix} = i\hbar\hat{\sigma}_z\end{aligned}$$

where $w = \hbar/2$. And we can also easily verify that $\underline{u}_+(s)$ and $\underline{u}_-(s)$ defined by

$$\underline{u}_+(s) = \begin{bmatrix} 1 \\ 0 \end{bmatrix} \text{ and } \underline{u}_-(s) = \begin{bmatrix} 0 \\ 1 \end{bmatrix} \tag{14.36}$$

are eigenstates of $\hat{\sigma}^2$ and $\hat{\sigma}_z$:

$$\begin{aligned}\hat{\sigma}^2\underline{u}_+ &= (3\hbar^2/4)\underline{u}_+, \quad \hat{\sigma}_z\underline{u}_+ = (\hbar/2)\underline{u}_+ \\ \hat{\sigma}^2\underline{u}_- &= (3\hbar^2/4)\underline{u}_-, \quad \hat{\sigma}_z\underline{u}_- = -(\hbar/2)\underline{u}_-]\end{aligned}$$

Let us next consider how the stationary states, that we have already obtained, are to be modified to accommodate the spin of the electron. When the Hamiltonian does not involve the spin, the modification is simple and straightforward. If the stationary states of the electron in the absence of spin are $\psi_v(\boldsymbol{r}; t) = \psi_v(\boldsymbol{r})exp(-iE_v t\hbar/)$, where v stands for a set of quantum numbers, the stationary states in the presence of spin will be twice as many. The degeneracy of each energy level, E_v, is doubled, with the corresponding states given by:

$$\begin{aligned}\psi_{v+}(\boldsymbol{r}, s; t) &= \psi_v(\boldsymbol{r})\underline{u}_+(s)exp(-iE_v t/\hbar) \\ \psi_{v-}(\boldsymbol{r}, s; t) &= \psi_v(\boldsymbol{r})\underline{u}_-(s)exp(-iE_v t/\hbar)\end{aligned} \tag{14.37}$$

where $\underline{u}_+(s)$ and $\underline{u}_-(s)$ are the spin states of Eq. (14.36).

When the Hamiltonian involves the spin the situation is different. It is easier understood through an example. This brings us back to our consideration of the fine structure of the energy levels of hydrogen.

The magnetic moment $\boldsymbol{\mu}$ associated with the spin, $\boldsymbol{\sigma}$, of the electron was found to be:

$$\boldsymbol{\mu} = -(e/m)\boldsymbol{\sigma} \tag{14.38}$$

While the magnetic moment $\boldsymbol{M}$ associated with the orbital angular momentum, $\boldsymbol{l}$, of the electron is given by (see exercise 14.8):

$$\boldsymbol{M} = -(e/2m)\boldsymbol{l} \qquad (14.39)$$

One can show that the interaction between $\boldsymbol{\mu}$ and $\boldsymbol{M}$ in the field of the atomic nucleus modifies the Hamiltonian of Eq. (14.25), to a good approximation, as follows:

$$\widehat{H} = -\frac{\hbar^2}{2m_r}\nabla^2 - \frac{Ze^2}{4\pi\varepsilon_0 r} + \frac{Z}{8\pi\varepsilon_0 m^2c^2r^3}(\widehat{\sigma}_x\widehat{l}_x + \widehat{\sigma}_y\widehat{l}_y + \widehat{\sigma}_z\widehat{l}_z) \qquad (14.40)$$

We note that $\widehat{l}^2$ commutes with the above Hamiltonian, but $\widehat{l}_z$ and $\widehat{\sigma}_z$ do not.[3] Therefore the eigenstates of $\widehat{H}$ can be also eigenstates of $\widehat{l}^2$, but can not be eigenstates of $\widehat{l}_z$ and/or $\widehat{\sigma}_z$ at the same time. However, one can easily verify that the square of the total angular momentum, $\widehat{j}^2$, and its z-component, $\widehat{j}_z$, defined in the parenthesis below, do commute with $\widehat{H}$, and with $\widehat{l}^2$. Therefore the eigenstates of $\widehat{H}$ can be also eigenstates of $\widehat{l}^2$, $\widehat{j}^2$, and of $\widehat{j}_z$. We denote the corresponding stationary states, in accordance with what we said above [see text following Eq. (14.32)] as follows: $\psi_{nljm_j}(r,\theta,\phi;s)exp(-iE_{nlj}t/\hbar)$. We note that the energy does not depend on m_j, since it can not depend on the arbitrary choice of the z-axis. And we expect, in accordance with Eq. (14.28), that E_{nlj} for the hydrogen atom may not depend explicitly on l.

[We define the total angular momentum as follows (See footnote 3):

$$\boldsymbol{j} = \boldsymbol{l} + \boldsymbol{\sigma} \qquad (14.41)$$

The operators for the components of $\boldsymbol{j}$ and the square of its magnitude defined by

$$\widehat{j}_x = \widehat{l}_x + \widehat{\sigma}_x,\ \widehat{j}_y = \widehat{l}_y + \widehat{\sigma}_y,\ \widehat{j}_z = \widehat{l}_z + \widehat{\sigma}_z \text{ and } \widehat{j}^2 = \widehat{j}_x^2 + \widehat{j}_y^2 + \widehat{j}_z^2 \qquad (14.42)$$

satisfy the commutation relations of Eq. (14.3) and therefore the eigenvalues of $\widehat{j}^2$ and $\widehat{j}_z$ should be given by Eq. (14.4a).]

---

[3] According to the rules of matrix multiplication (see Appendix A) we have: $\widehat{\sigma}_x\widehat{l}_x\begin{bmatrix}\alpha\\ \beta\end{bmatrix} = \widehat{\sigma}_x\begin{bmatrix}\widehat{l}_x & 0\\ 0 & \widehat{l}_x\end{bmatrix}\begin{bmatrix}\alpha\\ \beta\end{bmatrix} = \widehat{\sigma}_x\begin{bmatrix}\widehat{l}_x\alpha\\ \widehat{l}_x\beta\end{bmatrix}$ We note that when a spin-independent operator, such as $\widehat{l}_x$, operates on a spinor, it behaves like a 2×2 diagonal matrix with the diagonal elements equal to the given operator (as above). The same is implied in expressions involving summation of a spin-independent operator with a spin-dependent one as, for example, in Eqs. (14.40) and (14.41).

The question that now arises is: which values of the total angular momentum [of the quantum number j in Eq. (14.4a)] are compatible with the given orbital angular momentum (the given l). It turns out that for given l, the possible values for j are:

$$j = l + \frac{1}{2} \text{ and } j = l - \frac{1}{2} \tag{14.43}$$

Since for given j, there are $(2j + 1)$ states corresponding to $m_j = j,\ j - 1,\ j - 2,\ldots,\ -j$, we have in all $(2(l+\frac{1}{2})+1)+(2(l-\frac{1}{2})+1) = 2(2l+1)$ states. This is as many independent states [in the form of Eq. (14.37)] we would have, if the interaction term in the Hamltonian was not there. And because this interaction term is very small, we can write the $\psi_{nljm_j}(r,\theta,\phi;s)$ as a linear sum of the unperturbed states:

$$\psi_{nljm_j}(r,\theta,\phi;s) = \sum_{m=-l}^{l} C_{nl}^{+}(j,m_j;m)\psi_{nlm}(r)\underline{u}_{+}(s) + \sum_{m=-l}^{l} C_{nl}^{-}(j,m_j;m)\psi_{nlm}(r)\underline{u}_{-}(s) \tag{14.44}$$

where the C's are appropriate coefficients.

The corresponding energy levels, E_{nj}, when we include the relativistic kinetic energy correction of Eq. (14.30), are given by: $E_{nj} = E_n + \Delta E_r + \Delta E_s$, where E_n is the unperturbed eigenvalue (Eq. (14.28); ΔE_r is obtained from a modified version of Eq. (14.31) appropriate for spinors (ψ is replaced by $\psi_{nljm_j}(r,\theta,\phi;s)$); and ΔE_s is obtained from the same equation with $\widehat{V}_p$ given by the last term of Eq. (14.40). The relevant integrals are evaluated analytically and one obtains:

$$E_{nj} = E_n + \frac{Z^2|E_n|\alpha^2}{4n^2}\left(3 - \frac{4n}{j+1/2}\right) \tag{14.45}$$

where α is the so-called fine-structure constant, a pure number, given by:

$$\alpha = \frac{e^2}{4\pi\varepsilon_0\hbar c} = \frac{1}{137.0377} \tag{14.46}$$

The solid lines in Fig. 14.8 are in excellent agreement with the above formula (when $Z = 1$). We note that the orbital angular momentum (quantum number l) does not appear in the above formula [Eq. (14.45)]. Its effect is indirect, through Eq. (14.43), as evidenced in Fig. 14.8. This suggests that the magnitude of the orbital angular momentum, a conserved quantity for $\widehat{H}$ of Eq. (14.40), may not be so for the exact $\widehat{H}$.

Not long after the above result [Eq. (14.45)] was established in the manner we have described, Paul Dirac produced (in 1928) his celebrated theory of the electron (see Sect. 15.2), by establishing a Lorentz-invariant version of the Schrödinger equation for this purpose. It turns out that Dirac's equation admits solutions (stationary states of the electron) only in spinor form, and that this leads in the case of the hydrogen atom to the above formula for E_{nj}. In Dirac's theory, the spin of the electron and its magnetic moment are not postulated into the theory as we have done (in the manner of Pauli) but follow naturally from the requirement of relativistic invariance. Dirac's theory confirms the correctness of the results presented here.

14.3 The Radiation Field and its Interaction with Matter

14.3.1 The Linear Harmonic Oscillator

We have seen in Sect. 6.2.2 that, a particle of mass m will execute a harmonic motion about a centre (at $x = 0$) if the force acting on it has the form: $F = -kx$, where k is a constant. The force is of course a conservative one:$F = -\partial U/\partial x$ with $U(x) = kx^2/2$.

A quantum-mechanical study of the linear harmonic oscillator begins with the time-independent Schrödinger equation [Eq. (13.44)] which in the present case takes the form:

$$-\frac{\hbar^2}{2m}\frac{d^2u}{dx^2} + \frac{1}{2}kx^2u(x) = Eu(x) \tag{14.47}$$

Because $U(x) \to \infty$ as $x \to \pm\infty$, all energy eigenstates of the oscillator are bound states. Consequently, $u(x) \to 0$ as $x \to \pm\infty$. The situation is similar to that of a particle in a one-dimensional box (see Sect. 13.4.2). There is one eigenstate for each of the following (infinite many) energy eigenvalues:

$$E_n = \hbar\omega\left(n + \frac{1}{2}\right), \quad n = 0,\ 1,\ 2,\ 3, \ldots \tag{14.48}$$

where $\omega = (k/m)^{1/2}$. The ground state (the normalized eigenfunction corresponding to the minimum energy $E_0 = \hbar\omega/2$), is:

$$u_0(x) = \frac{\alpha^{1/2}}{\pi^{1/4}}\exp(-\alpha^2x^2/2) \tag{14.49}$$

where $\alpha = (km/\hbar^2)^{1/4}$. The eigenstates corresponding to $n = 1,\ 2,\ 3,\ \ldots$ are obtained from the following formula:

$$u_n(x) = \frac{1}{\sqrt{2n}}\left(-\frac{1}{\alpha}\frac{du_{n-1}}{dx} + \alpha x u_{n-1}(x)\right) \tag{14.50}$$

They are normalized as follows: $\int_{-\infty}^{+\infty} u_n^2(x)dx = 1$.

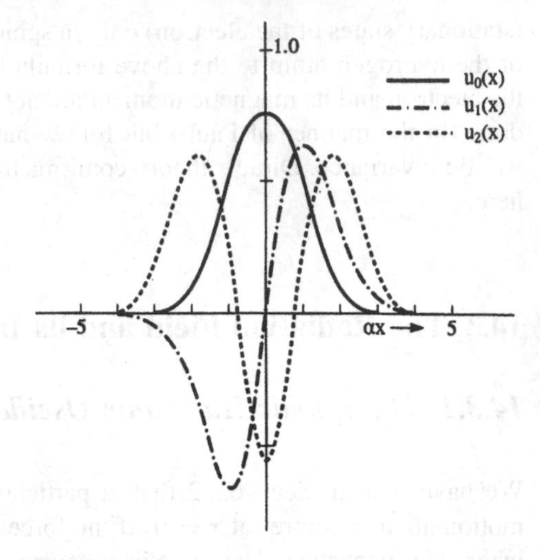

Fig. 14.9 The energy eigenfunctions of the harmonic oscillator for: $n = 0, 1, 2$

The ground state and the first two excited states are shown schematically in Fig. 14.9. We note that $u_0(x)$ has no zeros, $u_1(x)$ has one zero and so on: $u_n(x)$ has n zeros.

In what follows we shall make use of the following integrals:

$$\begin{aligned}
&\int_{-\infty}^{+\infty} u_{n-1}(x) x u_n(x) dx = \sqrt{\tfrac{n}{2}}\tfrac{1}{\alpha} \\
&\int_{-\infty}^{+\infty} u_{n+1}(x) x u_n(x) dx = \sqrt{\tfrac{n+1}{2}}\tfrac{1}{\alpha} \\
&\int_{-\infty}^{+\infty} u_m(x) x u_n(x) dx = 0, \quad \text{if } m \neq n-1,\ n+1
\end{aligned} \tag{14.51}$$

14.3.2 *Quantum-Mechanical Description of the Radiation Field*

We have seen (in Sect. 13.1.2) how Einstein was able to explain the photoelectric effect by *postulating* that the electromagnetic (EM) radiation field in a cavity consists of quantized 'oscillators'. In 1927 Paul Dirac developed a proper quantum-mechanical description of the radiation field.

Following Dirac we introduce the (classical) vector field (see Appendix A):

$$\boldsymbol{A}(\boldsymbol{r}, t) = \sum_{q,i} Q_{q,i}(t) \boldsymbol{e}_{q,i} \exp(i\boldsymbol{q} \cdot \boldsymbol{r}) + c.c. \tag{14.52}$$

where $Q_{q,i}(t)$ are functions to be determined. The sum is over all allowed values of $\boldsymbol{q}$ (given by Eq. (13.3), and for given $\boldsymbol{q}$, it includes the contributions from the two independent polarizations of the field normal to $\boldsymbol{q}$, defined by the unit vectors $\boldsymbol{e}_{q,i}$ ($i = 1, 2$). The second term, denoted by *c.c.*, is the complex conjugate of the first term, so that $\boldsymbol{A}(\boldsymbol{r}, t)$ is a real quantity. The electric field in the cavity is given by:

$$\boldsymbol{E}(\boldsymbol{r},t) = -\partial \boldsymbol{A}/\partial t \tag{14.53}$$

Using the above and Maxwell's equations (in the vacuum of the cavity) one obtains the following expression for the Hamiltonian (total energy) of the radiation field:

$$H_{rad} = \sum_{\alpha} \left(\left(\frac{1}{2\varepsilon_0} \right) P_\alpha^2 + \frac{\varepsilon_0 \omega_\alpha^2}{2} Q_\alpha^2 \right) \tag{14.54}$$

where $P_\alpha = \varepsilon_0 (dQ_\alpha/dt)$ and $\omega_\alpha = cq$, where c is the speed of light; and we remember that $\alpha \equiv \boldsymbol{q}, i$. The first term in the brackets gives the electrical energy (the integral of $\varepsilon_0 E_\alpha^2/2$ over the cavity), and the second term is the magnetic energy (the integral of $B_\alpha^2/(2\mu_0)$over the cavity) associated with the α eigenmode [see Eq. (10.52)]. We note that each term (corresponding to a certain α) of H_{rad} is formally identical with the Hamiltonian of a linear harmonic oscillator:

$$H = \left(\frac{1}{2m} \right) p^2 + \frac{k}{2} x^2 \tag{14.55}$$

P_α corresponds to the momentum p of the oscillator and Q_α to x; ε_0 takes the place of m, and $\varepsilon_0 \omega_\alpha^2$ that of k. Once this analogy has been established, going over to quantum mechanics is straightforward. Each oscillator, $\alpha = \boldsymbol{q}, i$, is described by a wavefunction $\psi(Q_\alpha, t)$ which satisfies Schrödinger's equation:

$$\left(-\frac{\hbar^2}{2\varepsilon_0} \frac{\partial^2}{\partial Q_\alpha^2} + \frac{\varepsilon_0 \omega_\alpha^2}{2} Q_\alpha^2 \right) \psi(Q_\alpha, t) = -\frac{\hbar}{i} \frac{\partial \psi}{\partial t} \tag{14.56}$$

Accordingly, the stationary states of this oscillator are given by:

$$\psi_{n_\alpha}(Q_\alpha, t) = u_{n_\alpha}(Q_\alpha) \exp(-iE_{n_\alpha} t/\hbar) \tag{14.57}$$

where E_{n_a} and $u_{n_\alpha}(Q_a)$ are the eigenvalues and eigenfunctions of the equation:

$$\left(-\frac{\hbar^2}{2\varepsilon_0} \frac{\partial^2}{\partial Q_\alpha^2} + \frac{\varepsilon_0 \omega_\alpha^2}{2} Q_\alpha^2 \right) u_{n_\alpha}(Q_\alpha) = E_{n_\alpha} u_{n_\alpha}(Q_\alpha) \tag{14.58}$$

Which is the analogue of Eq. (14.47); $u_{n_\alpha}(Q_a) \to 0$ as $Q_\alpha \to \pm\infty$. Therefore the energy levels of the oscillator α (=$\boldsymbol{q}$, i) are given by:

$$E_{n_\alpha} = \hbar\omega_\alpha (n_\alpha + \frac{1}{2}) \quad n_\alpha = 0, 1, 2, 3, \ldots \tag{14.59}$$

in complete agreement with Einstein's postulate. The $u_{n_\alpha}(Q_a)$ are obtained from the corresponding eigenstates, $u_n(x)$, of the harmonic oscillator via the substitutions: $n \rightarrow n_\alpha$, $x \rightarrow Q_\alpha$, and $\alpha = (\varepsilon_0\omega_\alpha/\hbar)^{1/2}$. For example: $u_0(Q_a) = \frac{\alpha^{1/2}}{\pi^{1/4}}\exp(-\alpha^2 Q_\alpha^2/2)$ with $\alpha = (\varepsilon_0\omega_\alpha/\hbar)^{1/2}$.

14.3.3 Interaction of the Radiation Field with Matter

We shall assume for the sake of simplicity that matter means a *hydrogen* atom. The extension to more complex systems is technically more laborious but in principle straightforward. In Sect. 13.4.5 we noted that the interaction of an atom with the radiation field can be treated by putting the atom in a time-varying electric field, $\boldsymbol{E}_0 \sin \omega t$, due to an incident plane wave, and this is how this problem is treated in many textbooks. In Dirac's treatment of the same problem (published in 1927), the system under consideration is: the atom *and* the electromagnetic field, the two interacting weakly with each other. Let me give you the Hamiltonian for this system without proof. It can be written as follows:

$$\widehat{H} = \widehat{H}_{atom} + \widehat{H}_{rad} + \widehat{H}_{\text{int}} \tag{14.60}$$

where the first term is the Hamiltonian for the atom [say Eq. (14.25)], the second term is the Hamiltonian of the radiation field, given by Eq. (14.54) with P_α replaced by $\widehat{P}_\alpha = -i\hbar\frac{\partial}{\partial Q_\alpha}$, and the last term represents the coupling between the two, which takes the following form (I shall not prove it here):

$$\widehat{H}_{\text{int}} = \frac{ie\hbar}{m}\mathbf{A}(\mathbf{r}, \mathrm{t}) \cdot \nabla$$

(we remember: $\boldsymbol{A} \cdot \nabla = A_x\partial/\partial x + A_y\partial/\partial y + A_z\partial/\partial z$). In the case of an atom, the electron wavefunction does not extend far out from the nucleus, and if the wavelength of the EM radiation is long compared with atomic dimensions (as it is the case for visible light), we have $\boldsymbol{q} \cdot \boldsymbol{r} \ll 1$ and we can put $exp(i\boldsymbol{q} \cdot \boldsymbol{r}) = 1$ in Eq. (14.52). In which case we obtain:

$$\widehat{H}_{\text{int}} = \frac{ie\hbar}{m}\sum_{q,i} Q_{q,i} \cdot \nabla \tag{14.61}$$

Let us, now, assume that in the absence of interaction between the atom and the radiation, the atom is in a state $\psi_{nlm}(\mathbf{r})$, as described by Eq. (14.24) (we disregard spin for the sake of simplicity). On the other hand, the oscillators of the radiation field are in one or the other of their stationary states: the α_{th} oscillator ($\alpha = \boldsymbol{q}$, i) in a state $u_{n_\alpha}(Q_a)$.

The following possibilities arise when $\widehat{H}_{\text{int}}$ is switched on:

1. *Absorption of light by the atom*: an atom in a state $v = n,\ l,\ m$ with energy E_v absorbs energy ΔE moving into a state $v' = n',\ l',\ m'$ with energy $E_{v'} = E_v + \Delta E$. At the same time one oscillator, say the α_{th} oscillator, moves from a state $u_{n_\alpha}(Q_a)$ with energy $\hbar\omega_\alpha(n_\alpha + 1/2)$ to a state $u_{n'_\alpha}Q_a)$ of lower energy $\hbar\omega_\alpha(n'_\alpha + 1/2) = \hbar\omega_\alpha(n_\alpha + 1/2) - \Delta E$ The probability of this transition occurring per unit time is small, and can be obtained using perturbation theory (in the manner of Sect. 13.4.5). It turns out to be proportional to a so-called *transition matrix element* defined by:

$$W(n'_\alpha, v'; n_\alpha, v) = |\boldsymbol{B}(n'_\alpha, n_\alpha) \cdot \boldsymbol{C}(v', v)|^2 \tag{14.62a}$$

where, $\boldsymbol{B}{\cdot}\boldsymbol{C}$ denotes as usual the scalar product of two vectors:

$$\boldsymbol{B}(n'_\alpha, n_\alpha) = \boldsymbol{e}_\alpha \int_{-\infty}^{+\infty} u_{n'_\alpha}(Q_a) Q_\alpha u_{n_a}(Q_\alpha) dQ_\alpha \tag{14.62b}$$

and
$$\begin{aligned} \boldsymbol{C}(v', v) &= \frac{ie\hbar}{m} \int \psi^*_{v'}(\boldsymbol{r}) \nabla \psi_v(\boldsymbol{r}) dV \\ &= -\frac{e}{\hbar}(E_{v'} - E_v) \int \psi^*_{v'}(\boldsymbol{r}) \boldsymbol{r} \psi_v(\boldsymbol{r}) dV \end{aligned} \tag{14.62c}$$

(the proof of the last step in Eq. (14.62c) need not concern us here). We note that, since $\hbar\omega_\alpha(n'_\alpha + 1/2) < \hbar\omega_\alpha(n_\alpha + 1/2)$, we must have $n'_a < n_\alpha$. And Eq. (14.51) tells us that $\boldsymbol{B}(n'_\alpha, n_\alpha) = 0$ except for $n'_\alpha = n_\alpha - 1$. We can say that the oscillator, α, emits energy in single steps: $n_\alpha \to n_\alpha - 1$. This in turn implies that $\hbar\omega_\alpha = E_{v'} - E_v$, which determines the magnitude of the wavevector $\boldsymbol{q}$ of the oscillator involved in the transition ($\omega_\alpha = cq$). Finally, Eq. (14.51) tells us that $\boldsymbol{B}(n_\alpha - 1,\ n_\alpha)$ is proportional to $\sqrt{n_\alpha}$, which means that:

$$W((n_\alpha - 1, v';\ n_\alpha,\ v) \propto n_\alpha,\ n_\alpha = 1, 2, 3, \ldots \tag{14.63}$$

Using the terminology of photons introduced in Sect. 13.1.2, we can say that the probability of the transition is proportional to the number, n_α, of α (= $\boldsymbol{q}$, i) photons in the radiation field. Finally the probability of the transition depends on the properties of the atom, through $\boldsymbol{C}(v',\ v)$. For example, a transition allowed energetically, will not occur if, as it happens occasionally, $\boldsymbol{C}(v', v) = 0$ for symmetry reasons[4].

2. *Emission of light by the atom*: an atom in a state $v = n,\ l,\ m$ with energy E_v emits energy ΔE moving into a state $v' = n',\ l',\ m'$ with energy $E_{v'} = E_v - \Delta E$. At the same time one oscillator, say the α_{th} oscillator, moves

[4] We have assumed that in the transition the energy emitted (absorbed) by the atom is absorbed (emitted) by a single oscillator. In a more detailed description of the theory this happens automatically, one does not need to assume it.

from a state $u_{n_\alpha}(Q_a)$ with energy $\hbar\omega_\alpha(n_\alpha + 1/2)$ to a state $u_{n'_\alpha}Q_a)$ of higher energy $\hbar\omega_\alpha(n'_\alpha + 1/2) = \hbar\omega_\alpha(n_\alpha + 1/2) + \Delta E$. The probability of this transition occurring per unit time is proportional to the *transition matrix element* defined by Eq. (14.62a), except that now: $n'_a > n_\alpha$. And, in this case, Eq. (14.51) tells us that $\boldsymbol{B}(n'_\alpha, n_\alpha) = 0$ except for $n'_\alpha = n_\alpha + 1$. We can say that the oscillator, α, absorbs energy in single steps. This $n_\alpha \to n_\alpha + 1$ in turn implies that $\hbar\omega_\alpha = E_v - E_{v'}$, which again determines the magnitude the wavevector $\boldsymbol{q}$ of the oscillator involved in the transition ($\omega_\alpha = cq$). Finally Eq. (14.51) tells us that $\boldsymbol{B}(n_\alpha + 1, n_\alpha)$ is proportional to $\sqrt{(n_\alpha + 1)}$, which means that:

$$W((n_\alpha + 1, v'; n_\alpha, v) \;\propto (n_\alpha + 1) \qquad (14.64)$$

We note that a transition is possible even if $n_\alpha = 0$; it corresponds to spontaneous emission (as described in Sect. 13.1.3). On the other hand, the term proportional to n_α in the above formula is proportional to the number of α (= $\boldsymbol{q}$, i)-photons in the radiation field; it corresponds to the induced emission of Sect. 13.1.3. Finally, the probability of the transition depends on the properties of the atom, through $\boldsymbol{C}(v', v)$, as in case (*i*).

Let me close this section by noting that the quantum-mechanical treatment of the absorption and emission of radiation by matter, which we summarized above, accords with Eq. (13.18) relating to the Einstein coefficients.

14.4 Indistinguishableness of Same Particles and Quantum Statistics

14.4.1 The Indistinguishableness of Same Particles

When dealing with systems of same particles, whether these are electrons in a many-electron atom, or same atoms in a gas, Heisenberg's uncertainty principle makes it impossible to treat any of them as if it were a particle with an identity of its own (distinguishable from the other particles). In classical physics we can, at least in principle, distinguish one from the other particles (let them be electrons) as follows: we call the electron which is at a certain position $\boldsymbol{r}_1$ at time $t = 0$, electron 1; and we call the electron which is at position $\boldsymbol{r}_2$ at time $t = 0$, electron 2; and so on. And, by following the paths of the electrons we can tell where electron 1 is, and where electron 2 is, and so on, at any moment of time. In quantum mechanics there is no such thing as the path of a particle. Assuming that we know the positions of the electrons at $t = 0$ (in principle possible), the best we can say is that, at time $t > 0$ there will be one electron at $\boldsymbol{r}_1$, a second one at $\boldsymbol{r}_2$, a third one at $\boldsymbol{r}_3$, according to some probability distribution, but we can not know which is which.

We must then consider how to write the wavefunction of a system of same particles to take into account their indistinguishableness. Let us begin with a system of two same particles. If the particles have no spin, the wavefunction for the system of the two particles will have the form: $\psi(\boldsymbol{r}_1, \boldsymbol{r}_2; t)$. The probability of finding a particle within a volume dV_1 about $\boldsymbol{r}_1$ and the other within a volume dV_2 about $\boldsymbol{r}_2$ is given by $|\psi(\boldsymbol{r}_1, \boldsymbol{r}_2; t)|^2 dV_1\, dV_2$; and it should be exactly the same with $|\psi(\boldsymbol{r}_2, \boldsymbol{r}_1; t)|^2 dV_1\, dV_2$, since it makes no difference which particle we denote by '1' and which by '2'. And the same argument applies to particles, like the electron which have spin. The wavefunction of a system of two electrons will have the form (an extension of $\psi(\boldsymbol{r}, s; t)$ introduced in Sect. 14.2.2): $\psi(\boldsymbol{r}_1, s_1, \boldsymbol{r}_2, s_2; t)$. The probability of finding an electron within a volume dV_1 about $\boldsymbol{r}_1$ with spin s_1 and the other within a volume dV_2 about $\boldsymbol{r}_2$ with spin s_2 is given by $|\psi(\boldsymbol{r}_1, s_1, \boldsymbol{r}_2, s_2; t)|^2 dV_1\, dV_2$; and it should be exactly the same with $|\psi(\boldsymbol{r}_2, s_2, \boldsymbol{r}_1, s_1; t)|^2 dV_1\, dV_2$, since it makes no difference which electron we denote by '1' and which by '2'. This requirement will be satisfied if:

$$\psi(\boldsymbol{1}, \boldsymbol{2}; t) = \psi(\boldsymbol{2}, \boldsymbol{1}; t) \tag{14.65a}$$

or

$$\psi(\boldsymbol{1}, \boldsymbol{2}; t) = -\psi(\boldsymbol{2}, \boldsymbol{1}; t) \tag{14.65b}$$

where **1** stands for $\boldsymbol{r}_1$, s_1 if the particle has spin, and for $\boldsymbol{r}_1$ if the particle has no spin; similarly **2** stands for $\boldsymbol{r}_2$, s_2 or $\boldsymbol{r}_2$. It was found empirically that of the particles met in nature, the ones with zero or integer spin (those with an intrinsic angular momentum given by Eq. (14.4a) with $j = 0, 1, 2$, etc.) obey Eq. (14.65a): they have a symmetric (with respect to the interchange of the particle coordinates) wavefunction; and those with half-integer spin (those with an intrinsic angular momentum given by Eq. (14.4a) with $j = 1/2, 3/2$, etc.) obey Eq. (14.65b): they have an antisymmetric (with respect to the interchange of the particle coordinates) wave function. The particles obeying Eq. (14.65a) are called *bosons*. Examples of bosons are: photons (spin 1), neutral helium atoms in their ground states (spin 0), and α-particles (spin 0). The particles obeying Eq. (14.65b) are called *fermions*. Examples of fermions are electrons, protons, neutrons (all spin ½). Finally, I should mention that Pauli was able to prove that the antisymmetry of fermions and the symmetry of bosons described above follows from the demands of special relativity.

I have, by now, referred to a number of important works by Pauli, and at this stage I should say something about the man himself. Wolfgang Ernst Pauli was born in 1900 in Austria. As a student in Munich, Pauli published three papers on relativity, following which, his professor Arnold Sommerfeld, asked him to write an encyclopaedia article on the same topic. When Einstein read the article, he was impressed[5]: *One wonders,* he wrote, *what to admire most … the complete treatment of the subject matter, or the sureness of critical appraisal.* After graduating from Munich in 1921, Pauli was appointed as an assistant to Max Born in Göttingen. There he met Niels Bohr, and thereafter became a frequent visitor of his

[5] The following extracts on Pauli appear in: P. Bizony, *Atom*, Icon Books Ltd, U.K., 2007.

Institute in Copenhagen. He taught at Heidelberg and, finally, at Zurich. In 1945 he was awarded the Nobel Prize for physics for his formulation (in 1925) of the *Exclusion Principle.* He died in 1958. Pauli's sharp remarks and put-downs became legendary. Heisenberg, a fellow-student in Munich never took offense. *He was extremely difficult,* Heisenberg remembered, *I don't know how many times he told me, 'You are a complete fool', and so on. That helped a lot.* On one occasion when Heisenberg expressed confidence with a theory he was working on, saying that *only the technical details are missing*, Pauli answered by drawing a jagged rectangle on a piece of paper, which he passed to Heisenberg with the remark: *This is to show that I can paint like Titian. Only the technical details are missing.* On another occasion he said to Heisenberg: *Only someone who lacks a thorough understanding of classical physics could think like you, so you have an advantage there. But remember, ignorance is no guarantee of success.* Pauli was, of course, his own worst critic, and this prevented him, perhaps, from making the great intuitive leaps into the unknown, that some of his contemporaries achieved.

14.4.2 Bosons

Let us assume that we have two *non-interacting* same bosons (of zero spin, and mass *m*) in a given potential field *U*(***r***). The Hamiltonian for the system of the two particles will be given by:

$$\hat{H} = \sum_{i=1}^{2} \widehat{H}(i) = \sum_{i=1}^{2} \left(-\frac{\hbar^2}{2m} \nabla_i^2 + U(\mathbf{r}_i) \right) \tag{14.66}$$

We assume that the stationary states (the eigenstates of the energy) of a single particle in the given field are known:

$$\widehat{H}(i)\varphi_v(\mathbf{r}_i) = \varepsilon_v \phi_v(\mathbf{r}_i) \tag{14.67}$$

where v stands for an appropriate set of quantum numbers. And we assume that:

$$\begin{aligned} \int \phi^*{}_{v_1}(\mathbf{r})\phi_{v_2}(\mathbf{r})dV &= 1, \quad \text{if } v_1 = v_2 \\ &= 0, \quad \text{if } v_1 \neq v_2 \end{aligned} \tag{14.68}$$

where the integral is over all space. We can easily verify that:

$$\widehat{H}\phi_{v_1}(\mathbf{r}_1)\phi_{v_2}(\mathbf{r}_2) = (\varepsilon_{v_1} + \varepsilon_{v_2})\phi_{v_1}(\mathbf{r}_1)\phi_{v_2}(\mathbf{r}_2) \tag{14.69}$$

However, $\phi_{v_1}(\mathbf{r}_1)\phi_{v_2}(\mathbf{r}_2)$ is not symmetric in respect to the coordinates of the particles. We obtain the corresponding symmetrical state [which satisfies Eq. (14.65a)] as follows:

$$\Phi_{v_1v_2}(\boldsymbol{r}_1,\boldsymbol{r}_2) = \frac{1}{\sqrt{2}}\{\phi_{v_1}(r_1)\phi_{v_2}(r_2) + \phi_{v_1}(r_2)\phi_{v_2}(r_1)\} \tag{14.70}$$

We have :
$$\hat{H}\Phi_{v_1v_2}(\boldsymbol{r}_1,\boldsymbol{r}_2) = (\varepsilon_{v_1} + \varepsilon_{v_2})\Phi_{v_1v_2}(\boldsymbol{r}_1,\boldsymbol{r}_2) \tag{14.71}$$

$$\int|\Phi_{v_1v_2}(\boldsymbol{r}_1,\boldsymbol{r}_2)|^2dV_1dV_2 = 1 \tag{14.72}$$

The latter equation follows, of course, from Eq. (14.68). We can read Eq. (14.70) as follows: we have two non-interacting particles in the one-particle states v_1 and v_2. And we note that it may happen that $v_2 = v_1$, in which case the two particles are in the same one-particle state. Extending the above to a system of N non-interacting identical (same) bosons is straightforward. The Hamiltonian is given by:

$$\hat{H} = \sum_{i=1}^{N}\hat{H}(i) = \sum_{i=1}^{N}\left(-\frac{\hbar^2}{2m}\nabla_i^2 + U(\boldsymbol{r}_i)\right) \tag{14.73}$$

A stationary state of the *N-particle* system will be described by a symmetrized wavefunction:

$$\begin{aligned}\Phi_{v_1,v_2\ldots v_N}(\boldsymbol{r}_1,\boldsymbol{r}_2,\ldots,\boldsymbol{r}_N) =& \frac{1}{\sqrt{N!}}\{\phi_{v_1}(r_1)\phi_{v_2}(r_2)\ldots r_N) \\ &+ \phi_{v_1}(r_2)\varphi_{v_2}(\mathbf{r}_1)\ldots r_N) + \ldots\}\end{aligned} \tag{14.74}$$

where the sum in { } includes all possible arrangements (in total $N!$) of $\boldsymbol{r}_1$, $\boldsymbol{r}_2,\ldots$, $\boldsymbol{r}_N$ in the N one particle states: v_1, $v_2,\ldots$, v_N. We read Eq. (14.74) as follows: we have N non-interacting particles in the one-particle states v_1, v_2, …, v_N. And we note again that, some of the v, or even all of them, can be the same: there can be any number of particles in a given one-particle state.

The energy of the system when in the above state is given by:

$$E_{v_1,v_2\ldots v_N} = \varepsilon_{v_1} + \varepsilon_{v_2} + \ldots + \varepsilon_{v_N} \tag{14.75}$$

Finally, we remember that the complete time-dependent wavefunction of the stationary state will be given by:

$$\Phi_{v_1,v_2\ldots v_N}(\boldsymbol{r}_1,\boldsymbol{r}_2,\ldots,\boldsymbol{r}_N)exp(-iE_{v_1,v_2\ldots v_N}t/\hbar) \tag{14.76}$$

In many practical applications we need only know how the N non interacting particles are distributed among the one particle states, φ_v, when the system (say a gas of bosons) is in thermal equilibrium at temperature T.

We remember [see Eq. (9.24)] that in classical physics the number of particles in the v_{th} state is proportional to $exp(-\varepsilon_v/kT)$, and that this is obtained by maximizing the entropy of the system, which is in turn determined by the number of the different arrangements (microscopic states) of the particles compatible with a given distribution. The procedure remains the same in quantum physics, but now

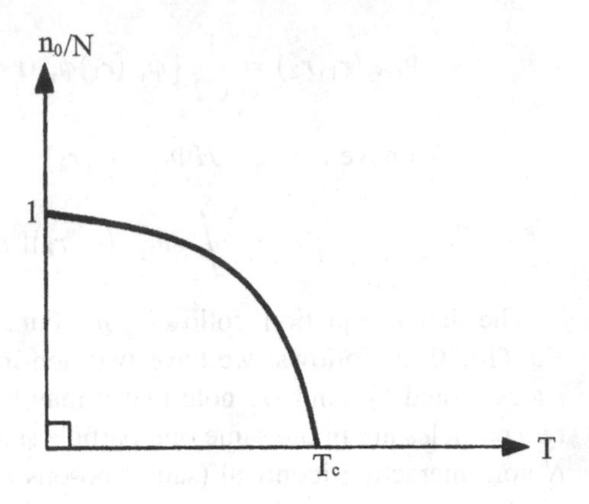

Fig. 14.10 The fraction of atoms (bosons) in their ground state as a function of the temperature. n_0 is the number of particles in their ground state and N is the total number of particles in the gas

we must bear in mind that arrangements that were counted as different in classical physics (e.g., *particle* 1 *in state* v_1 *and particle* 2 *in state* v_2 is different from *particle* 2 *in state* v_1 *and particle* 1 *in state* v_2) are not different in quantum physics. When the counting is properly done one arrives at the following formula for the number of particles (bosons) in the v_{th} state (at temperature *T*):

$$n(v) = \frac{1}{e^{(\varepsilon_v - \mu)/kT} - 1} \tag{14.77a}$$

where μ is a constant, known as the chemical potential of the system, determined, at any temperature, by the requirement:

$$\sum_v n(v) = N \tag{14.77b}$$

where the sum is over all one-particle states, and N is the total number of particles in the gas. We note that $\mu < \varepsilon_v$ for any v. The distribution of Eq. (14.77a) is known as the *Bose–Einstein* distribution in honour of the Indian physicist S. N. Bose who first proposed it in relation to a photon gas [putting $\mu = 0$ in the above equation one obtains Eq. (13.12)], and to Einstein who recognized that it applied to particles as well. The story goes like this: Bose, working in Dacca, posted to Einstein, in 1924, a short manuscript in English, where he demonstrated a new method of doing quantum statistics and applied it to give a new derivation of Planck's radiation law. Einstein was greatly impressed, he translated the paper into German, and submitted it for publication, on behalf of Bose, in the *Zeitschrift furPhysik*.

We note that, for ordinary gases, under ordinary conditions (small density of molecules at not very low *T*), the Bose–Einstein distribution relating to the translational motion of the molecules reduces to the classical: $n(v) \propto exp(-\varepsilon_v/kT)$, where ε_v is the translational (kinetic) energy of the molecule, since $n(v)$ is, in this case, very small for any v.

We note, in passing, a most interesting phenomenon concerning a gas of N non-interacting bosons (say He^4 atoms in their ground states) in a box of volume V at very low temperatures. We are concerned only with the translational motion of the atoms. We assume a cubic box ($V = L^3$) in which case the energy levels of the single atom (of mass m) are given by (see exercise 13.7):

$$E_{npl} = \frac{\hbar^2\pi^2}{2m}\left(\frac{n^2}{L^2} + \frac{p^2}{L^2} + \frac{l^2}{L^2}\right) \tag{14.78}$$

where n,p,l are any three positive integers. We find that EqS. (14.77) imply the following phenomenon: When $T \leq T_c$, where T_c is a critical temperature (determined by the atomic mass and the density, N/V, of the gas), a significant fraction of the atoms are found in their ground state ($n = p = l = 1$), with energy $E_{1,1,1}$ (which tends to zero as $V \to \infty$), as shown schematically in Fig. 14.10. We see that when $T > T_c$ the fraction of atoms which are in their ground state is practically zero. *Abruptly*, at $T = T_c$, it begins to rise as the temperature decreases; nature is forcing more and more atoms into a single state (their ground state) as the temperature decreases. The phenomenon is known as *Bose–Einstein condensation*. The only naturally occurring system which approximates a gas of bosons at low temperatures is liquid He^4. The specific heat of this substance, which depends of course on the energy distribution of the He^4 atoms, tends to infinity at $T = 2.8$ K, which is very near to $T_c = 3.14$ K [obtained from Eqs. (14.77)]; and it is reasonable to assume that this anomaly is a manifestation of Bose–Einstein condensation modified by interatomic forces, which though weak can not be neglected in the liquid state (He^4 liquifies at 4.25 K). This interpretation is supported by the fact that no such anomaly is observed in the specific heat of He^3 whose atoms are fermions.[6]

In a Bose–Einstein condensate the atoms loose their individuality and operate collectively. One manifestation of this collective behaviour is the phenomenon of *superfluidity*. Liquid helium (He^4) flows without friction below the critical temperature. The phenomenon is similar to that of superconductivity (see Sect. 14.7.7). It was discovered in 1938, by Peter Kapitza in Moscow, and independently, it seems, by Jack Allen and Donald Misener in Cambridge (U.K.). In the

[6] In 1995, Eric Cornell, Carl Weiman and coworkers at JILA (Joint Institute for Laboratory Astrophysics) at Boulder, Colorado (USA), were the first to observe the phenomenon of Bose–Einstein condensation, by cooling a gas of rubidium (Rb) atoms, trapped in a *magneto-optical trap* (MOT), to near absolute zero temperature. In their experiments cooling is achieved by slowing down the atoms by bombarding them with photons from several laser beams from different directions. This slowing and the presence of a magnetic field across the sample confine the atoms to a small region (trap). These traps can confine about 10^7 individual alkali or rare gas atoms (bosons), at densities of 10^{10} to 10^{12} atoms/cm^3, and at temperatures of about two microkelvin at which condensation occurs (we remember that the critical temperature depends on the density of the gas). Bose–Einstein condensation can be demonstrated as follows: one switches the trap off and allows the atoms to expand ballistically. A laser beam in resonance with an atomic transition is directed on the flow of atoms; the resulting absorption creates a 'shadow' which, recorded on a camera, indicates how much the atoms have expanded after being released from the trap, and in this way records the velocity distribution of the particles. The atoms condensed in the lowest energy state expand very little; their late arrival is seen as a sudden peak in the flow recorded by the camera!

1940s the Russian theoretical physicist L. D. Landau developed a phenomenological theory of the phenomenon which formed the basis for later work on the subject, which remains to this day an active field of research.

14.4.3 Fermions

Let us assume that we have two *non-interacting* same fermions (of spin ½ and mass m) in a given potential field. The Hamiltonian for the system of the two particles will be given by Eq. (14.66), with a possible addition of spin-dependent terms. Proceeding as in the derivation of Eq. (14.70), we write the normalized antisymmetrical [satisfying Eq. (14.65b)] wavefunction of a stationary state of the two-particle system as follows:

$$\Phi_{\nu_1\nu_2}(\boldsymbol{r}_1,s_1,\boldsymbol{r}_2,s_2)=\frac{1}{\sqrt{2}}\{\phi_{\nu_1}(\boldsymbol{r}_1,s_1)\phi_{\nu_2}(\boldsymbol{r}_2,s_2)-\phi_{\nu_1}(\boldsymbol{r}_2,s_2)\phi_{\nu_2}(\boldsymbol{r}_1,s_1)\} \quad (14.79)$$

We have : $$\hat{H}\Phi_{\nu_1\nu_2}(\boldsymbol{r}_1,s_1,\boldsymbol{r}_2,s_2)=(\varepsilon_{\nu_1}+\varepsilon_{\nu_2})\Phi_{\nu_1\nu_2}(\boldsymbol{r}_1,s_1,\boldsymbol{r}_2,s_2) \quad (14.80)$$

$$\sum_{s_1=1}^{2}\sum_{s_2=1}^{2}\int|\Phi_{\nu_1\nu_2}(\boldsymbol{r}_1,s_1,\boldsymbol{r}_2,s_2)|^2 dV_1 dV_2=1 \quad (14.81)$$

where ν stands for a set of quantum numbers specifying the one electron energies, ε_ν, and the corresponding eigenstates φ_ν (in the present case these are spinors, as defined in Sect. 14.2.2). We note that $\Phi_{\nu_1\nu_2}$ vanishes when $\nu_1=\nu_2$: the two particles can not be in the same state. The wavefunction of Eq. (14.79) can be written most conveniently in the form of a determinant (see Appendix A):

$$\Phi_{\nu_1\nu_2}(\boldsymbol{r}_1,s_1,\boldsymbol{r}_2,s_2)=\frac{1}{\sqrt{2}}\begin{vmatrix}\Phi_{\nu_1}(1) & \Phi_{\nu_1}(2)\\ \Phi_{\nu_2}(1) & \Phi_{\nu_2}(2)\end{vmatrix} \quad (14.82)$$

where we used the shorthand notation of Eq. (14.65a) to denote the coordinates of the particles. We note that the interchange of the coordinates ($1\rightarrow 2$, $2\rightarrow 1$) of the two particles is effected by the interchange of the two columns of the determinant, and we know that the interchange of any two columns of the determinant changes the sign of the determinant.

Extending the above to a system of N non-interacting identical fermions is straightforward. The Hamiltonian is given by Eq. (14.73) with the possible addition of spin terms. A stationary state of this system is described by a normalized antisymmetric wavefunction, which we can write in the form of an $N\times N$ *Slater determinant* as follows:[7]

[7] It is named so in honour of the American physicist John C Slater who first introduced it in the 1930s and who subsequently made important contributions to atomic and solid state physics.

$$\Phi_{v_1 v_2 \ldots\ldots v_N}(\mathbf{1},\mathbf{2},\ldots,\mathbf{N}) = \frac{1}{\sqrt{N!}} \begin{vmatrix} \phi_{v_1}(1) & \phi_{v_1}(2) & \cdot & \cdot & \phi_{v_1}(N) \\ \phi_{v_2}(1) & \phi_{v_2}(2) & \cdot & \cdot & \phi_{v_2}(N) \\ \cdot & \cdot & \cdot & \cdot & \cdot \\ \cdot & \cdot & \cdot & \cdot & \cdot \\ \phi_{v_N}(1) & \phi_{v_N}(2) & \cdot & \cdot & \phi_{v_N}(N) \end{vmatrix} \quad (14.83)$$

We note that the interchange of the coordinates of any two particles (for example: 1 → 2, 2 → 1) is effected by the interchange of the corresponding two columns of the determinant, and we know that the interchange of any two columns of the determinant changes the sign of the determinant. We also know that the determinant vanishes if any two rows are the same, which tells us that we can not have two particles (fermions) in the same state. This is of course what Pauli stated in his famous *Exclusion Principle*, before the formulation of quantum mechanics by Schrödinger and Heisenberg (see Sect. 13.2.3).

The energy of the system when in the above state is given by:

$$E_{v_1 v_2 \ldots\ldots v_N} = \varepsilon_{v_1} + \varepsilon_{v_2} + \ldots + \varepsilon_{v_N} \quad (14.84)$$

And we remember that that the complete time-dependent wavefunction of the stationary state will be given by:

$$\Phi_{v_1 v_2 \ldots\ldots v_N}(\boldsymbol{r}_1, s_1, \boldsymbol{r}_2, s_2, \ldots, \boldsymbol{r}_N, s_N) exp(-iE_{v_1 v_2 \ldots\ldots v_N} t/\hbar) \quad (14.85)$$

In many practical applications we need only know how the N non interacting particles are distributed among the one particle states, φ_v, when the system (say a gas of fermions) is in thermal equilibrium at temperature T. Proceeding as in the case of bosons (in the case of fermions one must also take into account the Exclusion Principle) one obtains the following formula for the number of particles (fermions) in the v_{th} state (we note that $n(v) \leq 1$):

$$n(v) = \frac{1}{e^{(\varepsilon_v - \mu)/kT} + 1} \quad (14.86)$$

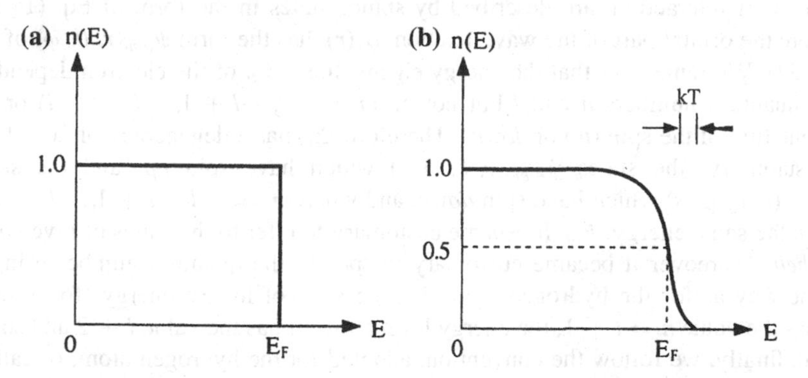

Fig. 14.11 The Fermi–Dirac distribution: (a) $T = 0$; (b) $T > 0$. We assume that $kT \ll E_F$, in which case $\mu(T) \approx \mu(0) = E_F$

where μ, the chemical potential of the system, is determined, at any temperature, by the requirement:

$$\sum_{\nu} n(\nu) = N \tag{14.87}$$

The above distribution is known as the Fermi–Dirac distribution in honour of Enrico Fermi and Paul Dirac who first proposed it (independently). In Fig. 14.11 we show n as a function of the energy of the particle when $T \to 0$, and when $0 < kT \ll E_F$, where E_F (known as the *Femi level*) is the value of μ as $T \to 0$. Figure 14.11a tells us that at $T = 0$, when the N-particle system is in its ground state, all one-particle states with energy below E_F are occupied, and those with energy above E_F are empty. If there are states with energy equal to E_F then only some of them will be occupied.

Finally we note that for ordinary gases, under ordinary conditions, and in relation to the translational motion of the molecules, Eq. (14.86) reduces to: $n(\nu) \propto exp(-\varepsilon_\nu/kT)$, as for bosons.

14.5 The Many-Electron Atom

14.5.1 The Periodic Table of the Elements

We assume (the reasonableness of the assumption is justified by the results) that the electrons of a many-electron atom move (to a good approximation) independently of each other in a spherically symmetric field $U(r)$, which depends of course on the charge of the nucleus (+Ze) and the distribution of the electrons about it. The way to obtain $U(r)$ in a self-consistent manner for any given atom will be summarized in the next section, but for now we shall assume that it is known. The stationary states of an (independent) electron in this field, if we disregard the small spin–orbit interaction, are described by spinor states in the form of Eq. (14.37), where the orbital part of the wavefunction, $\psi_\nu(\mathbf{r})$, has the form $\psi_{nlm}(r, \theta, \varphi)$ of Eq. (14.24). We remember that the energy eigenvalues, E_{nl}, of the electron depend on the quantum numbers n and l but not on m ($= -l, -l+1,\ldots,l-1, l$) or the orientation of the spin (*up* or *down*). Therefore, E_{nl} has a degeneracy of $2(2l + 1)$. In summary: the states $\psi_{nlm+}(r, \theta, \varphi, s)$ which have spin *up*, and the states $\psi_{nlm-}(r, \theta, \varphi, s)$ which have spin *down*, and where $m = -l, -l+1,\ldots,l-1, l$, have the same energy: E_{nl}. It became customary to refer to the states of given nl as a *shell*. Moreover it became customary to specify the quantum number n in the same way as for the hydrogen atom, i.e. the state of lowest energy (for a given l) has the value $n = l + 1$, the energy level above it has the value $l + 2$ and so on. And, finally, we follow the convention, adopted for the hydrogen atom, of calling the states with $l = 0$, s- states, those with $l = 1$, p-states and so on [see Eq.

(14.29)]. And, accordingly, we refer to the above mentioned shells as the 1*s* shell, the 2*s* shell, the 2*p* shell, and so on.

Let us now consider an atom with N electrons, which, we assume, move independently in a given field $U(r)$. Because the separation between the energy levels E_{nl} is much larger than kT at ordinary temperatures, the N electrons will occupy the first N one-electron states (ψ_{nlm+} and ψ_{nlm-}) in the given potential field in order of increasing energy. Let me give you some examples: The helium atom (He) has two electrons: these occupy the two states of the 1*s* shell, which has the lowest energy. We describe the He atom (in its ground state) as follows: He:$1s^2$. The lithium atom (Li) has three electrons: the first two are accommodated in the 1*s* shell of this atom; the third electron occupies one of the states in the second (in order of increasing energy) shell of this atom, which happens to be the 2*s* shell. We describe the Li atom (in its ground state) as follows: Li: $(1s^2)2s$. Which tells us that there are two electrons in an inner (the electron stays nearer to the nucleus; see below) fully occupied shell (shown in parenthesis) and one in a partly occupied outer shell (shown outside the parenthesis). [We should note that the energies and corresponding states of 1*s* shells vary significantly from atom to atom. And the same is true for 2*s*, 2*p* and every other type of shell (see, e.g. the table on the *nl* shells of carbon and sodium given below). The next atom in order of increasing *atomic number* (the number of electrons in the atom) is Beryllium (Be), which has four electrons. Its ground state is described by: Be: $(1s^2)2s^2$. Which tells us that Be has two electrons in an inner 1*s* shell and two electrons in an outer 2*s* shell. The next atom, Boron (B) has five electrons and the following configuration: B: $(1s^2)2s^22p$. There are two electrons in the fully occupied inner shell (1*s*), two in a fully occupied outer shell (2*s*) and a fifth one in a 2*p* outer shell, at an energy level above that of the 2*s* shell. A *p*-shell has $l = 1$, and can therefore accommodate up to $2 \times (2l + 1) = 6$ electrons. Therefore the atoms with atomic numbers $N = 5$, 6, 7, 8, 9 and 10 will have an outer 2*p* shell. These atoms are respectively: Boron (B), Carbon (C), Nitrogen (N), Oxygen (O), Fluorine (F) and Neon (Ne). The latter, with a configuration Ne: $(1s^2)2\ s^22p^6$, has two outer fully occupied shells (2*s* and 2*p*). We can go on in this way to describe the occupied (partly or fully) shells of all atoms. The results are most usefully presented in the so-called Periodic table of the elements, shown on page 388. We read the table as follows: Each atom is denoted by its conventional symbol (hydrogen by H, helium by He, etc.). The atomic number of the atom appears on the upper left of its symbol. A row in the table (or part of a row) defines a shell, noted in front of the row. When a shell is full we move to the next row (on the same line or the next line) representing the shell with an energy a step higher in the energy ladder. We read: after the 1*s* shell comes the 2*s* shell, after the 2*s* shell comes the 2*p* shell, after the 2*p* shell comes the 3*s* shell, followed by the 3*p* shell, which is followed by the 4*s* shell, and then comes the 3*d* shell and so on. Using the above we can write down the configuration of any particular atom. For example: the configuration of carbon (C) is: C: $(1s^2)2s^22p^2$, that of sodium (Na) is: Na: $(1s^22s^22p^6)3s$, where we put in parenthesis the inner shells. That of nickel (Ni) is: Ni: $(1s^22s^22p^63s^23p^6)4s^23d^8$ and so on. When an 'anomaly' occurs, as for example in the case of copper (Cu), this is noted under the symbol of the element (in copper an electron is missing from the $4s$ shell rather than from the $3d$ shell).[8]

We now come to the most important feature of the Periodic table, which is the following: the atoms appearing in the same column of the table have very similar chemical behaviour. And this is of course how a periodic table was conceived originally (on an empirical basis) by Dimitri Mendeleev, and improved upon by Henry Moseley, as we have already stated in Sect. 13.2.3. And we remember that Pauli proposed his Exclusion principle in relation to Bohr's efforts to explain this property of the Periodic table. However, it was only after the development of quantum mechanics that Bohr's ideas were put on a firm foundation.

We note that the atoms (elements) appearing in a given column of the table have an outer shell of the same type (s, p, d, etc.) and the same number of electrons in this shell, as noted below each column. And there lies the similarity of their chemical behaviour.

Periodic table of the elements

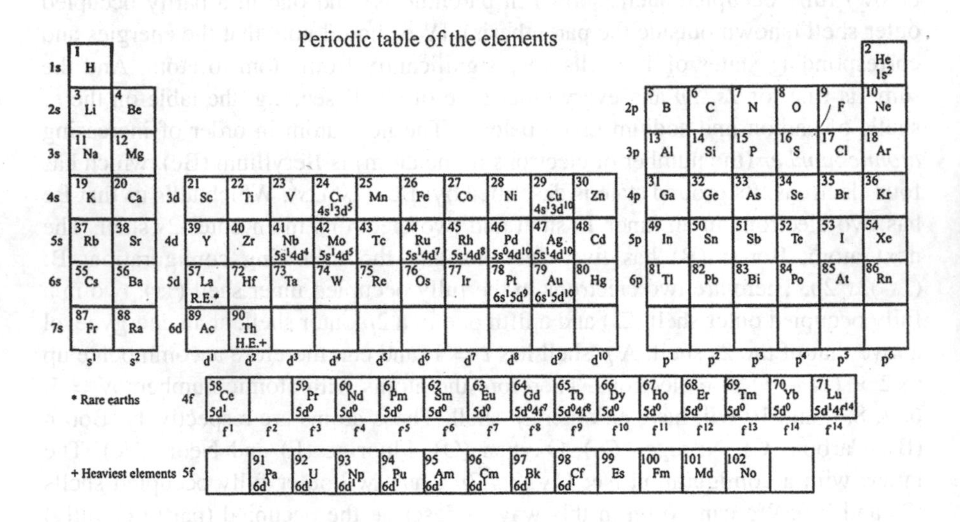

Let me, to begin with, clarify the meaning of the terms *inner* and *outer* that I used already in describing certain shells. In a many-electron atom the energy levels of the inner shells lie much below those of the outer shells, and the corresponding wavefunctions are concentrated much nearer to the nucleus in comparison with those of the outer shells. It follows from the form of the wavefunction [see Eqs. (14.37) and (14.24)] that an electron in any state of the nl shell will be in the space between two spherical surfaces, with radii r and $r + dr$ respectively, with a probability:

$$P_{nl}(r)dr = R_{nl}^2(r)r^2dr \tag{14.88}$$

[8] When one goes beyond the approximate model of independent electrons this anomaly is easily explained: the noted configuration is nearer to the actual ground state of the atom.

In Fig. 14.12 we show schematically $P_{nl}(r)$ for various shells. The *radius* $r_{max}(nl)$ of the shell, defined by the distance of the last maximum of $P_{nl}(r)$ from the atomic nucleus determines the spatial extension of the *nl* shell. We note that for given l the radius $r_{max}(nl)$ increases with n, i.e. as the energy increases. Which is what we expect by the analogy with the hydrogen atom. And we also remember that the effective barrier $U_l(r)$ the electron sees (Eq. 14.23a), makes the outward spread of the electron more difficult as l increases (see Fig. 14.12). In any case it is clear that in an atom some of the occupied shells (the so-called inner shells) will have a radius $r_{max}(nl)$ much smaller than the rest (the so-called outer shells). And we further note that while the inner shells of an atom are complete (fully occupied), the outer shells may be full, or partly occupied.

In the table below we give the calculated energies (E_{nl}) and the radii [$r_{max}(nl)$] of the inner (shown in parenthesis at the top of the table) and of the outer shells of two atoms: carbon (C) and sodium (Na).[9]

nl-shells of carbon and sodium**

C: $(1s^2)2s^22p^2$			Na: $(1s^22s^22p^6)3s$		
nl	E_{nl}	$r_{max}(nl)$	*nl*	E_{nl}	$r_{max}(nl)$
1*s*	−10.69	0.09	1*s*	−39.03	0.05
2*s*	−0.64	0.62	2*s*	−2.36	0.32
2*p*	−0.33	0.59	2*p*	−1.33	0.28
			3*s*	−0.19	1.17

**The E_{nl} are measured in Hartree units (1 Hartree = 27.21 eV) and $r_{max}(nl)$ *are measured in angstrom* (1 angstrom = 10^{-8} cm)

We note the great difference, in respect to the values of E_{nl} and $r_{max}(nl)$, between inner and outer shells in both atoms. The low energies of the inner shells arise, of course, from the fact that an electron in an inner shell lies closer to the nucleus with less electronic cloud between it and the nucleus to moderate the interaction between the two. It is also worth noting that in the case of carbon, the fully occupied 2*s* shell is rightly defined as an outer shell, because the radius of this shell is about the same (it is in fact a bit larger) than that of the 2*p* shell, in spite of the fact that E_{2p} lies well above E_{2s}.

[9] Taken from *Quantum Theory of Matter (2nd Edition)*, by John C Slater, McGraw-Hill Book Company, N.Y. 1968. The calculated values of the E_{nl} are generally in good agreement with the corresponding experimental data. One test of the deeper energy levels is provided by X-ray spectra such as the one shown in Fig. 11.3. When an electron is removed from a deep energy level by an incident radiation, an electron from a much higher energy level of the atom falls into the emptied state emitting in the process photons with well defined frequencies giving rise to the peaks in Fig. 11.3. The continuous part of the spectrum arises when the initial state of the electron lies in the continuum (the electron is not bound to a certain atom). Similar spectra are obtained from solid targets, because the deep energy levels of the atoms do not change significantly when the atoms bind together to form a molecule or solid.

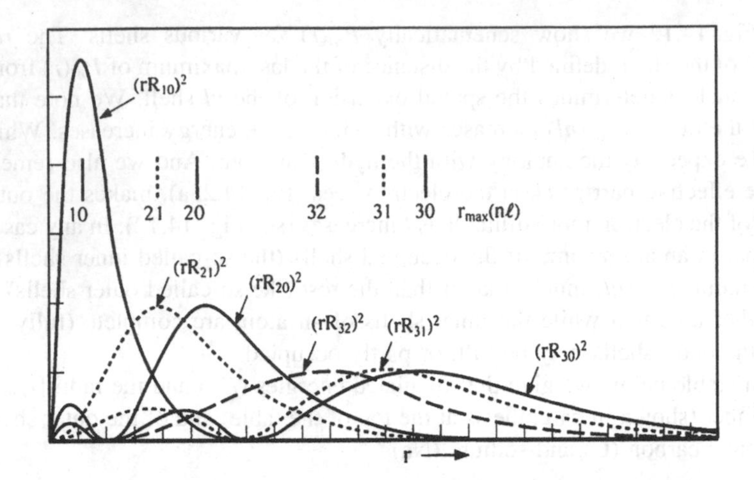

Fig. 14.12 Schematic description of $[rR_{nl}(r)]^2$ for various shells (nl). A *vertical line* marks the position of the last maximum, $r_{max}(nl)$, of $[rR_{nl}(r)]^2$

We expect, and this is confirmed by observation, the chemical behaviour of an atom to be determined by its outer shells. When two or more atoms bind together to form a molecule, the inner shells of the atom are affected only slightly. To a first approximation they remain the same. [We can not add electrons to these shells because they are already full (closed); and they remain full because the energy required to remove an electron from an inner shell to a partly occupied outer shell is usually quite large. Finally, the wavefunctions of the inner states of an atom, because of their small radial extension, are not significantly affected when the atom binds with other atoms to form a molecule.] Therefore by putting the atoms which have in their outermost shell the same number of electrons in a same type orbital, in the same column of the Periodic table, we have in any given column atoms with similar chemical behaviour.

Let me close this section with a brief note on the isotopes of each atom. We shall see (in Sect. 15.1.2) that, in general, an atomic nucleus consists of protons and neutrons. The proton has a positive charge (+e) but the neutron (with a mass practically the same as that of the proton) has zero charge. In a neutral atom the number of protons in the nucleus equals the atomic number (the number of electrons in the atom), but the number of neutrons may be zero or an integer smaller or greater than the number of protons. The atoms which have zero, one or more neutrons but the same number of protons in their nuclei are known as the *isotopes* of the given atomic number. In the case of hydrogen, for example, we have the common hydrogen (with just one proton), and two more isotopes, the deuterium and the tritium, which have, in addition to the proton, one and two neutrons respectively. The thing to note, in relation to our present discussion, is that all the isotopes of a given atomic number have the same chemical properties. In our treatment of the atom, the nucleus has, in effect, been replaced by a point-particle with a positive charge (+Ze).

14.5.2 The Self-consistent Field

In the previous section we have assumed that the N electrons of an atom move *independently* in a spherically symmetric field $U(r)$. Assuming that this is so, how does one determine $U(r)$ for a given atom? Hartree assumed that $U(r) = U_c(r)$, where $U_c(r)$ is the potential energy of an electron in the Coulomb field generated by the nucleus and the electrons of the atom[10]:

$$U_c(r) = -\frac{Ze^2}{4\pi\varepsilon_0 r} + \frac{e^2}{4\pi\varepsilon_0}\int \frac{n(\vec{r}')}{|\vec{r} - \vec{r}'|} dV' \tag{14.89}$$

where $+Ze$ is the charge of the nucleus (for a neutral atom $Z = N$) which is stationed at the origin of the coordinates, and $en(\mathbf{r})dV$ is the electronic charge in volume dV about $\vec{r} \equiv \mathbf{r}$. And we remember that $|\mathbf{r} - \mathbf{r}'|$ denotes the distance between $\mathbf{r}$ and $\mathbf{r}'$. The second term in the above equation represents the interaction of any one electron with the rest of the electrons in the atom. We note, to begin with, that the second term on the right of Eq. (14.89) will be spherically symmetric (depend only on r and not on the angles θ and φ) only if $en(\mathbf{r})$ is spherically symmetric. Now $n(\mathbf{r})$ may be taken as the sum of the electronic clouds corresponding to the *occupied* orbitals of the atom. But the electronic clouds of individual orbitals, excepting s-type orbitals, are not spherically symmetric (see Fig. 14.7). Only the electronic cloud corresponding to a fully *occupied* shell *is* spherically symmetric, because of Eq. (14.17). Therefore, for an atom with one or more partly occupied shells, $n(\mathbf{r})$ is not, strictly, spherically symmetric. However, it is not a bad approximation to replace the actual $en(\mathbf{r})$, by the following spherically symmetric contribution:

$$en(r) = e\sum_{n,l} q_{nl} R_{nl}^2(r)/4\pi \tag{14.90}$$

where $q_{nl}$is the number of occupied states in the nl shell. For an empty shell $q_{nl} = 0$, for a partly occupied shell $0 < q_{nl} < 2(2l + 1)$, and for a fully occupied shell $q_{nl} = 2(2l + 1)$. The above equation is obtained by dividing $P_{nl}(r)dr$ of Eq. (14.88) by $4\pi r^2 dr$ (the volume between two spherical surfaces with radi r and $r + dr$ respectively), and adding up the contributions from all occupied shells. In what follows we assume that $n(\mathbf{r})$ in Eq. (14.89) is given by Eq. (14.90). There is of course a second (and more serious) difficulty with Eq. (14.89): The $en(r)$ of Eq. (14.90) is made up from all the electrons in the atom, and therefore includes a contribution from the electron whose motion is being considered, and this constitutes an anomaly, because the electron does not 'feel' its own field. However, Hartree assumed that this anomaly can be disregarded in a first approximation, and he proceeded to calculate $U_c(r)$, as follows: he began by calculating numerically the energy levels E_{nl} (a sufficient number of them) and the corresponding wavefunctions [their radial parts R_{nl} (r)] for a more or less arbitrary potential field. He

[10] D. R. Hartree, *Proceedings of the Cambridge Philosophical Society,* ***24***, 89, 111, 426 (1928).

then used the calculated $R_{nl}(r)$ to obtain $en(r)$ from Eq. (14.90), which he then used to obtain a new $U_c(r)$ from Eq. (14.89). He then used the new $U_c(r)$ to calculate a new set of E_{nl} and R_{nl} and therefore a new $en(r)$, which he used to calculate a new $U_c(r)$ and so on. Hartree found that after a few iterations the output $en(r)$ coincided with impressive accuracy with the input $en(r)$; that it was, in other words, *self-consistent.*[11] And he claimed that the E_{nl} obtained for the self-consistent potential field are the right ones to be compared with the experimentally measured energy levels of the atom. [He tacitly assumed that $|E_{nl}|$ is equal to the minimum energy required to remove an electron from the *nl* shell to infinity.] And it is true that his calculations gave, if not accurate, reasonably good results when compared with the experimental data.[12] One thing is certain: the self-consistent method, that Hartree introduced, played and continues to play an important role in calculations of the electronic structure of atoms, molecules and solids.

Further progress along the same lines was made when an effective method was found to *correct* $U_c(r)$ by removing from it the contribution of an electron's own field:[13] Having in mind Pauli's exclusion Principle, which can be stated also as follows: *there can not be more than one electrons with the same spin at the same position*, we proceed as follows: We construct a sphere about the position ($\mathbf{r}$) of an electron with spin *up* (+) or *down* (−) with a radius $R_{\pm}$ such that:

---

[11] The effort that Hartree put in his numerical calculations can hardly be appreciated by the present day physicist. Hartree spent days and weeks on calculations that would take no more than a minute on a modern computer.

[12] Within a year or two from the publication of Hartree's work, an alternative and more reliable method of doing atomic calculations was proposed by V. Fock and refined by J. C. Slater working independently. The method, known as the Hartree–Fock method, assumes a wavefunction $\Psi(1, 2,\ldots, N)$, where **1** stands for $\mathbf{r_1}$, s_1, **2** for $\mathbf{r_2}$, s_2, etc., for the *N*- electron system in the form of a Slater determinant of *N* one-particle wave functions described by Eqs. (14.37) and (14.24), whose radial parts are determined by minimizing the total energy *E* of the atom given by: $E = \sum_{s_1,\ldots s_N} \int \Psi^*(1,2,\ldots N)\widehat{H}\Psi(1,2,\ldots N)dV_1\ldots dV_N$(a)

where $\widehat{H}$ is the Hamiltonian for the *N*-electron system described by:

$\widehat{H} = \sum_{i=1}^{N}(-\frac{\hbar^2}{2m})\nabla_i^2 - \sum_{i=1}^{N}\frac{Ze^2}{4\pi\varepsilon_0 r_i} + \sum_{i\neq j}\frac{e^2}{4\pi\varepsilon_0|\vec{r}_i-\vec{r}_j|}$(b)

Where, the first term stands for the kinetic energy of the electrons, the second for the attraction between the electrons and the nucleus (for a neutral atom $Z = N$), and the third term represents the repulsion between the electrons (see also Appendix A3). And we remember that the integrals in (a) extend over all $\mathbf{r}_1$, $\mathbf{r}_2$, etc, and the sums are over the two components of s_1, s_2, etc. The minimization of (a) leads to a set of integro-differential equations whose solution determines the energy levels E_{nl} and the radial parts $R_{nl}(r)$ of the corresponding one-particle states. It turns out that the Hartree–Fock values for the E_{nl} are in very good agreement with the available experimental data on the ionisation energies of at least the lighter atoms. They can not be bettered within an 'independent' electron approximation. Unfortunately the Hartee-Fock method, because of its complexity, can not be applied to atoms and molecules with a large number of electrons and, of course, it is impossible to apply to solids.

[13] The method, which came to be known as the $X\alpha$-method was elaborated by J.C. Slater in the 1950s. See his book *Quantum Theory of Matter,* Second Edition, McGraw-Hill Book Company, N.Y., 1968; see also the book cited in footnote 17.

$$\frac{4\pi}{3}R_{\pm}^{3}n_{\pm}(r) = 1 \text{which is the same as } R_{\pm} = \left(\frac{4\pi}{3}n_{\pm}(r)\right)^{-1/3} \quad (14.91)$$

Which tells that there is a total of one electron's charge of spin *up* (*down*) in the sphere of radius R_+ (R_-). By definition, $en_+(r)$ is the electronic charge density with spin *up*, and $en_-(r)$ is the electronic charge density with spin *down*. If there are no spin-dependent terms in the Hamiltonian of the atom, as we have assumed to be the case, we have: $n_+(r) = n_-(r) = n(r)/2$; and

$$R_+ = R_- = R = \left(\frac{2\pi}{3}n(r)\right)^{-1/3} \quad (14.92)$$

Now, it is reasonable to argue that by subtracting from $U_c(r)$ the interaction of the electronic charge within the above sphere with the electron (assumed to be at its centre) we remove (to a good approximation) the anomaly inherent in Hartree's potential field. The potential energy of an electron at the centre of the above sphere equals $3e^2/8\pi\varepsilon_0 R$ (see exercise 10.3). Substructing this from $U_c(r)$ we obtain the following expression for the effective potential field, $U(r)$, seen by an electron of the atom:

$$U(r) = U_c(r) + U_{ex}(r) \quad (14.93)$$

$$U_{ex}(r) = -\frac{3e^2}{8\pi\varepsilon_0 R} = -\frac{3\alpha e^2}{4\pi\varepsilon_0}\left(\frac{3}{8\pi}n(r)\right)^{1/3} \quad (14.94)$$

where $\alpha = (4\pi/3)^{2/3}/2$. The $U_{ex}(r)$ is known as the *exchange* correction to the potential energy.

Using the above potential field in self-consistent field calculations, and adjusting the value of α slightly for each atom, the theorists were able to reproduce with very good accuracy the ionization energy levels, $|E_{nl}|$, of many atoms.(See footnote 13) However, the method, which was also used in the calculation of the electronic structure of solids, remained a semi-empirical one. The limitations of the method were clarified with the development of the *density functional* formalism by W. Kohn and his co-workers in the 1960s.[14]

Kohn and his co-workers proved that the ground state energy, E, of a many-electron system (atom, molecule or solid) is uniquely determined by the electronic density distribution in the system, and though they could not determine the exact functional dependence of the ground-state energy on this distribution, they established that E is given with very good accuracy by the following formula:[15]

[14] .Hohenberg and W.Kohn, *Physical Review*, ***136***, *864-871 (1964)*; W. Kohn and L. J. Sham, *Physical Review*, ***140***, *1133-1138 (1965)*.

[15] We write the formula for an atom. The extension to molecules and solids (when the positions of the nuclei are fixed) is straightforward.

$$E = \sum_{nl} q_{nl} E_{nl} - \int U(r)n(r)dV + E_s + E_{ex} + E_{cr}$$
$$E_s = -\frac{Ze^2}{4\pi\varepsilon_0}\int \frac{n(r)}{r}dV + \frac{e^2}{8\pi\varepsilon_0}\int \frac{n(r)n(r')}{|\vec{r}-\vec{r}'|}dVdV' \tag{14.95}$$

where, E_{nl}, $U(r)$ and $n(r)$ are calculated self-consistently in the manner of Hartree, with $U(r)$ given by:

$$U(r) = U_c(r) + U_{ex}(r) + U_{cr}(r) \tag{14.96}$$

where $U_c(r)$ is given by Eq. (14.89), and $U_{ex}(r)$ by Eq. (14.94) with $\alpha = 2/3$. And $U_{cr}(r)$ is an additional term, known as the correlation term, which takes into account the fact that two electrons can not come as close together as the self-consistent field without this term would allow, because of the Coulomb repulsion between them. There are only approximate expressions for this small term and we need not reproduce them here.

The first two terms in Eq. (14.95) for E, taken together, give the kinetic energy of the electrons.[16] E_s represents the Coulomb interactions in the atom. E_{ex}, the exchange correction, derives from the second term in Eq. (14.93):

$$E_{ex} = -\frac{3}{2}\left(\frac{e^2}{4\pi\varepsilon_0}\right)\int \left(\frac{3}{8\pi}(n(r)\right)^{1/3} n(r)dV \tag{14.97}$$

And there is the final term, E_{cr}, represented by a similar expression, which takes into account the correlation correction.

We should note that, the ground-state energy calculated from Eq. (14.95) is generally a good approximation to the actual ground energy, but the one-electron energy levels, E_{nl}, used in its calculation are not necessarily a good approximation to the actual ionization levels. Interpreting them in that way does not have a sound theoretical basis, but physicists continue to do so, because quite often (especially in relation to crystalline solids) they reproduce the data reasonably well. Semi-quantitative arguments as to when and why this is so have been given by J. C. Slater.[17]

14.5.3 The Energy Levels of a Many-Electron Atom

Let us take carbon as an example (see the Table in Sect. 14.5.1). One of its two outer shells, the $2p$ shell, holds two electrons when it can accommodate six electrons. We have therefore 15 different states of the carbon atom corresponding

[16] It follows from: $E_{nl} = \int\psi_{nl} * (\mathbf{r})(-\hbar^2/2m)\nabla^2\psi_{nl}(\mathbf{r})dV + \int\psi_{nl} * (\mathbf{r})U(r)\psi_{nl}(\mathbf{r})dV$.

[17] *The Self-consistent Field for Molecules and Solids: Quantum Theory of Molecules and Solids*, Volume 4, McGraw-Hill Book Company, N.Y., 1974.

Fig. 14.13 The unperturbed energy level ε^0 splits into five energy levels

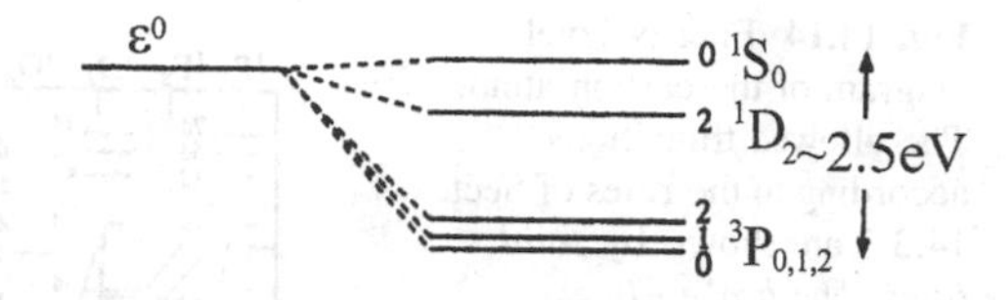

to the 15 different ways we can choose two out of the possible six one-electron states. The 15 states will be described by 15 different 6 × 6 Slater determinants. Four rows of the determinant [see Eq. (14.83)] are taken by the four states of the two closed shells (1*s* and 2*s*), the remaining two can be taken by any two of the six states of the 2*p* shell. And if there were no residual interactions to be taken into account the above states of the atom (we shall call them *unperturbed* states) would have the same energy: ε^0. But there are residual interactions due to the correlation between the electrons, and due to spin–orbit interactions. As a result of these interactions the orbital angular momenta and the *z*-components of the spins of the individual electrons are not conserved quantities. On the other hand the square of the total angular momentum, $\boldsymbol{J}$, and its z-component, J_z, are conserved quantities. And, when relativistic effects are not very strong (as is the case for the lighter atoms), the squares of the total orbital angular momentum $\boldsymbol{L}$, and of the total spin $\boldsymbol{S}$ are also conserved.[18] Accordingly, when the residual interactions mentioned above are taken into account, the 15-fold degeneracy of ε^0 is lifted and we obtain instead the five energy levels shown in Fig. 14.13.

The lowest level in the above diagram (the ground level) corresponds to a state of the atom (the ground state) with zero total angular momentum ($J = 0$) and it is not degenerate. The second level corresponds to a state with $J = 1$, and is three-fold degenerate (the three states corresponding to $M_j = -1, 0, 1$ have the same energy because the energy does not depend on the arbitrary choice of the *z*-direction). The third level corresponds to a state with $J = 2$, and is therefore five-fold degenerate. The value of J of the state corresponding to a given level is shown next to the level in the diagram. We note that the the fourth level has $J = 2$ and therefore a five-fold degeneracy, the fifth with $J = 0$ is not degenerate. We note that in total the number of states corresponding to the five levels is 15, as many as the independent unperturbed states with energy ε^0. We can in fact write approximate expressions for the eigenstates corresponding to the five energy levels shown

[18] The operators for the total orbital angular momentum $\boldsymbol{L}$ of the atom and its *z*-component L_z, are the sums of the corresponding operators for the N electrons of the atom, and they satisfy the same commutation relations [see Eq. (14.3)], and the same is true for the total spin $\boldsymbol{S}$ and its *z*-component S_z, and for the total angular momentum $\boldsymbol{J} = \boldsymbol{L}+\boldsymbol{S}$ and its *z*-component J_z. Therefore the eigenvalues of these quantities must accord with Eq. (14.4a). It turns out that the allowed values, for the square of the total orbital angular momentum of two *p*-electrons are: $\hbar^2 L(L+1)$ where $L = 0, 1, 2$. Similarly one finds that the allowed values for the square of the total spin of two electrons are: $\hbar^2 S(S+1)$ where $S = 0, 1$. Finally, the values for the square of the total angular momentum compatible with given given L and S are: $\hbar^2 J(J+1)$ where J is any integer in the range: $|L-S| \leq J \leq |L+S$. And we remember that for given J, the *z*-component of the total angular momentum takes the values: $J_z = M_j\hbar$, $M_j = -J, -J+1, \ldots, J$.

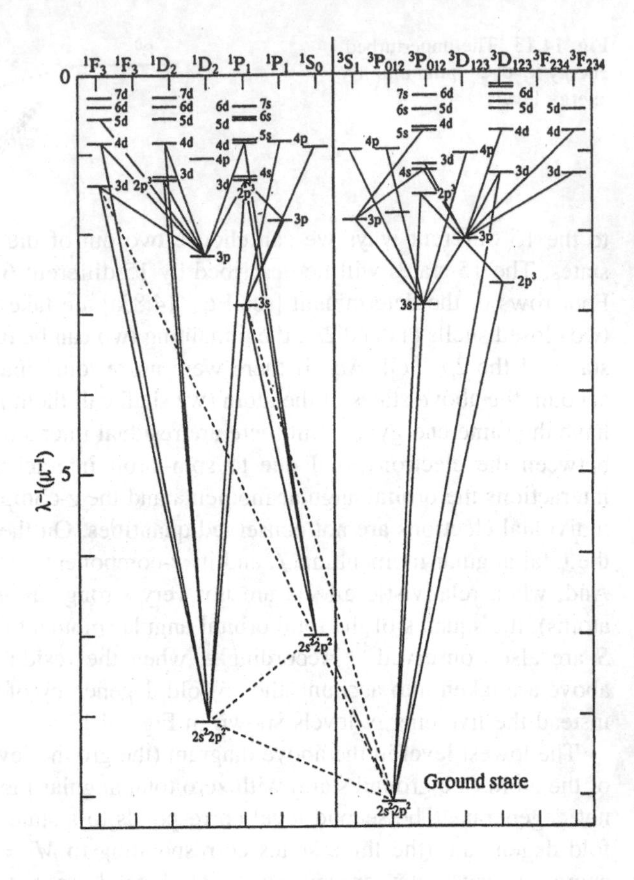

Fig. 14.14 Energy-level diagram of the carbon atom. The allowed transitions according to the rules of Sect. 14.3.3 are shown by *solid lines*. The *broken lines* indicate much weaker transitions. [From R.B. Leighton, *loc. cit.*]

in the diagram as linear sums of the unperturbed states (the 15 Slater determinants we mentioned above). We need not do so here. The symbols next to the energy levels in Fig. 14.13 have the following meaning. The letter tells us the magnitude of the total orbital angular momentum: the letters *S, P, D, F,* … stand for $L = 0$, 1, 2, 3,… respectively (See footnote 18). The upper index on the left of the letter equals $(2S + 1)$ where S is the quantum number for the total spin of the atom, and the lower index on the right of the letter is the value of J (for the total angular momentum).

The energy levels shown in Fig. 14.13 are not the only energy levels of the carbon atom. There are many others which we can introduce as follows. We note that the self-consistent calculation which gives the occupied one-electron energy levels (noted in the table of Sect. 14.5.1) produces at the same time a number of higher energy levels E_{nl} and their corresponding one-electron states (shells). Removing an electron from an occupied (in the ground state) shell to a higher energy one, produces a new configuration of the electrons which in turn leads (when residual interactions are taken into account) to a new collection of excited states of the atom. The corresponding energy level diagram is very rich indeed as can be seen in Fig. 14.14. We note that a multiplet splitting of an energy level, similar to that of $^3P_{0,1,2}$ in Fig. 14.13, can not be discerned in this diagram. Though an exact calculation of such a diagram is not possible, the theory allows us to classify the energy levels and determine the allowed transitions between them, using the formulae of Sect. 14.3.3 and appropriate extensions.

14.6 Molecules

14.6.1 The Born-Oppenheimer Approximation

A molecule consists of Λ nuclei ($\Lambda \geq 2$) with a complement of (say) N electrons. Here, we are interested in the motion of the electrons and the nuclei with respect to the centre of mass of the molecule which we assume to be at the origin of the coordinates. We note, to begin with, that the nuclei, being much heavier, will move very slowly compared with the electrons, under their mutual interaction. Classically, we would say that in the time taken by an electron to complete an orbit in the space of the molecule, the nuclei move so little that, to a good approximation we can assume they have not moved at all. With the nuclei motionless at the positions $\boldsymbol{R}$, we can calculate the energy $E(\boldsymbol{R})$ of the molecule at $\boldsymbol{R}$. [Here and throughout this section $\boldsymbol{R}$ stands for the coordinates of *all* the nuclei of the molecule.] $E(\boldsymbol{R})$ includes: the kinetic energy of the electrons, the (negative) potential energy of attraction between the electrons and the nuclei, the (positive) potential energy of repulsion between the electrons and, also, the (positive) potential energy of repulsion between the motionless nuclei. Finally, we 'allow' the nuclei to move, always very slowly compared with the electrons, to obtain the total energy, denoted by ε, of the molecule (with the nuclei in motion):

$$\varepsilon = K + E(\boldsymbol{R}) \tag{14.98}$$

where K denotes the kinetic energy of the Λ nuclei of the molecule. We can think of the above equation as the Hamiltonian (total energy) of a system of Λ nuclei in a potential energy field $E(\boldsymbol{R})$.

We can now proceed to a quantum-mechanical treatment of the molecule as follows.[19] For given $\boldsymbol{R}$ (motionless nuclei) we calculate the eigenvalues $E_n(\boldsymbol{R})$ (they are known as the *electronic terms* of the molecule), and the corresponding wavefunctions, $\psi_n(\boldsymbol{r}_1, \boldsymbol{r}_2,\ldots, \boldsymbol{r}_N; \boldsymbol{R})$, which describe the energy eigenstates of the N electrons when the nuclei are stationed at $\boldsymbol{R}$. The index n denotes the set of quantum numbers required for a complete description of $E_n(\boldsymbol{R})$ and $\psi_n(\boldsymbol{r}_1, \boldsymbol{r}_2,\ldots, \boldsymbol{r}_N; \boldsymbol{R})$. In Fig. 14.15 we show schematically the electronic terms for the simplest of molecules: the hydrogen molecular ion, H_2^+. Similar curves are

[19] We omit the spin coordinates of the electrons and the nuclei for the sake of simplicity.

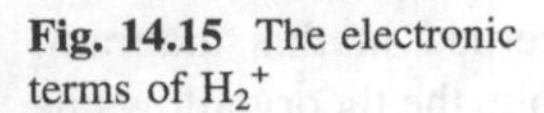
Fig. 14.15 The electronic terms of H_2^+

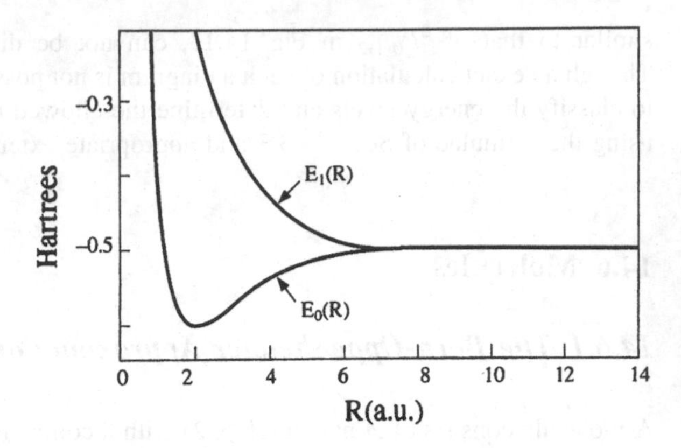

obtained for any diatomic molecule.[20] If the nuclei were indeed stationed at $\boldsymbol{R}$, we would have a stationary state of the molecule described by $\psi_n(\boldsymbol{r}_1, \boldsymbol{r}_2, \ldots, \boldsymbol{r}_N; \boldsymbol{R}) exp(-E_n(\boldsymbol{R})t/\hbar)$. But the nuclei are moving, however slowly, and $E_n(\boldsymbol{R})$ is not conserved (it is not a constant of the molecule's motion). Only the total energy of the molecule, ε, is conserved and this includes, beside $E_n(\boldsymbol{R})$, the kinetic energy of the nuclei. In accordance with Eq. (14.98), we identify the eigenvalues of ε corresponding to a given electronic term $E_n(\boldsymbol{R})$ with the energy eigenvalues of a system consisting of the Λ nuclei moving in the 'field' $E_n(\boldsymbol{R})$ defined above. To these eigenvalues, denoted by $\varepsilon_{n\mu}$, correspond, as usual, eigenstates of the nuclei which we shall denote by $\chi_{n\mu}(\boldsymbol{R})$, where μ is an appropriate set of quantum numbers. And because the state of the electrons (when the nuclei are at $\boldsymbol{R}$) is given by $\psi_n(\boldsymbol{r}_1, \boldsymbol{r}_2, \ldots, \boldsymbol{r}_N; \boldsymbol{R})$, and because in the approximation we have adopted the electrons adjust rapidly (instantly) to the variation in the position of the nuclei, the eigenstate of the molecule (electrons and nuclei) corresponding to $\varepsilon_{n\mu}$ is described by the wavefunction:

$$\Psi_{n\mu}(\boldsymbol{r}_1, \boldsymbol{r}_2, \ldots, \boldsymbol{r}_N; \boldsymbol{R}; t) = \chi_{n\mu}(\boldsymbol{R})\psi_n(\boldsymbol{r}_1, \boldsymbol{r}_2, \ldots, \boldsymbol{r}_N; \boldsymbol{R})\, exp(-\varepsilon_{n\mu}t/\hbar) \qquad (14.99)$$

The above description of the energy eigenstates (stationary states) of a molecule was proposed by Max Born and Robert Oppenheimer in 1927, and is known as the Born–Oppenheimer approximation.[21]

[20] For polyatomic molecules one obtains $E_n(\boldsymbol{R})$ as a function of the positions of the nuclei. While this function is difficult to present schematically, the physics is essentially the same with that of a diatomic molecule.

[21] Born and J. R. Oppenheimer, *Ann. Physik, 84, 457 (1927).*

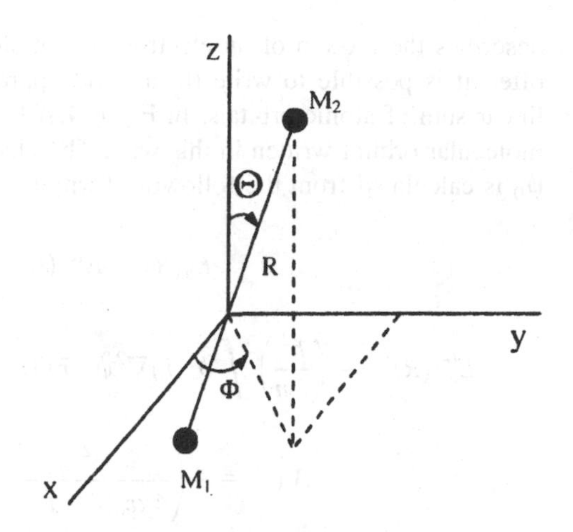

Fig. 14.16 The positions of the nuclei (of mass M_1 and M_2.) of a diatomic molecule, when its centre of mass is at the origin of the coordinates, is determined by the separation R between the atoms and the angles Θ and Φ [$0<R<\infty$; $0 \leq \Theta \leq \pi$; $0 \leq \Phi < 2\pi$]. For a homonuclear molecule, such as H_2^+, we have $M_1 = M_2$

14.6.2 Diatomic Molecules: Electronic Terms

The hydrogen molecular ion, H_2^+, consists of one electron and two same nuclei (protons) at $\boldsymbol{R}_1 = -\boldsymbol{R}/2$ and $\boldsymbol{R}_2 = \boldsymbol{R}/2$ respectively (see Fig. 14.16).

With the nuclei stationed at $\boldsymbol{R}_1$ and $\boldsymbol{R}_2$, the electron sees a potential field:

$$U(\boldsymbol{r}) = -\frac{e^2}{4\pi\varepsilon_0|\vec{r} - \vec{R}_1|} - \frac{e^2}{4\pi\varepsilon_0|\vec{r} - \vec{R}_2|} + \frac{e^2}{4\pi\varepsilon_0 R} \tag{14.100}$$

where $R = |\boldsymbol{R}_2 - \boldsymbol{R}_1|$, and $\boldsymbol{r} \equiv \vec{r}$ is the position of the electron. The last term is, of course, a constant when the nuclei do not move, and does not complicate the eigenvalue problem at hand: to find the eigenstates of the electron in the above field. It simply adds $e^2/4\pi\varepsilon_0 R$ to the energy eigenvalues [the electronic terms $E_n(R)$] obtained in the absence of this term.

We note, to begin with, that the $E_n(R)$ depend only on the separation R between the two nuclei (the orientation of the molecule in space can not possibly influence its energy). We may assume, for the sake of clarity, that the two nuclei are stationed on the z-axis at $z = -R/2$ and $R/2$ respectively. The above eigenvalue problem can be solved exactly, but here we shall present the essential results in a manner, which though approximate, is more informative. It turns out that the ground state of an electron in the field of Eq. (14.100) is described to a good approximation by: (See footnote 19)

$$\psi_0(r) = A_0(e^{-r_1/a_0} + e^{-r_2/a_0}) \tag{14.101}$$

where $r_1 = |\boldsymbol{r} - \boldsymbol{R}_1|, r_2 = |\boldsymbol{r} - \boldsymbol{R}_2|, a_0 \approx a_0'$ is the Bohr radius (see caption of Fig. 14.6), and A_0 is a normalization constant: $\int|\psi_0(r)|^2 dV = 1$. We note that the two terms in the above formula are identical with the ground state wave function of the hydrogen atom when the latter is placed at $\boldsymbol{R}_1$ and $\boldsymbol{R}_2$ respectively. We refer to these terms as *atomic orbitals*, and to a wavefunction, such as $\psi_0(\boldsymbol{r})$, which describes the motion of an electron in a molecule as a *molecular orbital*. Quite often it is possible to write (to a good approximation) a molecular orbital as a linear sum of atonic orbitals. In Eq. (14.101) we have the simplest example of a molecular orbital written in this way. The electronic term $E_0(R)$ corresponding to ψ_0 is calculated from the following formula (with $n = 0$):

$$E_n(R) = E_n^{(e)}(R) + \frac{e^2}{4\pi\varepsilon_0 R} \tag{14.102}$$

$$E_n^{(e)}(R) = -\left(\frac{\hbar^2}{2m}\right)\int \psi_n^*(\vec{r})\nabla^2\psi_n(\vec{r})dV + \int V(\vec{r})|\psi_n(\vec{r})|^2 dV \tag{14.102a}$$

$$V(\vec{r}) \equiv -\left(\frac{e^2}{4\pi\varepsilon_0|\vec{r}-\vec{R}_1|} + \frac{e^2}{4\pi\varepsilon_0|\vec{r}-\vec{R}_2|}\right) \tag{14.102b}$$

The first term of $E_n^{(e)}(R)$ represents the kinetic energy of the electron and the second term its potential energy in the field of the two nuclei. We can think of $E_n^{(e)}(R)$ as the ground-state energy of the electron in this field when the nuclei are at a distance R apart. The corresponding electronic term, $E_0(R)$, obtained from Eq. (14.102) for $n = 0$, is shown in Fig. 14.15. Its main feature is the existence of a minimum which secures the existence of the hydrogen molecular ion. We shall in what follows explain, briefly, how this minimum arises, but before doing so, it is worth noting that the next in energy term, $E_1(R)$, given by Eq. (14.102) with $n = 1$, corresponds to the following molecular orbital:

$$\psi_1(r) = A_1(e^{-r_2/a_0} - e^{-r_1/a_0}) \tag{14.103}$$

Let us look a bit more closely at how the energy levels $E_0^{(e)}$ and $E_1^{(e)}$, corresponding to the molecular orbitals $\psi_0(\boldsymbol{r})$ and $\psi_1(\boldsymbol{r})$, vary with R. These levels are shown in Fig. 14.17, by the horizontal lines marked $n = 0$ and $n = 1$, for two different values of R. Evidently $E_0^{(e)}$ is lower than $E_1^{(e)}$. We can understand this as follows: we remember that $E_0^{(e)}$ and $E_1^{(e)}$ are obtained from Eq. (14.102a), and we note that where V is smaller (more negative), i.e. in the space between the nuclei, $|\psi_0|^2$ is greater than $|\psi_1|^2$ (see Fig. 14.18), which in turn leads to $E_0^{(e)} < E_1^{(e)}$. Figures 14.17 and 14.18 tell us more. The difference $E_1^{(e)} - E_0^{(e)}$ decreases as R increases. In fact this difference vanishes as $R \to \infty$. And because the second term in Eq. (14.102) also vanishes in this limit, the corresponding electronic terms converge as $R \to \infty$ (as seen in Fig. 14.15). In this limit $E_1^{(e)} = E_0^{(e)} = -0.5$ Hartree, which is the ground state energy of a hydrogen atom. We can interpret the above as follows: at large separations the electron is found about nucleus '1'or about nucleus '2' (but not in between) whether in the ψ_0 or the ψ_1 molecular orbital: the overlap between the two terms vanishes in both Eqs. (14.101) and (14.103). We have then, in effect, a hydrogen atom in its ground state and a bare

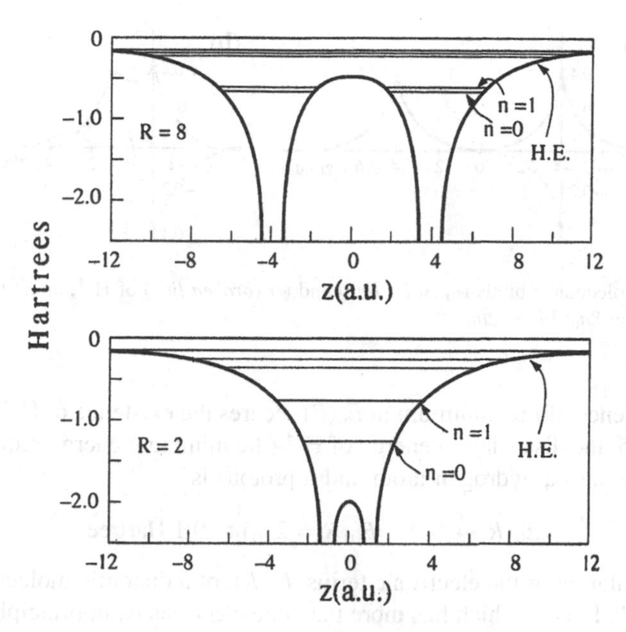

Fig. 14.17 The solid lines show the variation of V, defined by Eq. (14.102b), along the axis of the molecule (assumed to be along the z-direction, with the centre of mass of the molecule at $z = 0$) for two values of the distance R between the nuclei. The *top diagram* is obtained for $R = 8a_0$; the *bottom diagram* for $R = 2a_0$ (where the minimum in $E_0(R)$ occurs (see Fig. 14.15). We note the lowering of V in the space between the nuclei. The *horizontal lines* marked $n = 0$ and $n = 1$ give the energy levels $E_0^{(e)}$ and $E_1^{(e)}$ corresponding to the molecular orbitals $\psi_0(\mathbf{r})$ and $\psi_1(\mathbf{r})$ for the noted R; the levels denoted by H.E. are higher energy levels of the electron in the field of the two nuclei that need not concern us here. The separation of the energy level has been exaggerated in the top diagram

proton, the two of them not interacting with each other. As R gets smaller the electron is able to move between the two nuclei (tunneling plays an important role here) and the atomic level is split in two: $E_1^{(e)}$ and $E_0^{(e)}$ corresponding to the molecular orbitals we have described.

The difference between these two levels increases as R approaches the equilibrium separation, $R = 2a_0$ of the molecule. The variation of the corresponding electronic terms with R (shown in Fig. 14.15) is easy to understand. As R gets smaller the reduction in $E_0(R)$, given by Eq. (14.102), arising from the lowering of $E_0^{(e)}$ due to the accumulation of electronic cloud between the nuclei is opposed by the increase of the second term of Eq. (14.102), arising from the Coulomb repulsion between the nuclei. The two balance each other at the equilibrium separation, which corresponds to the minimum of $E_0(R)$. At smaller separations the Coulomb repulsion wins and $E_0(R)$ rises. In the case of $E_1(R)$ there is no significant accumulation of electronic cloud between the nuclei (see Fig. 14.18b) and Coulomb repulsion wins all the way.

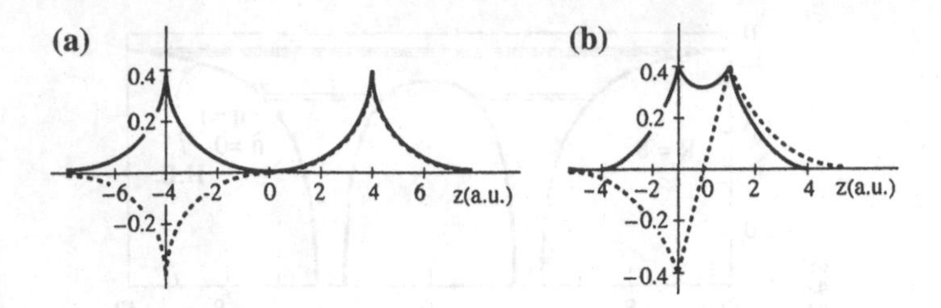

Fig. 14.18 Molecular orbitals ψ_0 (*solid line*) and ψ_1 (*broken line*) of H_2^+, along the axis of the molecule. **a**$R = 8a_0$; **b**$R = 2a_0$

The existence of the minimum in $E_0(R)$ secures the existence of H_2^+. According to Fig. 14.15, the dissociation energy of H_2^+ (the minimum energy required to split the molecule into a hydrogen atom and a proton) is:

$$E_0(R \to \infty) - E_0(R = 2a_0) \approx 0.1 \text{ Hartree}$$

The calculation of the electronic terms, $E_n(R)$, of a diatomic molecule XY, such as H_2, O_2, HCL, etc., which has more than one electrons is, in principle, similar to the calculation of the energy levels of a many-electron atom, but it is technically more difficult because of the reduced symmetry of the molecular field. It is possible nowadays to obtain self-consistently the electronic density and the corresponding energy, $E_0(R)$, of the molecule in its ground state. On the other hand, one may write the molecular orbitals of the electrons as sums of atomic orbitals centered on one or the other of the atomic nuclei and choose the coefficients of the atomic orbitals in the sums in such a way that the molecular orbitals, so obtained, are much the same with the orbitals one would obtain from a self-consisted field. It turns out that the binding of two atoms into a molecule is effected in more or less the same way as in the case of H_2^+: electronic cloud accumulating between the two nuclei, where the effective potential the electrons see is more negative, leads to a dip in the corresponding electronic-energy term when combined with the Coulomb repulsion between the nuclei. A bond of this kind is often called a *covalent bond*: the increased electronic cloud between the two nuclei 'belongs to both of them'.

There is another way of looking at covalent bonding which goes as follows:[22] the electronic charge between the two nuclei (calculated in the manner we have described) attracts the nuclei towards it, and therefore towards each other, while the repulsive force between the nuclei pushes them apart. At the equilibrium distance, the two forces balance each other. When the distance between the nuclei becomes smaller than the equilibrium one, electronic charge is removed from the region between the nuclei to the left of the left-hand nucleus and to the right of the

[22] This interpretation derives from the so-called Hellman-Feynman theorem proposed by H. Hellman in 1937, and independently by R. P. Feynman in 1939.

right-hand nucleus and from there helps to pull the nuclei apart acting in the same direction as the repulsive force between the nuclei.

14.6.3 Nuclear Motion of Diatomic Molecules

The positions of the nuclei of a diatomic molecule are determined by R, Θ, and Φ as defined in Fig. 14.16. Therefore the nuclear part, $\chi_{n\mu}(\mathbf{R})$, of the wavefunction defined by Eq. (14.99), takes the form: $\chi_{n\mu}(R, \Theta, \Phi)$. The energy eigenvalues $\varepsilon_{n\mu}$ of the molecule and the corresponding $\chi_{n\mu}(R, \Theta, \Phi)$, for a given electronic term $E_n(R)$, are obtained from the eigenvalue equation:

$$\widehat{H}\chi_{n\mu} = -(\hbar^2/2M_r)\nabla^2\chi_{n\mu} + E_n(R)\chi_{n\mu} = \varepsilon_{n\mu}\chi_{n\mu} \tag{14.104}$$

where $M_r = M_1M_2/(M_1 + M_2)$ is the reduced mass of the molecule (See footnote 1). We note that $\widehat{H}$ has exactly the same form as for the hydrogen atom (see Eq. 14.25). Therefore, we write [by analogy to Eqs. (14.24) and (14.22)]:

$$\chi(R, \Theta, \Phi) = (u(R)/R)Y_{lm}(\Theta, \Phi) \tag{14.105}$$

where $u(R)$ is to be determined [by analogy to Eq. (14.23)] from the equation:

$$-\frac{\hbar^2}{2M_r}\frac{d^2u}{dR^2} + U_{nl}(R)u(R) = \varepsilon u(R) \tag{14.106}$$

$$\text{where} \quad U_{nl}(R) = E_n(R) + \frac{\hbar^2 l(l+1)}{2M_rR^2} \tag{14.106a}$$

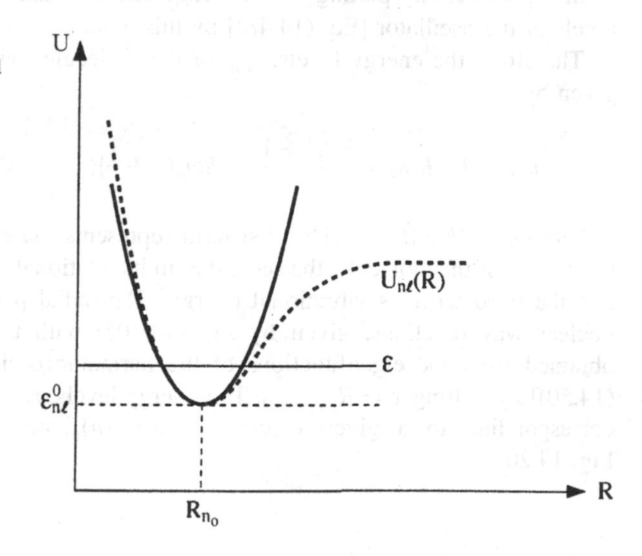

Fig. 14.19 The potential field $U_{nl}(r)$ (*broken line*) and its harmonic approximation of Eq. (14.107) (*solid line*)

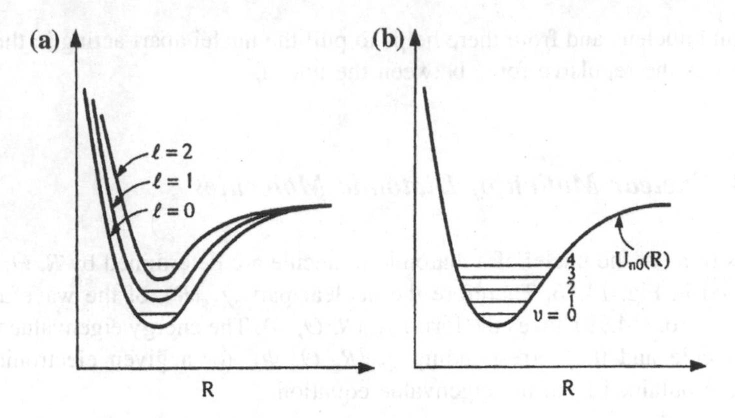

Fig. 14.20 **a** The *horizontal lines* represent the energy levels, $\varepsilon_{nl\upsilon}$, (for given n and $\upsilon = 0$) for $l = 0,1,2$. **b** The *horizontal lines* represent the energy levels, $\varepsilon_{nl\upsilon}$, (for given n and $l = 0$) for $\upsilon = 0, 1, 2, 3, 4$. We note that the energy scale in diagram **a** is about ten times larger than in diagram **b**

The $U_{nl}(R)$ is shown schematically in Fig. 14.19. The existence of a minimum of $U_{nl}(R)$ at R_{n0} allows us to approximate it by:

$$U_{nl}(R) = \varepsilon^0_{nl} + \frac{1}{2}k_n(R - R_{n0})^2 \tag{14.107}$$

$$\varepsilon^0_{nl} = E_n(R_{n0}) + \frac{\hbar^2 l(l+1)}{2M_r R^2_{n0}} \tag{14.107a}$$

and k_n is an appropriate constant. The above field (shown by a solid line in Fig. 14.19) is of course, the potential field of a harmonic oscillator. It is obtained from Eq. (14.47) by putting $x = R - R_{n0}$. The constant ε^0_{nl} only shifts the energy levels of the oscillator [Eq. (14.48)] by this amount.

Therefore: the energy levels, $\varepsilon_{n\mu}$, of the molecule (with μ replaced by $l\upsilon$) are given by:

$$\varepsilon_{nl\upsilon} = E_n(R_{n0}) + \frac{\hbar^2 l(l+1)}{2M_r R^2_{n0}} + \hbar\omega_n(\upsilon + \frac{1}{2}), \quad \upsilon = 0, 1, 2, \ldots \tag{14.108}$$

where $\omega_n = (k_n/M_r)^{1/2}$. The first term represents the electronic contribution to the energy of the molecule, the second term its rotational energy ($l = 0, 1, 2, 3, \ldots$) and the third term its vibrational energy. The radial parts of the corresponding nuclear wavefunctions, given by Eq. (14.105) with the $u = u_{n\upsilon}(R - R_{n0})$, are obtained from the eigenfunctions of the harmonic oscillator [Eqs. (14.49) and (14.50)] by putting $x = R - R_{n0}$. The energy levels, $\varepsilon_{nl\upsilon}$, of a diatomic molecule, corresponding to a given electronic term (n), are shown schematically in Fig. 14.20.

The diagrams of Fig. 14.20 provide a convincing explanation to the variation of the specific heat of a gas of diatomic molecules described in Sect. 9.2.4 (see Fig. 9.11). We recall that the specific heat of the gas is made of a contribution (3 $k/2$ per molecule) from the translational motion of the molecules, and contributions from the rotational and vibrational motions of the molecules. At low temperatures [when kT is smaller than the separation of the energy levels in diagram (a)] the molecule stays in the lowest level ($l = 0$, $\upsilon = 0$) and only the translational motion contributes to the specific heat. When kT exceeds the energy difference between the rotational energy levels, shown in diagram (a), molecules can absorb heat by moving into states of larger l, and this explains the first (at $T \approx 100$ K) upward change in the C_V of Fig. 9.11. And when kT exceeds the energy difference between the vibrational energy levels, shown in diagram (b), molecules can also absorb energy by moving into states of higher vibrational energy (larger υ), and this explains the larger value of the C_V of Fig. 9.11 at high temperatures. Apparently, at the very high temperatures the discreteness of the energy levels does not matter much and one obtains the classically expected result.

Finally, it is worth noting that the energy-level diagram of Fig. 14.20 is useful in understanding the optical properties of a diatomic molecule. Absorption and emission of radiation occurs in three different regions of frequency:

1. The region between microwaves and infrared corresponds to transitions between different l states: $nl\upsilon \rightarrow nl'\upsilon$.
2. The infrared region corresponds to transitions where rotational and vibrational levels change: $nl\upsilon \rightarrow nl'\upsilon'$.
3. The optical and ultraviolet region correspond to transitions where all three quantum numbers change: $nl\upsilon \rightarrow n'l'\upsilon'$.

14.6.4 Polyatomic Molecules

Figure 14.21 shows the equilibrium positions of the nuclei of methane: CH_4. The carbon nucleus sits at the centre of a regular tetrahedron and the hydrogen nuclei occupy the four corners of the tetrahedron. The shaded regions describe the electronic clouds associated with the occupied molecular orbitals of methane. These are combinations of the 1s orbitals of the hydrogen atoms with appropriate 2s and 2p orbitals of the carbon atom (see the table in Sect. 14.5.1). We need not describe the details of these molecular orbitals here. The thing to note is that the molecule is kept together by the accumulation of electronic charge between the carbon and the hydrogen atoms. And apparently the tetrahedral arrangement is the one among all possible arrangements that minimizes the total energy of the molecule.

In polyatomic molecules, in contrast to the case of diatomic molecules, it is practically impossible to calculate the spatial arrangement of the atomic nuclei that minimizes the energy of the molecule. It is usually obtained by meticulous experimentation and a careful analysis of the data. The larger the molecule the more difficult is this task. This is why it took so long to determine the structure of

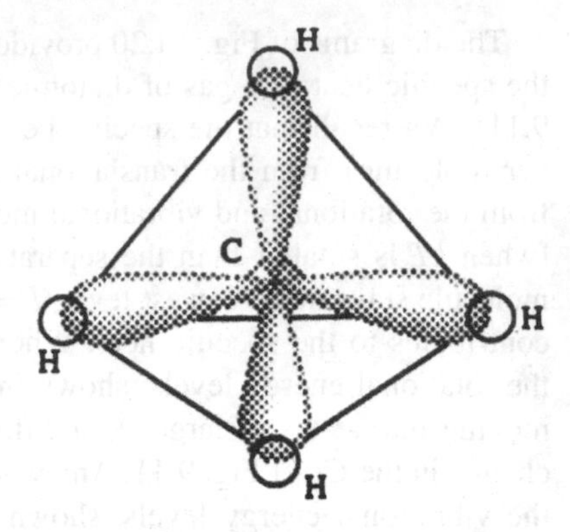

Fig. 14.21 Covalent bonding in CH_4

DNA. But because the structure of a molecule plays a critical role in defining its properties, the chemists need to know it. I shall not say anything more here. The advances made by chemists, in the years that followed the development of quantum mechanics, both in the understanding of naturally existing materials, and in the creation of new materials with specific properties are impressive, to say the least, but their description lies beyond the scope of the present book. The interested reader will find more information on pioneering work in atomic and molecular physics in the books of J.C. Slater.[13,17]

14.7 Solids

14.7.1 Symmetric Cells of Real and Reciprocal Lattices

The main characteristic of a crystalline solid is its periodic structure: unit cells of the same volume, shape and content fill up all the space occupied by the solid. A periodic structure is defined by a space lattice:

$$\boldsymbol{R}_n = n_1\boldsymbol{t}_1 + n_2\boldsymbol{t}_2 + n_3\boldsymbol{t}_3; \quad n_1, n_2, n_3 = 0, \pm 1, \pm 2, \pm 3, \ldots \tag{14.109}$$

where, $\boldsymbol{t}_1$, $\boldsymbol{t}_2$ and $\boldsymbol{t}_3$ are the primitive vectors of the lattice. There are in all 14 different space lattices, known as the Bravais lattices. The ones most common in nature are the bcc and the fcc lattices which we have already described in Sect. 11.4. The space lattice associated with a given crystal is determined experimentally in the manner we have described in that section.

The parallel piped of Fig. 11.5 is not the only way to choose the primitive cell of a lattice. We can construct a more symmetric primitive cell of the bcc lattice, defined by Eq. (11.3), as follows: we draw straight lines from a lattice point to all its neighbour points, and through the midpoint of each line and normal to it we draw a plane. In this way we obtain a closed volume of 14 faces about the central point as shown in Fig. 14.22a. By the way it was constructed, the volume of this

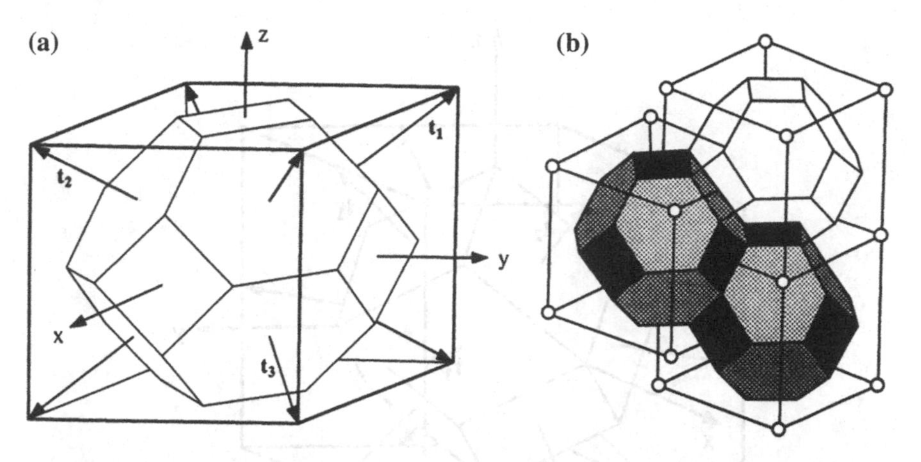

Fig. 14.22 **a** The Wigner–Seitz cell of the bcc lattice. **b** Wigner–Seitz cells fill up all the space of the ctystal

cell is that corresponding to one lattice point and, therefore, equals $a^3/2$, where a is the lattice constant (see Sect. 11.4). In Fig. 14.22b we show how a succession of such cells fills up all the space of the crystal.

A primitive cell constructed in the above manner is called a Wigner–Seitz cell.[23] What makes it special is its symmetry: rotations about symmetry axes, mirror reflections with respect to symmetry planes and other operations (48 in all) which leave a lattice point fixed (at the origin of coordinates) and bring the lattice upon itself, bring the Wigner–Seitz cell upon itself as well. The Wigner–Seitz cell has the cubic symmetry of the bcc lattice. No other primitive cell has this property.

The Wigner–Seitz cell of the fcc lattice, defined by Eq. (11.4), can be constructed in similar fashion by drawing planes normal to the vectors from a lattice point to its 12 neighbor points (at the midpoints of these vectors). The resulting closed volume, a dodecahedron of volume $a^3/4$, is shown in Fig. 14.23.

We have seen (in Sect. 11.4) that the reciprocal of the bcc lattice is an fcc lattice in $\boldsymbol{k}$-space (reciprocal space), and that the reciprocal of the fcc lattice is a bcc lattice in $\boldsymbol{k}$-space. The equivalent of the Wigner–Seitz cell in $\boldsymbol{k}$-space is known as the first Brillouin zone, or simply as the Brillouin zone,[24] and denoted by BZ. The Brillouin zones of the bcc and fcc lattices are shown in Fig. 14.24. The physical significance of the BZ will be made clear in the next section.

---

[23] In honour of Wigner and Seitz who introduced this cell in their calculations (the first of the kind) of the electron-energies in bcc and fcc metals: *Physical Review, **43**, 804 (1933); **46**, 509 (1934).*

[24] In honour of Brillouin who introduced the concept in his book: *Die Quantenstatistic*, Berlin, 1931.

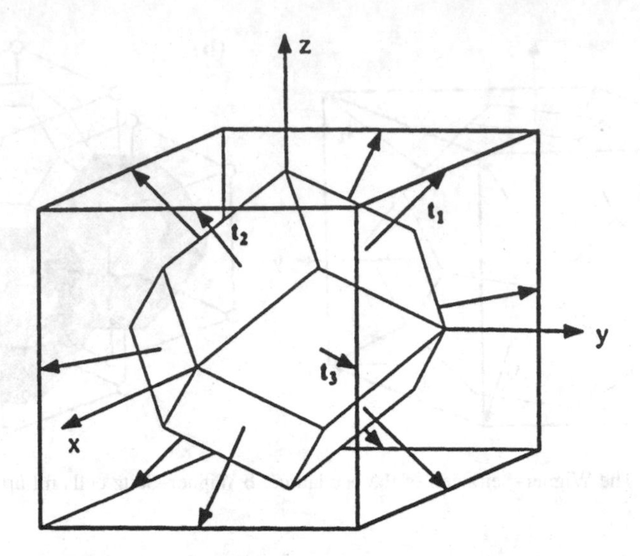

Fig. 14.23 The Wigner–Seitz cell of the fcc lattice

14.7.2 One-electron States in Crystalline Solids

Experience has shown that we obtain a very good description of many of the electronic properties of a crystalline solid by assuming that the electrons in the crystal move *independently* of each other in a potential field $U(\boldsymbol{r})$ which has the periodicity of the space lattice [Eq. (14.109)] associated with the given crystal[25]:

$$U(\boldsymbol{r} + \boldsymbol{R}_n) = U(\boldsymbol{r}) \qquad (14.110)$$

where $\boldsymbol{R}_n$ is any vector of the lattice. The above potential field can be calculated, in principle and nowadays in practice in many cases, self-consistently in essentially the same manner as for atoms (see Sect. 14.5.2).[26] In what follows we assume that $U(\boldsymbol{r})$ is known.

The German physicist Felix Bloch showed [*Zeits. f. Phys. 52, 555* (1928)] that the stationary states (energy-eigenstates) of an electron in the periodic field of Eq. (14.110) can always be written in the following form (Bloch's theorem):

[25] We assume that the nuclei are frozen at their equilibrium positions. At ordinary temperatures the vibration of the nuclei about their equilibrium positions is small and can be disregarded in the calculation of the one-electron levels. We have, of course, assumed the validity of the Born-Oppenheimer approximation.

[26] V. L. Moruzzi, J. F. Janak and A.R. Williams, Calculated Electronic Properties of Metals, New York: Pergamon, 1978. The self-consistent calculation of the one-electron energy levels and of the total energy of the crystal (for stationary nuclei) proceeds in the same way as for atoms (see section 14.5.2). The potential energy field is spherically averaged in a sphere about the nucleus inscribed in the unit cell and volume-averaged to a constant in the remainder of the unit cell.

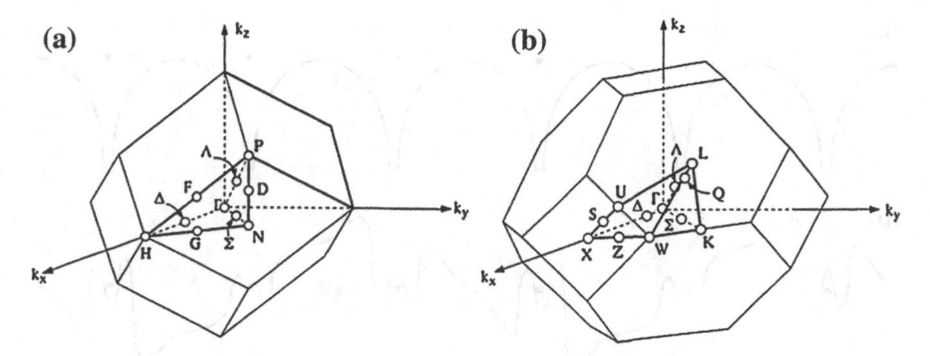

Fig. 14.24 **a** BZ of the bcc lattice. **b** BZ of the fcc lattice

$$\psi_{k\alpha}(\boldsymbol{r}, t) = Aexp(i\boldsymbol{k} \cdot \boldsymbol{r})u_{k\alpha}(\boldsymbol{r})exp(-iE_\alpha(\boldsymbol{k})t/\hbar) \quad (14.111)$$

where $u_{k\alpha}(\boldsymbol{r})$ is a periodic function: $u_{k\alpha}(\boldsymbol{r}) = u_{k\alpha}(\boldsymbol{r} + \boldsymbol{R}_n)$, where $\boldsymbol{R}_n$ is any vector of the given space lattice. A is a normalization constant such that:

$$\int_V |\psi_{k\alpha}(\boldsymbol{r}, t)|^2 dv = NV_0 \int_{V_0} |u_{k\alpha}(\boldsymbol{r})|^2 dV = 1 \quad (14.112)$$

where V_0 is the volume of the unit cell and N is the number of unit cells in the crystal under consideration. The eigenstate described by Eq. (14.111) is known as a Bloch wave. The index α stands for a set of quantum numbers which specify the Bloch wave and the corresponding energy eigenvalue $E_\alpha(\boldsymbol{k})$. The first thing to note is that only values of $\boldsymbol{k}$ within and on the boundary surface of the BZ need to be considered. A Bloch wave with $\boldsymbol{k}'$ outside the BZ is identical with a Bloch wave corresponding to a $\boldsymbol{k}$ within the BZ. In Fig. 14.25 we show schematically the variation of a Bloch wave along a line passing through the centers of a row of atoms.

Finally, we note that Bloch's theorem is strictly valid for an infinite crystal (only then is Eq. (14.110) strictly valid). We get around this difficulty arguing in very much the same way as we have done in our consideration of the modes of the electromagnetic field in a cavity (see Sect. 13.1.1). We replace the actual boundary condition, the vanishing of the wavefunction at the surface of the given crystal, by a periodic boundary condition.[27] We demand that:

$$\psi_{k\alpha}(x + L, y, z) = \psi_{\mathbf{k}\alpha}(x, y + L, z) = \psi_{\mathbf{k}\alpha}(x, y, z + L) = \psi_{\mathbf{k}\alpha}(x, y, z) \quad (14.113)$$

[27] The electronic properties of the bulk of the crystal (specific heat, electrical conducton, etc) are described perfectly well within this scheme. Obviously, when considering properties related to surface phenomena (e.g., electron emission) one must employ a different approach.

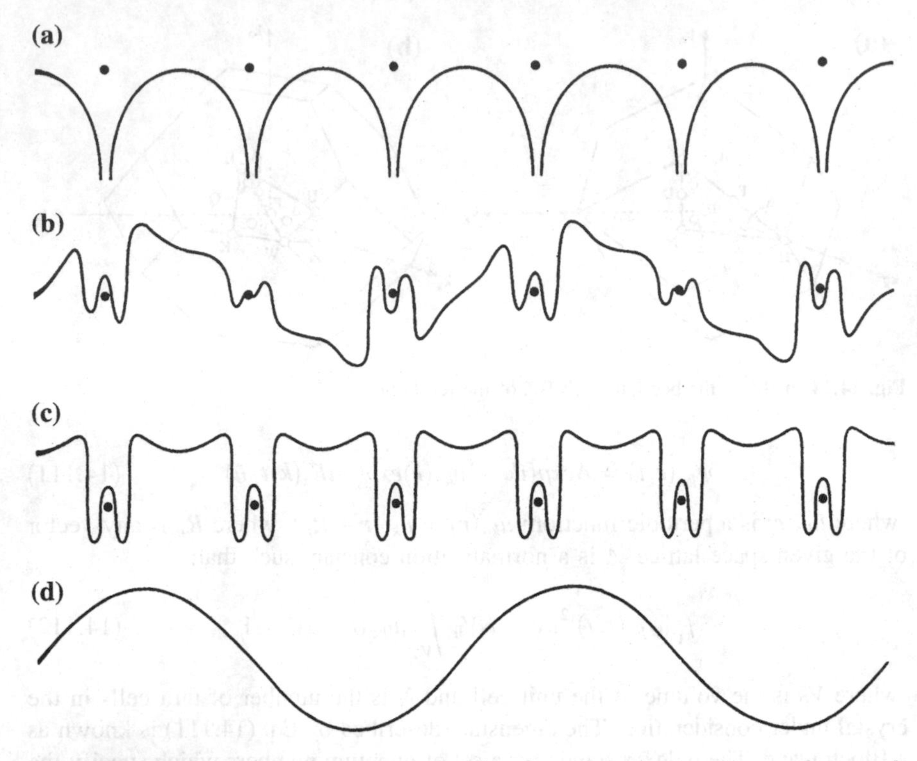

Fig. 14.25 Schematic variation along a line passing through the centres of a row of atoms of: **a** $U(\boldsymbol{r})$; **b** real part of $\psi_{\boldsymbol{k}\alpha}(r, t = 0)$; **c** real part of $u_{\boldsymbol{k}\alpha}(\boldsymbol{r})$; **d** real part of $A\exp(i\boldsymbol{k} \cdot \boldsymbol{r})$. [From: W. A. Harrison, *Solid State Theory*, McGraw-Hill, N.Y., 1970.]

where $L^3 = V$ is the volume of the crystal under consideration, which we assume to be sufficiently large (say a thousand unit cells or more). Eq. (14.113) are satisfied if:

$$\boldsymbol{k} = (k_x, k_y, k_z) = (2\pi/L)(n_x, n_y, n_z)$$

$$n_x,\ n_y,\ n_z = 0,\ \pm 1,\ \pm 2,\ \pm 3, \ldots \tag{14.114}$$

We find the number of allowed (satisfying the above equation) $\boldsymbol{k}$-points in the BZ by noting that there is one such point per volume (in $\boldsymbol{k}$-space) $(2\pi/L)^3 = (2\pi)^3/(NV_0)$, where V_0 is the volume of the unit cell and N the number of unit cells in the crystal under consideration. Given that the volume of the BZ equals $(2\pi)^3/V_0$, we conclude that there are N allowed $\boldsymbol{k}$-points in the BZ, as many as the number of unit cells in the crystal.

14.7.3 Metals

We shall take tungsten (W) as an example. It has a bcc lattice with one atom per unit cell. Its BZ is that of Fig. 14.24a. We know (see the Periodic table in Sect. 14.5.1) that the atom of tungsten has 74 electrons, 68 of which are in inner cells. These are practically unaffected. They do not contribute significantly to the binding of the crystal and do not play any role in the conduction of electricity and the other metallic properties of tungsten. The remaining 6 electrons of the atom are in outer 6*s* and 5*d* shells. These (the 6*s* and the five 5*d* atomic orbitals) develop into Bloch orbitals [see Eq. (14.116)] corresponding to the six *energy bands*: $E_\alpha(\boldsymbol{k})$, $\alpha = 1, 2, \ldots 6$, shown in Fig. 14.26. And we note that we have the same six bands for spin *up* and spin *down*. We often refer to these bands collectively as the conduction band of the metal. The figure shows the bands along certain symmetry directions of the BZ noted in Fig. 14.24a. The reader will note that there are six energy bands [six $E_\alpha(\boldsymbol{k})$] along, for example, the G-line (from point H to point N on the face of the BZ). Along the Δ-line (from the centre Γ of the BZ to point H) there seem to be only five bands. This is so because the Δ_5 band is doubly degenerate: there are two independent Bloch waves corresponding to the $E_\alpha(\boldsymbol{k})$ on this line. The second thing to note is that $E_\alpha(\boldsymbol{k})$ exist for $\boldsymbol{k}$ satisfying Eq. (14.114) anywhere in BZ; and it is only for clarity's sake that we plot $E_\alpha(\boldsymbol{k}i)$ only along the so-called symmetry directions of the BZ. And because L is large, the $\boldsymbol{k}$-points satisfying Eq. (14.114) are close together and appear as continuous lines in the BZ.

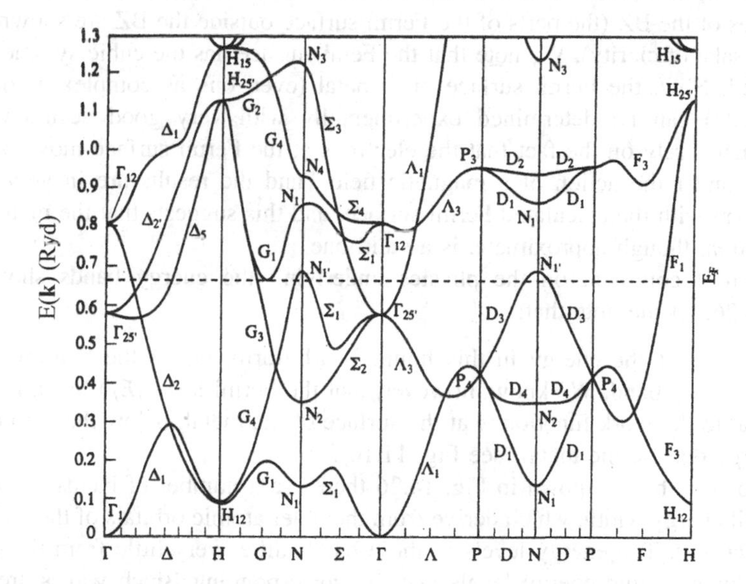

Fig. 14.26 Energy bands $E_\alpha(\boldsymbol{k})$ of tungsten, along various symmetry lines of the BZ (Fig. 14.24a). The energy is measured in Rydbergs (1 Ryd = 13.6 eV). [Calculation by L.F. Mattheiss, Physical Review, 139, A1893,1965).]

And it is also worth noting that the shown symmetry lines lie within a small region (1/48) of the BZ, known as the irreducible part of the BZ. The 48 symmetry operations (rotation about an axis, reflection with respect to a plane etc.) which keep the centre of a cube at the same point and bring the cube upon itself, bring the irreducible part of the BZ to the other regions of the BZ covering in this way its entire space. The bands $E_\alpha(\boldsymbol{k})$ along lines of the BZ, which are obtained from a given line in the irreducible part of the BZ by the above mentioned operations, are the same with the band along the given line.

It follows from the above that there are many one-electron states (Bloch waves) corresponding to different values of $\boldsymbol{k}$ with the same energy E. The $\boldsymbol{k}$-points which satisfy the equation:

$$E_\alpha(\boldsymbol{k}) = E \tag{14.115}$$

where α can be any band, define a surface of constant energy in the $\boldsymbol{k}$-space of the BZ. This can be a simple surface (e.g. a sphere) or a complex one consisting of different pieces that may cross each other. Of particular importance is the so-called Fermi surface of a metal, defined by: $E_\alpha(\boldsymbol{k}) = E_F$, where E_F is the Fermi level defined in Sect. 14.4.3. In a metal, E_F lies somewhere within the conduction band so that, at $T = 0$ all states with $E_\alpha(\boldsymbol{k}) < E_F$ are occupied and those with $E_\alpha(\boldsymbol{k}) > E_F$ are empty. The E_F of tungsten is shown by the horizontal broken line in Fig. 14.26. The Fermi surface of tungsten is shown in Fig. 14.27. We recognize a central piece which cuts into six spherical pieces with centres on the $\pm k_x$, $\pm k_y$, $\pm k_z$ axes, which are joined to the octahedra at the H points of the BZ, and finally we have the ellipsoids which are centred on the N points and are symmetric about the faces of the BZ (the parts of the Fermi surface outside the BZ are shown only for the sake of clarity). We note that the Femi surface has the cubic symmetry as expected. Now, the Fermi surface of a metal (even one as complex as that of Fig. 14.27) can be determined experimentally with very good accuracy (the experiments rely on the fact that the electrons at the Fermi surface move on this surface under the action of a magnetic field), and the results are in very good agreement with the calculated Fermi surface, and this suggests that the method of calculation, though approximate, is a valid one.

Before I comment on the physics underlying the energy bands shown in Fig. 14.26, let me note that:

1. The zero of the energy in this figure is arbitrarily put at the bottom of the conduction band. We know, however, that the Fermi level (E_F) lies a few eV, equal to the work function φ at the surface of the metal, below the zero of the energy outside the metal (see Fig. 11.10).
2. Below the bands shown in Fig. 14.26 there are a number of bands, of a very small energy width, which derive from the inner atomic orbitals of the atoms in the crystal. The energy levels in these bands differ very little from the corresponding atomic energy levels and the corresponding Bloch waves are very well approximated by Eq. (14.117) given below. The electronic cloud due to

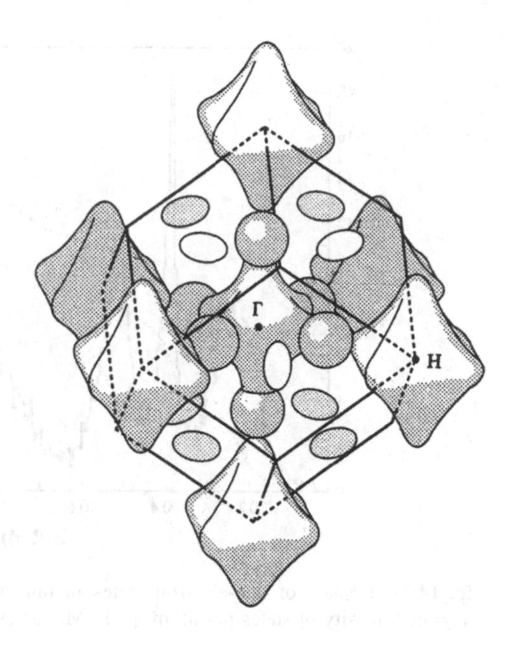

Fig. 14.27 The Fermi surface of tungsten. [L. F. Mattheiss, loc. cit.]

these bands is practically the same as in the free atoms. We need not say anything more about these bands here.

The Bloch waves corresponding to the bands shown in Fig. 14.26 can be written, approximately, as sums of 6*s* and 5*d* atomic orbitals as follows. We remember that the *d*-orbitals correspond to the orbital angular momentum quantum number $l = 2$ and, therefore, there are five such orbitals corresponding to the quantum number $m = -2, -1, 0, 1, 2$. Having the above in mind we write the $u_{k\alpha}(r)$ appearing in Eq. (14.111) as follows:

$$u_{\mathbf{k}\alpha}(\mathbf{r}) = \frac{A}{\sqrt{N}}\sum_{n} exp(-i\mathbf{k}\cdot(\mathbf{r}-\mathbf{R}_n))\varphi_\alpha(\mathbf{r}-\mathbf{R}_n) \qquad (14.116)$$

Where the sum is over all lattice sites of the crystal; we remember that in tungsten there is one atom centred on each site. The $\varphi_\alpha(r - R_n)$ is an appropriate combination of the 6*s* and the five 5*d* electron orbitals of the atom at $\boldsymbol{R}_n$, and accordingly there are six independent such combinations: $\alpha = 1, 2, \ldots, 6$. Substituting Eqs. (14.116) in (14.111), we obtain:

$$\psi_{\mathbf{k}\alpha}(\mathbf{r},t) = \{\frac{A}{\sqrt{N}}\sum_{n} exp(i\mathbf{k}\cdot\mathbf{R}_n))\varphi_\alpha(\mathbf{r}-\mathbf{R}_n)\}exp(-iE_\alpha(k)t/\hbar) \qquad (14.117)$$

Which tells us that an eigenstate of the electron in the crystal is given, approximately, by a linear sum of atomic orbitals centred on the nuclei at the lattice sites of the crystal. The above expression constitutes an extension of the molecular orbitals we used in the case of the hydrogen molecular ion (see Sect.

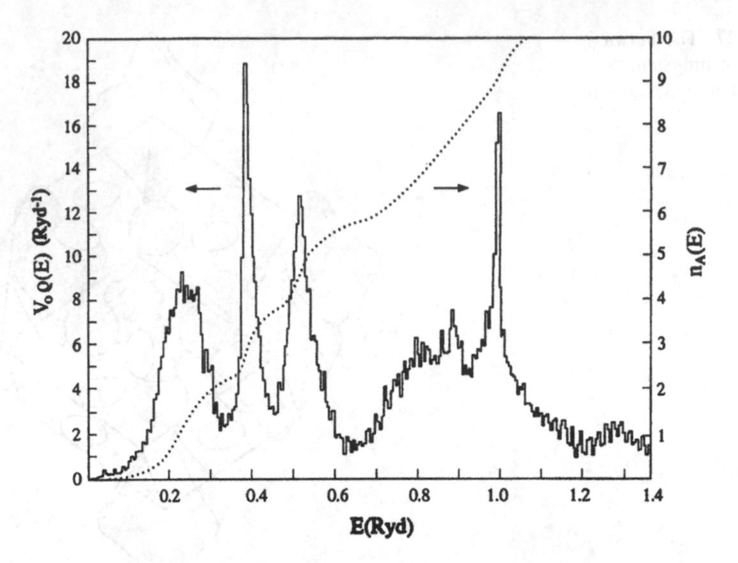

Fig. 14.28 Density of one-electron states in tungsten (*solid line*). The dotted line gives the integrated density of states per atom. [L.F. Mattheiss, loc.cit.]

14.6.2). We had then two molecular orbitals, given by Eqs. (14.101) and (14.103), corresponding to the two energy levels $E_0^{(e)}$ and $E_1^{(e)}$ (arising from the atomic levels) as shown in Fig. 14.17. We now have a band of 'molecular' orbitals represented by the Bloch waves of Eq. (14.117), corresponding to a band of energy levels (arising from the 6*s* and 5*d* energy levels of the tungsten atoms in the crystal). The above way of writing a Bloch wave clarifies its physical meaning, but it is not the best way to calculate $u_{\boldsymbol{k}\alpha}(\boldsymbol{r})$ and the corresponding energy bands (See footnore 26).

Finally, we remember that we have two states with the same $\psi_{\boldsymbol{k}\alpha}(\boldsymbol{r}, t)$ corresponding to spin *up* and spin *down* respectively.

The most important characteristic of the conduction band shown in Fig. 14.26 is that E_F 'cuts' through this band. At $T = 0$, all energy levels below E_F are occupied and those above it are empty. We have, therefore, empty one-electron states in the immediate vicinity, energy- wise, to occupied states. It is this fact that makes tungsten a metal. All metals share this property. In what follows I shall briefly describe how this property determines the electronic contribution to the specific heat of the metal and the electrical conductivity of the metal. To begin with, we define a density of one-electron states with spin *up* as follows: $\rho_\uparrow(E)dE$ is the number of one-electron states with spin *up* per unit volume of the crystal with energy between and E and $E + dE$; similarly $\rho_\downarrow(E)dE$ is the number of one-electron states with spin *down* per unit volume of the crystal with energy between E and $E + dE$. For a non- magnetic metal (like tungsten) we have:

$$\rho_\uparrow(E) = \rho_\downarrow(E) = \rho(E) \tag{14.118}$$

If we know the energy bands, $E_\alpha(\boldsymbol{k})$, we can calculate $\rho(E)$ without much difficulty [we need to remember that $\boldsymbol{k}$ is restricted according to Eq. (14.114)]. In the case of tungsten this is shown in Fig. 14.28. The solid line in this figure shows $V_0\rho(E)$, where V_0 is the volume of the unit cell; and since there is only one tungsten atom in the unit cell, $V_0\rho(E)dE$ is the number of one-electron states per spin, per atom, with energy between E and $E + dE$. The dotted line in the figure gives $n_A(E)$: the number of one-electron states per atom, with energy between zero (the bottom of the conduction band) and E:

$$n_A(E) = 2V_0 \int_0^E \rho(E)dE \tag{14.119}$$

where the factor of 2 takes into account the spin degeneracy. We know that there must be (in total) six electrons per atom in the bands shown in Fig. 14.26. At $T = 0$ these occupy all states up to E_F; therefore E_F is determined so that:

$$n_A(E_F) = 6 \tag{14.120}$$

which implies that $E_F = 0.675$ Ryd. We now know how the Fermi level shown in Fig. 14.26 has been determined. And in the same way one determines the Fermi level in any metal.

Specific heat: As the temperature rises above $T = 0$ the following happens (see Fig. 14.11). Some electrons with energy immediately below E_F take up energy ($\sim kT$) from the environment and move to states with energy above E_F. This is possible, of course, only if $\rho(E) \neq 0$ for E about E_F. The resulting increase, ΔE_{el}, in the *total* energy per unit volume of the metal is:

$$\Delta E_{el} \approx (2\rho(E_F)kT)kT = 2\rho(E_F)k^2T^2 \tag{14.121}$$

Therefore the electronic contribution to the specific heat (per unit volume of the metal) is given by:

$$c_{V,el} = \frac{d}{dT}\Delta E_{el} = 4\rho(E_F)k^2T = c_1T \tag{14.122}$$

where c_1 is a constant. The reader should note that the above result is very different from what we would obtain by treating the conduction band electrons as a Boltzmann gas, as one was bound to assume prior to the development of quantum mechanics. [We note that Eq. (9.19) leads to a specific heat independent of temperature.] The agreement of Eq. (14.122) with the experimental data was one of the first achievements of the quantum theory of solids.

Conductivity: In the same way that a superposition of plane waves produces a wavepacket that travels in space as shown in Fig. 13.14, a wavepacket of Bloch waves (made up of Bloch waves of about the same energy, $E_\alpha(\boldsymbol{k})$, and wavevectors within a narrow range of $\boldsymbol{k}$) describes an electron travelling in the space of the crystal with a velocity: $\boldsymbol{v}_\alpha(\boldsymbol{k}) = (1/\hbar)(\partial E_\alpha/\partial k_x, \partial E_\alpha/\partial k_y, \partial E_\alpha/\partial k_z)$, the terms in the

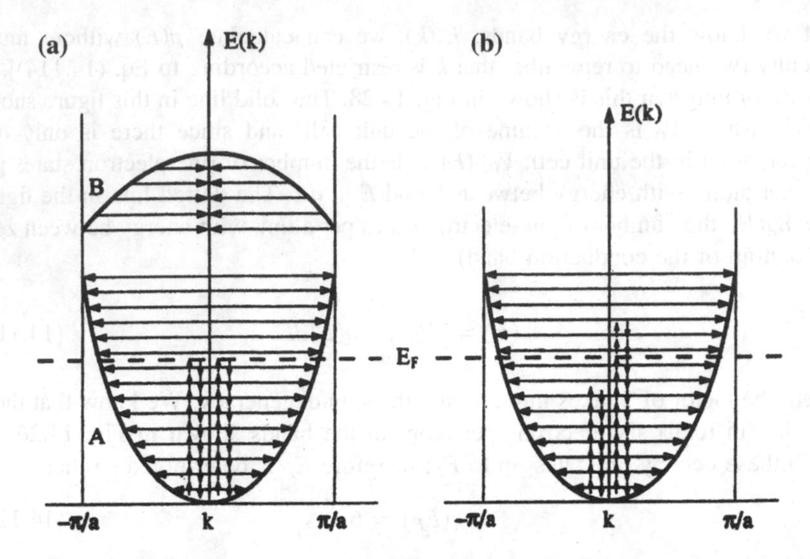

Fig. 14.29 **a** At equilibrium for every electron travelling to the right, there is one travelling with the same velocity to the *left*, so the total current equals zero. **b** More electrons travel to the *right* than to the *left*, so there is a net current different from zero. The *vertical arrow* ↑(↓) signifies the existence of an electron with spin up (*down*) at the given energy level. The velocity of the electrons is denoted by *horizontal arrows*

parenthesis corresponding to the components of the velocity along the x, y, and z directions. At equilibrium for every electron travelling to the right there is an electron travelling with the same velocity to the left, as shown schematically in Fig. 14.29a. It turns out that, when an electric field, F, exists in the metal (we assume it to be along the x-direction) the electron distribution changes to that of Fig. 14.29b: In time δt the k_x component of the electrons in the conduction band changes to $k_x + \delta k_x$ in accordance with the semi-classical formula:

$$\delta k_x = (-eF/\hbar)\delta t \tag{14.123}$$

At some moment, say when $\delta t = 2\tau$, collisions with impurities in the metal and with the vibrating lattice, bring the electrons back to their equilibrium distribution, so that on average (in the presence of the applied field) we have:

$$\delta k_x = (-eF/\hbar)\tau \tag{14.124}$$

where τ is known as the collision time. Its actual value depends on the concentration of impurities in the metal and on the temperature (since the amplitude of the vibrating nuclei depends on the temperature). A typical value at $T = 300$ K would be $\tau \approx 10^{-14}$ s. Accordingly, the average distance an electron (at the Fermi level) travels before suffering a collision, known as the *mean free path* is of the order of a few hundred angstrom for a typical metal at $T = 300$ K.[28] A detailed calculation of the current density corresponding to a given applied field, is somewhat technical and we shall not reproduce it here. One obtains the following formula for the conductivity, σ, of the metal (as defined in Eq. (10.19); we note that the applied field is denoted by E in that equation):

$$\sigma = \frac{2}{3} e^2 \overline{\upsilon^2}(\mathrm{E}_F)\rho(E_F)\tau \tag{14.125}$$

where $\overline{\upsilon^2}(E_F)$ is the average of the square of $\upsilon_\alpha(\boldsymbol{k})$, defined above, over all states at the Fermi surface. We note the presence of $\rho(E_F)$ in the formula. If $\rho(E_F)$ vanishes the conductivity vanishes. If $\rho(E_F)$ is different from zero (as it is the case for metals) we have a finite conductivity at any temperature (including $T = 0$). The rearrangement of the electrons shown in Fig. 14.29b, is possible only if there are empty one-electron states in the immediate vicinity, energy-wise, of occupied ones.

At this point I should mention that the first quantum-mechanical treatment of metals, similar in principle to the one we have described above, is due to Sommerfeld,[29] who in 1928 proposed that, a number of electrons, say N in a volume V of the metal are essentially free: the energy levels of an electron are those of a free particle in a box given by Eq. (14.78). When V is sufficiently large, these energy levels form a quasi-continuum like in the conduction band of a real metal. The corresponding density of states $\rho(E)$, defined by Eq. (14.118), is given by:

$$\rho(E) = (4\pi/h^3)\left(2m^3\right)^{1/2}\sqrt{E} \tag{14.126}$$

The number of *free* electrons (we can think of them as the conduction band electrons) per unit volume of the metal, given by $n = N/V$, is treated as an empirical parameter, which in turn determines the Fermi level of the metal from the formula:

$$n - 2\int_0^{E_F} \rho(E)dE \tag{14.127}$$

where the factor of 2 takes into account spin degeneracy. In Sommerfeld's model one chooses the value of n so that $\rho(E_F)$, when substituted in Eq. (14.122), gives the value of the specific heat determined experimentally. One may then use the same $\rho(E_F)$ in Eq. (14.125) for the conductivity, which in turn allows one to determine the collision time τ after measuring σ.

---

[28] The curvature in a macroscopic wire is negligible in comparison with the mean free path of the electron, and we are therefore justified in replacing the wire by a straight line along the x-axis, as we have done.

[29] A. Sommerfeld, *Z. Physik*, *47*, *1* (1928).

Sommerfeld's model was also employed in the quantum-mechanical treatment of electron emission from metal surfaces (see Sect. 11.5). But there are things that this model can not explain. A free electron model, can not account for a complicated Fermi surface, such as the one shown in Fig. 14.27. And it can not account for the observed optical transitions: When a photon with energy $\hbar\omega$ penetrates a metal, it can be taken up by an electron with energy $E_\alpha(\boldsymbol{k}) < E_F$ which then transits to a state of the same $\boldsymbol{k}$ with energy $E_\beta(\boldsymbol{k}) = E_\alpha(\boldsymbol{k}) + \hbar\omega > E_F$. The probability of such a transition is not zero when the corresponding states are described by Bloch waves, but vanishes identically when the initial and final states of the electron are those of a free particle as in the Sommerfeld model. In this respect Sommerfeld's model is not different from many other empirical models which are proposed at times in order to explain certain experimental data. The empirical model will explain these data, may be a few others, but it is unlikely that it will have a general validity. And though the independent electron model we introduced in Sect. 14.7.2 has a much wider applicability than Sommerfeld's model, it also is limited: there are phenomena which it can not explain (see, e.g., Sect. 14.7.8).

14.7.4 Ferromagnetic Metals

In the case of tungsten and other non magnetic metals, such as copper, silver, etc., the number of electrons in the conduction bands with spin *up* is equal to the number of electrons with spin *down*. But this need not be always the case. It may happen that the total energy of the metal in its ground state, calculated by a modified version of Eq. (14.95), is lower if there are more electrons with, say, spin *up* than there are electrons with spin *down*. In that case the potential field an electron 'sees' depends on its spin. $U_{ex}(\boldsymbol{r})$ of Eq. (14.94) is different for the two spins: $R = R_+$ (R_-) for spin *up* (*down*) in accordance with Eq. (14.91), and therefore the energy bands depend on the spin. And, of course, the potential fields for the two spins and the corresponding energy bands must be self consistent. Let us take ferromagnetic nickel (Ni) as an example. It has an fcc space lattice with one atom per unit cell. The electrons (18 per atom) from the inner shells of the atoms occupy lower energy bands and do not affect the magnetization (magnetic moment per unit volume) of the metal. The remaining ten electrons (per atom) are accommodated in conduction bands qualitatively similar to those of tungsten, but those of spin *up* (the so-called majority bands) are on the average displaced downward relative to those of spin *down* (minority bands) as shown in Fig. 14.30.

At $T = 0$ all states with $E < E_F$ are occupied, but because of the downward displacement of the spin *up* states, there will be more electrons with spin *up* than with spin down. We define, by analogy to Eq. (14.119), $n_{A\uparrow}(E)$ and $n_{A\downarrow}(E)$ to be the number of electrons per atom, with energy between zero (the bottom of the conduction bands shown in Fig. 14.30) and E, with spin *up* and *down* respectively. The E_F is determined (as in Eq. (14.120) for tungsten) by the requirement:

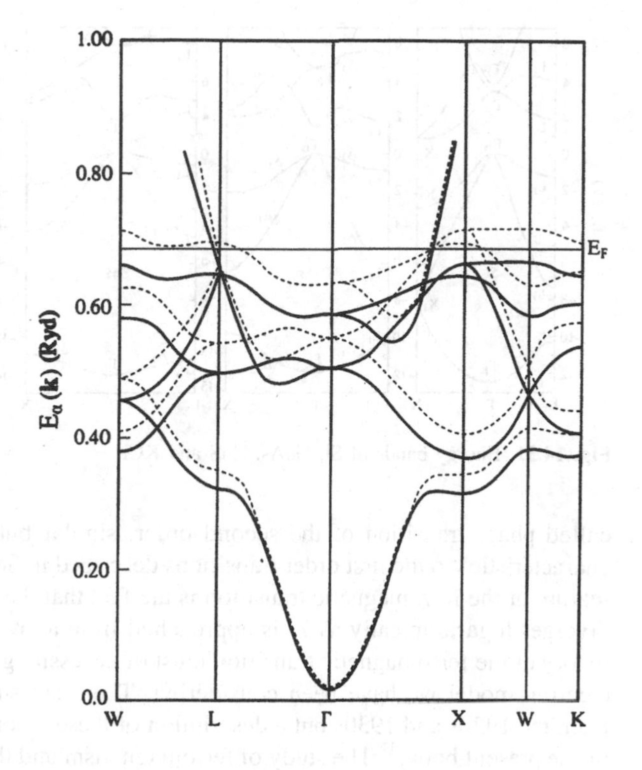

Fig. 14.30 Energy bands of ferromagnetic nickel along various *symmetry lines* of the BZ (Fig. 14.24b). Solid lines: spin *up*; broken lines: spin *down*. [Calculation by V. L. Moruzzi, J. F. Janak and A.R. Williams.(See footnote 26)]

$n_{A\uparrow}(E_F) + n_{A\downarrow}(E_F) = 10$. And, finally, the magnetization of nickel (the magnetic moment per atom) is given by $(n_{A\uparrow}(E_F) - n_{A\downarrow}(E_F))\mu_B$, where $\mu_B = e\hbar/2m$ is the magnetic moment associated with the spin of the electron known as the Bohr magneton [see Eqs. (14.38) and (Sect. 15.24)]. The calculated value of this quantity, $0.59\mu_B$ per atom is in very good agreement with the measured value of this quantity ($0.60\mu_B$ per atom as $T \to 0$).

I should emphasize that while the band theory of ferromagnetism described above gives the correct magnetization at $T = 0$, it can not describe the variation of the magnetization with the temperature. Observations tell us that the magnetization of a ferromagnetic metal, such as nickel and iron, is reduced as temperature is raised above $T = 0$, and vanishes altogether above a critical temperature T_c, known as the Curie temperature. For nickel, $T_c = 631$ K; for iron, $T_c = 1043$ K. At T_c the metal passes from its ferromagnetic to its paramagnetic phase,[30] in a so-

[30] In the paramagnetic phase ($T > T_c$) there is no spontaneous magnetization. However, the metal acquires a finite magnetization under the influence of an external magnetic field. The coupling of the spin magnetic moment with the external field leads to a small decrese or increase of the electron energy depending on whether its magnetic moment is parallel or antiparallel to the direction (z) of the magnetic field. This means, given that E_F is constant, that there will be more electrons with a magnetic moment parallel to the external field than antiparallel, and in this way an induced magnetization (***M***) is obtained proportional to the applied field (***B***). At low

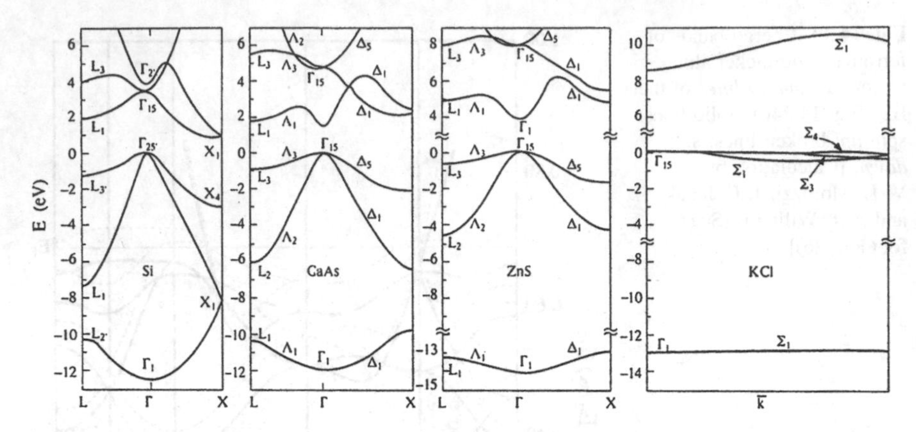

Fig. 14.31 Energy bands of Si, GaAs. ZnS and KCl

called phase transition of the second order, similar but different in some of its characteristics to the first order transitions described in Sect. 9.1.3. A characteristic feature of the ferromagnetic transition is the fact that the specific heat of the metal diverges logarithmically as T_c is approached from above and also from below. A theory of the ferromagnetic transition must of necessity go beyond the independent electron model we have been considering. There are such theories, some dating from the 1920s and 1930s but a description of these theories lies beyond the scope of the present book.[31] The study of ferromagnetism and the related phenomenon of antiferromagnetism, and of the corresponding phase transitions continues to the present day.

14.7.5 Semiconductors

In Fig. 14.31 we show sections of the energy bands of silicon (Si), gallium arsenide (GaAs), zink sulphide (ZnS) and potassium chloride (KCl). The four of them are non magnetic so we have the same bands for spin *up* and *down*. In each case, at $T = 0$, all one-electron states with energy below $E = 0$ (chosen arbitrarily) are occupied (we refer to the corresponding bands collectively as the *valence* band of the crystal; and all states with energy $E > 0$ are empty (we refer to the corresponding bands collectively as the *conduction* band. And we remember that below the valence band are a number of bands arising from the inner shells of

(Footnote 30 continued)
temperatures, this is given by: $\boldsymbol{M} = \chi \boldsymbol{B}$, where $\chi = 2\mu_{\mathrm{B}}^2 \rho(E_F)$, with $\rho(E)$ given by Eq.(14.118). This formula was first derived by Pauli.

[31] E. Ising, *Z. Physik, 31, 253* (1925); W. Heisenberg, *Z. Physik,* ***49****, 619* (1928); W. L. Bragg and E. J. Williams, *Proc. Soc. London Ser. A* ***145****, 699 (1934).*

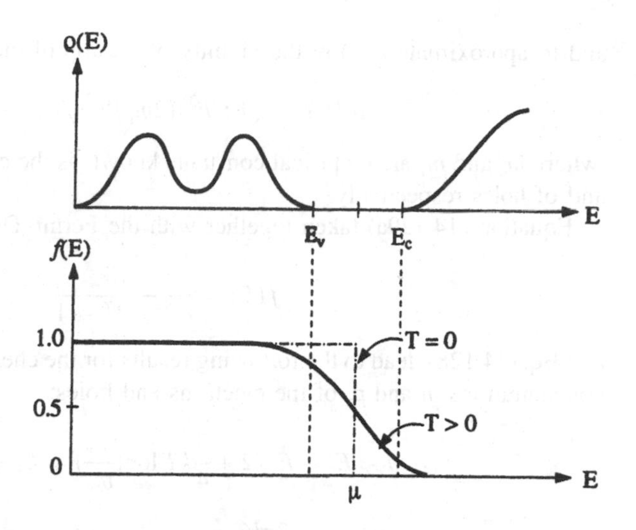

Fig. 14.32 Schematic description of the density of one-electron states per spin, $\rho(E)$, of a semiconductor; and of the probability $f(E)$ that a state of energy E will be occupied when $T = 0$, and when $T > 0$

the atoms which will not concern us here. We refer to the energy difference between the minimum of the conduction band and the maximum of the valence band as the energy gap, E_g, of the crystal. The energy gaps of the above crystals and that of Germanium (Ge), which has a crystal structure and an energy band-structure similar to that of silicon, are listed in the following table:

	Ge	Si	GaAs	ZnS	KCl
$E_g(eV)$:	0.67	1.11	1.43	3.56	8.2

Crystals with relatively small energy gaps are known as semiconductors, and those with larger energy gaps are known as insulators for reasons that will become apparent in what follows. Ge, Si, and GaAs are by this measure semiconductors, and KCl is an insulator. ZnS lies somewhere between the two; whether we call it an insulator or a semiconductor is unimportant.

In Fig. 14.32 we show schematically the density of one-electron states in a semiconductor, and the probability of occupation of these states at $T = 0$ and at $T > 0$ according to Eq. (14.86).

We see that when $T > 0$, a number, n, of electrons per unit volume move up from the valence band to the conduction band (to energy levels in the vicinity of the bottom, E_c, of this band), leaving a number, p, of empty states (so-called holes) per unit volume in the valence band (with energies in the vicinity of the top, E_v, of this band). Obviously:

$$n = p \tag{14.128}$$

It is usual, and reasonable, to approximate $\rho(E)$ in the vicinity of the bottom of the conduction band in the form of Eq. (14.126):

$$\rho_c(E) = (4\pi/h^3)(2m_c^3)^{1/2}\sqrt{(E - E_c)} \tag{14.129a}$$

and to approximate $\rho(E)$ in the vicinity of the top of the valence band by:

$$\rho_v(E) = (4\pi/h^3)(2m_v^3)^{1/2}\sqrt{(E_v - E)} \tag{14.129b}$$

where m_c and m_v are empirical constants known as the effective mass of electrons and of holes respectively.

Equation (14.129a) taken together with the Fermi–Dirac distribution,

$$f(E) = \frac{1}{e^{(E-\mu)/kT} + 1} \tag{14.130}$$

and Eq. (14.128), lead to the following results for the chemical potential, μ, and the concentrations, n and p, of the electrons and holes:

$$\mu \approx E_v + E_g/2 + \frac{3}{4}kT \log(\frac{m_v}{m_c}) \approx E_v + E_g/2 \tag{14.131}$$

$$n = p = 2(\frac{2\pi kT}{h^2})^{3/2}(m_v m_c)^{3/4} \exp(-E_g/2kT) \tag{14.132}$$

Obviously, the change ΔE in the total energy of the semiconductor due to the rearrangement of electrons as the temperature increases is given by $\Delta E \approx nE_g$, and since n varies exponentially with the temperature according to Eq. (14.132), the specific heat, $c_V = d(\Delta E)/dT$, will also vary exponentially with the temperature, and because $E_g \gg kT$ it will be very small; for an insulator is practically zero.

Let us consider the conductivity (σ) of a semiconductor. At $T = 0$ the valence band is completely occupied and the conduction band completely empty. There are no empty states next to occupied ones, and therefore no possibility of motion (see Sect. 14.7.3), and therefore $\sigma = 0$. As T increases above zero the situation gradually changes: next to the occupied energy levels in the conduction band there are empty ones and, similarly, next to the empty levels at the top of the valence band there are occupied ones, and therefore the basic requirement for electrical conductivity to occur is satisfied to some degree. This is indeed so, and we find that σ is given by:

$$\sigma = e\,\mu_n\,n + e\,\mu_p\,p \tag{14.133}$$

where n and p are given by Eq. (14.132), and μ_n and μ_p are certain constants (they depend on the material and the temperature) known as the mobility of the electrons and the mobility of holes respectively. Because of the smallness of n and p at ordinary temperatures, even for small-gap semiconductors, the conductivity is very much smaller than that of a metal (see Sect. 10.2.5), but sufficient for the purpose of carrying small signals. Moreover, it was soon discovered that one can generate electrons in the conduction band and/or holes in the valence band by means other than an increased temperature. It was discovered that introducing a relatively small amount of impurities in an intrinsic (pure) semiconductor (as the one we have

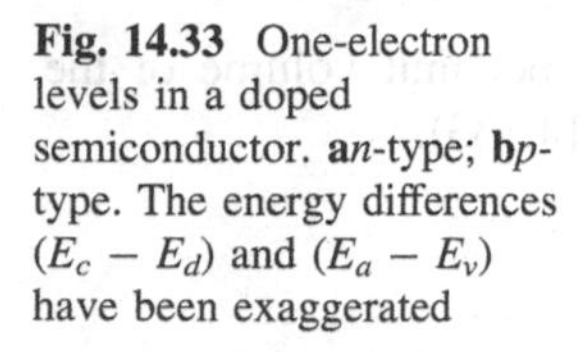

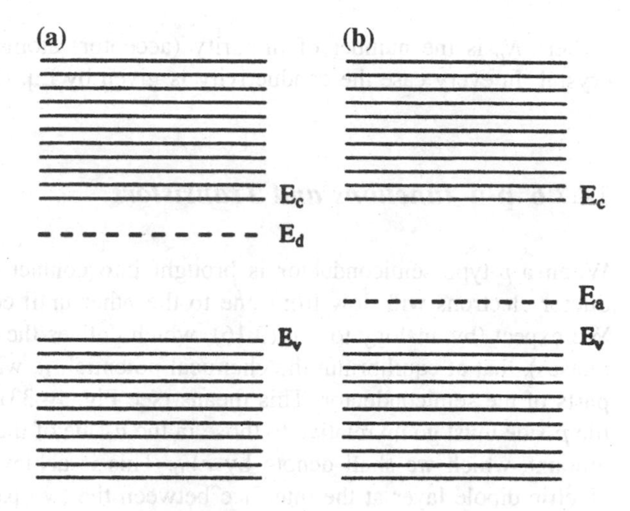

Fig. 14.33 One-electron levels in a doped semiconductor. **a** *n*-type; **b** *p*-type. The energy differences $(E_c - E_d)$ and $(E_a - E_v)$ have been exaggerated

described) is an effective way of achieving this end.[32] In what follows we shall assume that the semiconductor is either silicon or germanium. Both have an fcc space lattice, with two atoms per unit cell. And both atoms have *four* electrons in their outer shells. The eight electrons (four for each atom in the unit cell) fill up the valence bands of the corresponding crystals. Now, if a small number of the atoms in the pure crystal, say 10^{15} per cm^3 (we remember that there are about 10^{23} atoms per cm^3 in the pure crystal) are replaced by atoms which have *five* electrons in their outer shells, the '*fifth*' electron is accommodated in a localized state about the impurity atom (within a volume that contains many atoms of the host crystal), at an energy E_d a little below E_c as shown schematically in Fig. 14.33a. In this case the chemical potential, μ, at $T = 0$, lies just above E_d. At ordinary temperatures the electrons are removed from the above localized states into the conduction band, so that n is given by:

$$n \approx N_d \tag{14.134}$$

where N_d is the number of impurity (donor) atoms per unit volume of the crystal. Similarly, if an atom in the pure crystal, is replaced by an atom which has *three* electrons in its outer shells, an empty state is generated (an electron state is expelled from the valence band) localized about the impurity atom, at an energy level E_a a little above E_v as shown schematically in Fig. 14.33b. In this case the chemical potential, μ, at $T = 0$, lies just below E_a. At ordinary temperatures electrons are removed from the valence band into these localized states leaving an equal number of empty states (holes) in the valece band, so that p is given by:

$$p \approx N_a \tag{14.135}$$

32 The verification of the properties of intrinsic and doped semiconductors became possible with the development of appropriate methods of preparation in the 1940s.

where N_a is the number of impurity (acceptor) atoms per unit volume of the crystal. In every case the conductivity is given by Eq. (14.133).

14.7.6 p-n *Junctions and Transistors*

When a *p*-type semiconductor is brought into contact with an *n*-type semiconductor electrons will flow from one to the other until equilibrium is established. We expect (by analogy to Eq. (9.16), which defines the equilibrium between two phases), that at equilibrium the chemical potential, μ, will be the same in the two parts of the semiconductor. This means (see Fig. 14.33) that the energy levels in the *p*-side must go up relative to those in the *n*-side of the *p*-*n* junction by a certain amount, which we shall denote by eV_0. This is achieved by the creation of an electric dipole layer at the interface between the two parts: electrons move from the *n*-side to the *p*-side, creating an electric field between the two, similar to the one between the plates of a charged condenser; the width of this transition region (dipole layer region) is denoted by W in the top diagram of Fig. 14.34a. The bending of the bands across the transition region at equilibrium is shown in the bottom diagram of Fig. 14.34a.

At equilibrium we have so many electrons crossing the interface from the *n*-side to the *p*-side per unit time as there are electrons crossing the interface in the opposite direction. When an external voltage, V, is applied to the junction in the forward bias, as shown in Fig. 14.34b, the barrier the electrons have to overcome in crossing from the *n*-side to the *p*-side is reduced and the current in this direction increases exponentially, but the flow of electrons in the opposite direction remains the same (very small): the electrons on the *p*-side must be raised from the top of the valence band to the bottom of the conduction band, and this process, determined by thermal agitation, is not significantly affected by the applied voltage. As a result the *net* current flowing through the junction increases exponentially with the applied voltage for forward bias, as shown in Fig. 14.35. When an external voltage is applied to the junction in reverse bias, the energy diagram changes to that of Fig. 14.34c; as a result the flow of electrons from the *n*-side to the *p*-side practically vanishes, while that in the opposite direction is unchanged, which means that the current through the junction in reverse bias is very small, as shown in Fig. 14.35. The *I*–*V* characteristic presented in this figure shows that the *p*-*n* junction can be an effective rectifier of small a.c. currents/voltages. It does, for small currents, what the vacuum diode did in the past (see Sect. 11.5.1).

In Fig. 14.36a we show an example of a *p*-*n*-*p* transistor circuit. The emitter of the transistor (E) is *p*-type, the base (B) is *n*-type, and the collector (C) is *p*-type. In the absence of the a.c. voltages, υ_{BE} and υ_{CE}, the collector current, I_C, flowing through the 500 Ω resistance, is much greater than the base current, I_B, flowing through the 50 kΩ resistance. In our example we have $\beta = I_C/I_B = 100$. Adding a small a.c. variation, i_b, to the base current results in a much greater a.c. variation,

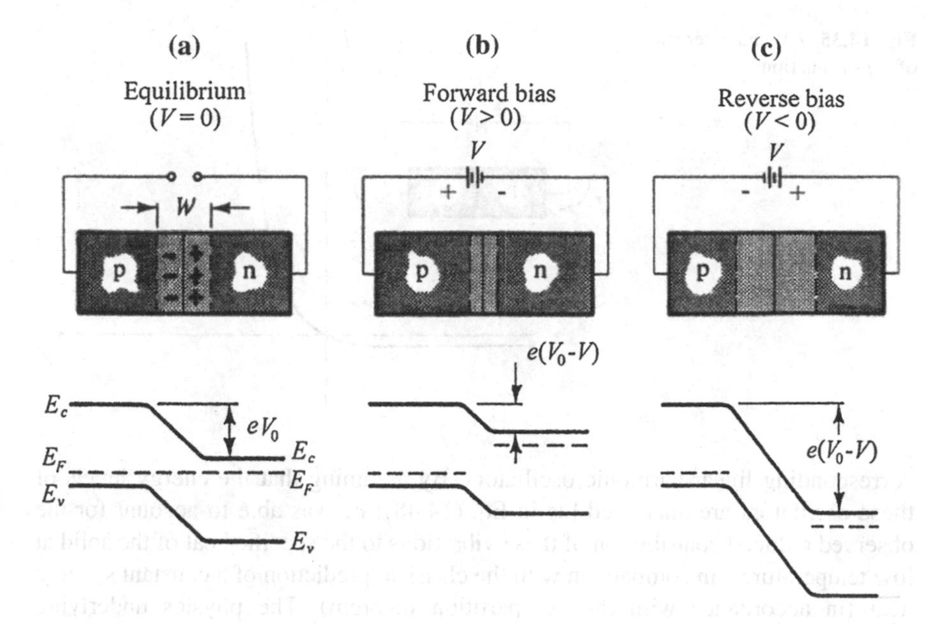

Fig. 14.34 Energy-band bending across a *p-n* junction (schematic). (**a**) *equilibrium* ($E_F = \mu$), (**b**) *forward bias*, (**c**) *reverse bias*

$i_c = \beta i_b$, in the collector current, as shown in Fig. 14.36b. Which demonstrates the efficiency of the transistor as an amplifier of small a.c. currents and voltages. We note that the operation of the transistor is analogous to that of the vacuum triode (see Sect. 11.5.1).

The first working transistor was invented by Walter Brattain, John Bardeen and William Shockley at the Bell Laboratories in December 1947. A discovery which earned them the 1956 Nobel Prize for Physics. The junction transistor shown in Fig. 14.36a, and the similar *n-p-n* transistor are known as bipolar transistors. There is another type of transistor, the field-effect transistor proposed by Shockley, which operates along different lines but with the same effect when it comes to the amplification of small signals. Needless to say that the discovery of the transistor and the subsequent development of integrated circuits by the Americans Jack Kilby and Robert Noyce in 1958 played and continues to play a central role in the development of modern day technologies.

14.7.7 Lattice Vibrations

The first quantum theory of lattice vibrations was proposed by Einstein in 1907. He assumed that each atomic nucleus in the crystal vibrates about its equilibrium position independently: its vibrations along the x, y and z directions represented by

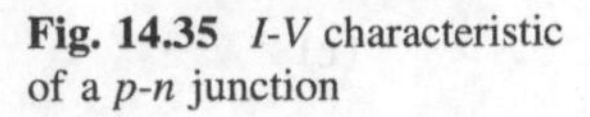
Fig. 14.35 *I-V* characteristic of a *p-n* junction

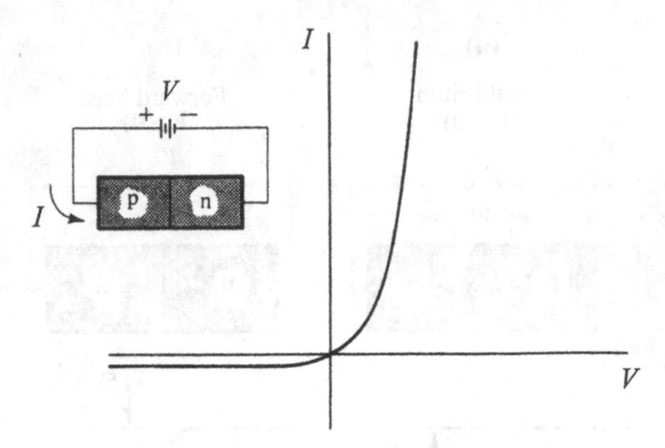

corresponding linear harmonic oscillators. By assuming that the energy levels of these oscillators are quantized [as in Eq. (14.48)] he was able to account for the observed reduced contribution of these vibrations to the specific heat of the solid at low temperatures, in comparison with the classical prediction of a constant specific heat (in accordance with the equipartition theorem). The physics underlying Einstein's theory is the same with that we described in relation to the vibration of a diatomic molecule (see Sects. 9.2.4 and 14.6.3).

In reality the nuclei in the crystal do not vibrate independently, they are coupled to each other and vibrate in the same way that particles connected with each other, via elastic springs, vibrate. The periodicity of the crystal allows us to write the displacement of the atom in the i_{th} unit cell (we assume that there is only one atom in a unit cell) as follows[33]:

$$\boldsymbol{R}_i^{(\mathbf{k},\alpha)}(t) = \mathbf{A}_0^{(\mathbf{k},\alpha)} exp(i\mathbf{k}\cdot\mathbf{R}_i) exp(i\omega_\alpha(\mathbf{k})t) \tag{14.136}$$

where $\boldsymbol{k}$ is a wavevector in the BZ given by Eq. (14.114); α is a band index whose meaning will be clarified by what follows, and $\boldsymbol{A}_0^{(\boldsymbol{k},\alpha)}$ is a constant. For a crystal with one atom per unit cell, we obtain, for given $\boldsymbol{k}$, *three* independent modes of vibration, described by the above formula, corresponding to certain frequencies: $\omega_\alpha(\boldsymbol{k})$, $\alpha = L, T_1, T_2$; the first (L) is a longitudinal vibration (along certain symmetry directions the nuclei vibrate parallel to $\boldsymbol{k}$) and the other two (T_1 and T_2) are transverse vibrations (along certain symmetry directions the nuclei vibrate normal to $\boldsymbol{k}$). In Fig. 14.37 we show the calculated frequency bands for lead (an fcc crystal with one atom per unit cell). We remember that there are N allowed $\boldsymbol{k}$-points in the BZ, as many as there are unit cells in the crystal. In the case of lead there is one atom per unit cell, so that the number of allowed $\boldsymbol{k}$-points in the BZ equals the number of atoms in the crystal. For each $\boldsymbol{k}$ we have three independent modes of vibration, so that there are three such modes per atom as expected. [When there are

[33] To be exact I should say that the displacement of the nucleus is given by a sum of the real and imaginary parts of the given expression.

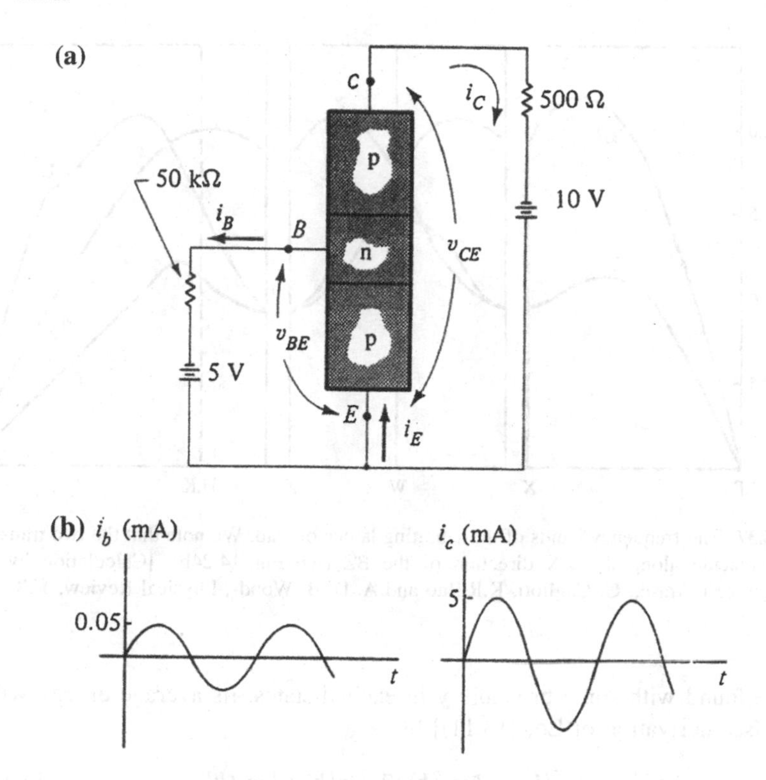

Fig. 14.36 An example of amplification in a common-emitter transistor circuit

two or more atoms per unit cell, there will be additional frequency bands at higher frequencies.]

I should emphasize that the calculation of the frequency bands, $\omega_\alpha(\boldsymbol{k})$, and of the corresponding displacements, $\boldsymbol{R}_i^{(\mathbf{k},\alpha)}(t)$, is entirely classical. Each atomic nucleus moves under the action of the force exerted on it by the other atoms, as a result of their elastic displacement from their equilibrium positions, according to Newton's law of motion. We need not go into the details of such a calculation here (see exercise 14.26). Once the frequency bands have been calculated in the above manner, each vibrational mode $(\boldsymbol{k},\alpha)$ is treated as an independent linear harmonic oscillator whose energy levels are, according to Eq. (14.48), given by: $E_n(\boldsymbol{k},\alpha) = \hbar\omega_\alpha(\boldsymbol{k})(n + 1/2)$, $n = 0, 1, 2, \ldots$

We observe that for $\boldsymbol{k}$ along any direction (defined by the angles θ, φ as in Fig. 14.1) every one of the above frequency bands varies linearly with k near $k = 0$ (the centre of the BZ): $\omega_\alpha(k) = c_\alpha(\theta,\varphi)k$, for k along the direction of (θ,φ) and small. These long-wavelength vibrational waves are acoustic waves and $c_\alpha(\theta,\varphi)$ is the corresponding sound velocity along the given direction.

The contribution of vibrations to the specific heat: at $T = 0$ every oscillator, $\omega_\alpha(\boldsymbol{k})$, is in its ground state. As the temperature is raised above zero, the oscillator

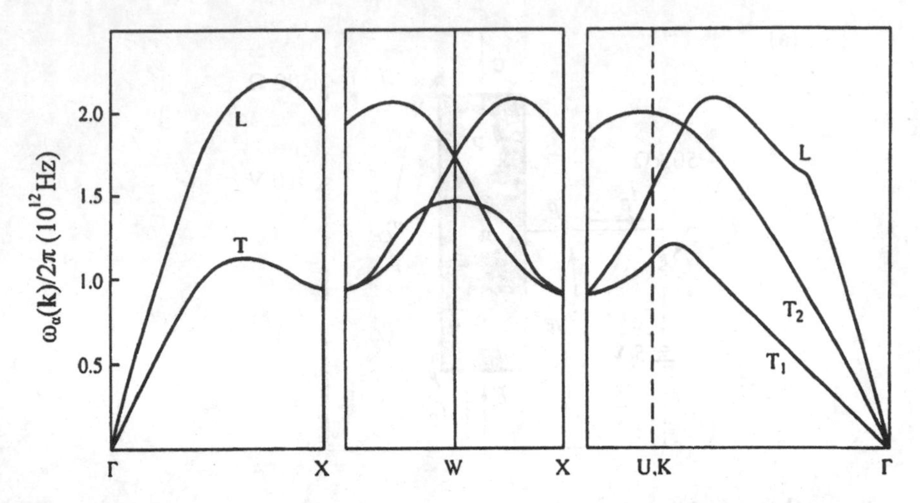

Fig. 14.37 The frequency bands of the vibrating lattice of lead. We note that the two transverse bands coincide along the ΓX direction of the BZ (see Fig. 14.24b). [Calculation by B.N. Brockhouse, T.Arase,, G. Caglioti, K.R.Rao and A. D. B. Woods, Physical Review, **128**, 1099, 1962]

will be found with some probability in excited states. Its average energy will be given [see derivation of Eq. (13.11)] by:

$$\overline{E_\alpha}(\boldsymbol{k}) = \hbar\omega_\alpha(\boldsymbol{k})/2 + n(\boldsymbol{k},\alpha)\hbar\omega_\alpha(\boldsymbol{k}) \tag{14.137a}$$

$$n(\boldsymbol{k},\alpha) = 1/(exp(\hbar\omega_\alpha(\boldsymbol{k})/kT) - 1) \tag{14.137b}$$

In evaluating the temperature variation of the energy associated with the lattice vibrations of the crystal, we can disregard the constant first term in Eq. (14.137a). In which case the contribution of the vibrations to the total energy per unit volume of the crystal is given by:

$$\Delta E_{vb} = \frac{1}{V}\sum_{\boldsymbol{k},\alpha} n(\boldsymbol{k},\alpha)\hbar\omega_\alpha(\boldsymbol{k}) = \int (\frac{\hbar\omega}{\exp(\hbar\omega/k\mathrm{T}) - 1})D(\omega)d\omega \tag{14.138}$$

where $V = L^3$ is the volume of the crystal and $D(\omega)d\omega$ is the number of vibrational modes, per unit volume, with frequency between ω and $\omega + d\omega$. The contribution of the vibrations to the specific heat per unit volume of the crystal is, accordingly, given by:

$$c_{V,vb} = \frac{d}{dT}\Delta \mathrm{E}_{vb} \tag{14.139}$$

Once the frequency bands have been obtained, one can calculate numerically $D(\omega)$ without much difficulty [we must remember that $\boldsymbol{k}$ is restricted according to Eq. (14.114)], and then calculate $c_{V,vb}$ using Eqs. (14.138) and (14.139). However, it turns out that for most crystals $c_{V,vb}$ is given to a very good approximation, by a

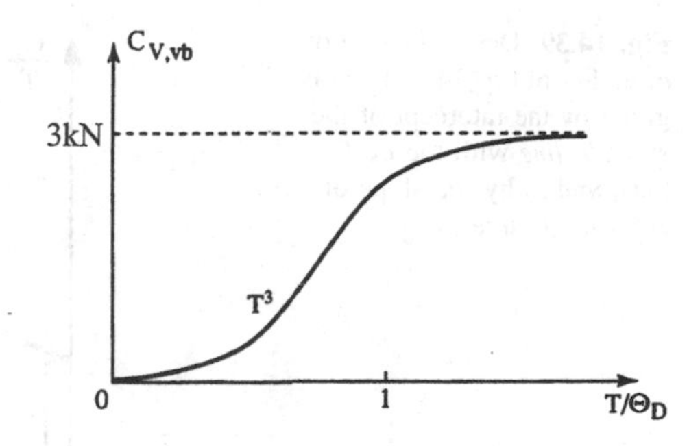

Fig. 14.38 The Debye law of specific heat

model of $D(\omega)$ proposed by the German physicist Peter Debye (1884–1966). In the Debye model one assumes that there are *three* acoustic-like bands described by:

$$\omega_\alpha(\mathbf{k}) = c_s k = c_s(k_x^2 + k_y^2 + k_z^2)^{1/2}, \quad \alpha = 1, 2, 3 \tag{14.140}$$

where c_s is an average velocity of sound, and $\boldsymbol{k}$ takes all values of Eq. (14.114) within a sphere of radius k_D, so chosen that there are so many $\boldsymbol{k}$-points in the sphere as there are atoms in the crystal. The density of vibrational modes in the Debye model is given by:

$$\begin{aligned} D(\omega) &= \frac{3\omega^2}{2\pi^2 c_s^3}, \quad \omega < \omega_D \\ &= 0, \qquad\quad \omega > \omega_D \end{aligned} \tag{14.141}$$

where $\omega_D = c_s k_D$. The contribution of the vibrations to the specific heat of a crystal according to the Debye model is shown schematically in Fig. 14.38. The temperature Θ_D is an empirical parameter, known as the Debye temperature defined by: $k\Theta_D = \hbar\omega_D$. [Typical values of Θ_D: 330 K for Cu, 390 K for Al, 224 K for As, etc.] We see that at high temperatures, $T \gg \Theta_D$, we obtain

$$c_{V,vb} = 3Nk \tag{14.142}$$

where N denotes the number of atoms per unit volume of the crystal. The above result is, of course, what we expect from the equipartition theorem: each of the $3N$ oscillators (three per atom) in the unit volume of the crystal contributes an amount of kT to its total energy. At low temperatures (when $T \ll \Theta_D$) we obtain:

$$c_{V,vb} = \frac{12\pi^4}{5} Nk(T/\Theta_D)^3 \tag{14.143}$$

Which tells that as $T \to O$ the contribution of the vibrations to the specific heat goes to zero proportionally to T^3.

We remember that the electronic contribution to the specific heat of a metal goes to zero proportionally to T (see Eq. (14.122). The experimentalist measures, of course, the sum, c_V, of the two contributions (electronic and vibrational):

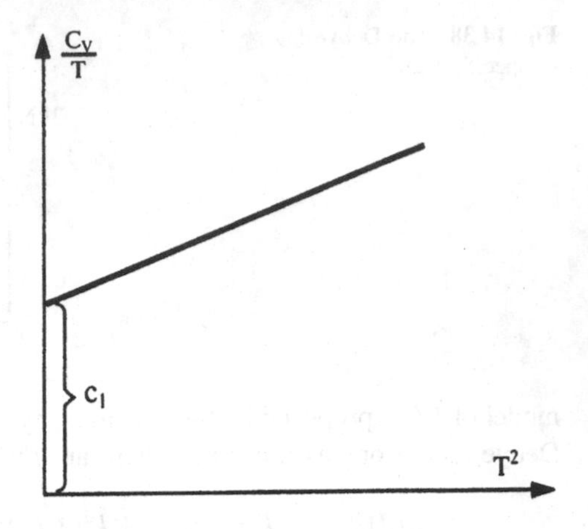

Fig. 14.39 Determination of c_1 and c_2 of Eq.(14.144): c_1 is given by the intercept of the *straight line* with the c_V/T axis, and c_2 by the slope of the *straight line*

$$c_V = c_{V,el} + c_{V,vb} = c_1 T + c_2 T^3 \tag{14.144}$$

Which can be written as:

$$c_V/T = c_1 + c_2 T^2 \tag{14.144a}$$

The experimentalist is able to determine c_1 and c_2 by plotting c_V/T as a function of T^2 in the manner of Fig. 14.39.

14.7.8 Superconductivity

An impressive phenomenon which the model of independent electrons can not explain is that of superconductivity. We expect the resistivity ($1/\sigma$) of a pure (without impurities) metal to be reduced to zero, as $T \to 0$, continuously. This is because the amplitude of the lattice vibrations (which are the cause of the resistance) are reduced to practically zero continuously as $T \to 0$. In many metals, however, the resistance vanishes abruptly at a temperature above zero, as seen in Fig. 14.40a, which shows the variation of the resistance of a mercury sample in the region of the so-called critical temperature (denoted by T_c). This is the temperature (a characteristic property of the metal) at which the transition occurs from the normal phase ($T > T_c$) to the superconducting phase ($T < T_c$). The data of Fig. 14.40a are due to Onnes who observed the phenomenon for the first time in 1911. Since then, the phenomenon has been observed in Aluminum ($T_c = 1.19$ K), Gallium ($T_c = 1.09$ K), Lead ($T_c = 7.18$ K), Vanadium ($T_c = 5.30$ K) and many other metals. The vanishing of the resistivity means that an electric current can be sustained in a metallic ring (in its superconducting phase) for a very long time without an electromotive force. There is experimental evidence of currents persisting, in some cases, for two years or so, which is indeed a long time. The

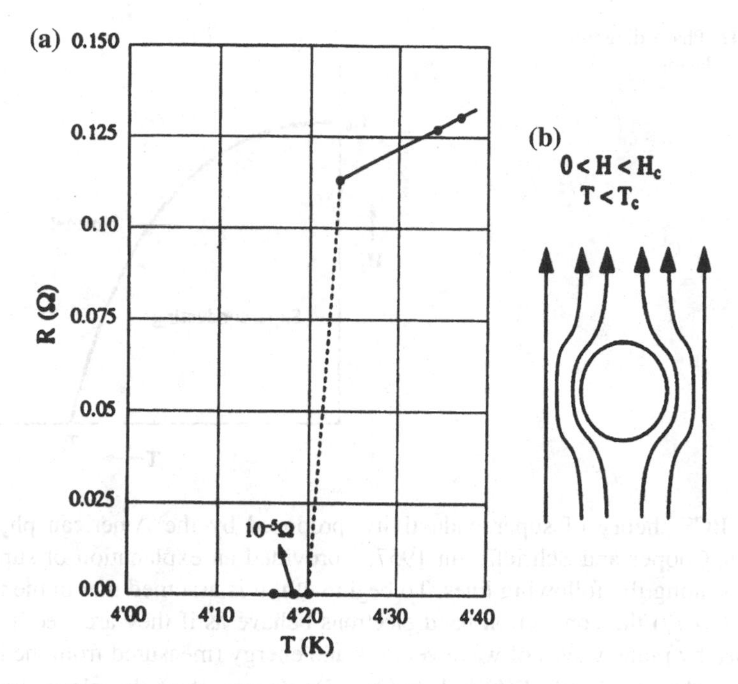

Fig. 14.40 (a) Variation of the resistance of mercury with the temperature in the region of T_c. (b) The Meissner effect

superconductor has also interesting magnetic properties: it will not allow a magnetic field in its interior. Placed in a magnetic field whose magnitude does not exceed a critical value (H_c), it modifies it in such a way as to expel it from its interior. The phenomenon, known as the Meissner effect, is demonstrated schematically in Fig. 14.40b.[34]

An understanding (in terms of a microscopic theory) of the vanishing resistivity, of the Meissner effect and other thermodynamic properties of a superconductor (for example, the variation with the temperature of its specific heat is quite different from that of a normal metal) turned out to be difficult and took a long time to develop. One may appreciate the difficulty of the problem by noting that at $T = 0$ the difference in the total energy (per atom) between the superconducting phase and the normal phase of a metal is orders of magnitude smaller than the total energy of either phase. This means, in fact, that one is looking for a mechanism of superconductivity sustained by a reduction $\delta\varepsilon$ in the total energy of the normal phase, when we can only estimate this energy within an error $\Delta\varepsilon \gg \delta\varepsilon$.

[34] It was first observed by W. Meissner and R. Ochsenfeld in 1933.

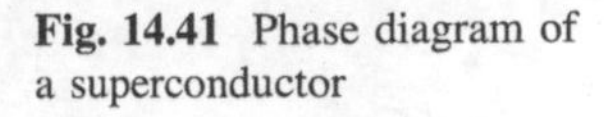

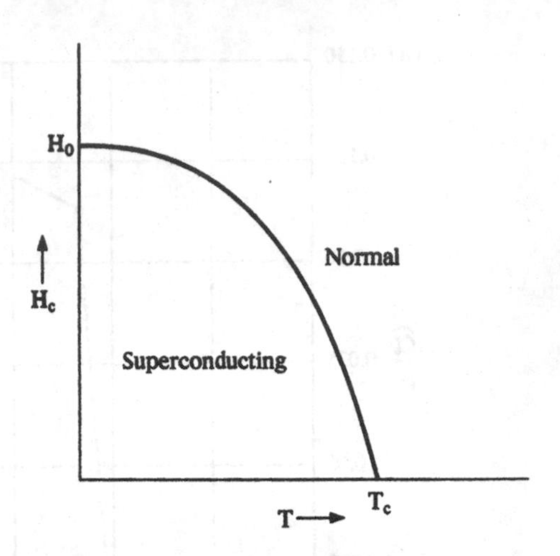

Fig. 14.41 Phase diagram of a superconductor

The BCS theory of superconductivity, proposed by the American physicists Bardeen, Cooper and Schrieffer in 1957,[35] provided an explanation of superconductivity along the following lines. To begin with, it is assumed that in the normal phase ($T > T_c$) the conduction band electrons behave as if they are free: they are described by plane waves of wave vector $\boldsymbol{k}$ and energy (measured from the bottom of the conduction band) $E(\boldsymbol{k}) = \hbar^2 k^2/2m$. (Replacement of the plane waves by Bloch waves does not affect the essential results of the theory.) It is further assumed that all interactions, except that which is responsible for superconductivity, produce the same corrections to the total energy in either of the two phases (normal and superconducting) and therefore, for the purpose of understanding superconductivity we can disregard them. In this way it becomes possible to estimate the energy difference between the two phases without it being necessary to know accurately the total energy of either phase. The interaction responsible for superconductivity according to the BCS theory can be described as follows: Two electrons with energy $E(\boldsymbol{k})$ and $E(\boldsymbol{k}')$, respectively, within the region:

$$E_F - \hbar\omega_D < E(\boldsymbol{k}), E(\boldsymbol{k}') < E_F + \hbar\omega_D \qquad (14.145)$$

where ω_D is the Debye frequency introduced in relation to Eq. (14.141), may be attracted to each other in spite of the residual Coulomb repulsion that exists between them. This is possible when the lattice of the atomic cores is polarized by a pair of electrons (within the above energy region) in such a way as to produce an effective attraction between them which is greater than their residual Coulomb repulsion. We shall not present here the evidence (experimental and theoretical) supporting the above statement. Suffice it to say that it has turned out to be quite a common occurrence in metals, as evidenced by the fact that so many of them become superconductors at sufficiently low temperatures. Let us assume then, that

[35] J. Bardeen, L.N. Cooper and J. Schrieffer, *Physical Review*, ***108***, *1175 (1957)*. The three authors shared a Nobel Prize for Physics for their discovery.

there exists a negative (attractive) interaction between any two electrons in the energy region of Eq. (14.145), and let us denote the corresponding potential energy term by U'. In the BCS theory U' is approximated by $U' = U_0'/L^3$, where L^3 is the volume of the metal and U_0' is a constant. This interaction favours the formation of weakly bound pairs of electrons (a bit like 'molecules' of two electrons loosely bound to each other and separated by a distance of the order of $\xi = 10^{-4}$ cm). Between and beyond the 'electrons of a given pair' lies the electronic cloud due to all other electrons, paired and unpaired, which are of course indistinguishable from the electrons of the given pair. It is obvious that we are dealing with the dynamics of a complex many-electron system, whereby the electrons exchange their roles in the collective action all the time; it is as if we are watching a formation dance by thousands of dancers and we perceive order to be there, because the dancing roles are given, though the dancers exchange their roles all the time.

We admit, then, that in thermodynamic equilibrium (at $T \approx 0$) the electrons that would (when $U' = 0$) be occupying the one-electron states with energy between $E_F - \hbar\omega_D$ and E_F, are now (when $U' < 0$) accommodated with certain probability in pairs $(\boldsymbol{k}, \boldsymbol{k}')$ with $E(\boldsymbol{k}), E(\boldsymbol{k}')$ in the region specified in Eq. (14.145). We assume that electrons with energy below $E_F - \hbar\omega_D$ are not affected by U'. Obviously, there are more possible pairs $(\boldsymbol{k}, \boldsymbol{k}')$ in the energy region of Eq. (14.145) than there are electrons to occupy them, and the question arises as to which pairs are actually occupied. It turns out that the total energy in the superconductive phase is minimized when the electrons in the energy region of Eq. (14.145), are in so-called Cooper pairs: $(\boldsymbol{k}\uparrow, -\boldsymbol{k}\downarrow)$. Which means that the electrons in a pair have opposite momenta and opposite spins. But still we have more Cooper pairs with $E(\boldsymbol{k})$ in the energy region of Eq. (14.145), than we have electrons to fill them. The probability, $0 < w(\boldsymbol{k}) < 1$, that a Cooper pair will be occupied at $T = 0$ is finally determined by the requirement that the total energy of the system be a minimum.

It turns out that the ground-state energy of the superconductor, determined in the above manner, is separated from that of the first excited state of the superconductor by a finite energy gap given by:

$$\Delta_0 = 4\hbar\omega_D \exp(-\frac{1}{\rho(E_F)|U_0'|}) \tag{14.146}$$

where $\rho(E_F)$ is the one-electron density, per spin, at the Fermi level of the normal phase. Δ_0 is the energy required to break a Cooper pair (but I should point out that because the electrons of different Cooper pairs interact with each other, the said energy can not be calculated by considering a Cooper pair in isolation). We find that $(\rho(E_F)|U_0'|)^{-1} \approx 3.5$ which means that $\Delta_0 << \hbar\omega_D$. The energy gap between the ground state and the first excited state of the superconductor plays a crucial role in the determination of its properties.

The momentum of a Cooper pair equals zero and therefore the total momentum of the electrons in the ground state of the superconductor vanishes, and so does the current in a superconducting ring in its ground state. The persistence of an electric current in a superconducting ring is explained as follows: To a given current corresponds a metastable state constituted in very much the same way as the ground state,[36] except that now all electrons in the energy region of Eq. (14.145) are to be found in Cooper pairs of the form: $(\boldsymbol{k}\uparrow, -\boldsymbol{k}'\downarrow)$, where $\boldsymbol{k} - \boldsymbol{k}' = \boldsymbol{K}$ is a non-zero vector, the same for all pairs. In this case the total momentum of the electrons is a multiple of $\hbar\boldsymbol{K}$, which implies a finite electric current, which persists for a very long time for the following reason. The state of the superconductor can change if one or more Cooper pairs break up, but for this a minimum energy is required equal approximately to the energy gap of Eq. (14.146), and the vibrating nuclei with which the bound pair might collide can not provide that much energy at temperatures below T_c. The state of the superconductor could also change if the momentum (the value of $\boldsymbol{K}$) of a large number of pairs could change in a collision, but the probability of this happening is negligible.

The BCS theory explains the Meissner effect as well: a state of mutually interacting Cooper pairs expels the magnetic field from the interior of the superconductor.

When $T > 0$, a fraction of the electrons in the energy region of Eq. (14.145) are not to be found in pairs, but moving independently like normal electrons (electrons in the normal phase). As the temperature rises this fraction increases, and at $T = T_c$ all electrons become normal. It can be shown that: $3.5\ kT_c \approx \Delta_0$. Is is also worth noting that as the temperature rises, the superconductor's energy gap is reduced from its $T = 0$ value (Eq. (14.146)) to zero at $T = T_c$. Finally it is worth noting that when $T < T_c$, the application of a magnetic field $H \geq H_c(T)$ destroys the superconducting phase. The function $H_c\ (T)$ is shown schematically in Fig. 14.41.

I shall close this section with a brief reference to the so-called high-temperature superconductors discovered in 1986 by Georg Bednorz and Karl Mülller. These superconductors are crystals with a chemical composition of the type: $YBa_2Cu_3O_7$, $Bi_2Sr_2CaCu_2O_8$, etc., and a correspondingly large unit cell. The critical temperatures T_c of these superconductors can be greater than a 100 K. The mechanism leading to superconductivity at these relatively high temperatures may be different, and more complicated, than that of the BCS theory. The study of high-temperature superconductivity is an active field of research, the review of which lies beyond the scope of the present book.

In closing the section on solids, I should say that there are many phenomena, of great theoretical, experimental and technological interest that I had not even mentioned. I said very little about surface phenomena and nothing about our ability to observe the atomic structure of solid surfaces, I said very little about magnetic phenomena, and nothing about non-crystalline solids and their very interesting electronic properties. All these have been the subject of very active research in the

---

[36] The energy of a metastable state lies above the ground energy and therefore it is not a stable state in the strict sense that thermodynamics attaches to this term. However it has a very long lifetime, because a transition to a state of lower energy is very difficult.

last fifty years or so, and it would be very difficult to do justice to them, without making the book too long. Suffice it to say that, in every field the application of quantum mechanics has led to satisfactory explanations of the observed facts, limited only by the complexity of the calculations involved, and our ability to conceive the right model (as demonstrated by the BCS theory discussed above).

Exercises

14.1. Verify the commutation relations [Eqs. (14.3)] of the orbital angular momentum operators.

14.2. Beginning with the definition of the spherical coordinates (see Fig. 14.1), verify the expressions for $\widehat{l}^2$ and $\widehat{l}_z$ given in Eq. (14.5).

Hint: To begin with, show that:

$$\widehat{l}_z = \frac{\hbar}{i}\frac{\partial}{\partial\varphi}; \quad \widehat{l}_x + i\widehat{l}_y = \hbar e^{i\varphi}\left(\frac{\partial}{\partial\theta} + i\frac{\cos\theta}{\sin\theta}\frac{\partial}{\partial\varphi}\right);$$
$$\widehat{l}_x - i\widehat{l}_y = \hbar e^{-i\varphi}\left(-\frac{\partial}{\partial\theta} + i\frac{\cos\theta}{\sin\theta}\frac{\partial}{\partial\varphi}\right)$$

Note that: $(\widehat{l}_x + i\widehat{l}_y)(\widehat{l}_x - i\widehat{l}_y) = \widehat{l}_x^2 + \widehat{l}_y^2 + \hbar\widehat{l}_z$.

14.3. Verify that the spherical harmonics given in Eqs. (14.14) satisfy (14.13).

14.4. Convince yourself that the variation of $(P_l^m(\cos\theta))^2$ with the angle θ is described correctly by the polar diagrams of Fig. 14.2.

14.5. Verify that the $R_{nl}(r)$ described in Fig. 14.6 satisfy Eq. (14.21) with $E = E_{nl}$ given by Eq. (14.28), when $U(r)$ is given by Eq. (14.26) with $Z = 1$. And verify that they also satisfy the appropriate boundary conditions and that they are indeed normalized according to Eq. (14.24a). [You may use published tables for the evaluation of the normalization integrals.]

14.6. Verify that the energy eigenfunctions, $\psi_{nlm}(r,\theta,\varphi)$, of the hydrogen atom have *even parity*: $\psi_{nlm}(-\boldsymbol{r}) = \psi_{nlm}(\boldsymbol{r})$, when l is even; and that they have *odd parity*: $\psi_{nlm}(-\boldsymbol{r}) = -\psi_{nlm}(\boldsymbol{r})$, when l is odd.

Note: $\boldsymbol{r} \to -\boldsymbol{r}$ is equivalent to: $\theta \to \pi - \theta$, $\varphi \to \varphi + \pi$.

14.7. Determine the classical turning point r_c of the electron in the ground state of the hydrogen atom (when $l = 0$, $U(r_c) = E$), and calculate the probability, P, for the electron to be in the classically forbidden region ($r > r_c$).

Note: $\int x^2 e^{-ax}dx = -e^{-ax}(\frac{x^2}{a} + \frac{2x}{a^2} + \frac{2}{a^3})$
Answer: $r_c = 2a_0$; $P = 0.238$.

14.8. The magnetic moment $\boldsymbol{M}$ due to a current i in a circular loop of radius r is normal to the plane of the loop and has a magnitude $M = iS$, where $S = \pi r^2$

is the area of the loop (see exercise 10.7). Assuming a circular motion of the electron as in Bohr's model and putting $i = -e/T$, where T is the period of the motion, show that there is a magnetic moment associated with the orbital motion of the electron given by: $\boldsymbol{M} = -\frac{e}{2m}\boldsymbol{l}$

where, $\boldsymbol{l}$ is Bohr's orbital angular momentum of the electron. In Schrödinger's quantum mechanics $\boldsymbol{M}$ and $\boldsymbol{l}$ are replaced by their respective operators.

14.9. (a) Determine, using first order perturbation theory, the correction to the ground-state energy of the hydrogen atom, when the latter is placed in a weak uniform electric field (such as the one between the plates of a condenser).

Hint: Putting the nucleus at the origin of the coordinates, we write the perturbing interaction as follows: $V = -eFr\cos\theta$.

(b) Consider what might happen in an electric field of large magnitude.

Answer: (a) $\Delta E = 0$. (b) Apart from any shift of the energy levels, electron escape by tunneling becomes possible.

14.10. (a) Verify that the Pauli matrices defined in Eq. (14.35) satisfy the commutation relations of Eq. (14.3).

(b) Verify that the operators for the total angular momentum given by Eq. (14.42) satisfy the commutation relations of Eq. (14.3).

14.11. The numerical results relating to the nl-shells of carbon and sodium (see the relevant table in Sect. 14.5.1) show that the radius $r_{max}(nl)$ of a shell depends on the energy and the angular momentum of the shell, and that it is possible for the radius of a shell of given angular momentum (given value of l) to be smaller than that of a shell of lower angular momentum, even if the energy of the former is higher than that of the latter. Provide a qualitative explanation of the above result on the basis of Eq. (14.23) and Fig. 14.5.

14.12. We have established that there are three independent p-orbitals which have the form: $\psi_{lm} = R(r)Y_{lm}(\theta,\varphi)$, where $l = 1$, $m = -1, 0, 1$ (A)

(a) Show that the combinations of the above defined by

$$p_x = -\frac{\psi_{11} - \psi_{1-1}}{\sqrt{2}} = (\frac{3}{4\pi})^{1/2}\frac{x}{r}R(r)$$

$$p_y = -\frac{\psi_{11} + \psi_{1-1}}{\sqrt{2}} = (\frac{3}{4\pi})^{1/2}\frac{y}{r}R(r) \qquad \text{(B)}$$

$$p_z = \psi_{10} = (\frac{3}{4\pi})^{1/2}\frac{z}{r}R(r)$$

are also normalized and orthogonal orbitals; which means that we can choose the set (B), instead of the set (A), to represent the p-orbitals.

(b) Verify that the probability densities, $|p_x(\mathbf{r})|^2$, $|p_y(\mathbf{r})|^2$, $|p_z(\mathbf{r})|^2$, are concentrated along the x, y, and z directions respectively.

14.13. (a) Describe the similarities and differences between the energy spectrum of a particle in a one-dimensional box [Eq. (13.47)] and that of a particle of the same mass in the potential field, $U(x) = kx^2/2$, of a harmonic oscillator.

(b) Using Eq. (14.50) derive the expressions for $u_n(x)$, $n = 1, 2$ of the harmonic oscillator.

(c) Verify Eq. (14.51) for one or two $u_n(x)$.

14.14. (a) Evaluate the ground-state energy of a system of five identical non-interacting particles (the mass of the particles equals that of an electron) in a box of dimensions (in angstrom): $a = 1$, $b = 1.5$, $c = 2$ (see exercise 13.9), (i) when the particles are fermions of spin ½ and (ii) when the particles are bosons of zero spin.

Answer: (i) 426.08 eV; (ii) 319.3 eV.

(b) Evaluate the ground-state energy of a system of five identical non-interacting particles (the mass of the particles equals that of an electron) in a box of dimensions (in angstrom): $a = b = c = 1.5$ (see exercise 13.9), (i) when the particles are fermions of spin ½ and (ii) when the particles are bosons of zero spin.

Answer: (i) 401.76 eV; (ii) 251.1 eV.

14.15. An electronic term $E(R)$ of a diatomic molecule which has a minimum, such as the $E_0(R)$ of Fig. 14.15, is described approximately by

$$E(R) = D\{\exp[-2\alpha(R - R_0)] - 2\exp[-\alpha(R - R_0)]\}$$

which is known as the Morse curve.

Show that (i) the minimum of $E(R)$ has the value $-D$ and occurs at $R = R_0$, and (ii) that by an appropriate choice for α one can obtain the observed frequency, ω_n, of the vibration spectrum corresponding to the term under consideration.

14.16. The space lattice of crystalline LiH is cubic with hydrogen atoms at the corners and at the centres of the faces of the cubic unit cell and lithium atoms at the centre of the cube and at the centres of its edges. Estimate the equilibrium separation between the atoms of the molecule LiH assuming it is the same with that between these atoms in crystalline LiH. We know that the density of crystalline LiH is 0.83×10^{13} kg/m^3. Using the above data estimate the wavelength of the transition: $l = 1 \rightarrow l = 0 (n = \upsilon = 0)$.

14.17. Convince yourself that the Fermi surface of tungsten shown in Fig. 14.27 is consistent with the energy bands shown in Fig. 14.26.

14.18. (a) Prove Eq. (14.126).

Hint: Remember the derivation of Eq. (13.4).

(b) Show that the Fermi level of a free-electron metal is given by: $E_F = \frac{\hbar^2}{2m}(3\pi^2 n)^{2/3}$, where m is the mass of the electron and n is the number of free (conduction band) electrons per unit volume of the metal.

14.19. The conduction band of a metal is approximated by a free electron model. The density of the metal is 9×10^3 kg/m^3, and every atom (of mass $M = 1.07 \times 10^{-22}$g) contributes one free electron to the conduction band. Calculate the Fermi energy E_F and the magnitude υ_F of the transport velocity of the electron at E_F. We remember that the transport velocity in band $E_\alpha(\boldsymbol{k})$ is given by:

$$\upsilon_\alpha(\boldsymbol{k}) = (1/\hbar)\left(\partial E_\alpha/\partial k_x,\ \partial E_\alpha/\partial k_y,\ \partial E_\alpha/\partial k_z\right)$$

14.20. When an electron in a metal gains energy due to an external electric field $\boldsymbol{F}$ it moves, in time dt, from an energy level $E_\alpha(\boldsymbol{k})$ to an energy level $E_\alpha(\boldsymbol{k} + \boldsymbol{dk})$. Using the expression for $\upsilon_\alpha(\boldsymbol{k})$ given above, show that

$$dE_\alpha/dt = \hbar\upsilon_\alpha(\mathbf{k}) \cdot d\mathbf{k}/dt$$

Noting that dE_α/dt equals the work done by the force $-e\boldsymbol{F}$ per unit time: $-e\boldsymbol{F} \cdot \upsilon_\alpha(\boldsymbol{k})$, show that: $d\boldsymbol{k}/dt = -e\boldsymbol{F}/\hbar$, which is of course the 3-dimensional form of Eq. (14.123).

14.21. The measured resistivity ($1/\sigma$) of copper at $T = 300$ K is $1.7 \times 10^{-8}\Omega m$.

(a) Estimate the relaxation time (τ) and the mean free path (L) of an electron with energy $E = E_F$ for the same temperature. Use the free-electron model approximation and assume one free electron per atom. The density of copper is 8930 kg/m^3.

(b) The resistivity of copper and other metals increases linearly with the temperature though not greatly (for copper it varies from about 1.7×10^{-8} Ωm at $T = 300$ K to about 2.7×10^{-8} Ωm at $T = 600$ K. Provide a qualitative explanation of the linear increase of the resistivity with T on the assumption that this is due to collisions with vibrating nuclei.

Hint: How does the amplitude of a vibrating nucleus vary with T over this region of temperature?

(c) One finds that the resistivity of a pure (without impurities) metal goes to zero like $1/\sigma \propto \mathrm{T}^5$ as $T \to 0$. Try to guess the reason for that.

Hint: Is one justified to treat the vibrating nuclei as independent classical oscillators as $T \to 0$? See Sect. 14.7.7.

14.22. Using the density of one-electron states shown in Fig. 14.28, estimate the electronic specific heat of tungsten when $T \to 0$, and evaluate the constant c_1 in Eq. (14.122) for tungsten.

Note: For tungsten: $V_0 = a^3/2$; $a = 3.16$ angstrom.

14.23. (a) Provide the missing steps in the derivation of Eq. (14.132) for the concentrations, n and p, of the electrons and holes in an intrinsic semiconductor.

(b) Explain how the energy gap of an intrinsic semiconductor can best be obtained from measurements of the conductivity of the semiconductor.

(c) Estimate the resistivity of intrinsic germanium (Ge) at $T = 300$ K. For Ge we have: $m_c = 0.55$ m, $m_v = 0.37$ m, where m is the mass of the electron. At $T = 300$ K, $\mu_n = 3900\ \text{cm}^2/\text{Vs}$ and $\mu_p = 1900\ \text{cm}^2/\text{Vs}$.

(d) Estimate approximately the resistivity of n-type germanium at $T = 300$ K, assuming a concentration of donor impurities equal to $10^{17}/\text{cm}^3$.

14.24. From the energy bands shown in Fig. 14.31 estimate the minimum frequency of an incident electromagnetic radiation capable of exciting an electron from the valence band to the conduction band of Si, GaAs, ZnS and KCl respectively.

Note: It can be shown that when the incident radiation has a wavelength much larger than the lattice constant of the crystal, only transitions between states (Bloch waves) of the same $\boldsymbol{k}$ are possible.

14.25. Assuming that the $\omega_\alpha(\boldsymbol{k})$ are given by Eq. (14.140), derive $D(\omega)$ of Eq. (14.141).

14.26. Let $x_i = ia$, $i = 0, \pm 1, \pm 2, \ldots$ be the equilibrium positions of a linear periodic chain of atoms. The classical equation of motion for the displacement X_i of the i_{th} atom from its equilibrium position is

$$M\frac{d^2X_i}{dt^2} = -K(X_i - X_{i+1}) - K(X_i - X_{i-1})$$

where M is the mass of the atom and K is an elastic constant. And we have assumed that an atom couples only with its nearest neighbours in the chain. Show that in this case, putting

$$X_i(t) = \text{Re}\{X_0^{(k)} \exp(ikx_i + i\omega(k)t)\}$$

we obtain the following equation:

$$\omega^2(k)MX_0^{(k)} = 2[1 - \cos ka]KX_0^{(k)}$$

and therefore, $\omega(k) = 2\omega_0|\sin(\frac{ka}{2})|$

where $\omega_0 = (K/M)^{1/2}$, and k takes the N values:

$k = 2\pi n/Na$, $n = 0, \pm 1, \pm 2, \ldots$ in the BZ: $-\pi/\text{a} < k \leq \pi/\text{a}$

where N is the number of atoms in the chain.

14.27. Assume that the electronic cloud due to *all* electrons in the conduction band of a metallic slab (parallel the *xy*-plane and of certain thickness Δz) is displaced by x in the x-direction. As a result of this displacement a layer of positive charge (due to the fixed positive ions in the slab) is created on the left face of the slab and a layer of negative charge due to electrons is created on its right face. The charges on the two faces of the slab looking much the same as the charges on the two plates of a parallel-plate capacitor. The surface charge density (apart from the sign) is the same on both ends of the slab: $\sigma = eNx$, where N is the number of conduction-band electrons per unit volume of the metal.

(a) Show that as a result a force $F = -(e^2N/\varepsilon_0)x$ is acting on every one of the conduction-band electrons in the slab.

(b) Assuming that the above force dominates over residual Coulomb interactions between the electrons and between the electrons and the vibrating nuclei, show that the electronic cloud of the conduction-band will vibrate *collectively*, as a single harmonic oscillator, whose quantized energy-levels are given by:

$$E_{P\upsilon} = \hbar\omega_P(\upsilon + \frac{1}{2}),\ \upsilon = 0, 1, 2, \ldots$$

$$\omega_P = (\mathrm{N}e^2/m\varepsilon_0)^{1/2}$$

Obviously, in order to excite the above oscillator, known as a plasmon oscillator, a minimum energy of $\hbar\omega_P$ is required. For most metals: $10 < \hbar\omega_P < 20\, eV$.

Note: The existence of such collective excitations has been confirmed experimentally as follows: A monoenergetic beam of electrons (E = *several thousand eV*) passes through a thin metal film. On exit some of the electrons have energies: $E - \Delta E$, where $\Delta E = \hbar\omega_P$, $2\hbar\omega_P$, $3\hbar\omega_P$, and so on. A proper quantum-mechanical theory of plasmons has also been developed, but its presentation lies beyond the scope of the present book.

14.28. When a light wave propagates through a crystal (say along the z-direction) the electric field associated with it (say along the x-direction) is described by:

$$E = \mathrm{Re}\{E_0 e^{i(qz-\omega t)}\} \approx \mathrm{Re}\{E_0 e^{-i\omega t}\}$$

The last equation assumes that the wavelength of the radiation is much greater than the lattice constant of the crystal. This is certainly true in the optical and ultraviolet part of the spectrum. The polarization P of the crystal [introduced in Eq. (10.64)] is in general given by:

$$P = \mathrm{Re}\{(\varepsilon_r(\omega) - 1)\varepsilon_0 E_0 e^{-i\omega t}\}$$

And we note that, $\varepsilon_r(\omega) = 1 + \chi(\omega)$ may be a *complex* number, which takes into account the absorption of light by the crystal. It can be shown that the reflection coefficient of *EM* radiation, of frequency ω, incident normally on a metal surface is:

$$R = \frac{(n-1)^2 + \kappa^2}{(n+1)^2 + \kappa^2}, \text{ where } n + i\kappa = \sqrt{\varepsilon_r(\omega)}$$

Now, for some metals (Cs, Rb, K, Na, Li), $\varepsilon_r(\omega)$ for frequencies in the optical and ultraviolet is well approximated by: $\varepsilon_r(\omega) = 1 - \frac{\omega_P^2}{\omega^2}$, where ω_P is the plasmon frequency introduced in exercise (14.27).

(a) Show that for $\omega < \omega_P$, normally incident light is totally reflected ($R = 1$) by these metals. Therefore show that by measuring the wavelength λ_0 above which total reflection occurs, we can determine *N*, the number of free (conduction-band) electrons per unit volume in the free-electron model of the metal.

(b) Using the experimental data shown below determine *N* for the given metals.

Metal	Cs	Rb	K	Na	Li
λ_0(angstrom)	4400	3600	3150	2100	2050

Chapter 15
The Very Small and the Very Large

15.1 Nuclear Physics

15.1.1 Radioactivity

The little town of Joachimsthal at the border between the Czech Republic and Germany was well known in the 16th century for its silver reserves. Along with the silver the miners had often found a shiny black mineral which they called *Pechblende*. It did not seem to be in any way useful. In 1789 an amateur German chemist, Martin Klaproth, examined it more carefully and found that it contained a 'strange kind of metal' which he named *uranium* in honour of the planet Uranus (at that time believed to be the last planet of the solar system). During the 19th century uranium was found in many places. It appeared to be the heaviest element on earth and its oxides and salts had vivid colours which made them useful in the creation of glass ware. Then, in 1896, a few months after the discovery of X-rays by Roentgen, the French physicist Antoine Henri Becquerel, whose earlier work related to optics, accidentally discovered that uranium salts radiated invisible rays which had similar properties to X-rays. In 1899 he showed that the radiation was deflected by a magnetic field, which implied that the radiation consisted, to a large degree at least, of charged particles.

Marie Curie was born Marya Skłodowska in Warsaw (Poland) in 1867. She was the youngest in a family with five children. Her father taught mathematics and physics at two *gymnasia* for boys. Her mother, who operated a Warsaw boarding school for girls before she became ill with tuberculosis, died when Marie was twelve, two years after Marie's oldest sister had died of typhus. As the family had very little money, Marie agreed to assist financially her sister Bronia to study medicine in Paris, and in return Bronia would assist Marie two years later. So she became a governess, first with a lawyer's family in Kraków, then with a landed family in Ciechanów, and for a third year with another family in Sopot on the Baltic Sea coast, all the while sending money to her sister. She then returned to her father in Warsaw, before leaving for Paris in 1891. In Paris, after staying with her sister and brother-in-law for a while, she rented a small attic room for herself.

A. Modinos, *From Aristotle to Schrödinger*,
Undergraduate Lecture Notes in Physics, DOI: 10.1007/978-3-319-00750-2_15,

During the day she studied physics, chemistry and mathematics at the Sorbonne, and in the evenings she tutored to earn her living. In 1893 she was awarded a degree in physics and began work in an industrial laboratory. At the same time she continued her studies at the Sorbonne and in 1894 she was awarded a degree in mathematics. In the same year she met the physicist Pierre Curie, a teacher and researcher at the School of Physics and Chemistry, who had already done useful work in magnetism. She began her scientific career with an investigation of the magnetic properties of various steels. And then she went back to Poland hoping for a university post. When she was rejected by the University of Kraków, merely because she was a woman, she returned to Paris. A year later, in 1895, she and Pierre Curie, married. They had a lot in common. They enjoyed long bicycle trips and travelling abroad, and they were both devoted to science, spending many hours together in their laboratory.

Pierre and Marie Curie were fascinated by the new rays discovered by Becquerel, which they called 'radio-actif' (radio-active), and decided to examine them more closely. Marie was to write her doctoral thesis on this topic. By 1898 they had secured the use of a laboratory space. It was no more than a disused room, unbelievably hot in the summer and very cold in winter. Here they began the grimy and labour-intensive effort to extract minute quantities of pure uranium from huge, boiling tanks of pitchblende solutions. Day after day sacks of pitchblende and torbenite (chalcolite) arrived; Marie would clean off the mud from the rough lumps, which then had to be ground into fine powder, and then boiled into a liquid which could be sieved and refined yet further. An electrolysis process followed, before a few grams of purified uranium might be extracted. *Sometimes I had to spend a whole day mixing a boiling mass with a heavy iron rod nearly as large as myself. I would be broken with fatique at the day's end*, she later wrote, ... *but it was in this miserable old shed that we passed the best and happiest years of our* life. And they made important discoveries.

Marie discovered, quite early, that the uranium rays caused the air around a sample to conduct electricity. Using an electrometer, devised by her husband and his brother some years earlier, she established that the activity of the uranium compound depended only on the quantity of uranium present. And then the Curie found to their amazement that the crude pitchblende was four times as active as pure uranium, and torbenite was twice as active as pure uranium. They concluded that these two minerals contain small quantities of some other substance that was far more active than uranium. In July 1898 the Curie published a paper together announcing the existence of a new element which they named *polonium*, in honour of her native Poland. And in December of the same year, they announced the existence of another element, which they named *radium* to emphasize its intense radioactivity.

In 1903, Marie Curie was awarded her doctorate from the University of Paris, and in the same year she, her husband and Henri Bequerel shared the Nobel Prize for Physics for their discoveries in radioactivity. The following year, the 45-year-old Pierre was appointed professor at the Sorbonne, and the 37-year-old Marie became his assistant. By this time they had two daughters, Irene and the newly born Eve. Suddenly, in April 1906, Pierre was killed in an accident; he was crushed under the wheels of a heavily loaded horse-drawn wagon which he failed to notice. Marie continued with her work and in 1908 became the first woman professor at the Sorbonne. And in 1911 she was awarded a second Nobel Prize, this time for Chemistry, in recognition of her discovery of radium. After the war she directed research relating to the treatment of neoplasms, using radioactivity. And in 1932 she founded the Radium Institute (now the Maria Skłodowska-Curie Institute of Oncology) in Warsaw, headed by her sister Bronia. She founded a similar Institute in Paris. Marie Curie died of leukaemia in 1934.

In August 1928 a young Ukrainian, George Gamow, arrived at Bohr's Institute in Copenhagen. He had spent a year in Göttingen absorbing the new theory of quantum mechanics, and was anxious to tell Bohr of his own explanation of the radioactive emission of α-particles (see Sect. 13.2.2). Bohr listened, liked what he heard and signed Gamow to the Institute for a year's work. About a month later, Gamow was shocked when he read in the magazine *Nature* a paper by two Princeton researchers, Ronald Gurney and Edward Condon, presenting a theory very much the same as his. Bohr acted swiftly to ensure that Gamow would still publish something worthwhile to get the recognition he deserved. The explanation of α-particle decay offered by Gamow and the two Americans is based on the realization that the phenomenon is essentially one of *tunnelling*. This is demonstrated in Fig. 15.1.

The broken-solid curve in this figure represents the repulsive Coulomb interaction between the positively charged ($+2e$) α-particle and the positively charged ($+Ze$) residual nucleus. For $r < R$, where R is the radius of the nucleus, the so-called strong nuclear interaction, which binds the nuclear particles into a compact nucleus, dominates. Very little was known at the time about this interaction, except that it was not electromagnetic in nature, that it was much stronger (holding the positively charged protons together) and that it had a very short range (about the size of the nucleus, $R \approx 10^{-14}$ m). It was reasonable to assume that within the

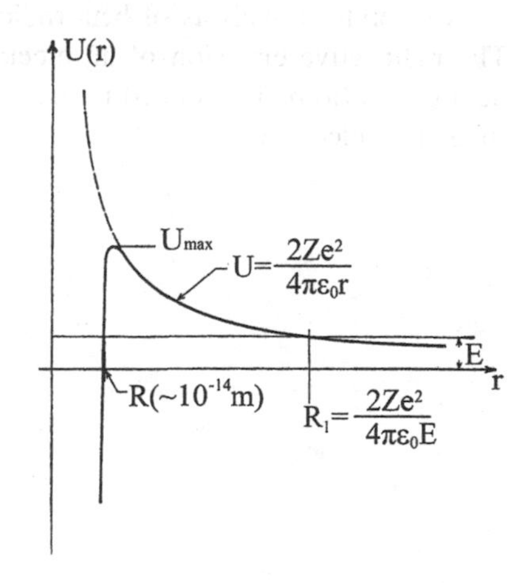

Fig. 15.1 Schematic representation of the potential energy of an α-particle near a nucleus

nucleus ($r < R$) the α-particle 'sees' an effective very deep potential well. It was further assumed that the α-particle has zero angular momentum, in which case its wavefunction (the radial part) satisfies a simple equation: Eq. (14.21) which reduces to Eq. (14.23) with $l = 0$, and a potential energy $U(r)$ as described above. However, for an explanation of α-particle decay, one does not require a full solution of Schrödinger's equation. One observes, to begin with, that the energy E of the ejected α-particle, ranging from 4.0 to 9.0 MeV, is lower than the maximum, U_{max}, of the potential barrier (see Fig. 15.1), and higher than the potential energy away from the nucleus ($r \gg R$). And this suggests that an α-particle can escape from the nucleus by *tunnelling* through the barrier of Fig. 15.1. Gamow started with the following fornula for the probability, W, of α-particle decay:

$$W = \nu T(E) \tag{15.1}$$

where $T(E)$ is the transmission coefficient for the barrier shown in Fig. 15.1, calculated using Eq. (13.74). We note that the potential barrier between the classical turning points at R (the radius of the nucleus) and R_1 is determined by the Coulomb repulsion alone. And ν is the times per second that an α-particle (in the nucleus) collides with the barrier: $\nu = N\upsilon/2R$, where N is the 'number' of α-particles in the nucleus (not a well defined quantity because one does not know if α-particles exist as such in the nucleus and if they do, how many they are), υ is a characteristic velocity, and $2R$ is the diameter of the nucleus. When the calculation of $T(E)$ is done, one finds that:

$$\log W \approx \log(N\upsilon/2R) + 2.97Z^{1/2}R^{1/2} - 3.95ZE^{-1/2} \tag{15.2}$$

where, *logW* is the natural logarithm of W. By 1928, *logW* had been measured experimentally for a number of radioactive nuclei, i.e. for different values of $ZE^{-1/2}$ and one could compare the above equation with measured data. It turned out the agreement between the two was good for values of R about 8.5×10^{-15} m, in excellent agreement with values of R obtained from Rutherford's and other scattering experiments.

A theoretical analysis of beta radioactivity (see Sect. 13.2.2) is more difficult. The radioactive emission of an electron implies the transformation of a neutral neutron (n) (to be introduced in the next section), into a positive proton (p) and a (negative) electron (e):

$$n \rightarrow p + e \tag{15.3a}$$

And similarly the radioactive emission of a positron (a positive electron) implies the transformation of a proton into a neutron and and a positron (e^+)[1]:

$$p \rightarrow n + e^+ \tag{15.3b}$$

The above equations make the point that β-particles (electrons and positrons) seem to be created *during* the ejection process, unlike α-particles which seem to pre-exist in the nucleus. The two processes differ in other ways as well: The energies of the ejected electrons and positrons range from a few keV to more than 15 MeV, whereas α-particles are rarely ejected with less than 4 or more than 9 MeV. Moreover, α-particles emerge from the nucleus with discrete well-defined energies, which suggests corresponding transitions between well-defined energy states in the parent and daughter nuclei. In contrast β-particles always possess a continuum spectrum from nearly zero to a maximum. Also, while α-particles, because of their relatively low energies, can be treated non-relativistically, a complete theory of β-decay must take into account relativistic effects, because β-particles often have energies comparable to m_0c^2.

And there is something else: In β-decay (Eqs. 15.3a), a particle of half-integer spin is transformed into two particles of half-integer spin, whose combined angular momentum can only be (according to the rules of adding angular momentum) an integral angular momentum! In 1930, Pauli suggested that the various observations in relation to the conservation of energy and angular momentum in β-decay can be explained, only *if* another particle were emitted beside the electron. This particle should be electrically neutral to conserve charge, it should have a spin of one-half (like the electron) to conserve angular momentum, and should have a rest mass very small (negligible) compared to that of the electron, because energy *is* conserved without it (if the more energetic electrons are used in the calculation). Finally, the interaction of this particle with matter must be much weaker than that of gamma-rays with matter, since it can apparently pass through large masses of solid matter without loosing energy. This hypothetical (at the time) particle was called *neutrino* (a little neutral). The existence of the neutrino was finally established in the 1960s.

An elementary theory of β-decay was formulated by Enrico Fermi in 1934. He postulated the existence of a short range (the same as that of the strong nuclear interaction) but very much weaker. This interaction which is known as the weak nuclear interaction is also weaker than the electromagnetic interaction, but it is of course stronger than the gravitational interaction. According to Fermi, the weak

[1] The existence of the positron was theoretically predicted by Paul Dirac in 1931 (see Sect. 15.2.1) and discovered in the cosmic radiation by Anderson in 1932. The positron differs from the electron only in that it carries a positive charge. In the presence of ordinary matter, the positron undergoes mutual annihilation with an electron producing EM radiation. Before annihilation the electron and the positron often form a 'neutral atom' called *positronium*. There are two states of the positronium, both unstable, with mean lifetimes of 8×10^{-9} and 7×10^{-6}s respectively. We should also note that the decay of the proton happens only within a nucleus. A free proton is practically stable, whereas the free neutron has a mean lifetime of about 18 min.

nuclear interaction is responsible for β-decay, which he treated as a transition from an initial state (i) of the nuclear system to a final state (f) consisting of the emitted particles (electron and neutrino) and the residual nucleus. He assumed a weak interaction of an extremely short range (described by a so-called delta-function), whose strength is desribed by a constant G with dimensions of energy times volume. And he evaluated the transition rate from (i) to (f) using perturbation theory in a manner similar to that described in Sect. 13.4.5 Fermi was able to account for the available data at the time, by putting $G \approx 1.41 \times 10^{-49} erg\, cm^3$. [1 erg $= 10^{-7}$ J].

15.1.2 Nuclear Reactions and The Discovery of The Neutron

Rutherford's early attempts to generate a nuclear reaction have been summarized in Sect. 13.2.2. In November 1927 Rutherford announced to the Royal Society his idea of a new scientific instrument: Its task would be to generate high electric fields that would accelerate charged particles (electrons, protons, α-particles) to high velocities, and then steer them with the use of electromagnets on nuclei (in a variety of targets) with the intent of smashing them into two or more pieces. He wanted to build a machine very different from the instruments he had used so far in his laboratory, one that could deliver up to 8 million volt of potential difference. Following Rutherford's idea, Ernst Walton, a young Irish physicist at his laboratory, suggested the following: he would accelerate the electrons a thousand times in succession by sending them round a ring-shaped track, contained there by magnetic coils. At a point in the circle the electron would go through a gap across which there is an a.c. voltage so tuned that every time the electron passed through it, it would receive a positive push increasing its energy. The idea was right, as later developments would show, but Walton failed in his attempts to implement it.[2] In 1927, he and John Cockcroft gained Rutherford's approval to build a *linear* accelerator. However, the scientists could not build this machine on their own. Private industry and the Department of Scientific and Industrial Research had to be involved. The era of 'big science' (meaning expensive) had arrived, and not only in England. Similar efforts were made at that time by Ernest Orlando Laurence at Berkeley in California.

In 1932 the linear accelerator of Walton and Cockcroft was ready. It was essentially a glass tube, about 3 m long, not much different from a cathode-ray tube. In the first experiment (April 1932), Lithium (Li) nuclei were bombarded with protons: a proton would enter a Li nucleus (with 3 protons) to create a nucleus with 4 protons, which would then split into two α-particles (two protons each). The α-particles would hit a scintillation screen and recognized as such. The number of

[2] A successful machine of this type, which came to be known as cyclotron, was first built in 1931 by the American physicist Ernest Lawrence in California.

scintillations observed was quite large, indicating that lots of α-particles were created, and all that with an accelerating voltage of about 200,000 V. The results were published in the magazine Nature under the title: *Disintegration of Lithium by Swift Protons*. When Walton and Cockcroft managed to attach properly calibrated detectors to their machine (in place of the scintillation screen), they established that the α-particles penetrated a long way into a Wilson cloud chamber (see parenthesis below), as if they had been accelerated by 8 million volt! That was in contrast to the energy required for the proton to enter the Li nucleus in the first instance, which was obtained by an accelerating voltage of 200,000 V. It seemed that the nuclear reaction by itself generated the extra energy. The Cavendish team soon established that the disintegration of the Li nucleus into two α-particles was accompanied by a two per cent loss of mass. After similar results were obtained by a team of German scientists (see next section), it became apparent that the energy ΔE generated by the nuclear reaction related to the loss of mass Δm according to Einstein's equation: $\Delta E = c^2 \Delta m$.

[The **Wilson cloud chamber** was invented in 1911 by C.T.R. Wilson. It is a container which contains air and ethanol vapour, which is cooled suddenly by adiabatic expansion which causes the vapour to become supersaturated. The excess moisture in the vapour is deposited as drops on the tracks of charged particles entering the chamber, which then become visible as a row of droplets, and they can be photographed. The tracks can, of course, be deflected by the application of an electric or magnetic field, and the deflection photographed.

The **Geiger counter** (Geiger-Müller counter) was devised by Hans Geiger (working with Rutherford) in 1908, and improved by Geiger and Müller in 1928. It consists of a tube containing a low pressure gas (usually a mixture of argon or neon with methane). In the tube there is a hollow cylindrical cathode with a fine-wire anode running through the centre of it. A voltage of about 1,000 V is applied between the two electrodes. An ionizing particle or photon passing through a window into the tube will cause an ion to be produced, which accelerated by the high electric field towards the one or the other of the electrodes, produces further ionizations (an avalanche of collisions). The resulting current *pulses* can be counted in electronic circuits, or amplified to work a small loudspeaker attached to the instrument].

By 1920 the Cavendish team, and other scientists, were convinced that atoms which are chemically the same (they interact in the same way with other atoms) do not always have the same mass. Francis Aston had constructed a *mass-spectrometer* in which ionized atoms or molecules were deflected by an electric or magnetic field, with the heavier ions deflected more than the lighter ones of the same charge. Since the mass of an ion is practically the same as that of its nucleus, he had a method of determining the mass of the nucleus. He soon found that there are a number of so-called *isotopes* of a chemical element: atoms of different mass with the same chemical behaviour. The chemical behaviour of an atom is determined (see Sect. 14.5.1) by the number of electrons in the atom, known as the *atomic number* (Z) of the atom. And because the atom is neutral, *Z* gives also the number of protons in the atomic nucleus. It was now apparent that atomic nuclei (except that of the common hydrogen atom which consisted of just one proton) contain something more than protons, and that this extra something was electrically neutral. The definite answer to this problem was obtained by Chadwick in 1932, a few weeks before The Cockcroft-Walton accelerator had its great success.

James Chadwick, born in 1891, was apparently so poor that, when (in 1908) he was accepted to study physics at the University of Manchester, he had to walk the four miles from his home to the University twice each day. In 1913 he travelled to Berlin to work with Rutherford's former coworker Hans Geiger. When the war came (in 1914), he was placed in an internment camp. There was little to eat, and the cold winters nearly killed him. On his return to England after the end of the war, Rutherford offered him a job at Manchester University, and when Rutherford left Manchester for Cambridge, Chadwick went along with him. And in 1921 he became Assistant Director of Research at the Cavendish Laboratory. Experimenting with radioactive materials, he observed that when the radiation collided with target substances the results, in some cases, could not be explained if one assumed that the radiation consisted only of electrons and α-particles. The results suggested that there was another component of radiation that carried too much momentum to be explained as X-rays or gamma-rays. He wrote the following in his notebook: *If we suppose that the radiation consists of particles of mass very nearly equal to that of the proton, all the difficulties connected with the collisions disappear.*

In 1932, when he bombarded beryllium targets with α-particles, he found that while some of the radiated particles on passing through a gas of hydrogen atoms would ionize some of them, there were also radiated particles (they certainly had mass) that passed right through the gas without leaving ionized trails, which suggested that these particles were electrically neutral. He had discovered the neutron. He established that neutrons could penetrate deep into all kind of materials including lead. Because it is neutral, the neutron does not interact with either the positive nuclei or the electrons around them and can therefore travel unhindered through any substrate.[3] For his discovery of the neutron, Chadwick was awarded the Nobel Prize for Physics in 1935, the same year in which he built Britain's first *cyclotron* at Liverpool University. He died in 1974.

15.1.3 Nuclear Fission and the First Nuclear Reactor

A young Austrian, Lise Meitner, arrived in Berlin in 1907, after becoming the first woman to gain a Ph.D. in physics from the University of Vienna. She teamed up with Otto Hahn, the talented chemist who trained with Rutherford at McGll in Montreal.

[3] Experiments similar to those of Chadwick had also been performed by Irene Joliot -Curie (daughter of Marie Curie) and her husband Pierre Joliot, who (in 1932) established that the new radiation (neutrons) was capable of knocking out of materials that contained hydrogen, protons of very high energy.

He would make the chemical separations to obtain the purest radioactive samples and she would investigate their physics. Their work was interrupted by the First World War, but after the war, in the 1920s and in the 1930s, they made significant contributions to the study of atoms, including the discovery of unknown elements in the heavy end of the periodic table. Then they entered the race, against rivals in Germany, Paris, and England to identify the new radiation, which ended with Chadwick's discovery of the neutron.

Meitner's Jewish origins put her in danger when Hitler came to power in 1933. Some of her rivals within the Chemistry Institute could now whisper a little more loudly: the Jewess endangers our Institute. And her old friend and coworker, Otto Hahn, was unwilling or afraid to help her. He apparently agreed with the administrators of the Institute that they should get rid of her. In 1938, a terrified Meitner escaped to Sweden, where she had no access to the kind of laboratory that she needed. She would write detailed notes to Hahn, who would conduct the experiments aided by Meitner's former assistant Fritz Strassman, and post the results back to Meitner for a detailed analysis. Sometime in 1938, Hahn and Strassman discovered that bombarding uranium with slowed neutrons (these had been discovered by Fermi in 1934; see below) produced new and lighter nuclei. After an elaborate chemical analysis they established that one of the products was most probably barium, which had a mass about half that of uranium. This was contrary to what they expected, i.e. uranium atoms transmuted into heavier isotopes by the insertion of extra neutrons. Hahn wrote to Meitner in late December of 1938, asking what his observations might mean, and promised that 'if there is anything you could propose that we publish, then it would still be work by the three of us. You will do us a good deed if you can find a way out of this.' And she found a way out of it. She argued as follows: The uranium nucleus (the rarest isotope, uranium-235) consists of 92 protons and 143 neutrons, held together by a strong nuclear force which wins over the repulsive Coulomb force between the protons. It seemed that the insertion of another neutron destroys this balance leading to the splitting of the nucleus. And she could explain why the insertion of a slow neutron led to a spectacular release of energy along with the splitting of the nucleus. This had nothing to do with the energy of the incident neutron, but much to do with Einstein's equation: $\Delta E = c^2 \Delta m$. Careful estimates of the masses involved showed that the sum of the masses entering the reaction (the uranium nucleus and neutron) was higher than the sum of the masses of the products of the reaction, by an amount sufficient to explain the released energy. The reader will remember that something similar had been observed (in 1932) by Cockcroft and Walton in their smashing of Li nuclei with α-particles. The possibility of generating energy by nuclear reactions was now an undisputed fact.

It might be of interest to note that Hahn never acknowledged Meitner's contribution to this discovery, not at that time and not when he received a Nobel Prize for his discovery of fission in 1944. Years later she wrote to a friend (P. Bizony, *Atom*, ibid): *I found it quite painful that in his interviews he did not say one word about me, or anything about the years we worked together.*

Enrico Fermi was born in Rome in 1901, the youngest of three children. His father was a railway official and his mother a schoolteacher. In 1915 his older brother died during a minor throat operation and his mother had a complete breakdown. A devastated Enrico took comfort in his books. And thanks to an observant teacher, who recognized his exceptional abilities, he was accepted as a student at Scuola Normale in Pisa, where some of the best brains in Italy were educated. In 1926 Fermi easily won the Chair of Theoretical Physics at the University of Rome. Fermi was an influential theoretical physicist who, independently of Paul Dirac, discovered the Fermi–Dirac statistics (see Sect. 14.4.3) and who, as we have already noted, postulated the weak nuclear interaction in his theory of beta-decay (the first theory on this subject). However, most people know of him as the man who led the team that built the first nuclear reactor in Chicago in 1942.

In 1934, working at his Physics Institute in Rome, Fermi succeeded by trial and error to produce a stream of slow (thermal) neutrons by impeding their progress with water and a shield of paraffin (about 6 cm thick). Lightweight hydrogen atoms in the paraffin slowed down the neutrons without actually blocking them. At the time, the great advantage of slowed neutrons seemed to be that they could be inserted effortlessly into nuclei: A neutron does *not* 'see' the Coulombic barrier a charged particle 'sees' at the periphery of the nucleus (see Fig. 15.1), so it reaches without hindrance the core of the nucleus where it is trapped by the strong nuclear force, creating a heavier isotope of the target nucleus. The heavier isotope can be unstable leading to automatic disintegration. Following Fermi's seminal discovery, many researchers around the world began to explore the investigative possibilities of slow neutrons. One of these teams was that of Otto Hahn in Berlin.

Until 1937 Fermi had no particular grievance with the Mussolini regime in Italy, so long as it was possible to carry on with his own work. But when Mussolini made an alliance with Hitler and agreed to abide with his racist policies, Fermi's Jewish-born wife Laura was in danger. In 1938, the year he was awarded the Nobel Prize for his work with slow neutrons, the Fermis moved to the United States. In the autumn of 1942 Fermi set about building the first nuclear reactor in Chicago, where he was a professor of physics at the University. In the centre of a 10 by 20 m room, lay a pile of black bricks and timber. Deeper inside the 'pile', as it was known, were layers containing blocks of uranium. It was extremely risky, but Fermi assured the 'National Academy of Sciences Committee to Evaluate the Use of Atomic Energy in War' that he could keep things under control. His grid of uranium blocks was interspersed at regular intervals by horizontal shafts into which control rods of cadmium-plated graphite could be inserted to block neutrons and damp the reactions. On 2nd December 1942 Fermi and his colleagues were ready to test their reactor. They listened amazed and almost in shock, as the various neutron counters clicked with increasing rapidity, and the microphones generated a roaring white noise. If the chain reaction was allowed to go on the reactor would have produced so much energy that it would have melted, killing everyone in the lab and poisoning the district around the University with intense radioactivity. After four and a half minutes, Fermii ordered the control rods to be inserted and the reaction stopped.

By 1943 two nuclear reactors were being built elsewhere in the U.S.A, and a third project was set in motion in Los Alamos (in the New Mexican desert) to produce a nuclear bomb, under the direction of Robert Oppenheimer. What followed is well known and we need not say anything about it here. Enrico Fermi died of cancer in 1954.

15.1.4 Nuclear Models

The nuclei are complex structures, like atoms, and as such they have a ground state and many excited states, the energy levels of which are determined experimentally in nuclear reactions and by the emission and absorption of gamma rays. To date there is not a reliable quantitative theory to describe the interaction between the particles that make up the nucleus, although much is known presently about the strong and weak interactions between the *elementary* particles that make up the proton and the neutron. However, several empirical models have been used to describe the energy states of the nucleus, of which the more commonly used are:

The *liquid-drop model*: In this model one assumes that a nucleon (proton or neutron) interacts only with its nearest neighbours. However, the nucleon moves about within the nucleus, it does not vibrate about a mean position as in a solid. In this respect it resembles a small drop of liquid. One can then visualize the excitation of a nucleus as a statistical 'heating' of the particles in it, and the emission of nuclear particles as a kind of evaporation. Moreover one can interprete certain low-lying excited states of the nucleus as excitation modes of the droplet regarded as a continuum fluid held together by surface tension.

In the *Fermi gasmodel*, the nucleons are treated as non-interacting particles confined in a spherical well of sufficient depth to give the uppermost occupied state an energy value of about 8 MeV below the zero level on the exterior of the nucleus. The radius of the well corresponds to the radius of the nucleus: $R = R_0A^{1/3}$, where A is the mass number (the number of protons + neutrons in the nucleus) and $R_0 \approx 1.2 \times 10^{-15}$ m.

The *shell model* of the nucleus is very similar to the shell model of the atom, but not as easy to justify because in the case of the nucleus we do not have a central particle (as is the nucleus in the atom) to sustain the model. However, a shell structure of the energy levels has been found to reproduce reasonably well the observed excitation energies of certain nuclei.

15.1.5 Elementary Particles and the Standard Model

In 1935 the Japanese theoretical physicist Hideki Yukawa suggested that the forces between nucleons could be the result of an *exchange* of particles (he called them mesons) between them, a process somehow similar to the way the atoms of a

Fig. 15.2 π^- capture. A negative pion comes to rest in the gas of the cloud chamber and is captured by an argon or carbon nucleus, from which a 65 MeV proton emerges

molecule are bound together by sharing some of their electrons (see Sect. 14.6.2). A proton or neutron would *emit* a meson which would then be *absorbed* by another proton or neutron. Such an exchange could take place only by a momentary (of duration Δt) violation of energy conservation, if the nucleons are at a distance (say R) from each other. Mesons were assumed to have a rest-mass energy $\Delta E = Mc^2$ which must be supplied in order to remove a meson from a nucleon; energy conservation is restored in time Δt when the meson is absorbed by another nucleon. Putting $\Delta t \approx R/c$, where R is the radius of the nucleus and c the speed of light, we can estimate ΔE from the 'uncertainty' relation $\Delta E \approx \hbar/\Delta t \approx \hbar\, c/R$, which suggests that the rest-mass of the meson should be about $M \approx \hbar/Rc \approx 275$ times the electron mass. A year after Yukawa's suggestion the so-called μ-meson or muon (with a mass about 200 times that of the electron) was discovered in cosmic radiation (high-energy particles that fall on the earth from space). The μ^- has a charge equal to that of the electron and its antiparticle μ^+ has a positive charge of equal magnitude. However their interaction with nuclei was too weak for them to be the Yukawa's mesons. In 1947 a family of mesons, similar to those postulated by Yukawa were discovered, and were named π- mesons or pions (see Fig. 15.2). But this was not the end of the story.

From about 1940 and, more so after the development of high-energy accelerators in the 1960s, many new and exotic particles were discovered. They were all unstable, existing for very small intervals of time of the order of 10^{-10} s, and their detection required highly sophisticated equipment. [This is one of the reasons that modern particle physics has become a very expensive research.]

The large number of these particles suggested that they do not constitute the elementary (basic) units of matter. In 1961 the American physicist Murray

Gell-Mann proposed that, from the particles that had been discovered up to that time, the leptons (these being the electron and the particles, listed below, which are governed by the *weak* nuclear force) are indeed elementary, but the rest, the ones governed by the so-called *strong* nuclear force, which are known as hadrons, are not elementary.[4]

THE LEPTONS

Name	Symbol	Rest-mass (MeV/c^2)	Charge
Electron	e	0.511	−1
Electron neutrino	ν_e	≈ 0	0
Muon	μ^-	105.7	−1
Muon neutrino	ν_μ	≈ 0	0
Tauon	τ^-	1784	−1
Tau neutrino	ν_τ	≈ 0	0

All particles have *one-half* (1/2) spin

The listed neutrinos appear to have an intimate connection with the electron, the muon and the tauon respectively, and in general they appear simultaneously with these particles. In 1985 a Soviet team reported a measurement of a non-zero neutrino mass, and in 1988–1989 Japanese and U.S groups presented theoretical arguments and corroborating experimental evidence to suggest that neutrinos do have mass.

We must also remember that for every particle there is an antiparticle with exactly the same properties except an opposite charge.

The *hadrons*, are in general much heavier than the leptons. There are two major types of hadrons: the *baryons*, which are fermions with half-integral spin (the most common examples are the proton and the neutron); and the *mesons*, which are bosons with integral spin. Both types are constituted from a set of six elementary particles (and their antiparticles), to which Gell-Mann gave the name: *quarks*. There are six of them, and some of their properties are listed in the table below. The masses shown are rough estimates, because quarks are never observed alone. It is worth noting that the *top* quark was found at CERN in Geneva in 1998, after its existence was predicted on theoretical grounds. We note that a quark has an electric charge (whether positive or negative) which is a fraction of the electron's charge. And we remember that for each quark there is an antiquark with exactly the same properties except an opposite charge.

[4] Murray Gell-Man was born in 1929 in New York, the son of Viennese Jewish immigrants. At fifteen years of age he entered Yale University. After graduating he spent short periods at Princeton University and the University of Chicago, before he moved to Caltech (The California Institute of Technology). In 1969 he was awarded the Nobel Prize for Physics for his contributions and discoveries concerning the classification of elementary particles and their interactions.

QUARKS			
Name	Symbol	Rest-mass (MeV/c^2)	Charge
Up	u	20	+2/3
Down	d	20	−1/3
Strange	s	200	−1/3
Charm	c	1,800	+2/3
Bottom	b	4,800	−1/3
Top	t	40,000	+2/3

All particles have *one- half* (1/2) spin

All baryons are made of three quarks. And all mesons are made of a quark and an antiquark. In Fig 15.3 we show schematically how a proton is made up of two u quarks and a d *quark* (resulting in a total charge of +1), and how a neutron is made up of two d quarks and a u quark (resulting in zero total charge). The fact that the neutron and the proton consist of charged particles going round inside them may explain why the two of them have slightly different magnetic moments, and why the neutron has a magnetic moment at all.

The quarks are held together by the strong interaction which is mediated by the exhange of *gluons* (so-called). These are particles of spin 1, zero charge and zero rest-mass, like the photons. And in the same way that the exchange of photons (mediators of the electromagnetic interaction; see Sect. 15.2.2) holds the electrons and the nucleus of an atom together, the exchange of gluons keeps the quarks of a baryon or a meson together. The situation is somewhat complicated by the fact that quarks are endowed also with a different kind of 'charge', known as *colour charge* or simply *colour* that plays an important role in the interaction process. At a particular time a quark can be *red* (R), *green* (G) or *blue* (B). These 'colours' have of course nothing to do with the ordinary colours (one can think of them as 'polarisations' of some kind). A gluon, on the other hand, can be one of 8 types depending on its effect on the colour of the quark, turning a red quark into a blue or green one etc. The way one calculates things in quark physics is modelled on quantum electrodynamics (see Sect. 15.2.2), which relies on successive approximations to obtain a final result. However, the existence of a number of particles, and the number of different ways these interact with each other (via different gluons), and finally because the strong interaction is much stronger than the

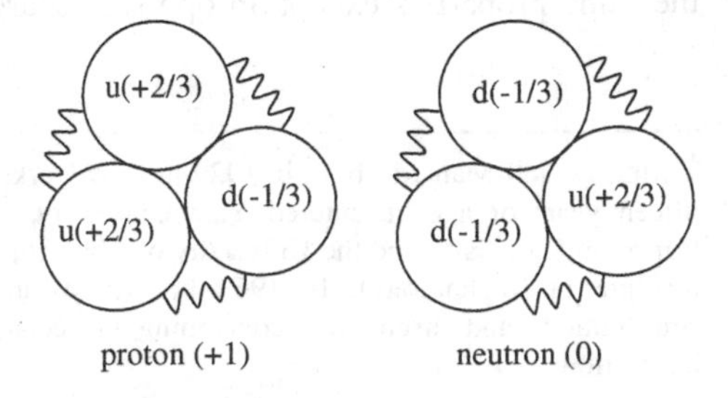

Fig. 15.3 The quarks in the proton and the neutron. The quarks are held together by gluons (depicted by the *wavy lines*)

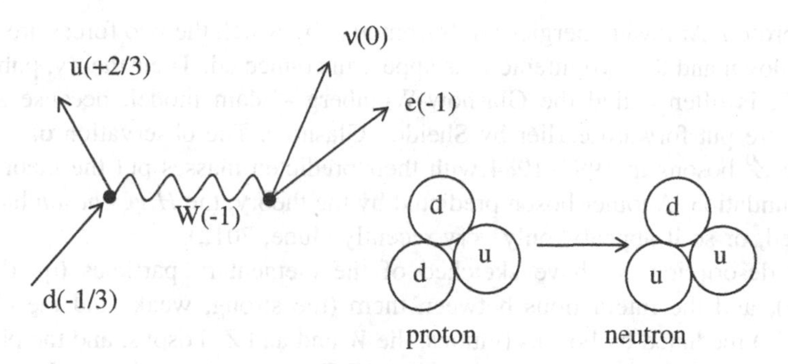

Fig. 15.4 The decay of a neutron into a proton, an electron and a neutrino is mediated by the W^- boson

electromagnetic interaction, the method (known as quantum chromodynamics) is not as efficient as quantum electrodynamics.

The weak interaction responsible for β-decay is desribed in similar fashion. Fermi's theory has been replaced by a relativistic theory in which the force is mediated by the exchange of one or the other of three particles called W^+, W^- and Z^0 bosons. The W^+ and W^- have a mass of 81×10^3 MeV/c^2 and a positive and negative charge, respectively, equal to that of an electron. The Z^0 boson is electrically neutral and has a mass of 93×10^3 MeV/c^2. We can, for example, explain the decay of a neutron, described by Eq. (15.3a), as follows (see Fig. 15.4): a W^- boson couples the d and u quarks on the left to the electron (e) and the neutrino (ν) on the right, and as a result of this interaction a d quark in the neutron changes into a u quark (turning the neutron into a proton) and at the same time an electron and a neutrino are emitted. Because W^- has a relatively high mass the process happens relatively slowly.

In processes where no change in the charge of any of the involved particles takes place the weak force is mediated by the Z^0 boson. Such processes are known as *neutral currents*. Finally, a coupling between a W^- and W^+ on the one hand and a Z^0 on the other hand is possible at relatively high energies (see Fig. 15.5), and it turns out that the strength of the *weak* interaction is in this case about the same as that of the *electromagnetic* interaction.

This led Stephen Weinberg and Abdus Salam to the idea that the W^-, W^+ and Z^0 bosons *and* the photon are facets of the same interaction (they called it *electroweak* interaction) at high energies or, equivalently, at distances smaller than the diameter

Fig. 15.5 Coupling between the W^- particles and Z^0

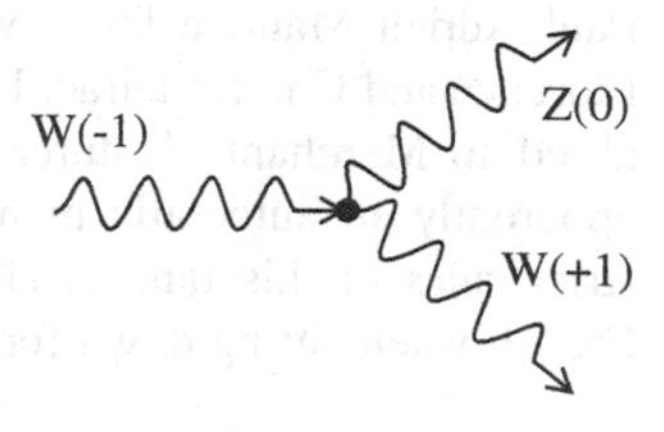

of the proton. At lower energies the 'symmetry' by which the two forces are united breaks down and the two interactions appear unconnected. Their theory, published in 1967, is often called the Glashow-Weinberg -Salam model, because similar ideas were put forward earlier by Sheldon Glashow. The observation of the W^-, W^+ and Z^0 bosons in 1983–1984 with their predicted masses put the theory on a firm foundation. Another boson predicted by the theory, the *Higgs boson* has been observed, or so it appears, only very recently (June, 2012).

The description we have sketched of the elementary particles (quarks and leptons), and the interactions between them (the strong, weak, and the electromagnetic) mediated by bosons (gluons, the W and and Z^0 bosons, and the photon), are known collectively as the *standard model.* The model seems to work very well as long as one uses the experimentally determined values of the masses of the particles and of the coupling constants (strengths) of the fundamental interactions.

We note that, for separations of about 10^{-15} m, the strong force is between ten and one hundred times stronger than the electromagnetic force, but it decreases rapidly with increasing separation and is negligible for separations greater than 10^{-14} m. The weak force is of similarly short range but much weaker (about 10^{10} times weaker) than the electric force.

There is of course a fourth fundamental interaction between material particles, that of gravity, but the gravitational attraction between elementary (or microscopic particles in general) is negligible (about 10^{40} times weaker than the electromagnetic force. However, because large masses are electrically neutral, the electrical force between them cancels out, while the gravitational force increases in proportion to their masses. It is for this reason that gravity dominates when we consider the motion of falling bodies and the motion of the stars.

However, many scientists believe that a quantum theory of gravity should be possible, and in such a theory the force will be mediated by a *graviton*. Moreover they expect to find that at the limit (when the whole universe was no more than a nucleus) all four fundamental interactions were facets of one unified force. It goes beyond the scope of this book to describe such attempts.

15.2 Dirac's Theory of the Electron and Quantum Electrodynamics

15.2.1 Dirac's Theory of the Electron

Paul Adrien Maurice Dirac was born in 1902. He was one of three children of Florence and Charles Dirac. His father was a French teacher at a secondary school allied to Merchant Venturers Technical College in Bristol. Charles Dirac was apparently an autocratic man with no interest in broader society, who imposed strict rules on his family. He would insist, for example, that every one spoke French when sitting down for dinner. Paul tried his best and was allowed to sit in the dining room while the rest of the family, his mother, his brother Reginald and his sister Beatrice, often ate in the kitchen. His father's attitude, who himself had run away from an unhappy childhood in Switzerland, no doubt contributed to Paul Dirac's extreme shyness. After finishing school Paul, like his older brother Reginald, studied Engineering at the University of Bristol, as was their father's wish. Reginald committed suicide at the age of 24. Paul Dirac moved to Cambridge where he devoted himself to theoretical physics. He shunned human company, and worked on his own. Of the many papers and books he wrote in his long career, mostly on quantum mechanics and relativity, only very few are co-authored. We have already noted his work on quantum statistics and his theory of the radiation field (see Sect. 14.3.3), but his fame derives mostly from his theory of the electron published in 1928, which is considered one of the greatest achievements of theoretical physics.[5] And it is for this work that Dirac shared the 1933 Nobel Prize for Physics with Heisenberg. In 1937 he married Margit Wigner (the sister of another famous physicist and Nobel laureate Eugene Wigner). She had two children from a previous marriage, and the couple had two children of their own. Following his retirement from Cambridge, Dirac moved to the United States where he stayed until his death in 1984.

Dirac set out to formulate a quantum theory of the electron which, in contrast to Schrödinger's theory, would be in accord with the demands of special relativity. His starting point, in relation to a free electron, is the invariant, under a Lorentz transformation, expression (see footnote 2 of Chap. 14):

$$p^2 - E^2/c^2 = -m^2c^2$$

which is the same as:

$$H^2 - c^2p^2 = m^2c^4 \tag{15.4}$$

where $\boldsymbol{p}$ is the momentum of the electron, E its energy, and m its rest mass. And we put $E = H$ (the Hamiltonian of the electron). When the electron is in an electromagnetic field, the above formula is replaced by the expression (invariant under a Lorentz transformation)[6]:

$$(H - e\varphi)^2 - c^2(\boldsymbol{p} - e\boldsymbol{A})^2 = m^2c^4 \tag{15.5}$$

where $-e$ denotes, as usual, the charge of the electron, and $\boldsymbol{A}(\boldsymbol{r},t)$ and $\varphi(\boldsymbol{r},t)$ are the electromagnetic potentials, vector and scalar respectively, defined in Appendix A. Dirac chose to write the above equation as follows:

$$H = c\left[(\boldsymbol{p} - e\boldsymbol{A})^2 + m^2c^2\right]^{1/2} + e\varphi \tag{15.6}$$

[5] P. A. M. Dirac, *Proc. Roy. Soc. Lond.* **A117**, 610 (1928); **A118**, 351 (1928).

[6] In the so-called Lagrangian and Hamiltonian formulations of Newtonian mechanics it leads to the correct equation for the motion of the electron in the electromagnetic field. We note that: $\boldsymbol{A}^2 \equiv \boldsymbol{A} \cdot \boldsymbol{A} = A^2$.

And he determined to write the square root on the right of the equation in a form symmetrical and linear in the momenta:

$$\left[(\boldsymbol{p}-e\boldsymbol{A})^2+m^2c^2\right]^{1/2}=\boldsymbol{\alpha}\cdot(\boldsymbol{p}-e\boldsymbol{A})+\beta mc \tag{15.7}$$

He then noted that the above expression can not be valid if $\boldsymbol{\alpha}$ and β are regarded as an ordinary vector and scalar quantities, because squaring both sides of Eq. (15.7) and equating the coefficients of similar terms leads to the apparently contradictory requirements:

$$\alpha_x^2=\alpha_y^2=\alpha_z^2=\beta^2=1 \tag{15.8a}$$

$$\begin{aligned}\alpha_x\alpha_y+\alpha_y\alpha_x&=\alpha_x\alpha_z+\alpha_z\alpha_x=\alpha_x\beta+\beta\,\alpha_x\\&=\alpha_y\alpha_z+\alpha_z\alpha_y=\alpha_y\beta+\beta\,\alpha_y\\&=\alpha_z\beta+\beta\alpha_z\\&=0\end{aligned} \tag{15.8b}$$

However, Eq. (15.7) is satisfied if α_x, α_y, α_z and β are treated as 4×4 matrix operators:

$$\widehat{\alpha}_x=\begin{pmatrix}0&0&0&1\\0&0&1&0\\0&1&0&0\\1&0&0&0\end{pmatrix}=\begin{pmatrix}\widehat{0}&\widehat{\sigma}'_x\\\widehat{\sigma}'_x&\widehat{0}\end{pmatrix}\quad\widehat{\alpha}_y=\begin{pmatrix}0&0&0&-i\\0&0&i&0\\0&-i&0&0\\i&0&0&0\end{pmatrix}=\begin{pmatrix}\widehat{0}&\widehat{\sigma}'_y\\\widehat{\sigma}'_y&\widehat{0}\end{pmatrix}$$
$$\widehat{\alpha}_z=\begin{pmatrix}0&0&1&0\\0&0&0&-1\\1&0&0&0\\0&-1&0&0\end{pmatrix}=\begin{pmatrix}\widehat{0}&\widehat{\sigma}'_z\\\widehat{\sigma}'_z&\widehat{0}\end{pmatrix}\quad\widehat{\beta}=\begin{pmatrix}1&0&0&0\\0&1&0&0\\0&0&-1&0\\0&0&0&-1\end{pmatrix}=\begin{pmatrix}\widehat{1}&\widehat{0}\\\widehat{0}&-\widehat{1}\end{pmatrix} \tag{15.9}$$

where $(\hbar/2)\widehat{\sigma}'_i=\widehat{\sigma}_i$, $i=x,\,y,\,z$ are the spin operators introduced by Pauli in 1927 (see Sect. 14.2.2); and $\widehat{1}$ and $\widehat{0}$ are the unit and null 2×2 matrices:

$$\widehat{1}=\begin{pmatrix}1&0\\0&1\end{pmatrix}\quad\widehat{0}=\begin{pmatrix}0&0\\0&0\end{pmatrix} \tag{15.10}$$

One can show that the matrices of Eqs. (15.9) satisfy (15.8a) using the rules of matrix multiplication (given in Appendix A).

It is then reasonable to write the equation of motion (the relativistic equivalent of Schrödinger's equation) of an elecrtron in the $\boldsymbol{A}$, φ field as follows:

$$-\frac{\hbar}{i}\frac{\partial\underline{\psi}}{\partial t}=\widehat{H}\underline{\psi} \tag{15.11a}$$

where $\widehat{H}$ is derived from Eq. (15.6), using Eq. (15.7) with the $\boldsymbol{\alpha}$ and β replaced by the operators defined in Eqs. (15.9) and (15.10), and the usual substitutions: $\boldsymbol{p} \to \frac{\hbar}{i}\nabla$ and $H \to \frac{\hbar}{i}\frac{\partial}{\partial t}$. One obtains:

$$\widehat{H} = c[\widehat{\alpha}_x(\frac{\hbar}{i}\frac{\partial}{\partial x} - eA_x) + \widehat{\alpha}_y(\frac{\hbar}{i}\frac{\partial}{\partial y} - eA_y) + \widehat{\alpha}_z(\frac{\hbar}{i}\frac{\partial}{\partial z} - eA_z)] + \widehat{\beta}mc^2 + \widehat{1}e\varphi \tag{15.11b}$$

where $\widehat{1}$ is the 4 × 4 unit matrix (its 4 diagonal elements equal unity and all its other elements equal zero).[7] The fact that $\widehat{H}$ is a 4 × 4 *matrix* operator, makes it necessary to look for the wavefunction in the form of a 4-component column matrix, $\underline{\psi}$, as follows:

$$\underline{\psi} = \begin{pmatrix} \psi_1 \\ \psi_2 \\ \psi_3 \\ \psi_4 \end{pmatrix}, \text{where} \psi_i = \psi_i(\boldsymbol{r}, t) \tag{15.11c}$$

We know (see Appendix A) that when an 4 × 4 matrix operates on a 4-component column matrix the result is another 4-component column matrix, as implied in Eq. (15.11a). We can look at $\underline{\psi}$ as a spinor of 4 components to be compared with the 2-component spinor introduced by Pauli (see Sect. 14.2.2). It is normalized in similar fashion by the requirement that:

$$\int \left(|\psi_1|^2+|\psi_2|^2+|\psi_3|^2+|\psi_4|^2\right)dV = 1 \tag{15.12}$$

[Introducing the row-matrix (see Appendix A)

$$\underline{\psi}^{\dagger} = (\psi_1^*, \psi_2^*, \psi_3^*, \psi_4^*) \tag{15.12a}$$

we can write Eq. (15.12) as follows (according to Eq.(A4.2):

$$\int \underline{\psi}^{\dagger}\underline{\psi}dV = 1 \tag{15.12b}$$

It is worth writing Dirac's equation for the electron (Eq. (15.11a, 15.11b) taken together) in component form:

---

[7] The 4 × 4 unit matix is often omitted, and one writes simply $e\varphi$ instead of $\widehat{1}e\varphi$. Similarly in what follows when a scalar quantity (or an operator such as $\widehat{l}_z$, etc.) is added to a 4 × 4 matrix, we imply that this quantity stands for a 4 × 4 diagonal matix whose diagonal elements are equal to the given quantity.

$$
\begin{aligned}
&(\frac{\hbar}{i}\frac{\partial}{\partial t} + e\varphi + mc^2)\psi_1 + c(\frac{\hbar}{i}\frac{\partial}{\partial z} - eA_z)\psi_3 + c[(\frac{\hbar}{i}\frac{\partial}{\partial x} - eA_x) - i(\frac{\hbar}{i}\frac{\partial}{\partial y} - eA_y)]\psi_4 = 0\\
&(\frac{\hbar}{i}\frac{\partial}{\partial t} + e\varphi + mc^2)\psi_2 + c[(\frac{\hbar}{i}\frac{\partial}{\partial x} - eA_x) + i(\frac{\hbar}{i}\frac{\partial}{\partial y} - eA_y)]\psi_3 - c(\frac{\hbar}{i}\frac{\partial}{\partial z} - eA_z)\psi_4 = 0\\
&c(\frac{\hbar}{i}\frac{\partial}{\partial z} - eA_z)\psi_1 + c[(\frac{\hbar}{i}\frac{\partial}{\partial x} - eA_x) - i(\frac{\hbar}{i}\frac{\partial}{\partial y} - eA_y)]\psi_2 + (\frac{\hbar}{i}\frac{\partial}{\partial t} + e\varphi - mc^2)\psi_3 = 0\\
&c[(\frac{\hbar}{i}\frac{\partial}{\partial x} - eA_x) + i(\frac{\hbar}{i}\frac{\partial}{\partial y} - eA_y)]\psi_1 - c(\frac{\hbar}{i}\frac{\partial}{\partial z} - eA_z)\psi_2 + (\frac{\hbar}{i}\frac{\partial}{\partial t} + e\varphi - mc^2)\psi_4 = 0
\end{aligned}
\tag{15.13}
$$

Let us now consider what the above equations imply for the energy and momentum eigenstates of a *free* electron. In this case $\boldsymbol{A}$ and φ equal zero and the equations simplify a lot.

We seek the eigenstates of the electron corresponding to a definite momentum $\boldsymbol{p} = (p_x, p_y, p_z)$ in the form:

$$
\underline{\psi} = \exp[(i/\hbar)(p_x x + p_y y + p_z z - Et)]\begin{pmatrix} C_1 \\ C_2 \\ C_3 \\ C_4 \end{pmatrix} \tag{15.14}
$$

where E denotes the energy of the electron. Substituting the above in Eq. (15.13), with $\boldsymbol{A} = \varphi = 0$, leads to a system of four algebraic equations for the four unknowns: C_1, C_2, C_3, C_4 as follows:

$$
\begin{aligned}
(-E + mc^2)C_1 + cp_zC_3 + c(p_x - ip_y)C_4 &= 0\\
(-E + mc^2)C_2 + c(p_x + ip_y)C_3 - cp_zC_4 &= 0\\
cp_zC_1 + c(p_x - ip_y)C_2 + (-E - mc^2)C_3 &= 0\\
c(p_x + ip_y)C_1 - cp_zC_2 + (-E - mc^2)C_4 &= 0
\end{aligned}
\tag{15.15}
$$

Now, a well known theorem (stated at the end of Appendix A) tells us that solutions of the above equations, where at least one of the unknowns is different from zero, are possible only if the determinant of the coefficients vanishes. In the present case this requirement is satisfied only if

$$
(E^2 - m^2c^4 - p^2c^2)^2 = 0
$$

which means that

$$
E = \pm(p^2c^2 + m^2c^4)^{1/2} \tag{15.16}
$$

This is of course an extraordinary and unexpected result because it tells us that a *free* particle can have a *negative* energy. According to Eq. (15.16), the particle may have any (positive) energy between its rest mass energy mc^2 and $+\infty$

(as expected from special relativity), but it may also have any (negative) energy from $-mc^2$ to $-\infty$. Dirac did not have a ready explanation for the *free particle with negative energy* that his equation predicted. It was in 1931, three years after the publication of his theory of the electron, that he came up with the *positron* idea. I will return to this aspect of Dirac's theory, after a brief description of some other aspects of the theory which are more readily accepted.

I begin by noting that there are four independent solutions of Eqs (15.15) corresponding to a given momentum $\boldsymbol{p}$ of the particle. Assuming that $\boldsymbol{p}$ is parallel to the z-direction, we have two eigenstates with positive energy given by:

$$\underline{\psi} = A\exp[(i/\hbar)(pz - Et)]\left\{\begin{pmatrix} pc \\ 0 \\ E - mc^2 \\ 0 \end{pmatrix} or \begin{pmatrix} 0 \\ -pc \\ 0 \\ E - mc^2 \end{pmatrix}\right\} \quad \text{for} E > mc^2 \tag{15.17a}$$

where $E = +\,(p^2c^2 + m^2c^4)^{1/2}$. And we have also two eigenstates with negative energy given by:

$$\underline{\psi} = A\exp[(i/\hbar)(pz - Et)]\left\{\begin{pmatrix} pc \\ 0 \\ E - mc^2 \\ 0 \end{pmatrix} or \begin{pmatrix} 0 \\ -pc \\ 0 \\ E - mc^2 \end{pmatrix}\right\} \text{for} E < -mc^2 \tag{15.17b}$$

where $E = -(p^2c^2 + m^2c^4)^{1/2}$.

In every case the normalization constant is determined according to Eq. (15.12a).

We note that for $p \ll mc$, we have $E - mc^2 \approx 0$ for positive energy, and the eigenstates of Eq. (15.17a), reduce to

$$\underline{\psi} = A'\exp[(i/\hbar)(pz - Et)]\left\{\begin{pmatrix} 1 \\ 0 \\ 0 \\ 0 \end{pmatrix} or \begin{pmatrix} 0 \\ 1 \\ 0 \\ 0 \end{pmatrix}\right\} \tag{15.18}$$

Which correspond to the spinors of the Pauli theory introduced in Sect. 14.2.2 (see Eq.(14.37); in the present case $\psi_v(\boldsymbol{r}) = exp[(i/\hbar)pz]$). In what follows we shall see that the relation between Dirac's theory and that of Pauli goes much deeper.

In the non-relativistic treatment of an electron (or any particle) in a central field (like that of a hydrogen atom) the orbital angular momentum operators $\hat{l}_z$ and $\hat{l}^2$ commute with the Hamiltionian of the motion and, therefore, the energy eigenstates of the electron are also eigenstates of $\hat{l}_z$ and $\hat{l}^2$ (see Sect. 14.1.2). This is *not* so with the Dirac Hamiltonian, which takes the following form in the case of a central field (we put $A_x = A_y = A_z = 0$ and $\varphi(r) = \varphi(r)$ in Eq. (15.11b):

$$\widehat{H} = c\frac{\hbar}{i}[\widehat{\alpha}_x\frac{\partial}{\partial x} + \widehat{\alpha}_y\frac{\partial}{\partial y} + \widehat{\alpha}_z\frac{\partial}{\partial z} + i\widehat{\beta}\frac{mc}{\hbar}] + e\varphi(r) \quad (15.19)$$

We shall demonstrate that the above does not commute with $\widehat{l}_z$. We remember [see Eqs. (14.2) and (14.5)] that the latter is given by:

$$\widehat{l}_z = \frac{\hbar}{i}(x\frac{\partial}{\partial y} - y\frac{\partial}{\partial x}) = \frac{\hbar}{i}\frac{\partial}{\partial\varphi} \quad (15.20)$$

Obviously, $\widehat{l}_z$ commutes with $\widehat{\alpha}_z(\partial/\partial z)$, $i\widehat{\beta}(mc/\hbar)$ and $\varphi(r)$, so these terms can be disregarded in the evaluation of $\widehat{l}_z\widehat{H} - \widehat{H}\widehat{l}_z$. We then have:

$$\begin{aligned}\widehat{l}_z\widehat{H} &= -c\hbar^2(x\frac{\partial}{\partial y} - y\frac{\partial}{\partial x})(\widehat{\alpha}_x\frac{\partial}{\partial x} + \widehat{\alpha}_y\frac{\partial}{\partial y})\\ &= -c\hbar^2(\widehat{\alpha}_x x\frac{\partial^2}{\partial x\partial y} + \widehat{\alpha}_y x\frac{\partial^2}{\partial y^2} - \widehat{\alpha}_x y\frac{\partial^2}{\partial x^2} - \widehat{\alpha}_y y\frac{\partial^2}{\partial x\partial y})\end{aligned}$$

and

$$\widehat{H}\widehat{l}_z = -c\hbar^2(\widehat{\alpha}_x\frac{\partial}{\partial x} + \widehat{\alpha}_y\frac{\partial}{\partial y})(x\frac{\partial}{\partial y} - y\frac{\partial}{\partial x})$$

$$= -c\hbar^2(\widehat{\alpha}_x\frac{\partial}{\partial y} + \widehat{\alpha}_x x\frac{\partial^2}{\partial x\partial y} + \widehat{\alpha}_y x\frac{\partial^2}{\partial y^2} - \widehat{\alpha}_x y\frac{\partial^2}{\partial x^2} - \widehat{\alpha}_y\frac{\partial}{\partial x} - \widehat{\alpha}_y y\frac{\partial^2}{\partial x\partial y})$$

Therefore:

$$\widehat{l}_z\widehat{H} - \widehat{H}\widehat{l}_z = -c\hbar^2(\widehat{\alpha}_y\frac{\partial}{\partial x} - \widehat{\alpha}_x\frac{\partial}{\partial y}) \neq 0 \quad (15.21)$$

And similarly one can show that $\widehat{l}_x$, $\widehat{l}_y$ and $\widehat{l}^2$ also do not commute with $\widehat{H}$. However, one finds that the quantity defined by

$$\widehat{j}_z = \widehat{l}_z + \begin{pmatrix}\widehat{\sigma}_z & \widehat{0}\\ \widehat{0} & \widehat{\sigma}_z\end{pmatrix} \quad (15.22)$$

where $\widehat{\sigma}_z$ is the Pauli matrix defined in Eq. (14.35) and $\widehat{0}$ the 2×2 null matrix, *commutes* with $\widehat{H}$. We define the operators for the x and y components of $\boldsymbol{j} = (j_x, j_y, j_z)$, which we know as the total angular momentum of the electron, by analogy to Eq. (15.22) (see Footnote 7):

$$\widehat{j}_x = \widehat{l}_x + \begin{pmatrix}\widehat{\sigma}_x & \widehat{0}\\ \widehat{0} & \widehat{\sigma}_x\end{pmatrix} \quad \text{and} \quad \widehat{j}_y = \widehat{l}_y + \begin{pmatrix}\widehat{\sigma}_y & \widehat{0}\\ \widehat{0} & \widehat{\sigma}_y\end{pmatrix} \quad (15.23)$$

They also commute with $\widehat{H}$. But the operators of the different components of j do not commute with each other; they do commute with the operator for the square of its magnitude: $\widehat{j}^2 = \widehat{j}_x^2 + \widehat{j}_y^2 + \widehat{j}_z^2$. They do, in summary, satisfy the commutation relations expected of angular momenta introduced in Sect. 14.1.1. Finally one can show that $\widehat{j}^2$ also commutes with $\widehat{H}$. It follows from the above that the energy eigenstates of the electron in a central field are also eigenstates of $\widehat{j}^2$ and one of its components (we usually choose $\widehat{j}_z$). In conclusion: we established that the intrinsic angular momentum (spin) of the electron, postulated by Pauli (see Sect. 14.2.2), is a property of the electron which derives automatically from Dirac's equation, whicn in turn derives from the requirements of special relativity.

Moreover, the solution of Dirac's equation for an electron in a magnetic field shows that the allowed energies include a term which corresponds to the known orientation energy of a particle with a spin of ½ and an associated magnetic moment equal to the so-called *Bohr magneton*:

$$\mu_{\mathrm{B}} = \frac{e\hbar}{2m} \tag{15.24}$$

A result in complete agreement with Eq. (14.38) introduced in Sect. 14.2.2.

Ideally, at this point one should present Dirac's solution of the hydrogen atom: the energy eigenvalues and the corresponding eigenstates (4-component spinors). That involves some rather complicated mathematics and I shall not do it here. Suffice to say that the results are essentially the same with those obtained in Sect. 14.2.2, as we have already pointed out at the conclusion of that section.

Let me conclude this section with a brief discussion of the negative-energy eigenstates of the free electron and Dirac's prediction the *positron*. His theory was so successful in explaining the observed properties of the electron, that one had to believe that there was some hidden physics behind the negative-energy states predicted by this same theory. After much hesitation Dirac proposed that these states imply the existence of a new kind of particle, which he called *anti-electron*, which differs from the electron only in this respect: it carries a positive charge, *+e*, instead of the negative charge, $-e$, of the electron. Dirac's idea of the anti-electron can be summarised as follows: The infinitely many negative-energy states, described by Eq. (15.17b), corresponding to the different values of the momentum ($\boldsymbol{p}$) are *all* occupied by electrons. Therefore the exclusion principle does not allow any more electrons to fall into this infinitely deep sea of negative-energy electrons. It is further assumed that the impossibility of motion of these electrons [the situation is analogous to that of electrons in the valence band of a semiconductor at $T = 0$; see text leading to Eq. (14.133)] makes them undetectable. An anti-electron is then regarded as a *hole* left by the removal of an electron from a negative-energy state into a positive-energy state, realized by the absorption of a *photon* whose energy is greater than $2mc^2$. Such a hole would appear to move in the opposite direction to that of an electron under the action of an electric field, because all the

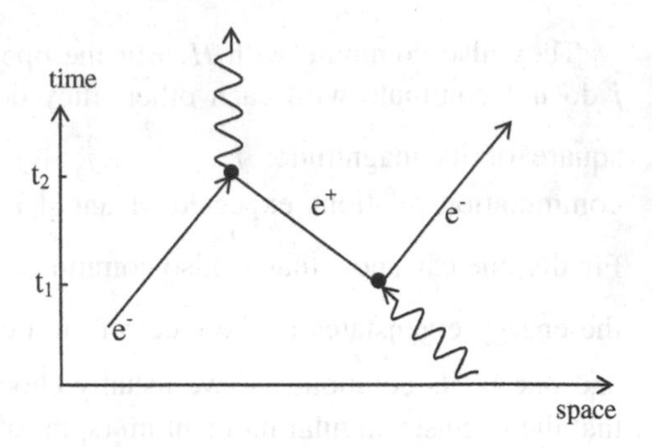

Fig. 15.6 Life history of a positron according to Feynman

negative-energy electrons will move in the opposite direction. As we have already noted, in 1932, a year after Dirac's postulation of the anti-electron, Carl Anderson discovered such a particle experimentally in cosmic radiation and named it positron. Nowadays, positrons can be easily produced in the laboratory (for example, in collisions of two photons) and can be kept for weeks in a magnetic field.

An alternative to the *hole* model of the positron was proposed br R. P. Feynman in 1949. It can be described schematically as shown in Fig. 15.6. At times $t < t_1$ we have an electron (represented by a straight line) and a photon (represented by a wavy line) moving towards each other. At $t = t_1$ the photon disintegrates into an electron (e^-) and a positron (e^+); at $t = t_2$ the positron annihilates with the original electron to create a new photon, while the electron created by the earlier photon continues forward in space-time . The above sequence of events is in fact what is observed in the laboratory.

We can, also, interprete the above diagram as follows: The original electron travels in space-time until $t = t_2$; at $t = t_2$ it emits a photon, then travels *backwards in time* to absorb a photon at $t = t_1$, and then travels forward again. In an actual experiment, the path of a *backwardin time* moving electron is perceived by us as the *forward in time* path of an anti-electron (which we call positron). We must remember that travelling *backwards in time* does not in any way violate the classical or quantum-mechanical laws of motion! It is only that, due to the second law of thermodynmics, which governs the development of large macroscopic systems, we are accustomed to *see* things as moving forward in time.

15.2.2 Quantum Electrodynamics

In Sect. 14.3.3 we have seen how the interaction between light and an atom can be treated as a perturbation in the manner of Sect. 13.4.5. That theory applies equally well to molecules and solids, and it is more than adequate for most practical purposes. But if one attempts to improve the accuracy of the calculation by going beyond the first order approximation (in the manner suggested at the end of Sect. 13.4.5), things become worse rather than better. Supposedly small corrections come out as infinities! One would not be satisfied with such a situation even if the experimental data could be accounted for satisfactorily by first order approximation theory. But in fact there are phenomena which this theory can not explain. Perhaps the most impressive of these phenomena is the so called Lamb shift in the spectrum of the hydrogen atom, discovered by Lamb and Retherford in 1947. Looking at the fine structure of the energy levels of the hydrogen atom (Fig. 14.8), we note that the energy levels of the two states with $n = 2$ and $j = ½$: the one obtained from $l = 0$ and spin *up* (and denoted by $^2S_{1/2}$), and the other obtained from $l = 1$ and spin *down* (and denoted by $^2P_{1/2}$) are the same. And this is true not only in Schrödinger's theory, but also in Dirac's theory of the hydrogen atom. But Lamb and Retherford found that there is a small energy difference between the levels of these two states of about $\Delta E = 4.37 \times 10^{-8}$ eV. The Lamb shift was rightly attributed to the interaction of the atom with the EM (electromagnetic) field, which had to be treated more accurately than to first order approximation.

Another physical quantity which is to a small degree affected by the interaction of the electron with the EM field in a way that can not be accounted for by ordinary perturbation theory relates to the magnetic moment of the electron. As we have already noted, Dirac's theory endows the electron with a magnetic moment, μ_B, given by Eq. (15.24). But in 1948 it was discovered experimentally that the actual value of this quantity is nearer to $\mu = 1.00118\ \mu_B$!

A theory for the calculation of the above, and other phenomena relating to the interaction of matter with light, with the required accuracy was eventually developed at about 1948 by three scientists, working independently: Julian Schwinger who was a professor at Yale University, the Japanese Sin-Itiro Tomonaga, and Richard P. Feynman who was working at Cornell University in Hans Bethe's department. Schwinger's approach to the problem was based on seemingly more rigorous mathematics in comparison to that of Feynman, whose method seemed to rely a lot on intuitive arguments and a series of diagrams to represent complicated integrals. Apparently, Schwinger, who was an excellent mathematician, did not think much of Feynman's unorthodox methods. The third man, Tomonaga, had developed his method in pre-war Japan but did not communicate it to the West until about 1948. He was then invited to Princeton by its director Robert Oppenheimer, and he participated in the discussions on this subject going on at that time. Tomonaga's approach was nearer to that of Feynman, without the diagrams. And then it happened that a young Englishman, Freeman Dyson, who at that time was a graduate student at Cornell University, was able to show, in a paper published in the Physical Review in 1948, that the methods of Schwinger and Feynman were actually equivalent! It is worth noting that Dyson's paper appeared before Schwinger's and Feynman's formal papers on the subject were published. Dyson championed Feynman's method, and so did Bethe. It was easier to understand and easier to work with. On meeting Tomonaga, whose method as we have already noted was nearer to that of Feynman, Dyson was impressed by his uncomplaining dignity; he would later write: *Tomonaga is more able than either Schwinger or Feynman to talk about ideas other than his own. And has enough of his own too. He is an exceptionally unselfish person.* Schwinger, Feynman and

Tomonaga were awarded the Nobel Prize for Physics for the year 1965 for their fundamental work in quantum electrodynamics. In what follows I shall desribe in qualitative terms their theory using Feynman's language, but perhaps I should first say a few words about the man who became, as time went by, one of the best known scientists of the 20th century.

Richard P. Feynaman was born in 1918 in New York. His father, Melville, was a clever and articulate salesman and entrepreneur, who had an interest in science and wanted to know how nature works, and he apparently passed his fascination with the physical world to his son. After completing his first degree at the Massachusetts Institute of Technology, Richard Feynman was encouraged by the head of his deparment, John Slater, to continue his studies at Princeton, where he arrived in the autumn of 1939. It was while at Princeton that he first developed his *path integral* formulation of quantum mechanics. A novel and most interesting way of looking at quantum mechanics, which space does not allow us to present here. John Wheeler, his mentor at Princeton, had this to say about Feynman's idea:

He treats on a footing of absolute equality every conceivable history that leads from the initial state to the final one, no matter how crazy the motion in between.The contribution of these histories are weighted in according to the classical principle of the least action integral.[8]*How could one ever want a simpler way to see what quantum theory is all about.*

After the war, during which Feynman collaborated, along with Bethe and many others, in the production of the first atomic bomb at Los Alamos, he was offered a job by Bethe at Cornell University, where he arrived in 1945. In June of that year his first wife, Arline Greenbaum, died of tuberculosis. Feynman, who had the reputation of a womanizer would marry again. And he had two children. In 1951, after a long and refreshing trip to Brazil, he took a professor's position at the California Institute of Technology, where he stayed until his death from cancer in 1988. Feynman is remembered, not only for his fundamental contribution to quantum electrodynamics, and other important contributions he made to particle physics and other branches of science, but also, and perhaps more so, for his enthusiasm and curiosity about the world, and the joy of doing physics, which he was able to express as a lecturer and writer on physics. Most students of physics will be familiar with the three volumes of the *Feynman Lectures on Physics*, which are basically edited transcripts by Robert Leighton and Mathew Sands of lectures that Feynman gave in the academic years 1961–1962 and 1962–1963, which is about the only time that he lectured to undergraduates. Any one who reads these volumes can not fail to be impressed by the originality of Feynman's approach to seemingly well 'understood' phenomena, and his endeavour to reformulate priciples or rules of physics so as to be understood by young students. And there are a number of other books by him, based on lectures that he gave on different

[8] The *principle of least action* is an important princible of analytical mechanics discovered by William Hamilton a century earlier, which states that the actual (classical) path of a particle is that which minimizes a certain integral.

occasions to varied audiences, and they are equally impressive.[9] Feynman enjoyed life, enjoyed solving problems of any kind, he liked music, he painted (he had an exhibition once), he liked telling stories about himself and others.[10] He liked to entertain his audience and played the buffoon at times.

Any theory of quantum electrodynamics, as the theory of the interaction of matter with the electromagnetic field came to be known, begins with Dirac's equation for the electron [Eq. (15.13)] and Maxwell's equations for the elctromagnetic potentials $\boldsymbol{A}(\boldsymbol{r},t)$ and $\varphi(\boldsymbol{r},t)$ defined in Appendix A. Disregarding, to begin with, the contribution to $\boldsymbol{A}$, φ from static fields (e.g., due to static nuclei) if such exist, we obtain $\boldsymbol{A}(\boldsymbol{r},t)$ and $\varphi(\boldsymbol{r},t)$ by solving Eqs. (A3.4), with ρ and $\boldsymbol{j} = (j_x, j_y, j_z)$ replaced by:

$$\rho(\boldsymbol{r},t) = e\,\underline{\psi}^{\dagger}\underline{\psi} = e\left(|\psi_1|^2+|\psi_2|^2+|\psi_3|^2+|\psi_4|^2\right) \tag{15.25}$$

where $\underline{\psi}^{\dagger}\underline{\psi}$ is the probability of findinng an electron at $\boldsymbol{r}$ at time t, in accordance with Eq. (15.12). The current density given by $\boldsymbol{j} = \rho\boldsymbol{u}$, where $\boldsymbol{u}$ is the electron velocity, takes the following form (when the electron is described by a Dirac spinor):

$$j_i(\boldsymbol{r},t) = ec\underline{\psi}\dagger\widehat{\alpha}_i\underline{\psi}, i = x,y,z \tag{15.26}$$

where $\widehat{\alpha}_i$ are the matrices of Eq. (15.9) and c denotes as usual the speed of light. Therefore Maxwell's equations take the form:

$$\nabla^2 A_i - \frac{1}{c^2}\frac{\partial^2 A_i}{\partial t^2} = -\mu_0 ec\underline{\psi}\dagger\widehat{\alpha}_i\underline{\psi}, i = x,y,z \tag{15.27a}$$

$$\nabla^2\varphi - \frac{1}{c^2}\frac{\partial^2\varphi}{\partial t^2} = -\frac{e}{\varepsilon_0}\underline{\psi}\dagger\underline{\psi} \tag{15.27b}$$

The above equations for $\boldsymbol{A}$, φ must of course be solved simultaneously with Dirac's Eq. (15.13). And we remember that, when an electron moves in an external static field (in the case of the hydrogen atom, for example, the external field is that of the static nucleus), this must be added to $\boldsymbol{A}$, φ in Eq. (15.13). It is obvious that an exact solution of the combined equations is practically impossible. The only way to proceed is by perturbation theory. The basic idea can be summarized as follows: In a system of non interacting particles, the different particles will go their own way, and in this case it is easy to write down the wavefunction for the total

[9] *The Character of Physical Law* (Penguin, 1992), *SixEasy Pieces* (Penguin, 1998), *Six Not-So-Easy Pieces* (Penguin, 1999) and others. From the books he wrote for a general audience, the most relevant to the subject matter of the present section is his 'popular exposition' of quantum electrodynmics entitled: *QED The strange theory of light and matter* (Princeton University Press, 1985).

[10] His memoirs, *Surely You're Joking, Mr. Feynman,* were published in 1985.

system. When a weak interaction between the particles is switched on, we can obtain the wavefunction of the system as a linear sum of its unperturbed eigenstates ($\psi_v(q)$), to a *first order approximation*, in the manner of Sect. 13.4.5. The probability of a transition from a given (initial) state, $\psi_i(q)$, to a final state $\psi_f(q)$, is proportional to $|T_{fi}|^2$, where T_{fi} is a complex amplitude defined by:

$$T_{fi} = \int \psi_f^*(q)\widehat{H}_{\text{int}}\psi_i(q)dq \tag{15.28}$$

where q is a set of coordinates which determines the positions of all particles in the system; i and f are the values of an appropriate set of quantum numbers which specify the eigenstates of the unperturbed system. $\widehat{H}_{\text{int}}$ is an interaction potential, and the integration is over all values of q. If a first-order perturbation treatment is not satisfactory, one can in principle calculate the wavefunction to second-order approximation, and then to a higher order, in the manner indicated in Sect. 13.4.5, but the process of doing so gets very complicated.

In the 1940s Feynman proposed an alternative treatment, according to which the interaction $\widehat{H}_{\text{int}}$ acts *discontinuously* on the system. One may look at the integrand of Eq. (15.28) as the probability amplitude for a sequence of events, each term in the integrand representing the probability amplitude for one of the events: The system arrives at a point q in *space-time* (the position of any particle in the above formula acquires a fourth component: time) with a probability amplitude $\psi_i(q)$; at the point q it is acted upon by the interaction potential, the probability amplitude of this happening is given by $\widehat{H}_{\text{int}}$; finally the system leaves point q in the state f, and the probability amplitude of this happening is given by ψ_f^*. The probability amplitude for the total effect is obtained by integrating over all points q of space-time at which the interaction potential may act. One great advantage of this scheme is that, it allows one to write expressions for higher-order perturbation effects simply, as follows: The system arrives at some point in space-time where it is acted upon by the interaction potential, and leaves this point in a new state; it can then go on to a second point in space-time , where it is acted upon a second time by the interaction potential, and leaves it in a third state, and so on. In what follows I can only describe qualitatively the application of this method to quantum electrodynamics. The complexity of the mathematics involved is such, that I can not summarise it in a meaningful manner.

Let me introduce the method through an example: the mutual scattering of two electrons. In second-order approximation it is described by the diagrams of Fig. 15.7. Diagram (a) corresponds to a matrix element T_{fi}, similar to the one defined above, where:

1. $\psi_i(q)$ is the probability amplitude that electrons a and b arrive at space-time points 1 and 2 respectively with momenta p_a and p_b respectively. We remember that p denotes the 4-component momentum-energy vector of a particle (electron or photon): $p \equiv (\boldsymbol{p}, E)$, and we note that the electrons would be described by appropriate 4- component spinors [similar to those of Eq. (15.17a)].

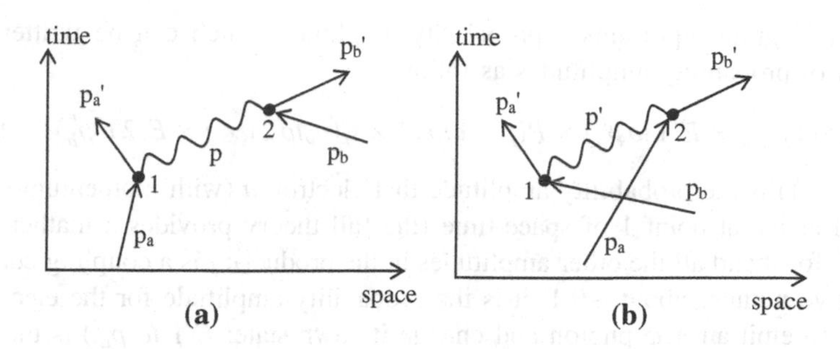

Fig. 15.7 Feynman diagrams for electron scattering. Three-dimensional space is represented by a single axis for the sake of simplicity. Electrons are represented by solid lines and photons by wavy lines. The number of coupling points in space-time (1 and 2 in the above diagrams) determines the order to which the interaction is treated; the above are second order diagrams. We note that p denotes the 4-component momentum-energy vector of a particle (electron or photon): $p \equiv (\boldsymbol{p}, E)$ In **a**: $p = p_a - p'_a$ at point 1, and $p + p_b = p'_b$ at point 2. Therefore $p_a + p_b = p'_a + p'_b$. In **b**: $p\prime = p_b - p'_a$ at point 1, and $p_a + p\prime = p'_b$ at point 2. Therefore $p_a + p_b = p'_a + p'_b$. These results demonstrate a general rule: The probability amplitude for any observable process vanishes unless the energy and the momentum are conserved

2. $\hat{H}_{\text{int}}$ is the probability amplitude that the two electrons will interact via an electromagnetic field $\boldsymbol{A}, \varphi$ in accordance with the equations of quantum elecrodynamics.
3. $\psi_f^*(q)$ is the probability amplitude that the electrons a and b leave the points 1 and 2 with momenta $p_a{}'$ and $p_b^{'}$ respectively.
4. To obtain the total amplitude for the event (two electrons of given initial momenta, p_a and p_b, change into two electrons with certain final momenta, $p_a{}'$and $p_b^{'}$) one must of course integrate over all possible coupling points (any values for 1 and 2) in space-time .

According to Feynman there is a more efficient way of looking at the diagrams of Fig. 15.7, and any other such diagrams, which now bear his name. A Feynman diagram tells a story about electrons propagating in space-time emitting and absorbing $\boldsymbol{A}, \varphi$ photons at coupling points along the way. The solid lines (arrows) in a diagram represent electrons [described by 4-component spinors as in Eq. (15.17a)]; the wavy lines represent 4-component $\boldsymbol{A}, \varphi$ photons (the term is explained in the parenthesis below). The coupling points are denoted by dots which may be marked 1, 2, etc. For example, we read diagram (a) of Fig. 15.7 as follows: An incident electron a, with momentum-energy p_a, arrives at point 1 in space-time ; there it emits a photon of momentum-energy p and it (the electron) changes into a state of momentum $p_a{}'$; the photon propagates to point 2 in space time; there it is absorbed by an incident electron b with momentum-energy p_b; the electron b changes into a new state with momentum-energy $p_b{}'$. More specifically, the said diagram represents a probability amplitude which can be written as a product of probability amplitudes as follows:

$$E(p_a to\, 1) \times j \times E(1\, to\, p'_a) \times P(p:\ 1\, to\, 2) \times E(p_b to\, 2) \times j \times E(2\, to p'_b) \quad (15.29)$$

$E(p_a to$ 1) is the probability amplitude that electron a (with momentum-energy p_a) will arrive at point 1 of space-time (the full theory provides a mathematical formula for it and all the other amplitudes in the product); j is a coupling constant, a negative number about -0.1, it is the probability amplitude for the electron at point 1 to emit an $\boldsymbol{A},\varphi$ photon and change its own state; $E(1\ to\ p_a')$ is the probability amplitude for the electron to change into a state of momentum p_a'; $P(p$: 1 *to* 2) is the probability amplitude for the photon (of momentum-energy p) emitted at point 1 to arrive at point 2 in space-time ; $E(p_b to$ 2) is the probability amplitude that electron b (with momentum-energy p_b) will arrive at point 2 of space-time ; the second j in the product is the probability amplitude that the electron at point 2 will absorb a photon at point 2 (we note that the probality amplitude for absorption of a photon is the same as that for emission of a photon); finally $E(2\ to\ p_b')$ is the probability amplitude that electron b will have a momentum-energy p_b' after absorbing the photon.

[The 4-component $(\boldsymbol{A},\varphi)$ photon (of momentum-energy $p \equiv (\boldsymbol{p}, E)$), usually referred to simply as photon, is a particle of zero mass and zero charge with an intrinsic angular momentum (spin) whose magnitude squared is $s(s+1)\,\hbar^2$ with $s = 1$. We say that photons have spin 1. This means that the component of the spin parallel to $\boldsymbol{p}$ takes three values: $m\hbar$ with $m = +1, -1, 0$. The +1 and -1 components correspond to *right* and *left* circularly polarized light (these can be regarded as linear combinations of the two transverse polarizations of light we are familiar with (see Sects. 10.4.3 and 13.1.1). The $m = 0$ component corresponds to a longitudinal polarization (it derives from a longitudinal component of $\boldsymbol{A}$, and φ) and exists only in the presence of charged particles, as when a photon is exchanged (emitted by one and absorbed by another electron). The $(\boldsymbol{A},\varphi)$ photons mediate the interaction between electrons in accordance with the equations of quantum electrodynamics. We should note, by the way, that the longitudinal component of the photons takes into account that interaction between electrons which we, previously, treated as an instantaneous interaction between two electrons (in the form: $e^2/|\boldsymbol{r}_1 - \boldsymbol{r}_2|$); see also Appedix A). $(\boldsymbol{A},\varphi)$ photons exchanged between charged particles are often called virtual photons.

In our qualitative desription [Eq. (15.29)] of the mathematics behind diagram (a) of Fig. (15.7), and the same applies to the other Feynman diagrams presented in this book, we disregard the fact that *solid lines* representing electrons of given p are 4-component spinors, and *wavylines* representing photons of given p are 4-component $(\boldsymbol{A},\varphi)$ vectors, treating them as if they were scalar quantities attached to a 4-component momentum-energy vector p. In the actual calculations one takes into account the way each of the four components of a photon couples with each of the four components of a spinor and vice versa, and this complicates the algebra. But it does not change the essence of the simple picture presented here. In this respect, it is worth noting that the coupling between any two components, is proportional to the same coupling constant, *j*, if not zero.]

The various component amplitudes appearing in the product represented by a Feynman diagram possess certain important symmetry properties with respect to the space-time axes which we shall summarize below. But let us first consider diagram (b) of Fig. 15.7. This diagram represents the amplitude for a process whereby an electron with momentum p_b changes, at point 1 in space-time , into a state with momentum $p_a{}'$, emitting a photon p'; the photon propagates to point 2 in space-time where it is absorbed by an electron with momentum p_a, the electron changes into a new state of momentum $p_b{}'$. We note that the initial and final states of the electrons are the same as in diagram (a). And because the electrons are indistinguishable, the two diagrams are probability amplitudes for the same event, and therefore to obtain the total probability amplitude for the event we must add the two amplitudes. In fact, because of the antisymmetry property of the wave-function describing two electrons, the amplitude of the second diagram is subtracted from that of the first. But this is a detail of the calculation that need not concern us here.

The next thing to note is that points 1 and 2 in the diagrams of Fig. 15.7*can be anywhere in space-time* , and therefore the probability amplitudes for *all* these points must be calculated and added together to obtain the total amplitude in second order approximation (this is defined by the number of *j's,* one per coupling point, in the diagrams). This is a laborious process but in principle a straightforward one. One finds that the various component amplitudes possess certain symmetry properties with respect to the space-time axes which are very important in this respect. These symmetries are:

1. The emission of a photon is equivalent to the absorption of a photon of opposite momentum.
2. The propagation of a photon between two points in space-time is equivalent to propagation in the reverse direction with reversed momentum.

Accordingly, in the diagrams (a), (b) and (c) of Fig. 15.8 the photons can be regarded as being emitted at point 1 and absorbed at point 2, even though the photon in (b) is absorbed at the same time as it is emitted, and in (c) it is emitted later than it is absorbed. As far as the calculation is concerned the three diagrams are equivalent. We can of course say, in relation to diagram (c), that a photon is emitted at point 2 and absorbed at point 1. It is all the same, so we can simply say that a photon is exchanged, and insert the locations of the two points in the formula for *P*(*p:* 1 to 2) appearing in the amplitude [Eq.(15.29)] for this diagram.

3. The arrival of an electron is equivalent to the departure of a positron, and vice versa. We have used this property in our discussion of the positron (preceding section).

Finally we remember that, always, the probability of an event happening (in the above example: two electrons of given initial momenta, p_a and p_b, change into two

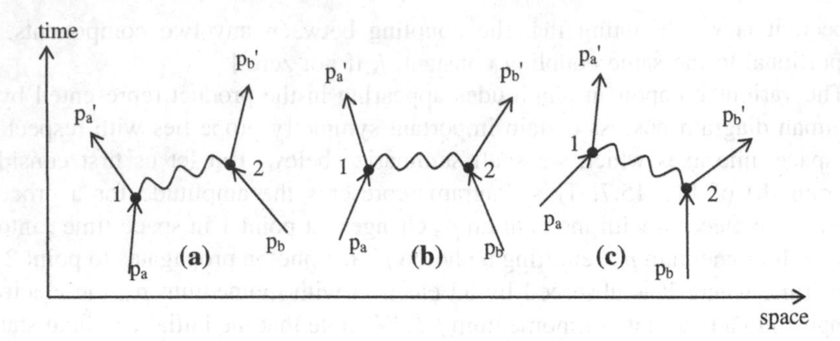

Fig. 15.8 *a*, *b* and *c* are equivalent

electrons with certain (final) momenta, p_a' and p_b') is given by the square of the magnitude of the (complex) total probability amplitude for the event.

If we wish to calculate the mutual scattering of two electrons more accurately, we must include in the evaluation of the total probability amplitude for the event the contribution of higher order diagrams, such as the one of Fig. 15.9, which involve four coupling points. The rules for writing down the amplitudes corresponding to these and higher order diagrams are the same as those of the second order diagrams we have described. Therefore the amplitudes corresponding to diagrams with four coupling points are proportional to j^4, and are therefore about 100 times smaller than those of second order diagrams which are proportional to j^2 (we remember that $j \approx -0.1$).

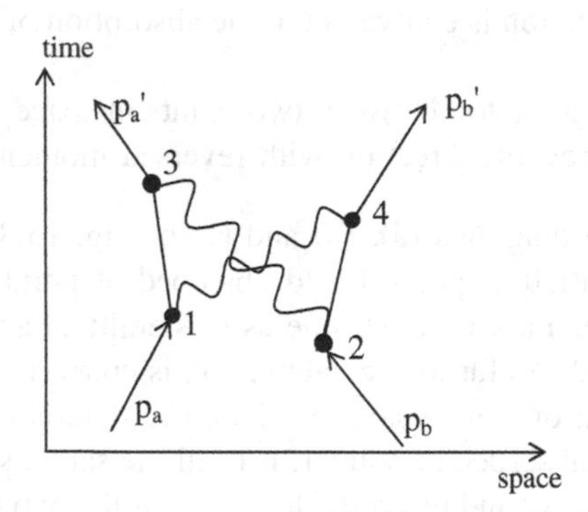

Fig. 15.9 Fourth order diagrams for electron–electron scattering. We note that the intermediate states of electrons (between points 1 and 3, and between points 2 and 4), and of photons (between points 1 and 4, and between points 2 and 3), can have any momentum $\boldsymbol{p}$ and any energy E. At each coupling point, momentum and energy are formally conserved, but the relation $E^2 = \boldsymbol{p} \cdot \boldsymbol{p}c^2 + m^2c^4$, where m is the rest mass of the particle (zero for photons), is *not* valid for the intermediate (unobserved) states of the electrons and photons involved. Integrating over all possible coupling points in space-time is equivalent to summing over all values of $\boldsymbol{p}$ and E for the intermediate states (of electrons and photons) in the diagrams, consistent with the above-mentioned conservation of momentum and energy at the coupling points

And of course, if one wants to, he can include amplitudes (diagrams) which involve six coupling points, which will be ten thousand times smaller than those of Fig. 15.7, and then go on to include diagrams with eight coupling points, and so on. And because coupling points can be anywhere in space-time one must always integrate (add up) the contribution from all these points (see caption of Fig. 15.9).

Before I tell you about the difficulties encountered in the implementation of the above procedure, let me describe briefly some phenomena, other than electron–electron scattering, which one can analyze using the same method.

One such phenomenon is that of *Compton scattering*, which is described diagrammatically in Fig. 15.10.[11] The two diagrams of this figure are probability amplitudes (to second order approximation) for the same event: an electron of momentum p_1 collides with a photon of momentum p_2; the result of the collision is: an electron with momentum p_1' and a photon of momentum p_2' (we have $p_1 + p_2 = p_1' + p_2'$). In diagram (a): An electron of momentum p_1 and a photon of momentum p_2 arrive at point 1 in space-time ; they couple; the electron absorbs the photon and changes into a state of momentum $p_1 + p_2$; the electron propagates to point 2; a second coupling; the electron emits a photon of momentum p_2' and it (the electron) changes into a state of momentum $p_1' = p_1 + p_2 - p_2'$. The probability amplitude for the above sequence of events is obtained as a product of probability amplitudes in the same way as in formula (15.29). Diagram (b) of Fig. 15.10 differs from that of diagram (a) in that the incident electron (of momentum p_1) emits the final photon (of momentum p_2') at point 1, then propagates (with momentum $p_1 - p_2'$) to point 2, where it absorbs the incident photon (of momentum p_2) and changes into a state of momentum $p_1' = p_1 + p_2 - p_2'$. The total probability amplitude for Compton scattering (to second order approximation) is obtained by adding the amplitudes corresponding to the two diagrams and integrating over all possible positions in space-time of the two coupling points 1 and 2.

---

[11] The first observations of Compton scattering were made by the American Arthur H Compton in 1923 while on faculty at Washington University in St. Louis. In his experiments a beam of X-rays was scattered by the electrons in a block of graphite. Compton found that the wavelength of the scattered X-ray had a wavelength λ slightly longer than the wavelength λ_0 of the incident ray, and that $\Delta\lambda = \lambda - \lambda_0$ was a function of the scattering angle θ (the angle between the directions of the scattered and the incident ray). Assuming that X-rays consist of photons of momentum $\hbar q$ and energy $\hbar cq$ (as suggested by Einstein; see Sect. 13.1.2) and that the total energy and the total momentum are conserved (do not change) in the collision of a photon with an electron, Compton derived the following formula for $\Delta\lambda$:

$$\Delta\lambda = (\hbar/mc)(1 - \cos\theta)$$

where m denotes the rest mass of the electron. We remember, in relation to the above, that the momentum p and the energy E of the electron are given by the 4-dimensional vector of Eq. (12.33), and that the total momentum (energy) is the sum of the momenta (energies) of the electron and the photon. Finally, it is worth noting that the above formula, which agrees with the experimental results, shows that the Compton effect is a purely quantum–mechanical phenomenon: $\Delta\lambda \to 0$ when $\hbar \to 0$. For this discovery, Compton was awarded the Nobel Prize for physics for the year 1927.

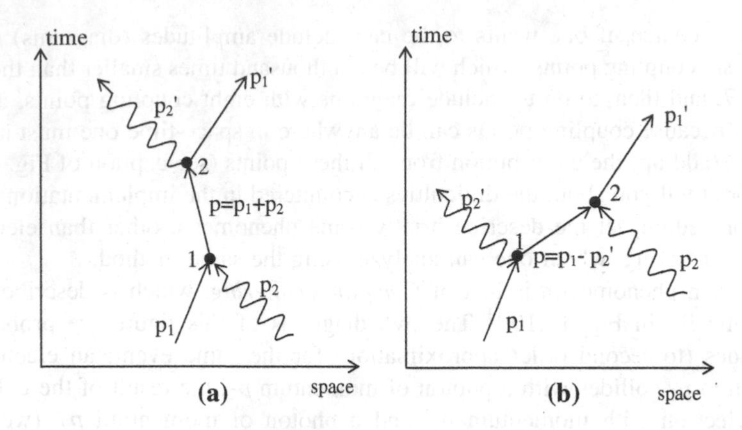

Fig. 15.10 Feynman diagrams for Compton scattering

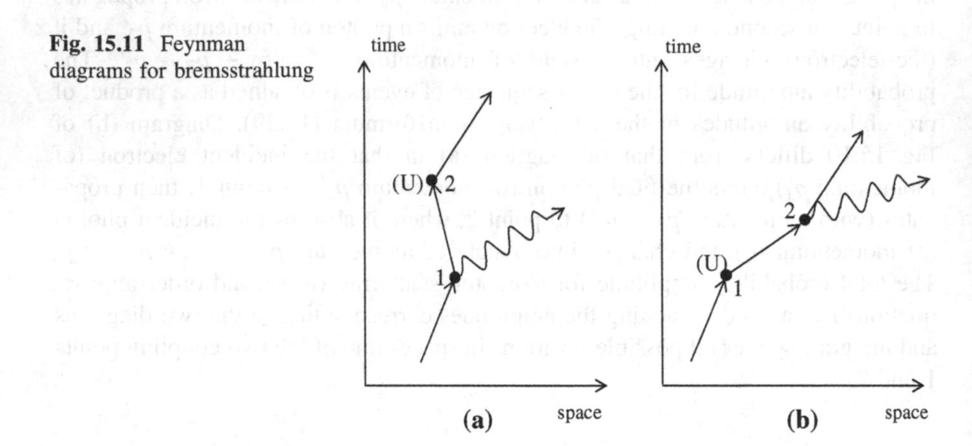

Fig. 15.11 Feynman diagrams for bremsstrahlung

Another phenomenon which one can describe in the above manner is that of *bremsstrahlung*, in which an electron emits a photon as it passes near a nucleus. The phenomenon is described in Fig. 15.11.

The two diagrams are probability amplitudes for the same event. In diagram (a): an electron of given (initial) momentum arrives at point 1 in space-time , emits a photon and it (the electron) changes into a new momentum; it then propagates to point 2 where it is scattered by the potential U of a nucleus into a new (final) momentum state. In diagram (b) the electron is scattered by the nucleus at point 1, then propagates to point 2, where it emits the photon and changes its own momentum accordingly.

Another phenomenon described by the same method is that of pair production, demonstrated in Fig. 15.12. In diagram (a) of this figure, a photon annihilates into a pair, an electron and a positron, at point 1 in space-time ; the electron propagates

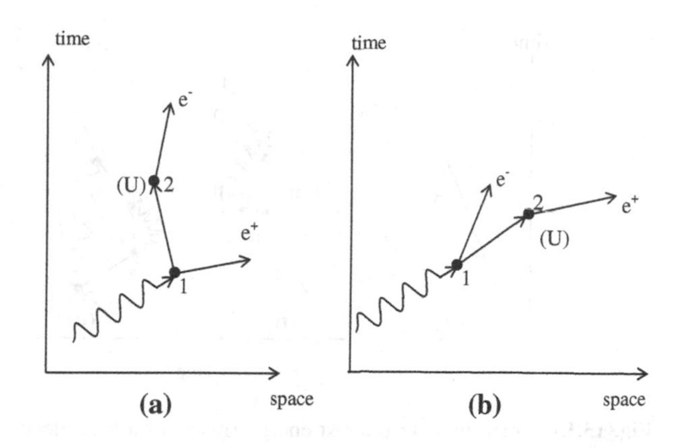

Fig. 15.12 Feynman diagrams for pair production

to point 2 in space-time where it is scattered by the potential (U) of a nucleus.[12] A second probability amplitude for the same event is given by diagram (b): the photon annihilates into an electron and a positron, at point 1 in space-time ; the positron propagates to point 2 in space-time where it is scattered by the potential U of a nucleus. [We can always think of the positron as an electron moving backward in time: at point 1 in space-time it absorbs a photon and changes into an electron moving forward in time; we can also say, following Dirac, that at point 1 the photon excites an electron from the 'sea' of the occupied negative-energy electron states into a positive-energy state creating at the same time a 'positive hole', which is the positron. The mathematics behind the diagrams is the same.]

I must now tell you about the difficulties met in the implementation of the perturbation method I described. We have noted that the probability amplitudes corresponding to the various Feynman diagrams, corresponding to any order of approximation, must be integrated over all possible coupling points in space-time and then added up to obtain the total scattering amplitude for an observable event. But it turns out that these integrals diverge to infinity! We can go around this difficulty by stopping the integration when the distance between coupling points is very small, say 10^{-30} cm, which is many orders of magnitude smaller than presently 'observable' distances between elementary particles within the nucleus. [Stopping the integration when the distance between coupling points is very small is equivalent to cutting off virtual photons with very small wavelength (large $|\mathbf{p}|$); see caption of Fig. 15.9]. One then obtains finite numbers at the end of the calculation; this is how the first calculations relating to the phenomena we have described were performed. But then inconsistencies appeared which could not easily be dealt with. The difficulties arise from the so called self-interactions of the electron and the photon, which we have not cited so far, but which are required by the theory nevertheless.

[12] We could say that the electron exchanges a photon with the nucleus. That would correspond to a description of the static field about the nucleus as a 'cloud' of photons about the nucleus.

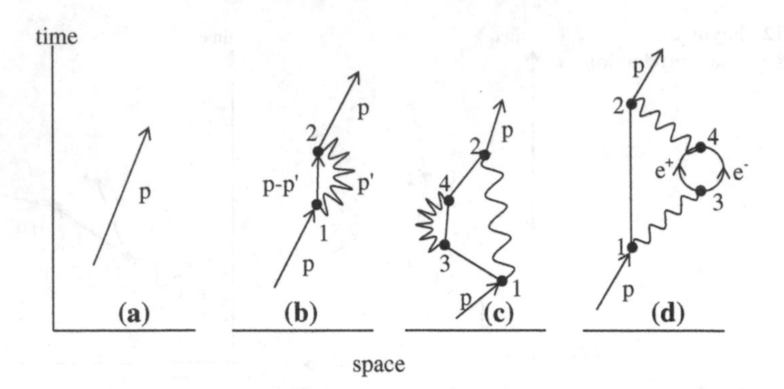

Fig. 15.13 Corrections to the rest-energy (mass) of a bare electron

Consider the Feynman diagrams of Fig. 15.13. Diagram (a) represents the probability amplitude for the propagation (in space-time) of a *bare* electron of momentum-energy p. Diagram (b) corresponds to the probability amplitude for the following sequence: at point 1 in space-time the electron emits a photon of momentum p' and it changes into a state of momentum $p - p'$; at point 2 in space-time the electron absorbs the photon it had emitted and returns to its initial state of momentum p. We remember that the two coupling points can be anywhere in space-time , so that an integration over all possible positions of the coupling points is implied. The result of this integration is infinity unless we stop the integration when the distance between the coupling points is very small. [This is equivalent to: p', the momentum of the (virtual) photon and that of the electron in its intermediate state, can take any values compatible with momentum-energy conservation at the coupling points. Therefore an integration over all values of p' is attached to diagram (b), and the result of this integration is *infinity*, unless we cut off photons of very short wavelength (very large $|\boldsymbol{p}|$).] Diagram (b) is a second order diagram; diagrams (c) and (d) are fourth order diagrams which are interpreted in similar fashion. We note that in diagram (d), at point 3 in space-time , a virtual photon changes into an electron–positron pair which at point 4 recombine into a single photon again. And since in all diagrams the momenta of intermediate photons (and electrons) can take any value, the corresponding integrals are all infinity, unless we cut off the photons of very small wavelength. Suppose we have done so, that we have established a rule which tells us at which distance between coupling points to stop the space-time integrations or, equivalently, which photon wavelengths to cut off in the integrals over the momenta of intermediate states. How are we then to interpret the diagrams (b) to (d) of Fig. 15.13, and similar higher order corrections to the amplitude of the bare electron represented by the straight line of diagram (a)? The total amplitude (the sum of the four diagrams of Fig. 15.13 and all other higher order corrections) represents the propagation of a *real* electron in space-time , whereas the straight line represents what we called a *bare* electron (which is not the thing we observe in the laboratory). But what exactly is the physical meaning of this amplitude?

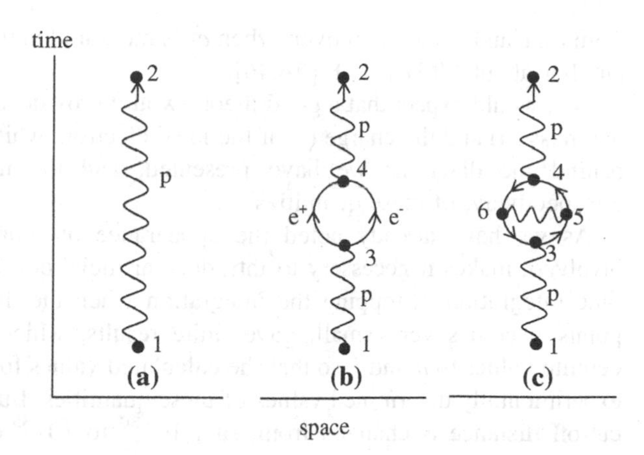

Fig. 15.14 Self-interactions change the ideal photon into a real photon. Diagram (**a**) represents the probability amplitude for an ideal photon to propagate from point 1 to point 2 in space time; diagram (**b**) represents a second-order correction to the probability amplitude of (**a**); diagram (**c**) represents a fourth-order correction

When the final state, ψ_f, of the electron (or any system) is different from its initial state, ψ_i, we have a clear physical meaning for the probability amplitude T_{fi}, it is: $|T_{fi}|^2$ is the probability that the electron, acted upon by the specified interaction will change from state ψ_i to state ψ_f. How are we to interpret T_{fi} when $\psi_f = \psi_i$? We can guess the answer by reference to Eq. (15.28). The first-order matrix element defined by this equation when $\psi_f = \psi_i$ gives the first-order correction to the energy of the system in the state ψ_i due to $\hat{H}_{\mathrm{int}}$. Therefore we surmise that the diagrams (b) to (d) of Fig. 15.13, and similar higher-order diagrams, represent corrections to the rest-mass energy, nc^2, of the bare electron, where n is the rest-mass of the *bare* electron. And because the speed of light is a universal constant, we can simply say that the above diagrams represent corrections to the rest mass of the bare electron due to its interaction with the electromagnetic field emanating from it (a self interaction which is always there). We can put it as follows: the mass m of a real electron (the one we measure) is obtained by adding to n the corrections corresponding to diagrams (b–d) of Fig. 15.13, and all similar higher-order diagrams. And it is annoying, to say the least, that these corrections come out to be infinity in the limit of zero separation between coupling points. But if no contradictions arise by stopping the above mentioned space-time integrations at minutely small separations between coupling points, the situation would be at least tolerable.

And then there are more peculiar diagrams: those shown in Fig. 15.14, and similar higher-order ones. The probability amplitude for a *real* photon to propagate from point 1 of space-time to point 2, is obtained by adding to the amplitude for the *ideal* photon, corresponding to diagram (a), the amplitudes corresponding to the diagrams (b) and (c) and similar higher order diagrams. We remember that the coupling points (3 and 4 in (b); 3, 4, 5 and 6 in (c)) can be anywhere in space-time, which implies corresponding integrations in space-time , which lead to infinities if not restricted as explained above.

The diagrams of Fig. 15.14 imply that the charge of an *ideal* electron, which we can identify with the coupling constant j, is different from the measured charge e of a real electron. [We note that e is the only constant in Maxwell's equations apart from ε_0, c and $\hbar$ and, moreover, when e^2 is measured in units of $4\pi\varepsilon_0\hbar c$, it is a pure number about 1/137 [see Eq.14.46].

One would expect that a good theory would provide a definite relation between the mass (n) and the charge (j) of the ideal electron, which appear in the formulae behind the diagrams we have presented, and the measured values, m and e respectively, of these quantities.

As we have already noted the appearance of infinities in the calculations involved, makes it necessary to introduce artificial cut-offs in the relevant space-time integrations. Stopping the integration when the distance between coupling points becomes very small, gave finite results, which allowed one to ascribe definite values to n and j, so that the calculated values for m and e agreed with the experimentally determined values of these quantities. But unfortunately, when the cut-off distance is changed from, say, 10^{-30} to 10^{-35} cm, the same calculation gives different values for n and j! After many unsuccessful attempts to remove this difficulty, Hans Bethe and Victor Weisskopf, at about 1948, noticed the following: If one uses the values of n and j, obtained for a certain cut-off distance, to calculate the outcome of a certain experiment, he obtains a certain answer: A. Now if he repeats the calculation using the values n' and j', obtained for a different cut-off distance, to calculate the outcome of the same experiment, he obtains a certain answer: B. Now, provided the two calculations are performed to the same order of approximation, the two results, A and B, are not exactly the same, but they are nearly the same; and more so if the cut-off distances are smaller! This seemed to suggest that the only quantities that depend on the small distances between coupling points (or equivalently on the exchange of very small wavelength photons) are the values of n and j, which can *not* be measured directly anyway. This observation by Bethe and Weisskopf opened the way to the development of a working theory of quantum electrodynamics by Schwinger, Tomonaga and Feynman.[13] They developed, each in his own way, a theory which is relativistically invariant and which 'removes' the divergencies in a systematic way in all orders of perturbation theory, by a process known as the *renormalisation of the charge and mass of the electron.* We note, however, that a cut-off of the interaction, when the distance between two coupling points is very small, remains an essential part of the theory. Looked at from a mathematician's point of view, the theory is not an exact theory (such a theory should allow one to go to the limit of *zero* distance between coupling points); and there are a few scientists (Paul Dirac was one of them) that do not like it. But the theory works!

The theory gives the correct answer to all relevant experiments, but it does not give us the value of either the mass or the charge of the electron. But the theory is able to calculate the observable *difference* between the value of the mass of the electron when free and its value when bound to a nucleus. And one can evaluate the effective magnetic moment for a free or bound electron.

---

[13] Feynman says so (see page 128 of his book on *QED* cited in footnote 9).

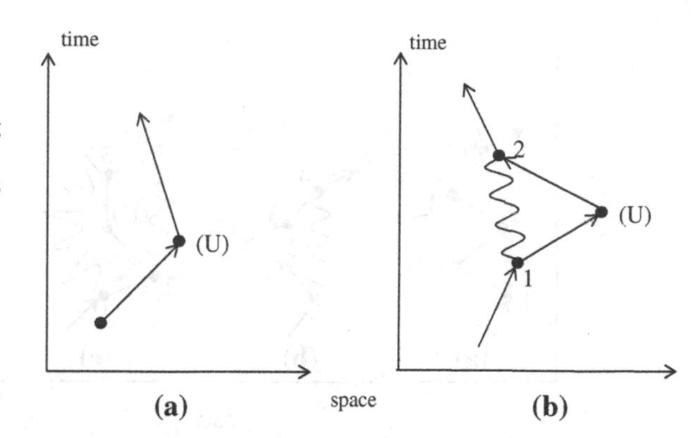

Fig. 15.15 **a** corresponds to the probability amplitude for the scattering of an electron by a magnetic field according to Dirac's theory.
b corresponds to Schwinger's second-order correction to this amplitude

We know that Dirac's theory endows the electron with a magnetic moment, μ_B, given by Eq. (15.24). But as we have already noted, the actual value of the magnetic moment is closer to $\mu = 1.0018\mu_B$. The first explanation of the deviation from Dirac's value was provided in terms of quantum electrodynamics by Schwinger, who in 1948 obtained the following formula for μ:

$$\mu = \mu_B(1 + \alpha/2\pi) \qquad (15.30)$$

where α is the fine-strucure constant of Eq. (14.46).

The physics behind Schwinger's formula is described diagrammatically in Fig. 15.15. Diagram (a) of Fig. 15.15 corresponds to the probability amplitude for the scattering of an electron by a magnetic field according to Dirac's theory and leads to a magnetic moment μ_B for the electron. We read it as follows: the electron is scattered by the field, *U,* of a magnet. In diagram (b) of Fig. 15.15: the electron goes along for a while, at point 1 in space-time it emits a photon and its own state changes accordingly; it is then scattered by the magnetic field (*U*), and then at point 2 in space-time it reabsorbs the photon it had emitted and changes into a new state accordingly. The corresponding probability amplitude must be added to that of diagram (a), and this accounts for Schwinger's correction to the magnetic moment of the electron. There are two coupling points in diagram (b) and therefore the corresponding probability amplitude is proportional to j^2. As usual the coupling points can be anywhere in spacetime and corresponding integrations with appropriate cut-offs are implied.

In order to obtain a better approximation to the measured value of the magnetic moment one must include higher order corrections to the probability amplitude for the scattering of an electron by the magnetic field. In Fig. 15.16 we show examples of fourth-order corrections: there are four coupling points in these diagrams and therefore the corresponding corrections are proportional to j^4. The way to read these diagrams should by now be obvious to the reader and not worth repeating here.

In order to compare with the experimentally observed value of the magnetic moment of the electron, which has become more accurate over the years, theorists

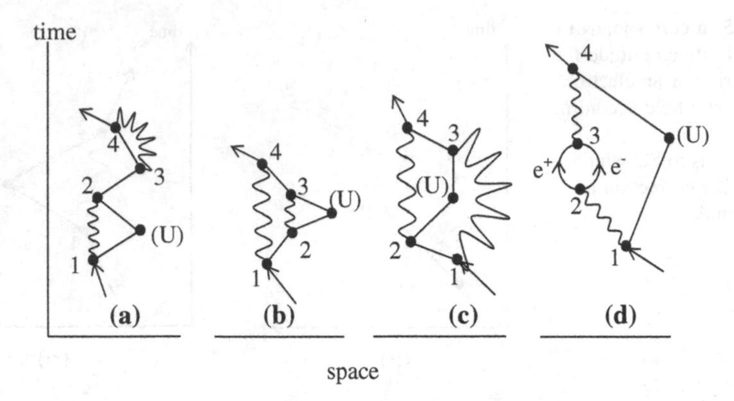

Fig. 15.16 Fourth-order Feynman diagrams involved in the evaluation of the magnetic moment of the electron

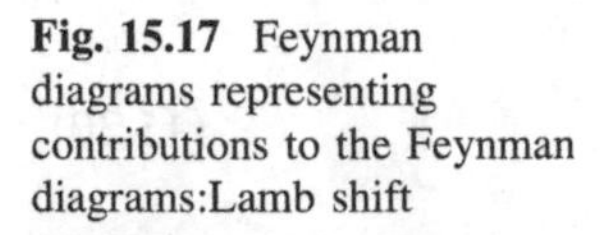

Fig. 15.17 Feynman diagrams representing contributions to the Feynman diagrams:Lamb shift

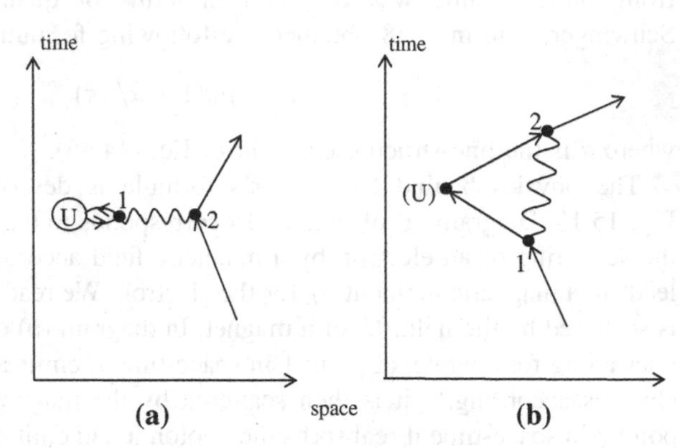

have included corrections up to the sixth order (proportional to j^6) in their calculations of this quantity. In his book on *QED*, published in 1985, Feynman compares with pride the theoretical value ($1.00115965242\mu_B$) with the experimental value of this quantity ($1.00115965221\mu_B$) as they stood at that time.

An equally important phenomenon in relation to quantum electrodynamics is the Lamb shift mentioned at the beginning of this section. The Lamb shift is known to arise from second and higher-order self-interacion effects, and in particular those corresponding to the diagrams of Fig. 15.17.

According to diagram (a), the Coulomb field, U, of the nucleus (which can be regarded as a 'cloud' of vitual photons) instead of acting directly upon the electron, it may create a virtual electron–positron pair which is subsequently annihilated to create a virtual photon which is absorbed by the electron. This leads to a small reduction in the effect of U, which is equivalent to a smaller effective charge of the electron. This shielding effect of apparently empty space is, appropriately, called *vacuum polarization*. The interpretation of diagram (b) is straightforward and requires no further comment. We need not go into the details of the calculation of the Lamb shift. Suffice it to say that the calculated value of this shift is in excellent agreement with the measured value of this quantity.

15.3 Cosmology

15.3.1 Galaxies

As a result of the vastly improved telescopes constructed in the 20th century (some of them radio-telescopes) and our better understanding of the laws of nature, both at the macroscopic level (Newton's law of gravity and Einstein's theory of General Relativity) and at the microscopic level (quantum mechanics), and also because present-day electronic computers allow us to perform previously unimaginably difficult calculations, we now know a lot about the universe, about how stars come into being, what they are made of, and how they move relative to each other. But there are still questions to which we can not give definite answers. In this, the last section of the book, I shall briefly tell you some of the things we know, and what some of the unanswered questions are.

Let me begin by telling you how vast the universe, as we know it, is. Our solar system is but a small neighborhood in our galaxy of stars which is known as the Milky Way. In it there are some 10^{11} to 10^{12} stars, some are smaller, and some are about the same size or larger than the Sun. There are many regions in our galaxy, the nearest is the Orion Nebula, consisting of clouds of dust, where new stars are now being born. The diameter of the Orion Nebula is about 15 light years; a light year is the distance travelled by light in a year, which is about 10^{12} km. The distance of the Orion Nebula from our solar system is about 100 times its diameter (about 1,500 light years), and the over-all diameter of the Milky Way is nearly 100 times greater still, about 100,000 light years.

And there are many galaxies. The nearest is the *M*31, known as the Andromeda Galaxy. Like our own galaxy, it is shaped like a flattened circular disk bulging at its centre (see Fig. 15.18). The distance from our galaxy to the Andromeda is about 10^7 light years, which is an enormous distance. However, when this distance is considered on a galactic scale it is rather small. By which I mean that, if we take the diameter of our galaxy as the unit of distance and call it the 'yard', then the average distance between galaxies is about 100 *yards*. In other words, on a galactic scale, stars are widely spaced, whereas galaxies are rather closely spaced.

The density of stars in our galaxy becomes smaller and smaller as the periphery of the galaxy is reached. But when astronomers looked for a similar reduction in the density of galaxies on a universal scale, they did not find one. The distribution of galaxies (there are about 10^9 of them of various shapes) goes on undiminished

Fig. 15.18 The Andromeda Galaxy. The stars we see scattered about it belong to our own galaxy. We look through our own galaxy to see the stars that lie beyond it

as far as the astronomers can detect with their largest telescopes. The galaxies (and the stars within the galaxies) are not static but in a continuous motion. For many practical purposes (when we are dealing with the rotation of the earth and its orbit about the sun) we may assume them stationary, but *not* when we look at the universe as a whole. The astronomers and the mathematicians have a lot to say about the motion of the galaxies, and how it relates to an assumed 'beginning' and an 'end' of the universe. But before I summarise for you some of the facts concerning the motion of the universe, and some of the theoretical ideas concerning their interpretation, let me tell you something about how stars are generated, how they live and die.

15.3.2 Generation and Death of Stars

At the beginning there is only a cloud of dust: a gas of hydrogen atoms, which is by far the most abundant material in the universe. When due to gravity lots of hydrogens come and stay together a star begins to form. As its mass contracts under the influence of gravity the interior of the star gets very hot. It is estimated, for example, that the temperature inside the sun rises to about 15 million degrees. When the protons (the nuclei of hydrogen) collide with each other at this temperature, the weak interaction changes some of them into neutrons (an *up* quark within the proton changes into a *down* quark [(see Fig. 15.4) in the process described by Eq. (15.3b)]. The neutron then joins another proton to create a *deuteron* (so called). When two deuterons collide with each other at these high temperatures, some of them join up to form the nuclei of helium atoms, releasing energy in the process. Not surprisingly, helium is the second most abundant material in the universe (about ten times less than hydrogen). The energy released in the above nuclear reaction (the burning of hydrogen into helium) in the interior of the sun, makes up the radiation loss from the surface of the sun (which keeps life going on Earth). [We note in passing that, if hydrogen could be converted to helium under controlled conditions, the burning of a few tons of water would provide enough energy to operate the world's industries for a year or so.] I should also tell you, that according to reliable estimates, the hydrogen still remaining inside the sun is sufficient to maintain its present power output for a further 10,000 million years.

Stars have different masses and they differ in the rate by which they emit energy. A star with ten times the mass of the sun emits energy a thousand times faster than the sun. This means that the hydrogen in it will be depleted sooner, may be within a few tens of millions of years, which is much less than the age of the Milky Way, estimated to be about 12,000 million years. One would like to know what happens after the hydrogen in the star is exhausted. According to the astrophysicists who studied this problem,[14] the further contraction of the central region of the star, which follows the exhaustion of hydrogen, raises further the temperature inside the star and other nuclear reactions become possible. Helium undergoes reactions which produce carbon and oxygen, also in great abundance in the universe. At still higher temperatures, carbon and oxygen undergo reactions which generate sodium, magnesium, aluminum, silicon, sulfur, and calcium. And when the temperature is raised further the common metals appear: iron, nickel, chromium, manganese, and cobalt. Elements of low abundance are also generated in the stars, but only in minor by-product reactions. One assumes that in the most massive stars the evolution from hydrogen to the iron-like elements is by now complete, but in stars of relatively small mass it is likely that this process is still going on.

Finally, one may inquire as to what happens when all nuclear fuel in a star has been exhausted. It is reasonable to assume that the star will cool off, and will not shine thereafter. Calculations show that this is indeed the case when the mass of the star is about the mass of the sun or smaller, but not so for the more massive stars. These, in order to cool off, eject matter into space. Some, the not so massive,

---

[14] A major contribution to our understanding of how the various elements are created in the interior of the stars was made by Fred Hoyle, a British astronomer (at Cambridge), and by Edwin Salpeter, an American astronomer (at Cornell University) in the 1950s. For a popular exposition see: Fred Hoyle, *Ten Faces of the Universe*, W. H. Freeman and Company, San Francisco, 1977.

do it gently, throwing off a cloud, a so-called planetary nebula, similar to the clouds they were made of in the first instance, except that now the clouds contain complex atoms with many protons and neutrons in their nuclei. So that when a new generation of stars is born from these clouds, they will contain some of these complex atoms. What is left of the original star cools further to become a *white dwarf*. If the white dwarf has a mass about that of the sun, its diameter will be about that of a planet with the material at its centre having a density of about a ton per cubic centimeter.

The more massive stars are likely to die more violently, exploding like a nuclear bomb. Such explosions happen about once in a century in our own galaxy. For a few days after the explosion, known as a *supernova*, the star is exceptionally bright, and then it fades away. The residues of these explosions, known as *neutron stars*, are quite remarkable. Neutron stars with a mass about that of the sun are much smaller in size than the planets. The mass-density at their centre may be as high as 100 million tons per cubic centimeter. Because of their small diameters, neutron stars spin around very quickly, most of them about once in a second, some even faster, and a lighthouse effect seems to operate in which all kinds of radiation (from radio waves to visible light and X-rays) sweep periodically across an observatory. When this happens the star is known as a *pulsar*.

Neutron stars are not far removed from what is known as a *black hole*, named so because the gravitation field about it is so strong that even light can not escape from it, rendering it invisible (see below). There is in fact evidence that supernovae do sometimes produce a black hole rather than a neutron star.

Initially it was thought that a complete collapse of a massive star to atomic size under the influence of gravity would not be possible because of the so-called *degeneracy* pressure arising from Pauli's exclusion principle (the electrons could not all be accommodated in the quantum state of the smallest radius, and the same would apply to other spin-half particles). However, in 1931, the Indian physicist Subrahmanyan Chandrasekhar showed that degeneracy pressure does not suffice to resist collapse when the mass of the star exceeds a certain limit (about 1.4 times the mass of the sun). Chandrasekhar's theory uses quantum mechanical arguments in conjunction with general relativity, but one may argue that a final answer to this question can only be obtained from a *quantum theory ofgravity*, which we do not yet have.

It appears that black holes may also appear when stars at the centre of a galaxy coalesce to form a super-massive black hole with a mass 10^6 to 10^9 times the mass of the sun.

Finally, I should mention a remarkable discovery made in 1974 by the British theoretical physicist Stephen Hawking, that a black hole is not simply an absorber of energy, that itself emits radiation through a quantum mechanical process by which particle—antiparticle pairs are created (by quantum fluctuations of the vacuum state) at the periphery of the black hole with a corresponding probability that before their mutual annihilation occurs one of the pair is pulled into the black hole and the other escapes outward to make up the Hawking radiation. As a result the black hole slowly evaporates, in proportion to its surface area, at a rate given by:

$$\frac{d(Mc^2)}{dt} = \sigma T^4 (surface\ area) \propto M^{-3}$$

where M is the mass of the black hole, and σ is the Stefan-Boltzmann constant of Eq. (13.1). One finds that the lifetime of a black hole is approximately given by:

$$\tau = (\frac{M}{10^{11} kg})^3 \times 10^{10} \text{ years}$$

According to the above formula a black hole with a mass equal to that of the sun would not have had enough time to evaporate since the beginning of the universe (which probably happened about 10^{10} years ago).

15.3.3 The Earth

The geologists assert that two or three hundreds of millions of years ago Africa was joined to America, until they separated by forces arising from a kind of nuclear engine within the Earth, using uranium as fuel. The motion produced by the heat generated by nuclear reactions within the Earth is rather complicated: the crust is made by a number of *plates* which move relative to each other. At some places the rocks of a plate emerge from the interior of the Earth, as in the Mid-Atlantic ridge, elsewhere the rocks of one plate may dip below another plate, below the rocks of a continent. The pressing together of plates has caused mountains to be pushed up, from the Alps in the west, through Turkey and Caucasus, to the Himalayas in the east. The motion of the plates against each other causes the eruption of volcanoes (due to the heat generated), and it is the cause of earthquakes. But without this turning over of the Earth's crust there would, probably, be no mineral deposits on the surface of the Earth, since these minerals derive from the interior of the Earth.

Measurements of seismic vibrations, when compared with data on the properties of materials under high shock pressures, strongly suggest that, the Earth has a metallic core with a radius of about 3,400 km, which is a bit more than half of the radius of the total Earth (about 6,370 km). Apparently the core of the Earth consists mostly of molten iron, at a temperature of about 5,000°, mixed with smaller quantities of titanium, vanadium, chromium, manganese, cobalt and nickel. Electric currents flowing in its metallic core generate the magnetic field of the Earth. Outside the core lies a mantle of rock, the crust of which we described above. Of course, extended regions of the Earth's surface are covered by the oceans, and if the great mountains were not there, most of the Earth would probably be covered by water.

One may ask: How did the Earth, and the other planets, come into being? Astronomers agree that they were generated during the formation of the sun, about 4,500 million years ago. They claim that during the early solar condensation, a rapidly rotating mass lost some of its angular momentum to the planets formed by the matter ejected by the sun. And this is the reason, according to this theory, that the planets have large angular momenta compared to that of the present-day sun. Another argument in favour of the above theory goes as follows: One observes that certain elements which are gases or form gaseous compounds, at temperatures of a few hundred degrees, are largely absent from Earth and the other inner planets (Mercury, Venus and Mars) but they exist in the outer planets (Saturn, Uranus and Neptune), which suggests that the inner planets were formed from solid aggregates, while the outer planets were formed from the gases which continued their motion to the outer region of the solar system.

15.3.4 The Universe: Expanding Universe

In 1965 the American physicists Arno Penzias and Robert W. Wilson, working at the Bell Telephone Laboratories in New Jersey, discovered radio waves arriving at Earth uniformly from all directions in space.[15] The only way to explain this microwave radiation is to assume that it originated at a time much more remote than any which is observable by the available astronomical methods.

As far as the astronomers can see (to a distance of about 5,000 million years) the galaxies appear to be uniformly distributed in space. One can put it like this: if at a given time *t*, *N* galaxies are placed on the *N* points (corners) of a polygon, then at a later time the *N* points will be different but the shape of the polygon will not change. Only the scale, *Q*, of the polygon changes. Remarkably, the Russian physicist Alexander Friedman was able to show in 1922 that, if one assumes that the universe looks the same in which ever direction we look and independently of wherever in the universe we stand, then Einstein's theory of general relativity leads to the conclusion that the universe is not static. It is worth noting that Friedman proposed the above before Hubble's discovery of the expanding universe mentioned in Sect. 12.4.3. In any case, the work of Friedman and others has shown that the scale *Q*, introduced above, may change with time in only one of two ways, represented by *A* and *B* in Fig. 15.19; the theory can not tell which of the two curves the actual universe obeys.

According to curve *A*, the expansion of the universe never stops. According to curve *B*, gravitation eventually stops the initial expansion. This happens when *Q* reaches the maximum of curve *B*. After this point expansion is replaced by contraction. Observations tell us that, if the universe obeys *B*, then we must be

[15] This discovery won for Penzias and Wilson the 1978 Nobel Prize for Physics.

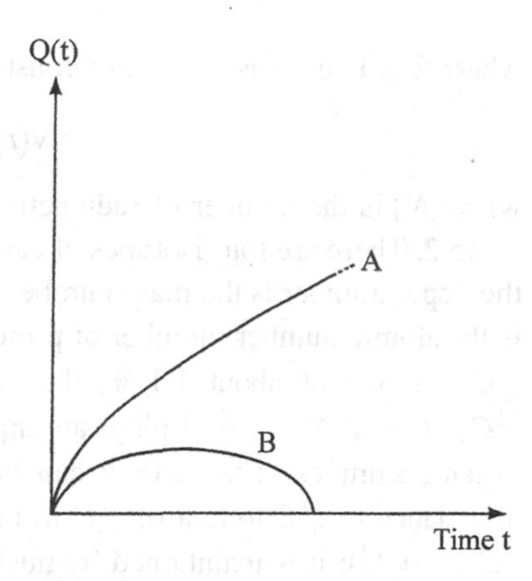

Fig. 15.19 The scale $Q(t)$ changes in the manner of either A or B

living before the maximum of the curve.[16] We note that in both curves, Q was zero at some point in the past (the astronomers put it between 12 and 15 billion years ago). It was at that point when the universe began!

We can now return to the microwave radiation mentioned at the beginning of this section. The origin of the universe (the point of $Q = 0$) is the point in space-time from which the microwave radiation is derived. In the first few seconds of the history of the universe, the radiation's temperature was about 10,000 million degrees. Its present-day temperature is only 3 degrees, as a result of cooling brought about by the expansion of the universe.

What was there, if anything, before the universe was born? Many say that the question is meaningless, space-time was not existent before the whole thing arose out of nothing. Others disagree. There are a number of hypotheses. Nobody knows. And nobody knows how the universe will end. Only one thing is certain, it will be a very long time from now!

Exercises

15.1. Let N be the number of certain radioactive nuclei in some material at time t, then the decay rate (rate of change) of N is:

$$\frac{dN}{dt} = -\lambda N \tag{1}$$

[16] A quantitative analysis of the astronomical data on the rate of expansion of the universe must accord with the observed distribution of matter in the universe and with the distribution of energy in the universe. And there arose great discrepancies between theory and observations, which led astronomers to postulate the existence of such things as dark (not visible) matter and dark energy, that brought back into being Einstein's cosmological constant (see Sect. 12.4.3) but with a different physical meaning attached to it. It appears that about 70 % or more of the energy (including rest-mass energy) in the universe is dark!.

where λ is known as the decay constant. Verify that $N(t)$ is given by:

$$N(t) = N_0 e^{-\lambda t} \tag{2}$$

where N_0 is the number of radioactive nuclei at $t = 0$.

15.2. There are four isotopes of carbon: ${}^{11}_{6}C$, ${}^{12}_{6}C$, ${}^{13}_{6}C$, ${}^{14}_{6}C$. We remember that the upper number is the mass number (protons + neutrons), and the lower number is the atomic number (number of protons). ${}^{12}_{6}C$ has a natural abundance of 98.9 %; ${}^{13}_{6}C$ has one of about 1.1 %, the others much less. However, the radioactive ${}^{14}_{6}C\left({}^{14}_{6}C \rightarrow {}^{14}_{7}N + e + \nu\right)$ plays an important role in the so-called *carbon-dating* of organic samples. The method can be summarized as follows: The ratio of the abundance of ${}^{14}_{6}C$ to that of ${}^{12}_{6}C$ in the CO_2 of the atmosphere is constant (about 1.3×10^{-12}); it is maintained by nuclear reactions induced by cosmic rays in the upper atmosphere. Accordingly, all *living* organisms have the same ratio of ${}^{14}_{6}C$ to ${}^{12}_{6}C$, since they receive their CO_2 from the atmosphere. But when the living organism dies the ratio between the two is reduced, according to Eq. (2) of the previous exercise, as time passes. Therefore: by comparing the decay rate of ${}^{14}_{6}C$ in the dead organic material to that in the living organism, we can estimate the time passed since its death. The method has been particularly successful in dating objects, such as old testament scrolls (which turned out to be about 2000 years old).

Given the above, solve the following problem: If 25 *gr* of charcoal shows a ${}^{14}_{6}C$ radioactivity of 250 decays per minute, how long has the tree from which the charcoal originated been dead. The decay constant for ${}^{14}_{6}C$ is: $\lambda = 3.843 \times 10^{-12} s^{-1}$.

Note: Remember that there are 6.02×10^{23} nuclei of ${}^{12}_{6}C$ in 12 g of carbon (Avogadro's number).

Answer: 3370 years.

15.3. Verify that $\widehat{\alpha}_x, \widehat{\alpha}_y, \widehat{\alpha}_z$ and $\widehat{\beta}$, as defined by Eqs. (15.9) satisfy Eqs. (15.8).

15.4. Verify Eqs. (15.15), and that the condition to be satisfied for a nonzero solution of these equations leads to Eq. (15.16).

15.5 (a) Verify that $\widehat{l}_x, \widehat{l}_y$ and $\widehat{l}^2$ do *not* commute with $\widehat{H}$ as defined by Eq. (15.19).

(b) Verify that $\widehat{j}_x, \widehat{j}_y, \widehat{j}_z$ and $\widehat{j}^2$ commute with $\widehat{H}$; that $\widehat{j}_x, \widehat{j}_y, \widehat{j}_z$ commute with $\widehat{j}^2$ and $\widehat{H}$ but do *not* commute with each other.

15.6. Prove Compton's formula for $\Delta\lambda$ given in footnote 11 of Chap. 15.

Appendix A
Notes on Mathematics

The Basics of Trigonometry

(a) The angle between two (straight) line-segments AB and AΓ is equal to the angle between A′B′ and A′Γ′, if A′B′ and A′Γ′ are parallel respectively to AB and AΓ (see Fig. A1.1a).
(b) The angle between two line-segments AB and AΓ is equal to the angle between A′B′ and A′Γ′, if A′B′ and A′Γ′ are normal (at 90°) respectively to AB and AΓ. (See Fig. A1.1b): we observe that a rotation of 90° about A′(A) of the primed triangle brings it into coincidence with the unprimed triangle.
(c) The angles of a triangle add up to two right angles (180°). This follows from (a) and (b) above (see Fig. A1.1c).
(d) It follows from (c) that a triangle is completely determined if one of its sides (its length) and two of its angles are known. It is also completely determined if two of its sides and the angle between them are known.
(e) Two triangles are obviously the same if the three sides of the first one are respectively equal to the sides of the second one. But this is not necessarily true when all three angles of a triangle are equal respectively to the angles of another one. Such triangles are said to be *similar*. Similar triangles differ in scale (the one being a magnified version of the other).

A triangle is said to be orthogonal if one of its angles is a right angle (90°).

Pythagoras's Theorem

Pythagoras's theorem (see Fig. A1.2a): The square of the hypotenuse of an orthogonal triangle (the area enclosed by ABEΔ) is equal to the sum of the squares of its two other sides (the areas enclosed by BZHΓ and AΓNM respectively).

Proof of the theorem (see Fig. A1.2b): The square ABEΔ of Fig. A1.2a has been turned on its side (AB) going into ABEΔ of Fig. A1.2b, and the square AΓNM has been replaced by KHΛΔ which can be shown to be equal to it as follows: we must show that KΔ = KH = AΓ. We do so by comparing the

Undergraduate Lecture Notes in Physics, DOI: 10.1007/978-3-319-00750-2,

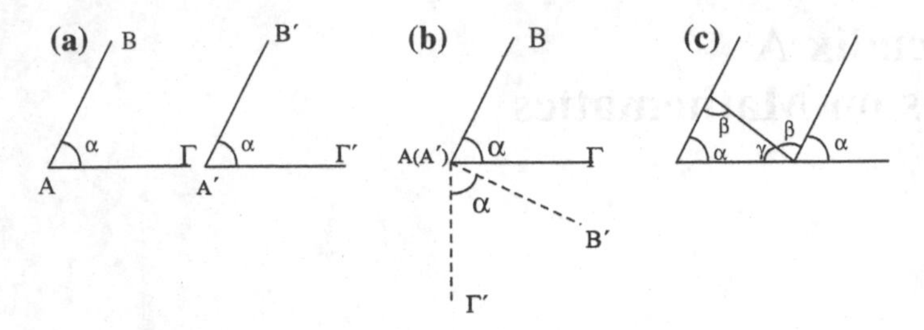

Figure A1.1 The angles of a triangle add up to 180°

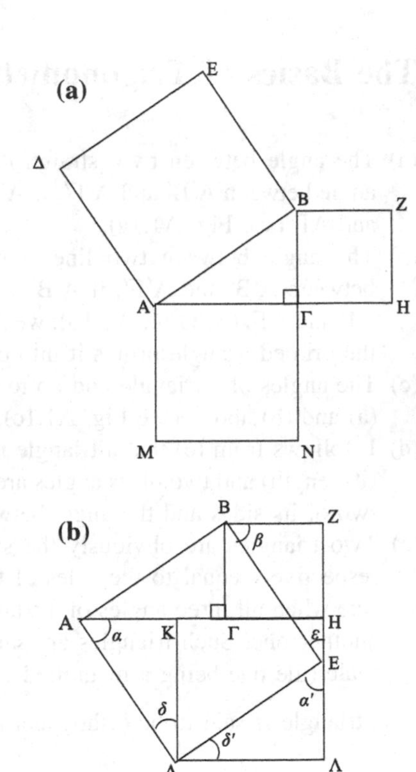

Figure A1.2 **a** Pythagoras's theorem: The square of the hypotenuse is equal to the sum of the squares of its two other sides. **b** Proof of the theorem (see text)

orthogonal triangles AKΔ and BZE: these are the same because they have the same hypotenuse (AΔ = BE) and the same angles ($\alpha = \beta$, $\delta = \varepsilon$), therefore AK = BZ = ΓH. Therefore (see figure): KH = AΓ. Similarly we establish that the orthogonal triangles AKΔ and EΛΔ are the same, therefore KΔ = ΔΛ = KH. Which proves that KΔ = KH = AΓ and that therefore the square KHΛΔ of Fig. A1.2b is the same with that of AΓNM of Fig. A1.2a. We must now convince ourselves that the area enclosed by ABEΔA (the square of the hypotenuse of the orthogonal triangle ABΓ) is equal to the area enclosed by KΓBZHΛΔK (the sum of the squares of its two other sides). We note that the two areas have in common the area enclosed by KΓBEΔK in Fig. A1.2b. The area of ABEΔA contains in addition to the common area the orthogonal triangles ΑΒΓ and ΑΚΔ, while the area ΚΓΒΖΗΛΔΚ contains in addition to the common area the triangles ΔΕΛ and ΒΖΕ. Noting that triangle ΑΒΓ is equal to ΔΕΛ, and that triangle ΑΚΔ is equal to ΒΖΕ, completes the proof of Pythagoras's theorem.

The $\sqrt{2}$ Is Not A Rational Number

We shall prove that the $\sqrt{2}$ can not be written as the ratio of two integers. The method we shall use is that of *reductio ad absurdum*: we assume that $\sqrt{2}$ is rational, i.e. that $\sqrt{2}$ = a/b where a and b are integers. We can safely assume that a and b have no common factor which means that the fraction a/b can not be further simplified by dividing numerator and denominator by an integer (the common factor). This is a trivial assumption because if a and b had a common factor we could divide numerator and denominator with this factor and by doing so remove it. So let us assume that $\sqrt{2} = a/b$ and that a and b have no common factor, and go on to show that this leads to a contradiction, and that therefore $\sqrt{2}$ can not be rational.

$$\text{If } \sqrt{2} = a/b, \text{ then } a^2 = 2b^2 \tag{A1.1}$$

Therefore a^2 is an even number.* But if a^2 is an even number, a must be an even number too, for the square of an odd number (an odd number times an odd number) is always an odd number. Therefore we can write:

$$a = 2c, \text{ where } c \text{ is an integer.} \tag{A1.2a}$$

Therefore, using (A1.1) we obtain:

$$2b^2 = a^2 = 4c^2; \text{ therefore: } b^2 = 2c^2 \tag{A1.2b}$$

Therefore b^2 is an even number, and therefore b is an even number, for as we have already said the square of an odd number is always odd. Therefore we can write:

$$b = 2d, \text{ where } d \text{ is an integer,} \tag{A1.2c}$$

Equations (A1.2a) and (A1.2c) tell us that a and b have a common factor: 2, in contradiction to our starting point (them not having a common factor), the contradiction arising from the assumed rationality of the $\sqrt{2}$. Therefore the $\sqrt{2}$ is not a rational number.

* A number is even if it can be written as twice some other integer. Every other of the natural numbers of Eq. (1.5), i.e. the numbers: 2, 4, 6, 8,… are even; every one of them can be written as twice some other integer. The rest, i.e. the numbers: 1, 3, 5, 7,… are odd, meaning that they can not be written as twice some other integer.

Complex Numbers

The idea of the square root of a negative number appeared for the first time in the 16th century, in the work of three Italian mathematicians working independently. The first was Nicolo Fontana, better known as Tartaglia, born at Brescia in 1499. He came from a humble background, did not attend University, but taught himself enough mathematics to become a mathematics teacher. He apparently devised a method of solving cubic equations (of the form: $ax^3 + bx^2 + cx + d = 0$), but would not reveal what the method consisted of. A well known mathematician of that time, Girolamo Cardano, persuaded Tartaglia to tell him the method, promising that he would not reveal it to any one else. Cardano generalized Tartaglia's method and published it in a book of his. Tartaglia was fully credited in Cardano's book, but he was furious nevertheless, and never forgave Cardano for breaking his promise. In Cardano's book there is a short description of a solution involving the square root of a negative number. A devise which Cardano describes as being *as subtle as it is useless*. The idea was further advanced by Rafael Bombeli who was a contemporary of Tartaglia and Cardano. Bombeli was the first to attribute to *negative* numbers an equal status to that of positive numbers (rather than viewing them as positive numbers to be subtracted from other positive numbers); and he went on, in the same spirit, to attribute to *imaginary* numbers, such as $\sqrt{(-1)}$, an independent existence as well. The introduction of *negative* imaginary numbers by Gauss completed the picture of imaginary numbers (and complex numbers) as we know them today. For the purposes of this book we need only know the basic properties of these numbers which we summarize below.

We define, to begin with, the imaginary unity:

$$i \equiv \sqrt{(-1)}, \tag{A2.1}$$

From which follows that

$$i^2 = -1 \tag{A2.2}$$

We remember, and this is very important that there is no real number which satisfies the equation: $x^2 = -1$. The imaginary numbers are defined by:

$$yi, -\infty < y < +\infty \tag{A2.3}$$

They can be thought as points on a straight line (the so-called axis of imaginary numbers) normal to the axis of real numbers which it crosses at the origin (see Fig. A2.1). To a certain point on the axis we correspond the imaginary unity (i). With the zero at the origin and i at the said point, we obtain a one-to-one correspondence between the imaginary numbers and the points of the imaginary axis.

The complex numbers are defined by:

$$z = x + yi \tag{A2.4}$$

Figure A2.1 The complex plane

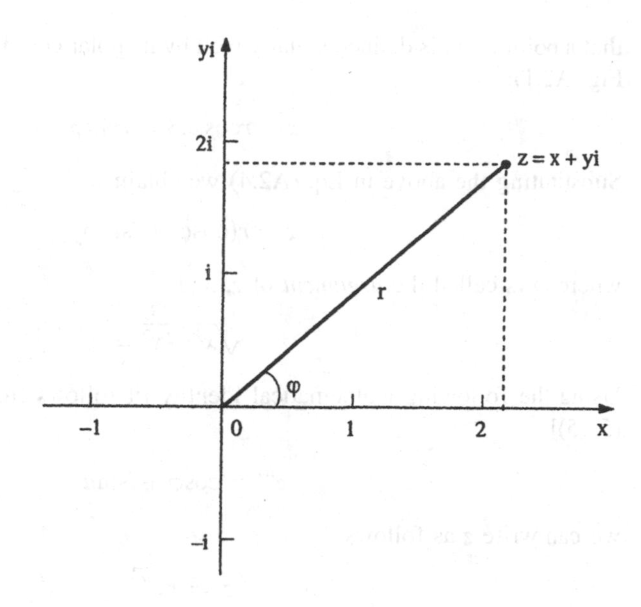

Where x and y are real numbers ($-\infty < x, y < +\infty$). We call x the real part of the complex number, and y its imaginary part, and write:

$$x = Re\{z\} \tag{A2.5a}$$

$$y = Im\{z\} \tag{A2.5b}$$

We add and multiply (and correspondingly subtract and divide) complex numbers using the familiar rules of addition and multiplication of real numbers, and Eqs. (A2.1) and (A2.2). Therefore, if

$$z_1 = x_1 + y_1 i \text{ and } z_2 = x_2 + y_2 i \tag{A2.6}$$

$$\text{Then: } z_1 + z_2 = (x_1 + x_2) + (y_1 + y_2)i \tag{A2.7}$$

$$z_1 z_2 = (x_1 + y_1 i)(x_2 + y_2 i) = (x_1 x_2 - y_1 y_2) + (x_1 y_2 + y_1 x_2)i \tag{A2.8}$$

To every complex number $z = x + yi$ there corresponds a complex conjugate number defined by:

$$z^* = x - yi \tag{A2.9}$$

The absolute value, $|z|$, of a complex number is a positive (real) number defined by:

$$|z| = \sqrt{zz^*} = \sqrt{x^2 + y^2} \tag{A2.10}$$

It is obvious from Fig. A2.1 that there exists a one-to-one correspondence between the complex numbers [defined by Eq. (A2.4)] and the points of the complex plane x-y (x: axis of the real numbers; y: axis of the imaginary numbers). We further note that a point (x, y) is defined equally well by its polar coordinates: r, φ (as shown in Fig. A2.1):

$$x = r\cos\varphi, y = r\sin\varphi \tag{A2.11}$$

Substituting the above in Eq. (A2.4) we obtain:

$$z = r(\cos\varphi + i\sin\varphi) \tag{A2.12}$$

where φ is called the *argument* of z, and

$$r = \sqrt{x^2 + y^2} = |z| \tag{A2.13}$$

Using the following mathematical identity [it follows from Eqs. (5.7), (5.8) and (5.15)]

$$e^{i\varphi} = \cos\varphi + i\sin\varphi \tag{A2.14}$$

we can write z as follows:

$$z = r\, e^{i\varphi} \tag{A2.15}$$

which is very useful when one multiplies complex numbers:

$$z_1 z_2 = (r_1 e^{i\phi_1})(r_2 e^{i\phi_2}) = r_1 r_2 e^{i(\phi_1+\phi_2)} \tag{A2.16}$$

Electromagnetic Potentials

Maxwell's equations [Eqs. (10.45)] allow us to write the electric field, $\boldsymbol{E}(\boldsymbol{r}, t)$, and the magnetic field, $\boldsymbol{B}(\boldsymbol{r}, t)$, in the following form:

$$\boldsymbol{B} = \nabla \times \boldsymbol{A} \tag{A3.1}$$

$$\boldsymbol{E} = -\partial \boldsymbol{A}/\partial t - \nabla\varphi \tag{A3.2}$$

Where $\boldsymbol{A}(\boldsymbol{r}, t)$ and $\varphi(\boldsymbol{r}, t)$ are known as the vector and scalar potential respectively. And they can be chosen in such a way that

$$\nabla \cdot \boldsymbol{A} = -\mu_0\varepsilon_0(\partial\varphi/\partial t) \tag{A3.3}$$

Substituting Eqs. (A3.1 and A3.2) in Maxwell's equations and using Eq. (A3.3), we obtain the following equations for A_x, A_y, A_z and φ:

$$\nabla^2 A_i - \mu_0\varepsilon_0 \frac{\partial^2 A_i}{\partial t^2} = -\mu_0 j_i, \quad i = x,\ y,\ z, \tag{A3.4a}$$

$$\nabla^2\phi - \mu_0\varepsilon_0\frac{\partial^2\phi}{\partial t^2} = -\frac{\rho}{\varepsilon_0} \tag{A3.4b}$$

Where ρ and $\boldsymbol{j} = (j_x, j_y, j_z)$ are the charge and current density respectively. Now $\boldsymbol{A}$ and φ can be combined into a 4-vector (it transforms like ds of Eq. (12.32) under a Lorentz transformation):

$$A_\mu = (A_x, A_y, A_z, i\varphi/c) \tag{A3.5}$$

Where $c = 1/\sqrt{(\varepsilon_0\mu_0)}$ denotes, as usual, the speed of light. Using A_μ we can write Eqs. (A3.4) in a Lorentz invariant form (it retains its form under a Lorentz transformation of the coordinates) as follows:

$$\left(\nabla^2 - \frac{1}{c^2}\frac{\partial^2}{\partial t^2}\right)A_\mu = -\mu_0 J_\mu, \quad \mu = 1, 2, 3, 4 \tag{A3.6}$$

Where J_μ is the 4-vector defined by Eq. (12.49).

Note: Eq. (A3.3) is known as the Lorentz condition. Another choice often made (when the relativistic invariance of Maxwell's equations need not be explicitly stated, is: $\nabla \cdot \boldsymbol{A} = 0$. When $\nabla \cdot \boldsymbol{A} = 0$, we have [from Eq. (A3.2)]: $\nabla^2\varphi = -\nabla \cdot \boldsymbol{E}$, and therefore using Eq. (10.45a) we obtain: $\nabla^2\varphi(\mathbf{r}, t) = -\rho(\mathbf{r}, t)/\varepsilon_0$, which implies that there is no *retardation* in the determination of $\varphi(\boldsymbol{r}, t)$ corresponding to $\rho(\boldsymbol{r}, t)$. The former is obtained from the latter as if they were constant in time. In other words: the interaction between charged particles is in effect an instantaneous one. This is how we treated the interaction between electrons in atoms and molecules in the relevant chapters of this book. And we emphasize that this does not constitute an approximation.

Matrices

A matrix consists of a rectangular array of numbers (real or complex) of *m rows* and *n columns*. Examples:

1. $\begin{pmatrix} a_{11} & a_{12} & a_{13} \\ a_{21} & a_{22} & a_{23} \end{pmatrix}$ and $\begin{pmatrix} 1 & 4 & 2 \\ -3 & 0 & 5 \end{pmatrix}$ are matrices having two rows and three columns. The matrix element a_{ij} is found at the i_{th} row and j_{th} column.

2. $\begin{pmatrix} a_1 \\ a_2 \\ a_3 \\ a_4 \end{pmatrix} = \underline{a}$ and $\begin{pmatrix} 0.3 \\ -2 \\ 0 \\ 5 \end{pmatrix}$ are one-column matrices having four rows. When no ambiguity arises we use an underlined small letter, $\underline{a}$ in this case, to denote a column matrix.

3. $(a_1 \quad a_2 \quad a_3 \quad a_4) = \underline{a}^T$ and $(1 \quad 4 \quad -2 \quad 0.5)$ are matrices having only one row and four columns. When no ambiguity arises we use an underlined small

letter with a superscript T, $\underline{a}^T$ in this case, to denote a row matrix. We say that $\underline{a}^T$ is the transpose of $\underline{a}$.

4. Square matrices have the same number (n) of rows and columns. The following are 4 × 4 matrices:

$$\begin{pmatrix} a_{11} & a_{12} & a_{13} & a_{14} \\ a_{21} & a_{22} & a_{23} & a_{24} \\ a_{31} & a_{32} & a_{33} & a_{34} \\ a_{41} & a_{42} & a_{43} & a_{44} \end{pmatrix} = \underline{A}; \quad \begin{pmatrix} 1 & 4 & 0.5 & 1 \\ -2 & 6 & 7 & 3 \\ 8 & 0.6 & 4 & 0 \\ 24 & -0.9 & 8 & 5 \end{pmatrix}$$

When no ambiguity arises we shall use an underlined capital letter, $\underline{A}$ in this case, to denote a matrix.

Let a_{ij} and b_{ij} be the elements of two $n \times m$ matrices. The *sum* of the two matrices is an $n \times m$ matrix whose elements are given by:

$$c_{ij} = a_{ij} + b_{ij} \tag{A4.1}$$

Accordingly, if λ is a number and $\underline{A}$ an $n \times m$ matrix with elements a_{ij}, $\lambda\underline{A}$ is an $n \times m$ matrix with elements λa_{ij}. And the same applies when λ is replaced by an operator. For example:

$$\frac{d}{dx}\begin{pmatrix} f_{11}(x) & f_{12}(x) \\ f_{21}(x) & f_{22}(x) \end{pmatrix} = \begin{pmatrix} df_{11}/dx & df_{12}/dx \\ df_{21}/dx & df_{22}/dx \end{pmatrix}$$

(a) *The product* $\underline{AB}$ of two matrices, where $\underline{A}$ is an $m \times p$ matrix and $\underline{B}$ an $p \times n$, is an $m \times n$ matrix, $\underline{C}$, whose elements c_{ij} $(i = 1,\ldots,m; j = 1,\ldots,n)$ are given by:

$$c_{ij} = \sum_{k=1}^{p} a_{ik} b_{kj} \tag{A4.2}$$

where a_{ik} and b_{kj} are the elements respectively of $\underline{A}$ and $\underline{B}$.

In particular, the product of a one-row matrix $\underline{a}^T$, with elements a_i $(i = 1,\ldots,n)$, with a column matrix $\underline{b}$, with elements b_i $(i = 1,\ldots,n)$, is a number: $c = a_1b_1 + a_2b_2 + \ldots + a_nb_n$. Example:

$$\begin{pmatrix} 2 & 0 & 4 & 1 \end{pmatrix} \begin{pmatrix} 3 \\ 2 \\ 0 \\ 1 \end{pmatrix} = 7$$

(b) *The product* $\underline{A}\,\underline{B}$ of two $n \times n$ matrices A and $\underline{B}$ is another $n \times n$ matrix $\underline{C}$, whose elements c_{ij} $(i, j = 1, 2, \ldots n)$ are given by

$$c_{ij} = \sum_{k=1}^{n} a_{ik} b_{kj} \tag{A4.3}$$

where a_{ik} and b_{kj} are the elements respectively of $\underline{A}$ and $\underline{B}$. We note that in general

$$\underline{AB} \neq \underline{BA} \tag{A4.4}$$

Example: $\underline{A} = \begin{pmatrix} 1 & 2 \\ 1 & 0 \end{pmatrix}$ and $\underline{B} = \begin{pmatrix} 1 & 0 \\ 0 & 2 \end{pmatrix}$

Then: $\underline{AB} = \begin{pmatrix} 1 & 4 \\ 1 & 0 \end{pmatrix}$ and $\underline{BA} = \begin{pmatrix} 1 & 2 \\ 2 & 0 \end{pmatrix}$

In some cases it happens that $\underline{AB} \neq \underline{BA}$. We then say that $\underline{A}$ and $\underline{B}$ commute.

Example: $\underline{A} = \begin{pmatrix} 1 & 0 \\ 0 & 2 \end{pmatrix}$ and $\underline{B} = \begin{pmatrix} 3 & 0 \\ 0 & 1 \end{pmatrix}$; then $\underline{AB} = \underline{BA} = \begin{pmatrix} 3 & 0 \\ 0 & 2 \end{pmatrix}$.

Applications: The following system of four equations with four unknowns (x_1, x_2, x_3, x_4):

$$\begin{aligned} a_{11}x_1 + a_{12}x_2 + a_{13}x_3 + a_{14}x_4 &= b_1 \\ a_{21}x_1 + a_{22}x_2 + a_{23}x_3 + a_{24}x_4 &= b_2 \\ a_{31}x_1 + a_{32}x_2 + a_{33}x_3 + a_{34}x_4 &= b_3 \\ a_{41}x_1 + a_{42}x_2 + a_{43}x_3 + a_{44}x_4 &= b_4 \end{aligned} \tag{A4.5a}$$

can be written in compact form as follows:

$$\underline{Ax} = \underline{b} \tag{A4.5b}$$

Where $\underline{A}$ is a 4 × 4 matrix of known elements (a_{ij}), $\underline{b}$ a column matrix of 4 known elements (b_i), and $\underline{x}$ is a column matrix of the 4 unknowns (x_i).

In general, Eq. (A4.5a) stands for a system of n equations with n unknowns: $\underline{A}$ is a given $n \times n$ matrix, $\underline{b}$ a given column matrix (with n rows), and $\underline{x}$ is a column matrix of n unknowns:

$$\sum_{j=1}^{n} a_{ij} x_j = b_i, \; i = 1, 2, \ldots, n \tag{A4.6}$$

The *determinant*, *det* $\underline{A}$, of an $n \times n$ matrix $\underline{A}$ is a number given by the sum:

$$\det \underline{A} = \sum_{\{j_1, j_2, \ldots j_n\}} (-1)^j a_{1j_1} a_{2j_2} a_{3j_3} \ldots a_{nj_n} \tag{A4.7}$$

where $\{j_1, j_2, \ldots j_n\}$ is any possible arrangement of the numbers 1, 2, …, n. There are $n! = 1 \times 2 \times 3 \times \ldots \times (n-1) \times n$ of them and each one is obtained from the natural sequence (1, 2, 3, …,$n-1$, n) by interchanging the position of two numbers and repeating the process a number (j) of times. For example: $\{j_1, j_2, \ldots j_n\} = \{2, 1, 3, 4, \ldots, n-1, n\}$ is obtained from the natural sequence by just one ($j = 1$) interchange (1↔2); $\{j_1, j_2, \ldots j_n\} = \{2, 1, 4, 3, 5, \ldots, n-1, n\}$ is obtained from the natural sequence by two ($j = 2$) interchanges (1↔2 and 3↔4), and so on. The determinant of an $n \times n$ matrix is often denoted by:

$$\det \underline{A} = \begin{vmatrix} a_{11} & a_{12} & . & . & . & a_{1n} \\ a_{21} & a_{22} & . & . & . & a_{2n} \\ . & . & . & . & . & . \\ . & . & . & . & . & . \\ . & . & . & . & . & . \\ a_{n1} & a_{n2} & . & . & . & a_{nn} \end{vmatrix}$$

Example:

$$\begin{vmatrix} a_{11} & a_{12} & a_{13} \\ a_{21} & a_{22} & a_{23} \\ a_{31} & a_{32} & a_{33} \end{vmatrix} = \det \underline{A} =$$

$$a_{11}a_{22}a_{33} - a_{11}a_{23}a_{32} - a_{12}a_{21}a_{33} + a_{12}a_{23}a_{31} + a_{13}a_{21}a_{32} - a_{13}a_{22}a_{31}$$

We note that:

$$\begin{vmatrix} a_{11} & a_{12} & a_{13} \\ a_{21} & a_{22} & a_{23} \\ a_{31} & a_{32} & a_{33} \end{vmatrix} = a_{11}\begin{vmatrix} a_{22} & a_{23} \\ a_{32} & a_{33} \end{vmatrix} - a_{12}\begin{vmatrix} a_{21} & a_{23} \\ a_{31} & a_{33} \end{vmatrix} + a_{13}\begin{vmatrix} a_{21} & a_{22} \\ a_{31} & a_{32} \end{vmatrix}$$

The above allows us to calculate a 3×3 determinant in terms of 2×2 determinants. A 2×2 determinant is by definition given by:

$$\begin{vmatrix} a_{11} & a_{12} \\ a_{21} & a_{22} \end{vmatrix} = a_{11}a_{22} - a_{12}a_{21}$$

A 4×4 determinant can be calculated in terms of 3×3 determinants, and so on, which facilitates the calculation of any given determinant.

We note the following important property of a determinant: the interchange of any two rows or of any two columns of a determinant changes the sign of the determinant. It follows from this property that a determinant equals zero if any two rows or any two columns of the determinant are the same.

A useful theorem: Consider the following system of equations:

$$\underline{A}\underline{x} = \underline{0} \qquad \text{(A4.8)}$$

where $\underline{A}$ is an $n \times n$ matrix, $\underline{x}$ is a column matrix of n unknowns we wish to determine, and $\underline{0}$ is a column matrix whose n elements are all zero. For example, for $n = 4$, the above stands for:

$$\begin{aligned} a_{11}x_1 + a_{12}x_2 + a_{13}x_3 + a_{14}x_4 &= 0 \\ a_{21}x_1 + a_{22}x_2 + a_{23}x_3 + a_{24}x_4 &= 0 \\ a_{31}x_1 + a_{32}x_2 + a_{33}x_3 + a_{34}x_4 &= 0 \\ a_{41}x_1 + a_{42}x_2 + a_{43}x_3 + a_{44}x_4 &= 0 \end{aligned} \qquad \text{(A4.9)}$$

The theorem tells us that a solution of the above equations, where at least one of the unknowns is different from zero, is possible only if the determinant of the coefficients vanishes, i.e., if

$$det\underline{A} = 0 \tag{A4.10}$$

Appendix B
Physical Constants

Planck's constant: $\hbar = h/2\pi = (1.05459 \pm 0.000001) \times 10^{-34}$ J s
Speed of light: $c = (2.997925 \pm 0.000001) \times 10^{8}$ m/s
Electronic charge: $e = (1.6021892 \pm 0.0000046) \times 10^{-19}$ C
Electron mass: $m = (9.109534 \pm 0.000047) \times 10^{-31}$ kg
Proton mass: $M_p = (1.672648 \pm 0.000006) \times 10^{-27}$ kg
Permittivity of free space: $\varepsilon_0 = 8.85434 \times 10^{-12}$ F/m
Permeabiliy of free space: $\mu_0 = 4\pi \times 10^{-7}$ H/m
Boltzmann's constant: $k = 1.38044 \times 10^{-23}$J/K
Atomic unit of length (*a.u.*): $a_0 = 0.52917 \times 10^{-10}$ m
Atomic unit of energy: *Hartree* = 2 *Rydbergs* (*Ryd*) = 27.21 eV
1 $eV = 1.6 \times 10^{-19}$ J
Avogadro's number: 6.0221367×10^{26} (kg mole)$^{-1}$

A. Modinos, *From Aristotle to Schrödinger*,
Undergraduate Lecture Notes in Physics, DOI: 10.1007/978-3-319-00750-2,

Subject Index

A
Absolute temperature, 145
Absorption of radiation
 by atoms, 376
 by crystals, 420
 by molecules, 405
Acceleration, 48, 88, 95
 in circular motion, 96
Accelerators, 452
Acoustic waves, 427
Action and reaction, 92
Action at a distance, 15, 58, 101, 220, 282
Adiabatic process, 159
Air, 19, 134, 136
Alchemy, 133
Alpha (α) particle, 310, 443
Ampere (A), unit of electric current, 200
Ampere's law, 199, 218
Analytical geometry, 59
Analytical mechanics, 119
Angstrom, 389
Angular momentum
 in classical mechanics, 107
  conservation, 122, 124
 in quantum mechanics
  commutation relations, 357
  orbital, 355
  spin, 318, 368
  total, 370, 395
Anode, 244
Archimedes's principle, 20
Astrolabe, 30
Atmospheric pressure, 65
Atom (many-electron), 386
 chemical behaviour, 290
 energy levels, 394
 ground state energy, 393
 one-electron energy levels (E_{nl}),, 360–363
 self-consistent field of, 390
Atom, model of, Boyle, 66, 67
 Dalton, 141
 Democritus, 15
 Rutherford, 310
 Thomson, 309
Atomic nucleus, 313, 450
Atomic number, 317, 386, 449
Atomic orbital, 399
Atomic shells, 386
 inner, 388
 outer, 388
 radii of, 388
Atomic theory of matter, 15, 67, 139, 263
Atomic unit of length (a.u.), 364
Atomic weight, 142
Atom (one-electron), 55
Avogadro's law, 144
 number, 145

B
Balmer's formula, 315
 series, 315
Barometer, 65
Baryons, 453
Battery, discovery of, 193
Bending of light by gravity, 282, 291, 292
Bernoulli's theorem, 124
Beta (β) radioactivity, 310, 446
Biot and Savart law, 197
Black-body radiation, 297
 Stefan-Boltzmann law, 297
 spectral distribution, 297
 Planck's formula, 301
  Einstein's derivation, 308
 Rayleigh-Jeans (classical) theory, 300
Black hole, 486
Bloch's theorem, 408
Bloch wave, 409

A. Modinos, *From Aristotle to Schrödinger*,
Undergraduate Lecture Notes in Physics, DOI: 10.1007/978-3-319-00750-2,

Body-centred cubic (bcc) lattice, 251, 406
reciprocal of, 252, 407
Bohr magneton, 463
Bohr's model of the hydrogen atom, 315
Boltzmann's, constant (*k*), 145
equation, 171
H-theorem, 172
Born-Oppenheimer approximation, 397
Bose-Einstein condensation, 382, 383
Bose-Einstein distribution, 381
Bosons, 379, 380
Boyle's law for gases, 66
Bragg's law of X-ray diffraction, 253
Brillouin zone (BZ), 409
Brownian motion, 179

C

Calculus, 72
Capacitor, 216
capacitance of, 216
Carbon dating, 490
Carbon dioxide, discovery of, 134
Carnot engine, 158
Cartesian coordinates, 59, 355
Cathode, 243
rays, 243
Celsius scale of temperature, 140
Central (centripetal) force, 96, 108, 121
Centre of mass (gravity), 20, 92
Charles' law for gases, 140
Chemical element, 133, 139
Chemical potential, 163, 381, 385
Circular motion, 96
Clausius's theorem, 160
Collision time, 171, 416
Commutation, of operators, 357
of matrices, 496
Complex conjugate of a number, 495
Compressibility, isothermal, 155
Compton scattering, 475
Conductivity, 202
of doped semiconductors, 422, 423
of intrinsic semiconductors, 422
of metals, 202, 416
Configuration of electrons in atoms, 386
Conservation, of energy, 150, 151
of mass, 138
of mass-energy, 447, 449
Constant-energy surface, 411
Continuity equation, 223
Convective acceleration, 123
Cooling of the earth, 159
Cooper pairs, 432
Copernicus' heliocentric system, 31
Correlation energy, 394
Correspondence principle, 344
Cosine, *cos(x)*, 62
Cosmological constant, 291
Coulomb (C), unit of electric charge, 200
Coulomb's law, 193
Covalent bond, 402
Covariant derivative, 287
Criterion of convergence, 2, 76
Critical temperature (T_c) for
Bose-Einstein condensation, 382
ferromagnetic transition, 419
superconducting transition, 430, 433
Crystalline solids, 408
Cupellation, 134
Curie temperature, 419
Curvature of, a two-dimensional surface, 288
spacetime, 289
Curvilinear tensors, contravariant, 285
covariant, 287

D

Dalton's law of partial pressures, 140
Dark energy and dark matter, 489
Darvinian evolution, 21, 159
Dating via radioactivity, 419, 490
Davy's miner's lamp, 204
de Broglie's hypothesis, 319
de Broglie's wave, 488
Debye, model of lattice vibrations, 425
temperature, 428
Dedekind cut, 55
Degeneracy of energy levels, 362, 365, 370, 396
Density functional formalism, 393
Density of one-electron states
in a metal, 413, 416
in a semiconductor, 420
Density of states, 176
Derivative of a function, 74, 76
partial, 114
Descartes' analytical geometry, 55
Determinant, 497
Deuterium, 390
Deuteron, 483
Differential calculus, 74
Diffusion, 181
Diffusivity, 186
Diode, 254, 256
Dirac's equation, 461

Dirac's theory, of the electron, 458
of the hydrogen atom, 465
of the positron, 234
Displacement current, 223
Distillation, 133
Divergence theorem, 215
Doppler shift, 294
Dynamo, 209

E

Earth, age of, 485
axis of rotation, 34
circumference, 16, 25
interior, 485
precessional motion, 34
orbit, 17
shape, 16, 100
Eclipse of, the moon, 7
the sun, 27, 292
Ecliptic, 29
Effective mass, of electrons, 421
of holes, 421
Eigenstates (eigenfunctions) of, energy, 326
momentum, 324, 325
physical quantity, 338
Eigenvalues of, energy, 326
momentum, 324, 325
physical quantity, 338
Einstein coefficients, 306, 307
Electric charge, 190
unit of, 192, 201
Electric current, 195, 196, 200
and magnetism, 195
unit of, 200
Electric dipole, moment, 228
oscillating, 228
Electric displacement vector, 232
Electricity and chemistry, 203
Electric susceptibility, 233
Electrolysis, 204, 205
Electrolyte, 194
Electromagnetic (EM) fields in material media, 231
Electromagnetic induction, 207
Electromagnetic potentials, 496
Electromagnetic radiation (EM waves), 226–228
from accelerated charge, 231
from an oscillating dipole, 227
interaction with matter, 265, 343, 465
quantum-mechanical description of, 374
Electromagnetic theory of light, 220, 224
Electromagnetic units, 200
Electromotive force (*e.m.f*), 320
Electron diffraction, 419
Electron, discovery of, 458
Dirac's theory, 458
Electron emission, 254
Electronic cloud, 359
Electronic (thermionic) valves, 254
Electroscope, 191
Electrostatics, 190
Electroweak interaction, 457
Elementary particles, 455
antiparticles, 455
Ellipse, 61
Emission of radiation
by atoms, 422
by crystals, 320
by molecules, 404
induced, 307
spontaneous, 306
Energy, 148
kinetic, 120
potential, 120, 121
unit of, 149
Energy bands, 140
of ferromagnetic nickel, 418
of Si, GaAs, ZnS, KCl, 420
of tungsten, 411
Energy eigenvalues and eigenstates of a particle
in a central field, 360
in a cubic box, 349, 382
in a one-dimensional box, 436
in a periodic field, 408
in a potential field, 329
in a rectangular potential well, 323
in a spherical well, 363
Energy gap, in a semiconductor, 419
in a superconductor, 431
Enthalpy, 152, 153
Entropy, 159, 161, 172, 176
Epicycles (in planetary motion), 35
Epicycloid, 29, 34
Equation of state, 152
of a real gas, 154
Equipartition theorem, 177, 299, 425
Ether, 15, 58, 67, 102, 213, 220, 235, 264–266
Euclidian, geometry, 25
space-time, 287, 288
Evaluation of π
Exchange energy, 393
Exclusion principle (Pauli's), 317, 384, 387, 486
Expansion of a gas, adiabatic, 162
isothermal, 162

Expansion of the universe, 489
Exponential function, e^x, 64
Extensive quantity, 152

F

Face-centred cubic (fcc) lattice, 251, 407
  reciprocal of, 252, 407
Faraday's, cage, 209
  field lines, 210, 220
  law of EM induction, 207
  rotor, 205
Fermi-Dirac distribution, 385
Fermi, level (E_F), 385
  surface, 411
Fermi's theory of β-radioactivity, 446
Fermions, 379, 383
Ferromagnetic metals, 417
Feynman diagrams, 471, 472
  bremsstrahlung, 476
  Compton scattering, 476
  electron-electron scattering, 474, 475
  Lamb shift, 482, 483
  magnetic moment of electron, 480
  pair production, 476
  propagation of an electron, 478
  propagation of a photon, 479
Field emission, 257
Fine-structure constant, 372
First law of thermodynamics, 151
Fluctuations, 170, 178
Fluid dynamics, 123
Force, 89
  conservative, 120, 122
  unit of, 90
Fourier series, 114, 117
Fowler-Nordheim equation, 257
Free electron model of metals, 417
Free energy, 162
Friction, 129, 164
Functions (of one variable), 58, 61, 63
  complete set of, 117, 324, 326
  complex, 322
  derivative of, 74, 77
  integral of, 72, 78
  orthogonal set of, 117, 325, 326
Functions of more than one variables, 115
  partial derivatives of, 115

G

Galaxies, 483
  Andromeda, 483
  our galaxy, the Milky Way, 483
Galileo, astronomical observations, 51
  experiments relating to falling bodies, 47
  law of inertia, 50
  pendulum, 44
  properties of matter, 53
  telescope, 38, 50
Galilean, relativity, 102
  transformation, 103
Galvanometer, 199
Gamow's theory of α-radioactivity, 445
Gaussian, coordinates, 284
  distribution, 166, 167
  units, 200
Gauss's, law, 214
  theorem, 215
Gay-Lussac's law for gases, 140
Geiger counter, 449
General relativity, principle of, 281
  Einstein's equation, 290
    and cosmology, 291
  bending of light, 282, 292
  gravitational waves, 291
  planetary motion, 291
Generation and death of stars, 483
Geocentric model of the universe, 17
Geodesic line, 288
Geometry and numbers, 6
Gibbs, energy, 163
  statistical physics, 174
Gluons, 456
Gradient of a scalar field, ∇f, 121
Gram (g), 89
Gravitational constant (G), 101
  measurement, 101
Gravitational waves, 291
Gravity, Aristotle's theory, 16
  Descartes' theory, 58, 59
  Einstein's theory, 283, 287
  Newton's theory, 99
  quantum theory of, 486
Gravity, centre of, 20, 92

H

Hadrons, 453
Hamiltonian, 322, 374, 380, 383
  relativistic, 459
Harmonic motion, 95
Harmonic oscillator, 372
Hartree-Fock method, 392
Hartree (self-consistent) method, 391
Hartree unit of energy, 364
Hawking radiation, 486

Heat, 147
    a form of energy, 150, 151
    latent, 135
    unit of, calorie (cal), 149, 150
Heisenberg's uncertainty principle, 333, 378
Hellmann-Feynman theorem, 402
Henry (H), 201
Hertzian waves, 227–228
Hydrogen atom
    energy eigenvalues and eigenstates, 362–365
    fine structure of the energy levels, 367, 317
Hydrogen gas, discovery of, 135
Hydrogen molecular ion, 398

I
Ideal gases, Avogadro's law, 143
    Charles' law, 140
    Dalton's law, 140
    Gay-Lussac's law, 140
    law of, 145, 153
Image potential energy, 237, 255
Imaginary numbers, 494
Indistinguishableness of same particles, 378
Induced (radiation) emission, 307
Infinite series and sequences, 1
Infinitesimal calculus, 76
Infinity, Aristotle's definition, 4
Insulator, 191, 202, 420
Integral, 72
    definite, 78
    indefinite, 79
Integral calculus, 72
    table of integrals, 79
Intensive quantity, 152
Interference, optical, 70, 212
    particle-wave, 320
Irreversible process, 156
Isotopes, 390, 449

J
Joule (J), unit of energy, 149

K
Kepler's, laws of planetary motion
    first law, 37
    second law, 37
    optics of human eye, 38
    Rudolphine Tables, 37, 40
    telescope, 38
    third law, 59
Kinetic energy, 102
    operator, 322, 361
Kinetic theory of gases, 67, 165
k-space, 252, 407

L
Lamb shift, 465, 480
Laplace operator (∇^2), 126
Laser, 308
Latitude, 18
Lattice, space, 250
    body-centred cubic, 251, 407
    face-centred cubic, 251, 407
    reciprocal, 252, 407
    simple cubic, 251
Lattice vibrations, 425
Least action principle, 468
Legendre polynomials, 358
Lenz law of electromagnetic induction, 207
Leptons, 455
Light, nature of, 68, 302
    see, velocity of light
Linearity (of an equation), 116
Logarithm, *log(x)*, 65
Lorentz, force, 231
    transformation, 266, 268

M
Mach's principle, 334
Magnetic moment, 238
Magnetic moment of the electron
    orbital, 318, 370
    spin, 370, 465, 467
Magnetic poles, 189
    field of earth, 189
Magnetic susceptibility, 234
Magnetization, 234, 418
Magnet (natural), 189
Mass, 89, 100
    centre of, 20, 92
    relation to energy, 271
    spectrometer, 449
    unit of, 889
Matrices, 497
    as operators, 369, 460, 461, 463
Matrix mechanics, 331
Maxwell-Boltzmann distribution, 166
Maxwell's equations, 219
    in material media, 231
        in relativistic (tensor) form, 280
    in vacuum, 224
Maxwell's model of ether, 221

Mean value of, position, 336
  momentum, 337
Mean free path, 171, 416
Meissner effect, 431
Meridian, 18
Mesons, 454
Metals, 202, 410
  conductivity, 415
  specific heat, 415
Metastable state, 433
Methane, 405
Metric tensor, $g_{\mu\nu}$, 285–287, 288
Michelson-Morley experiment, 264
Microwave radiation, cosmic, 488
Minkowski, force, 277
  spacetime, 272
M.K.S. units, 201
Mobility, of electrons, 422
  of holes, 422
Molecular orbital, 399
Molecules, diatomic
  covalent bond, 402
  dissociation energy, 401
  electronic terms, 399
  energy levels, 403
  molecular orbitals, 399
  nuclear motion, 402
  optical properties, 405
  specific heat, 404
Molecules, polyatomic, 405
Moment of inertia, 111
Momentum, 89, 93
  conservation, 122
  operator, 324
Momenum-energy vector, 277
  conservation, 474
Motors, 209

N
Navier-Stokes equations, 125
Neutrino, 445
Neutron, 390, 446
  decay, 446, 457
  slow, 450
Neutron star, 486
Newton rings, 70
Newton's, experiments with light, 85
  Principia, 87
    examples of motion, 93
    laws of motion, 89, 92
    theory of gravity, 99
Nuclear fission, 448
Nuclear models
  Fermi gas, 453
  liquid drop, 453
  shell, 453
Nuclear reactions, 313, 448
Nuclear reactor (first), 450
Numbers, 6
  and geometry, 7
  complex, 494
  continuum of, 55
  fractional, 6
  imaginary, 494
  irrational, 8, 55, 493
  large, 10
  natural (positive integers, counting), 6
  negative, 56
  rational, 7
  real, 56
Numerals, Attic (Greek), 9
  Roman, 9
  Hindu-Arabic (decimal), 9

O
Ohm's law, 202
Oxygen, discovery of, 137

P
Parallax, 28, 34
Paramagnetism, 498
  Pauli's theory, 419
Partial derivative, 114
Partial differential equations, 115
Pendulum, 44, 112
Periodic boundary conditions, 299, 410
Periodic potential field, 408
Periodic table of the elements, 317, 386, 388
Permeability, of vacuum (μ_0), 201
  of a material medium, 233
Permittivity, of vacuum (ε_0), 201
  of a material medium, 233
Perturbation theory, 342, 376, 469
Phase diagram, 154
  of a superconductor, 434
Phase transition (first order), 156, 164
  ferromagnetic, 419
  superconductivity, 430
Phlogiston, 135
Photoelectric effect, 244
Photon, 303, 305, 379, 466, 471
  polarization of, 302, 472
  virtual, 472
Planck's constant, 301, 314, 321, 333
Planetary motion according to

Copernicus, 34
general relativity, 290
Kepler, 37, 39
Ptolemy, 29
Plane wave, electromagnetic, 227
free particle, 323
p-n junction, 423
as a rectifier, 424
Polarization, of light, 213, 302, 303
of a material medium, 233
Pole Star, 17, 18
Polonium, 444
Positivism, 331
Positron, 445, 465, 466
Potential (energy) field, 120
central, 355
electrostatic, 193
Poynting vector, 230
Proper time interval, 276, 454
space-like, 276
time-like, 276
Proton, 313, 390, 456
Pulsar, 486
Pythagoras's theorem, 7, 491

Q
Quantum chromodynamics, 457
Quantum electrodynamics, 466
Quantum mechanics
Copenhagen interpretation, 335
relation to classical mechanics, 343
relativistic, 460
Quarks, 456
Quaternions, 222

R
Radioactivity, 310, 443
decay constant, 490
Radium, 444
Reflection and refraction of light, 68, 70, 234
Refractive index, 68, 69, 71
Relativistic invariance, 275, 280
Renormalisation of charge and mass of electron, 480
Resistivity, 202
Reversible process, 156
Reynold's number, 126
Rotation of a solid body, 109
Rotations, in space, 272
in spacetime, 274
Ruhmkorff induction coil, 244

S
Scalar quantity, 119, 276
Scattering states, 338
Schrödinger's equation, 322
time-independent, 326
Schrödinger's theory, 330
Schwarzchild metric, 290
Self-consistent field, 391
Self-inductance, 238
Semiconductors, 419
n-type, 422
p-type, 422
Simultaneity, 266, 269
Sine, *sin(x)*, 62
Slater determinant, 384
Snell's law of refraction, 68, 71, 236
Spacetime, 274, 284
Special relativity, 263
law of motion, 271
in tensor form, 277
length contraction, 269
time dilation, 269
Specific heat, 149
C_v, 152
C_P, 152
of a gas, 178, 404
of a metal, 415, 429
of a semiconductor, 421
Spherical coordinates, 355
Spherical harmonics, 359
Spin, discovery of, 318
Pauli matrices, 369
see also, Dirac's theory of the electron, 459
Spin-orbit interaction, 386
Spinors, 369, 465, 471
Standard model of elementary particles, 453, 458
Stationary states, 313, 326, 337, 338
of a diatomic molecule, 399
of a particle in a central field, 360
of the hydrogen atom, 363
Stefan-Boltzmann law, 297
Stern-Gerlach experiment, 318
Stokes theorem, 218
Strong interaction, 456
Sublimation, 133
Sun, distance from earth, 28
radiation, 483
Superconductivity, 430
BCS theory, 431
high-temperature, 434
Superfluidity, 383
Supernova, 486

T
Telescope, 38, 50
Tensors, Cartesian, 275–278
  curvilinear, 286
Thermal conductivity, 186
Thermionic emission, 254
Thermodynamic, equilibrium, 152
  quantities, 152–154
  variable, 152
Thermodynamics, 147
  First law, 149, 152
  Second law
    and direction of time, 164
    Clausius's statement, 160, 161
    Kelvin's statement, 159
Time, according to
  Aristotle, 13
  general relativity, 283
  Newton, 88
  special relativity, 266, 283, 287
Time reversal, 164, 466
Torque, 111
Transistor, 424
  characteristic of, 430
Transitions, 342
Transmission coefficient, 241
Triode, 256
Tunnelling, 257, 339, 341, 435, 445
Turbulence, 127

U
Uncertainty Principle
  see Heisenberg's Uncertainty Principle
Universe
  according to, Aristarchus, 11
  Aristotle, 13
  Copernicus, 33
  expanding, 488
Uranium atom, 451

V
Vacuum polarization, 483
van der Waals' equation, 154
Vectors, 90
  addition of, 91
  differentiation of, 92
  scalar product of two vectors, $\boldsymbol{A} \bullet \boldsymbol{B}$, 119
  vector product of two vectors, $\boldsymbol{A} \times \boldsymbol{B}$, 108
Vector field $\boldsymbol{A(r)}$
  $\nabla \times \boldsymbol{A}$ (= $\boldsymbol{curlA}$), 218
  $\nabla \cdot A$(= $\boldsymbol{divA}$), 216
Vector 4-dimension, 274, 277
Velocity (speed) of light, in vacuum, 71, 72, 225, 263, 283
  in a material medium, 71, 234
Vibrating string, 114
  normal modes of, 116
Vibrations, of a molecule, 403
  lattice, 427
Viscosity, 125
  coefficient, 126
Voltage, 194

W
Wavefunction, 322
Wave nature of light, 70, 211, 220, 235
Wavepacket, 339
Wave-particle duality, 319
Weak interaction, 446, 455, 456
Weierstrass's criterion of convergence, 3, 76
White dwarf, 484
Wigner-Seitz cell, 407
Wilson chamber, 449
Work, done by a force, 119
  by an expanding gas, 157
  unit of, 149
Work function, 255, 412

X
$X\alpha$- method, 393
X-ray, diffraction, 249
  spectrum, 249, 389
X-rays, discovery of, 245

Z
Zeeman effect, 310
Zeno's paradox of motion, 4

Name Index

A

Allen, J., superfluidity, 383
Ampere, A. M., 195–196
  Ampere's law, 199
Anaxagoras, physical processes, 21
  shape of the earth, 16
Anderson, C., discovery of positron, 447, 466
Antiphon, evaluation of π, 5
Archimedes, 19–20
  centre of gravity, 20
  evaluation of π, 6
  on large numbers, 10
  principle, 20
Aristarchus, model of the universe, 11
  size of moon, 26
Aristotle, 13, 14
  definition of infinity, 5
  nature acts for an end, 21
  space, time, and matter, 14
  theory of gravity, 16
  the Universe, 16
Aston, F., mass spectrometer, 449
Avogadro, L. R. A. C. B., 143
  theory of gases, 144

B

Bacon, R., invention of spectacles, 31
Balmer, J. J., emission spectra of hydrogen atom, 317
Bardeen, J., invention of transistor, 424
  theory of superconductivity, 431
Barkla, C., X-rays, 249
Becquerel, H., radioactivity, 443
Bednorz, G., superconductivity, 434
Bernoulli, D., fluid dynamics, 124
  kinetic theory of gases, 67
Berzelius, J. J., notation of atoms, 142
Bethe, H., quantum electrodynamics, 467, 480
Biot, J. B., electric current and magnetism, 197
Black, A.
  discovery of carbon dioxide, 134
  latent heat, 135
Bloch, F., electron states in a periodic field, 408
Bohm, D., interpretation of quantum mechanics, 335
Bohr, N., 312, 315
  Copenhagen interpretation of quantum mechanics, 317, 334
  model of hydrogen atom, 313
  periodic table of atoms, 317
Boltzmann, L., 172, 173
  Boltzmann equation, 171
    constant, 145
  H-theorem, 172
  Maxwell-Boltzmamm distribution, 166
Bombeli, R., complex numbers, 492
Born, M., Born-Oppenheimer approximation, 397
  quantum mechanics, 331, 336
Bose, S. N., Bose–Einstein distribution, 381
Boyle, R., 67
  gas law, 66
  gas models, 66
Bragg, W. H., X-ray diffraction, 253
Bragg, W. L., X-ray diffraction, 253
Brahe, T., planetary motion, 37
Brattain, W., invention of transistor, 424
Brillouin, L., Brillouin zone, 407
Brown, R., Brownian motion, 179

C

Cardano, G., complex numbers, 494
Carnot, S., Carnot engine, 157
Cavendish, H., 135, 136
  gravitational constant, 100, 101

A. Modinos, *From Aristotle to Schrödinger*,
Undergraduate Lecture Notes in Physics, DOI: 10.1007/978-3-319-00750-2,

properties of hydrogen gas, 135
Celsius, A., temperature scale, 140
Chadwick, J., 448
discovery of neutron, 450
Chandrasekhar, S., generation and death of stars, 486
Charles, J., law of gases, 140
Clausius, R, 160
Clausius's theorem, 160
entropy, 161
kinetic theory of gases, 165
second law of thermodynamics, 160, 161
Cockcroft, J., accelerators, 446
Compton, A. H., Compton scattering, 475
Condon, E., α-particle radioactivity, 445
Cooper, L. N. theory of superconductivity, 431
Copernicus, N., 31–36
heliocentric system, 34
Cornell, E., Bose-Einstein condensation, 382
Crookes, W., cathode rays, 243
Curie, M., radioactivity, 443–445
Curie, P., radioactivity, 440

D
Dalton, J., 139–140
atomic theory of matter, 142
law of partial pressures, 145
Dante, 45
Davisson, C. J., electron diffraction, 319
Davy, H., 148, 203
heat, 148
electrolysis, 204
miner's lamp, 204
de Broglie, L., wave-particle duality, 319
de Coulomb, C. A., Coulomb's law, 192
de Maricourt, P., magnetism, 190
Debye, P., 332
lattice vibrations, 428
Dedekind, R., continuum of numbers, 55
Democritus, atomic theory of matter, 15, 139
Descartes, R., 58–59
analytical geometry, 59–63
Dirac, P. A. M., 458, 459
Fermi–Dirac distribution, 385
radiation field, 374
interaction with matter, 375
theory, of the electron, 458
of the hydrogen atom, 465
of the positron, 466
Dulong, P. L., explosives, 204
Dushman, S., thermionic emission, 254
Dyson, F., quantum electrodynamics, 467

E
Eddington, A., bending of light by gravity, 211, 291, 293
Einstein, A., 261–264, 292–294
Brownian motion, 179
coefficients, 306
general relativity, 281
Einstein's equation, 287
photoelectric effect, 304
special relativity, 263
specific heat of crystals, 428
statistics, 382
Empedocles, atomic theory of matter, 15
cause of physical processes, 21
Eratoshenes, circumference of earth, 25
Euclid, 25
Eudoxos, continued fractions, 8
Euler, L., 107–108
analytical mechanics, 107
fluid dynamics, 123
motion of solid bodies, 107
Ewald, P., X-ray diffraction, 249

F
Faraday, M., 204–205
Faraday's, cage, 209
field lines, 210, 220
law of EM induction, 207
rotor, 205
Fermi, E., 450–451
Fermi–Dirac distribution, 385
Fermi, level (E_F), 385
surface, 411
nuclear reactor (first), 450
slow neutrons, 452
theory of β-radioactivity, 447
Feynman, R. P., 466–467
diagrams, 470–475
Hellmann-Feynman theorem, 402
model of positron, 466
quantum electrodynamics, 466
Fock, V., Hartree-Fock method, 392
Fontana (Tantanglia), N., complex numbers, 494

Forbes, J., heat, 148
Fourier, J. B. J., 117
  Fourier series, 117
Fowler, R. H., field emission, 257
Franklin, B., electric charges, 190
Fresnel, A. J., wave theory of light, 213, 236
Friedman, A., expanding universe, 488

G
Galileo, 43–54
  astronomical observations, 52
  experiments relating to falling bodies, 47
  law of inertia, 50
  pendulum, 44
  properties of matter, 53
  telescope, 38, 50
Galvani, L., discovery of battery, 193
Gamow, G. α-particle radioactivity, 445
Gauss, J. C. F., 214–215
  complex numbers, 494
  Gaussian, coordinates, 284
    distribution, 167
    units, 200
  Gauss's theorem, 215
Gay-Lussac, J. L., law of gases, 140
Geiger, H., atomic structure, 311
  Geiger counter, 449
Gell-Man, M., 455
  elementary particles, 453
Gerlach, W., Stern-Gerlach experiment, 318
Germer, L. H., electron diffraction, 319
Gibbs, J. W., statistical physics, 174
Gilbert, W., earth's magnetic field, 189
Giorgi, E., M.K.S. units, 200
Goudsmit, S., discovery of spin, 318
Guericke, O., atmospheric pressure, 65
Gurney, R. α-particle radioactivity, 445
Gutenberg, J., printing press, 31

H
Hahn, O., nuclear fission, 450
Halley, E., astronomy, 87
Hamilton, W., principle of least
    action., 457
Hartree, D. R., self-consistent field, 390, 391
Hawking, S., Hawking radiation, 486
Heaviside, O., electromagnetism, 222
Heisenberg, W., 331–333
  matrix mechanics, 334
  uncertainty principle, 334, 378
Hellmann, H., Hellmann-Feynman theorem,
    402
Helmholtz, von H., free energy, 163
Henry, J., electromagnetic induction, 207
Herapath, J., kinetic theory of gases, 165
Hershel, W., heat, 148
Hertz, H., electromagnetic waves, 228
  photoelectric effect, 304
Hipparchus, 26
  earth's, axis of rotation, 33
    distance to moon, 26
Hooke, R., 86, 87
Hoyle, F., generation and death of stars, 485
Hubble, E., expanding universe, 292
Humason, M., expanding universe, 292
Huygens, C., theory of light, 70
  pendulum as a clock, 46

J
Jeans, J., black-body radiation, 297, 299
Joule, J. P., 152–153
  equivalence of heat and mechanical work,
    149
  kinetic theory of gases, 165

K
Kapitza, P., superfluidity, 383
Kepler, J., 36
  laws of planetary motion, 37, 40
  on astrology, 39
  optics of human eye, 38
  Rudolphine Tables, 37, 40
  telescope, 38
Kohn, W., density functional
    formalism, 393
Krönig, A., kinetic theory of gases, 165

L
Lagrange, J. L., analytical mechanics, 135–136
Lamb, W., Lamb shift, 467, 482
Landau, L. D., superfluidity, 383
Laplace, P. S., essays on probabilities, 166
Lau, von M., X-ray diffraction, 249
  thermionic emission, 254
Laurence, E. O., accelerators, 448
Lavoisier, A., 138–139
  conservation of mass, 138
Legendre, A. M., Legendre polynomials, 358
Leibniz G. W., 72–73
  invention of calculus, 72
Lenard, P., cathode rays, 244
  photoelectric effect, 304
Lenz, H., electromagnetic induction, 207, 262

Leonardo of Pisa, Hindu-Arabic numerals, 9, 10
Lilienfield, J. E., electron emission, 257
Lippershey, H., invention of telescope, 31, 50
Lord Kevlin (Thomson, W.), 160
absolute temperature, 144, 145, 145
cooling of the earth, 159
second law of thermodynamics, 159
Lorentz, H., 265
Lorentz transformation, 265, 268
Mach, E., Mach's principle, 334
Malus, E., wave nature of light, 213
Marconi, G., radio transmission, 229
Marsden, E., atomic structure, 311
Maxwell, J. C., 219–221
Maxwell–Boltzmann distribution, 166
Maxwell's equations, 224, 232
model of ether, 221
Mayer, J. R., 149, 150
equivalence of heat and mechanical work, 149
Meissner, W., Meissner effect, 430
Meitner, L., nuclear fission, 451
Melloni, M., heat, 148
Mendeleev, D., periodic table of elements, 317
Michelson, A., Michelson-Morley experiment, 264
Millikan, R., measurement of the electron's charge, 248
Minkowski, H. 4-dimensional spacetime, 272
Misener, D., superfluidity, 383
Mitchell, J., magnetism, 189
Morley, E., Michelson-Morley experiment, 264
Moseley, H., periodic table of elements, 387
Müller, K., superconductivity, 434

N

Navier, C. L., fluid dynamics, 125, 125
Nernst, W., entropy, 161
Newton, I., 83–87, 103
experiments with light, 85
invention of calculus, 72
Principia, 87
laws of motion, 89, 92
gravity, 99
Nordheim, L. W., field emission, 257
Oersted, H., magnetism due to electric current, 195
Ohm, G. S., Ohm's law, 201
Oppenheimer, R., 465
Born-Oppenheimer approximation, 397
Ostrogradsky, M., divergence theorem, 215

P

Pauli, W. E., 379
bosons and fermions, 378
exclusion principle, 384, 384, 387, 392, 486
neutrino hypothesis, 447
paramagnetism, 419
spin-matrices, 369
Penzias, A., cosmic radiation, 488
Perrin, J. B., Brownian motion, 180
Planck, M., 300
black-body radiation, 297
Planck's constant, 301, 313, 323, 333
Plato, 13
Plücher, E., cathode rays, 243
Poincare, H., special relativity, 266
Priestley, J., 137
discovery of oxygen, 137
Ptolemy, planetary motion, 28–30
Pythagoras, 7
theorem, 7, 491

R

Rayleigh, J. W. S., black-body radiation, 297, 299
Rheticus, G., heliocentric system, 35
Richardson, O. N., thermionic emission, 254
Riemann, B., 285
Riemann's curvature tensor, 289
Roemer, O., speed of light, 72
Roentgen, W. C., 244–246
discovery of X-rays, 244
Rumford, Count, (Thompson, B.), 147
theory of heat, 147
Rutherford, E., 310, 311
atomic structure, 309
nuclear reactions, 313, 446
radioactivity, 310

S

Salam, A., electroweak interaction, 457
Salpeter, E., generation and death of stars, 485
Salviati, F., sunspots, 51
Salvinus de Amatus, invention of spectacles, 31
Savart, F., electric current and magnetism, 199
Schiller, F., 173
Schrieffer, J., theory of superconductivity, 431
Schrödinger, E., 321–322
Schrödinger's equation, 321
time-independent, 326
Schrödinger's theory, 330

Schotky, W., electron emission, 355
Schwarzchild, K., Schwarzchild metric, 290
Schwinger, J., quantum electrodynamics, 467, 480
Seitz, F., Wigner–Seitz cell, 407
Shockley, W., invention of transistor, 424
Slater, J. C., Hartree-Fock method, 391
  Slater determinant, 384
  $X\alpha$-method, 392
Snell, W., refraction of light, 68, 236
Sommerfeld, A., 247, 317
  free electron model of metals, 247, 417
  quantum theory, 317
Stefan, J., black-body radiation, 297
Stern, O, 174
  Maxwell–Boltzmann distribution, 173
  Stern-Gerlach experiment, 318
Stokes, G. G., 125
  fluid dynamics, 123
  Stokes theorem, 217
Strassman, S., nuclear fission, 451

T

Thomson, G. P., electron diffraction, 320
Thomson, J. J., discovery of electron, 247
  static model of the atom, 309
Tomonaga, S. I., quantum electrodynamics, 464
Torriccelli, E., 54
  barometer, 65
  nature of air, 66

U

Uhlenbeck, G., discovery of spin, 318

V

van der Waals, J., van der Waals equation, 154
Volta, A., discovery of battery, 193

W

Walton, E., accelerators, 448
Weber, W. W., electromagnetism, 200, 214
Weinberg, S., electroweak interaction, 457
Weierstrass, K. W., criterion of convergence, 2, 77
Weiman, C., Bose–Einstein condensation, 383
Weisskopf, V., quantum electrodynamics, 383
Wigner, E., Wigner–Seitz cell, 407
Wilson, C. T. R., Wilson cloud chamber, 449
Wilson R.W., cosmic radiation, 445

Y

Young, T., 197
  optical interference, 212
Yukawa, H., mesons, 451

Z

Zeeman, P., Zeeman effect, 309
Zeno, paradox of motion, 3